Digital Transformation in Cloud Computing

Digital Transformation in Cloud Computing

Top-level Design, Architecture, and Applications

Alibaba Cloud Intelligence GTS

CRC Press
Taylor & Francis Group
Boca Raton London New York

CRC Press is an imprint of the
Taylor & Francis Group, an **informa** business

HZ BOOKS
华章IT

First edition published in English 2022
by CRC Press
6000 Broken Sound Parkway NW, Suite 300, Boca Raton, FL 33487-2742

and by CRC Press
4 Park Square, Milton Park, Abingdon, Oxon, OX14 4RN

CRC Press is an imprint of Taylor & Francis Group, LLC

© 2022 Alibaba Cloud Intelligence GTS

Library of Congress Cataloging-in-Publication Data
Names: Alibaba Cloud Intelligence GTS, author.
Title: Digital transformation in cloud computing : top-level design, architecture, and applications / Alibaba Cloud Intelligence GTS.
Description: First edition. | Boca Raton, FL : CRC Press, 2022. | Includes bibliographical references. | Summary: "With the rapid development of cloud computing and digital transformation, well-designed cloud-based architecture is always in urgent need. Illustrated by project cases from the Chinese technology company Alibaba Group, this book elaborates how to design a cloud-based application system and build them on the cloud. Cloud computing is far from being just a resource provider; it offers database, storage and container services that can help to leverage key advantages for business growth. Based on this notion, authors from the Alibaba Cloud Intelligence Global Technology Services introduce new concepts and cutting-edge technology in the field, including cloud-native, high-availability and disaster tolerance design on cloud, business middle office, data middle office, and enterprise digital transformation."— Provided by publisher. Identifiers: LCCN 2021054021 (print) | LCCN 2021054022 (ebook) | ISBN 9781032223018 (hbk) | ISBN 9781032225326 (pbk) | ISBN 9781003272953 (ebk) Subjects: LCSH: Cloud computing.
Classification: LCC QA76.585 .D544 2022 (print) | LCC QA76.585 (ebook) |
 DDC 004.67/82—dc23/eng/20211222
LC record available at https://lccn.loc.gov/2021054021
LC ebook record available at https://lccn.loc.gov/2021054022

ISBN: 978-1-032-22301-8 (hbk)
ISBN: 978-1-032-22532-6 (pbk)
ISBN: 978-1-003-27295-3 (ebk)

DOI: 10.1201/9781003272953

Typeset in Minion
by codeMantra

Contents

Foreword

TODAY'S DIGITALIZATION WAVE HAS swept the world. Major countries and economies around the world are not willing to lose the initiative in this transformation, which is no less than electrification 100 years ago: the European Union released the digitalization strategy "Shaping Europe's Digital Future" at the beginning of 2020, and in September 2020, the United States released "Designing America's Digital Development Strategy". In the 14th Five-Year Plan of the People's Republic of China for National Economic and Social Development and the Outline of 2035 (hereinafter referred to as "the 14th Five-Year Plan"), which was released in March 2021, three chapters are devoted to elaborate on Digital China.

The global COVID-19 pandemic has further intensified the process of social digitization. People and enterprises have to rely on cloud technology to build digital survival tools, and the government has to rely more on the power of cloud and big data for epidemic prevention and control and national governance. Among the uncertainties of today, the digital trend is certain, and more people and organizations will embrace it in the future. We believe that the upsurge of digital transformation is actually the anxiety of human beings for the future development, and the application and development of digital technology can help solve this problem.

"Digitization" itself is a kind of productive force, closely combined with production tools and production relations. Zhang Jianfeng, president of Alibaba Cloud, believes that whoever has mastered the advanced productive forces will have a greater say, and digitalization and technological innovation are the advanced productive forces of this era. At the organizational and management level, digitalization drives internal processes, decisions, organizational design, and performance management to be more scientific and efficient. At the business level, digitalization drives the efficiency of end-to-end business processes and accelerates integration and collaboration. In terms of digital creation of new products and value, digitalization offers the possibility of more new tracks, reduces the cost of trial and error and the upgrading, and accelerates the incubation of new businesses. What enterprises and governments need is digital transformation with visible paths and effects, but there is no need to exaggerate our anxiety about "digital transformation". Digital innovation is generated by the development of enterprises' business to a certain stage or the needs of enterprise business development itself. We provide enterprises and governments with cloud-based digital technology services and solutions, as well as subsequent technical service guarantee, so that they have more confidence to do what they want to do.

We find that in the process of enterprises embracing digitalization, information system is no longer a simple business flow problem, but also data flow, mobility, artificial intelligence to deal with big data, and Internet of everything, which is not encountered in the previous information construction. In other words, digital transformation is not simply to purchase a digital software or tool, it requires top-level planning and design, overall consideration of the enterprise's entire digital business process, and even the internal organizational structure of the enterprise should be adjusted accordingly. So, we want to bring the methodology and best practices of digital transformation in cloud computing to our readers.

What kind of people are we? We are from Alibaba Cloud Intelligent global technology services (GTS), we are China's digital builders, and we are active in major technology speech exhibitions and frequent travellers on planes and trains. We may study production problems with workers and farmers in factory workshops and fields, or we may be the people who silently provide support and guarantee for you when you browse micro blogs and short videos on your mobile phone. We are deeply involved in the ten digital application scenarios proposed in the 14th Five-Year Plan, from intelligent transportation, intelligent energy, intelligent manufacturing to intelligent government affairs. Alibaba Cloud's technology and services, as well as its engineers, have become part of China's digital infrastructure, which is a very important part of the industrial structure. Through technology driving and business innovation, we are striving to become a practitioner of the digital transformation of the government and enterprises. We hope to cultivate 100,000 ecological developers and help 10,000 enterprises complete the digital transformation, and build ourselves into a dream team of digital transformation. Not only in China, but our business has also expanded overseas. Our company footprint has expanded to opening data centers in East Asia, Southeast Asia, West Asia, Europe and North America.

In 2018, Forrester, an authoritative research institute in the United States, selected Alibaba Cloud as the world's "Digital Transformation Expert", and as the only Chinese manufacturer, it was evaluated as "outstanding" and the best market performance. At the recent 2021 Cloud Computing Conference, Gartner jointly released the report "Cloud Innovation Digital Services 2.0" with a third party, defining the cloud service industry standard in the era of digital economy for the first time, and Alibaba Cloud became the only cloud service vendor selected into L4 (governance level). With this recognition, the author's team also feels the responsibility on their shoulders, and we hope to share our understanding and insights on digital transformation with the market and industry. In 2017, the author's team published the first edition of *Engineering Practice of Cloud Migration*. At that time, the cloud migration had just become a common consensus, that book focused on IaaS, involving the cloud migration of applications, databases, big data, etc. It helped enterprises smoothly migrate to the cloud. In 2019, we published the second edition of *Engineering Practice of Cloud Migration*, on the basis of the first edition, updated the product technology content, and combined with the technology trends and hot spots, it introduced topics such as Business Middle Office, database and big data system cloudization to help enterprises carry out application and architecture reconstruction on the cloud, so as to make the system more adapt to the technical characteristics and requirements of the cloud.

This book is the evolution of the first two books. We start with the evolution of cloud computing and introduce the concept of digital transformation. Under the guidance of uniform top-level design and business consulting, we make full use of Alibaba's methodology, technology, products, and best practices. They include application cloudization and cloud native, Business Middle Office, Data Middle Office, and AIoT platform. Finally, we use three comprehensive cases of digital transformation of large-scale customers to explain the whole process and final effect of digital transformation enabled by Alibaba Cloud. We mainly focus on the digital transformation of enterprises, which is also of reference significance to the government.

It is strongly recommended that enterprise business decision makers, technical decision makers, industry experts, architects and technical directors read this book. I hope this book can provide valuable information for everyone, give sufficient reference, help the pace of digital transformation move forward with speed and stability, and achieve better results. Those in the industry who hope to understand cloud computing and Alibaba Cloud will also benefit from this book.

The compilation and publication of this book is inseparable from the care and support of many leaders and colleagues of Alibaba Group and Alibaba Cloud Intelligence. We have listed the names of authors and workers separately in the list of staffs. Thank you!

In particular, although all codes and cases in this book were successfully run or implemented in the author's environment at that time, it does not mean that readers will succeed in their own environment. Please do not test them randomly in the production environment. At the same time, in view of the rapid development of technology and the author's limited knowledge, there are inevitable omissions and deficiencies in this book. Readers are warmly welcome to criticize and correct!

ZhangRui
GM of Alibaba Cloud Intelligence
GTS Delivery Technology Department

Preface I

SINCE THE END OF 2019, in the fight against COVID-19, a "health code" and a "consumption voucher" have enabled society to recover in an orderly manner, and "digitization" has gone from a professional term to a general term.

How to ensure the success of enterprise digital transformation and reform, and what kind of entry point is the most appropriate, are the topics that every enterprise person in charge will ponder and consider?

Through the exploration and practice summary in the digital field of recent years, Alibaba Group chose the most difficult thing, which is the easiest breakthrough to open the work situation. What is the "most difficult thing"? "Break" and "Establish"!

What does it mean to "break":

Break the "Empiricism": We have thought and tried this many times before, but historical experience tells us it is not feasible. It is like digital old wine in a new bottle.

Break the "fear of difficulty": Digitalization is not a department's business, process remodeling, system transformation, breaking departmental walls and benefit chains, and short-term results are not easy to measure, but rather it's too difficult. Is it better to do more than less?

What to "establish"?

Establish the "development view": Layout of the current work with the vision of development. Will the next 5–10 years continue to be as strong and leading as today? Can you keep up with your clients' demands? Do you run faster than your competitors? Will the growth of employees match the development of the market? Facing the uncertain future, digitalization should be the most certain job in many layouts.

Establishing the "values": Promoting future investment from the perspective of value. Where you have the most people, where you have the most money, where you have the most risk, where you have the least efficiency, is where you should start. Under digital governance, value can be designed and managed so that enterprises can see the current situation, discover the unknown and create the future. Digitalization is the key to intellectualization.

How is this done?

Each enterprise belongs to a different industry and stage. The book *Digital Transformation in Cloud Computing* summarizes Alibaba's own experience and the experience of helping many enterprise customers to migrate to the cloud. Some of them "break first and then establish", some of them "establish first and then break", and some of them "establish while breaking", which is a rare practical book in the field of digital construction. How can enterprise leaders make correct decisions to make the enterprise develop better and longer? Only with the forerunner, and practice of peer!

Peng Xinyu
VP of Alibaba Group

Preface II

As COVID-19 pandemic raged, the digital economy has become the engine of the whole economy, the development of digital economy is so fast, and digitalization has become the only certainty in an uncertain world. It is also because digital transformation has become so urgent that all industries are beginning to get digital anxiety. They want to embrace the digital economy, but for various reasons, they are hesitant to break from their current situation.

How does one solve digital anxiety, and is there a clear definition or method to guide digitalization? We believe that the answer is yes; first and foremost a strategic determination. Digital transformation is not a project, but a prime strategy. The goal of digitalization is closely related to 3 aspects, they are: cost reduction and efficiency; improved collaboration for the organization and the people in it; combination of business, data and intelligent technology. Starting from these three goals, through computing, data, connectivity and collaboration, digital transformation can be achieved. Especially for a large enterprises, they all have to go through the necessary stages of Internet technology(including cloud infrastructure), digitalization of core business and at last closed-loop data and business.

The first step of digital transformation is to build the innovation base through the cloud of infrastructure, and the computing power and storage efficiency of enterprises will also be greatly improved. Next, the core business processes are digitized. The most important thing is to identify what the core processes of the enterprise are and what are the most digitized at present, instead of digitizing all the processes. The next step is to achieve the closed-loop data business, gradually turning data into the production elements of the enterprise, to achieve the process from business to data, and then back to business, which is what we often say: the digitalization of all business and the businessization of all data. Digitization also helps enterprises restructure their organizations and processes, making decisions faster and making organizations more interconnected and agile. However, realizing digitalization is not the ultimate goal. The fundamental driving force of digitalization lies in intelligence, and the pattern of pursuing innovation on the basis of intelligence. Only more innovation can promote the continuous development of enterprises, then radiate and expand their digital capabilities.

Take FeiHe Dairy as an example, which is a transforming star company in recent years having in-depth cooperation with Alibaba Cloud Intelligence to promote the enterprise's comprehensive digital transformation, open up data islands in the enterprise, continue to develop capacity building in Middle Office, support business decisions and feedback business operations with the help of large data capabilities, so as to move from "manufacturing" to "smart manufacturing", and inject innovation power into the enterprise. The successful construction of the Data Middle Office has inspired customers' expectations for the construction of the Business Middle Office. Through the landing of the Business Middle Office in Feihe Dairy Group, it is expected to drive the business of the enterprise to better meet the needs of consumers and walk out of its own digital transformation road.

Who are the group of people who wrote this book? They are such a group of deliverers who are challenged by the implementation of business contracts. They aspire to be the dream team of enterprise digital transformation. They mentor customers as teachers, and fully immerse themselves into each industry, grow with customers and ecological partners, and help customers set up a system and leave a team for each project. This book is the result of the combination of Alibaba's own experience and the best practices of the digital transformation of many corporate customers. We hope readers will benefit from it.

Hu Chenjie
VP of Alibaba Cloud Intelligence

Author

Alibaba Cloud Intelligence GTS (Global Technology Services) is a technical commitment assurance team that provides complete life-cycle services for Alibaba Cloud Intelligence customers. Through professional delivery and implementation, systematic middle office support and standardized service product capabilities, we use the power of cloud and data intelligence to unite ecological partners to help customers achieve business value.

Book Staff

General Advisors
Li Jin, Peng Xinyu

Advisory and Expert Committee (No ranking):
Zhang Rui, Huang Huanhuan, Jin Jianming, Hu Chenjie, Huo Jia, Wang Sai, Ning Xiaoming, Wang Jiayi, He Long, Li Meng, Sun Lei, Fang Xiang, Yang Peng

Editorial Board (No ranking):
Jin Jianming, Zhuang Longsheng, Li Yichao, Cao Lin, He Qiang, Shen Shunhou

The authors of this book (by chapter):

- **Chapter 1. Cloud Computing and Digital Transformation**
  Jin Jianming
- **Chapter 2. The Top Level Design of Enterprise Digital Transformation:**
  Zhuang Longsheng, Huang Peidong, Li Yuchen
- **Chapter 3. Application Cloudization and Cloud Native**
  Li Yichao, Wang Keqiang, Cheng Zheqiao, Zhang Yuchen, Liu Xiao, Liu Xun
- **Chapter 4. Business Middle Office on Cloud**
  Qi Gaopan, Cao Lin, He Wei, Shi Jinliang, Sun Zhongpeng, Zheng Fa, Yu Kang, Song Wenlong, Yuan Tieqiang

- **Chapter 5. Data Middle Office on Cloud**
  He Long, Li Linyang, HeQiang, Dai Shaoqing, Ma Xin, Wang Yaqing, Xu Fu, Pu Cong, Song Chunying, Chen Li

- **Chapter 6. AIoT on Cloud**
  Shun Shunhou, Yang Peng

- **Chapter 7. Comprehensive Cases of Digital Transformation**
  Zhuang Longsheng, Li Chengqiang, Hu Zhiqiang, Caolin

Whole Book Planning and Overall Planning:
He Qiang

Special Thanks (No ranking):
Su Chen, Li Ting, Ma Peng, Wang Peijie, Liu Ming, Wang Luanfeng, Jia Linwei, Li Yali, Ya Lizhe, Wang Jiajia, Meng Xiangjun, Wang Yanping, Shao Wei

CHAPTER 1

Cloud Computing and Digital Transformation

Jin Jianming

THERE HAS NEVER BEEN an era like today that we can use information and data resource like hydropower and coal, and no era can use data mining and intelligence as much as today, in human history. The rapid arrival of the era of digital intelligence makes many people feel unprepared, but they also fully enjoy the convenience and innovation experience brought by the era of digital intelligence. Digital intelligence has a profound impact on all aspects of society and is constantly reshaping the way of life. More and more government institutions and enterprises put 'digital innovation' and 'governance modernization' on the agenda.

What is digital innovation? Digital innovation is to drive the reconstruction of business, organization and process with digital technology (Figure 1.1).

Computing: Centralized (Mainframe) → Distributed (Virtualization) → Cloud Computing (Using on demand) → Data Intelligence (Digitalization and intellectualization). This is key development milestone after human beings entered the computer age.

Data: File Store → Database → Big Data Platform → Data Analysis. Since the electronic storage of data, the data have evolved from simple storage to classified query, and then to in-depth analysis, which can help people make better and more efficient decisions.

DOI: 10.1201/9781003272953-1

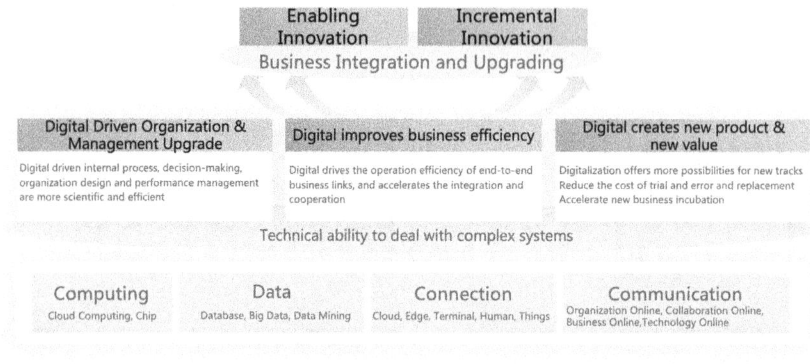
Enabling Innovation
Incremental Innovation
Business Integration and Upgrading
Digital Driven Organization & Management Upgrade
Digital driven internal process, decision-making, organization design and performance management are more scientific and efficient
Digital improves business efficiency
Digital drives the operation efficiency of end-to-end business links, and accelerates the integration and cooperation
Digital creates new product & new value
Digitalization offers more possibilities for new tracks
Reduce the cost of trial and error and replacement
Accelerate new business incubation
Technical ability to deal with complex systems
Computing
Cloud Computing, Chip
Data
Database, Big Data, Data Mining
Connection
Cloud, Edge, Terminal, Human, Things
Communication
Organization Online, Collaboration Online, Business Online,Technology Online

FIGURE 1.1 The value of digital innovation.

Connection: Cloud computing is not only a data center, a bunch of devices, a cloud operating system and some cloud computing, storage and network services, but also forms the interconnection of cloud, edge and terminal with people and everything. Cloud is in the center, providing unlimited computing power and massive data processing capacity, global control and macro analysis; edge is distributed in various regions, local computing and analysis, liberating transmission constraints, and local fast decision-making; the terminal connects all things, activates all kinds of devices, and touches all aspects of society like tentacles, so as to establish a connection between people and things. In the near future, when you go home, you may know at any time whether the street light on the road you want to cross is good or bad, whether you want to change the road, and so on, but this is just a very small use scenario of connection (Figure 1.2).

Communication: The biggest change brought by the Internet and digital age is the change of communication mode. New technologies such as IM (Instant Message) timely communication, video interaction, live shopping and face recognition are constantly impacting the original way of life, and life is becoming more and more convenient. Therefore, many organizations have begun to try to communicate and manage online, organize online management (paperless office, electronic travel application, real-time display of organizational effectiveness, online training, etc.), office coordination, online data-driven business, and communication of information flow

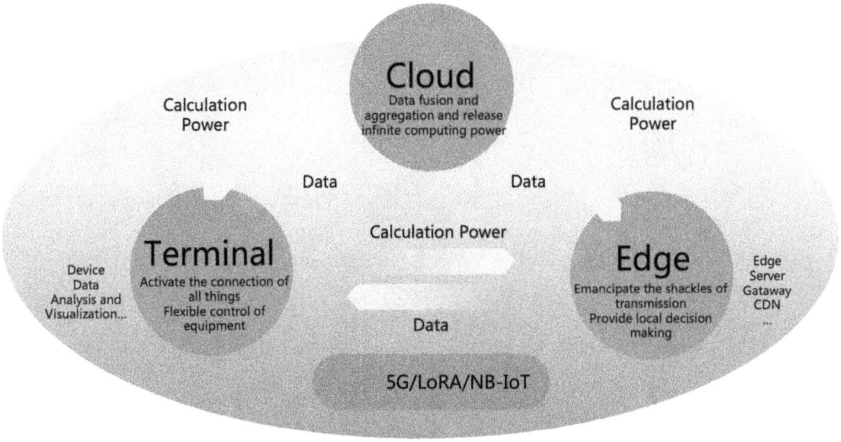

FIGURE 1.2 Connection.

and data flow of ecological partners are all advancing by leaps and bounds. In the traditional communication mechanism, information transparency has always been a challenging problem. Many isolated islands of information lead to fragmentation and incompleteness of information, which will further lead to information distortion and inaccurate decision-making when the whole organization makes many judgments and decisions. Personal experience becomes a crucial factor. When there are still a lot of policy instructions that need to be uploaded or sent to the management from bottom to top, it is actually inefficient in the traditional enterprise communication mechanism. People may send documents layer by layer, and it has been a long time since one thing has been conveyed. However, if the whole organization can be connected by data flow, the distribution of an instruction may take a few minutes. Especially in the face of enterprises with hundreds of thousands of employees, if we want to hold a full staff meeting, it used to be totally unthinkable. Now online video live conference, exhibition on cloud and other forms of online communication technology have become mature, and began to be widely used in all walks of life. Similarly, in the past, if there were dozens or hundreds of people modifying the same document in a project, it was unimaginable, let alone dozens or even a few people. It was also a troublesome version management. Now, some collaborative platforms can realize the ability of online multiple people editing at the same time. Therefore, in the era of digital intelligence, there are many innovations in communication (Figure 1.3).

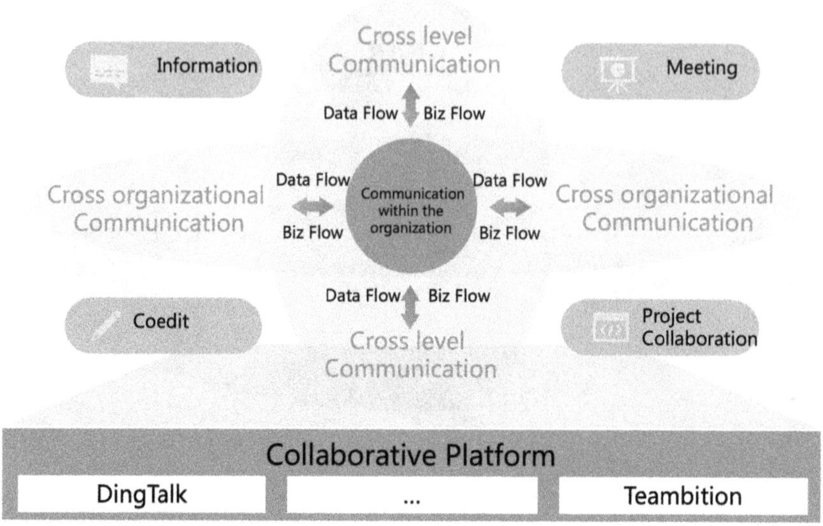

FIGURE 1.3 Communication.

In the big demand of digital innovation and governance modernization, cloud computing has entered the 2.0 era. From providing basic computing, storage, network, security and other resource services in 1.0, it has upgraded to a three-dimensional cloud service of infrastructure cloud & digital native operation system & industry solutions & ecological business system. From the cloud of 1.0 to the data, IoT (Internet of Things) and mobile, and then to the industrial and ecological development. Cloud based on 2.0 can achieve new business scenarios more flexibly, promote the landing of more innovative businesses through data and promote business change (Figure 1.4).

Alibaba has been trying to build a digital native operating system. Based on the cloud computing operating system, through the Internet of things and mobile technology, Alibaba has built a Cloud-Edge integration and Cloud-DingTalk integration technology system, so that all businesses can be digitized. With data as the blood, it connects all aspects of social life, so as to form a digital native operating system and promote business through digital innovation (Figure 1.5).

For any government, institution or enterprise, governance is the top priority. Good governance can ensure long-term stability and healthy operation, and vice versa. So, what is the modernization of governance?

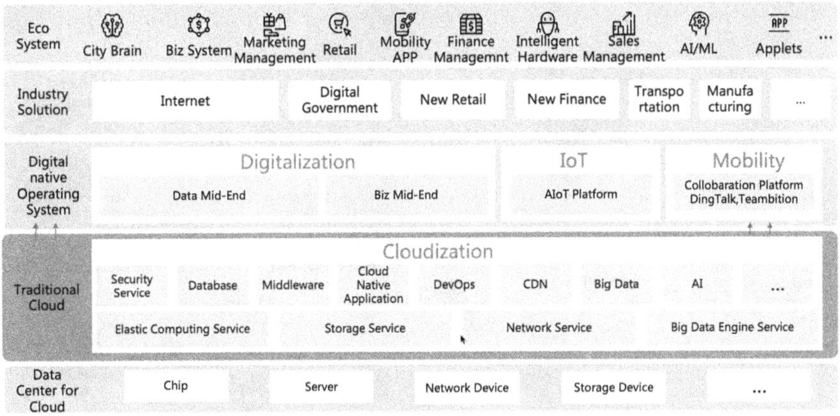

FIGURE 1.4 Cloud computing 2.0.

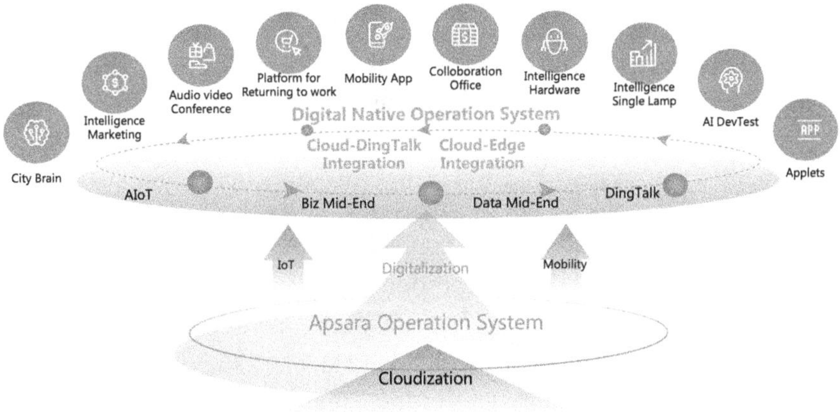

FIGURE 1.5 Digital native operation system.

Governance is nothing more than internal and external. The core goals of internal governance is to improve the efficiency of the organization, reduce operating costs, and ensure that the organization is safe and controllable; externally, take government institutions as an example, how to realize the refinement, source and rapid response of social governance, realize the people-centered service, and release the data dividend in the critical period of economic transformation, rather than continue to focus on the demographic dividend. Therefore, it can be summarized into two key words: 'dual online, internal and external communication', where

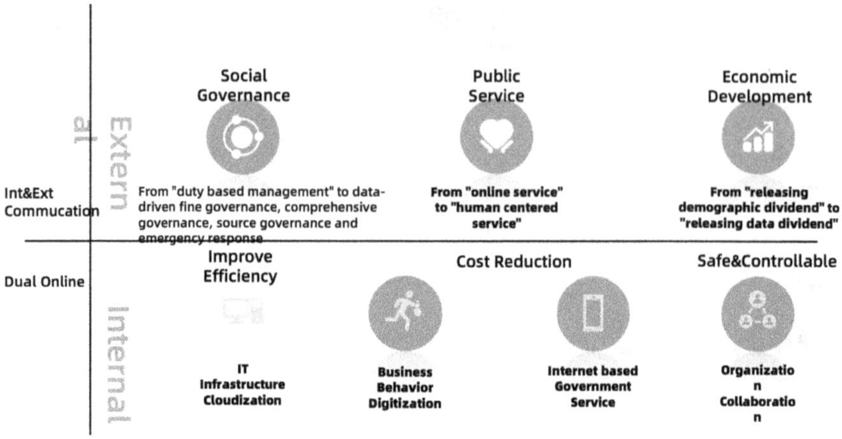

FIGURE 1.6 Governance modernization.

'dual online' is to realize online organization and online business, and 'internal and external communication' is to be proficient in internal governance and good at external management (Figure 1.6).

The above gives an overview of cloud computing and digitization. Next, let's uncover the mystery of cloud computing and digitization.

1.1 PAST AND PRESENT OF CLOUD

1.1.1 The Evolution of Cloud Computing

1.1.1.1 Born in the Era of Extreme Single Computer Computing

On August 9, 2006, Eric Schmidt, CEO of Google, first proposed the concept of 'cloud computing' at SES San Jose 2006. But in fact, the real external service of cloud computing comes from the elastic computing cloud (EC2) service launched by Amazon in March 2006. The earliest concept of cloud computing originated from 'the network is the computer' proposed by Sun Microsystems in the 1980s. In 2006, Microsoft, IBM, HP, Oracle, SAP and other companies are still in the ascendant era. Why did the cloud computing, which is now popular all over the world, come into being in the era of extreme stand-alone computing and single software? I think you may have found an interesting phenomenon that Google and Amazon are both studying cloud computing, but there is a big difference in the direction. First of all, let's look at the respective businesses of Google and Amazon. Google's main business is search business, and Amazon's main business is e-commerce business. When it comes to Google's cloud computing, the most famous papers are FS, MapReduce and BigTable. Google has been focusing on the research of distributed file system and big data-related technologies in its early stage, mainly because Google's business is related to how to process data on a large scale and efficiently. Amazon's early research direction of cloud computing is also very interesting. Because Amazon's e-commerce enterprises have strong seasonal sales characteristics and short-term seasonal characteristics, when the sales peak comes, they need to quickly invest a lot of computing resources, and these resources will become a lot of waste after the seasonal sales peak has passed, so it makes Amazon incubate EC2, the earliest cloud computing service. However, these requirements cannot be solved by traditional software and hardware, which forces new companies such as Google and Amazon to start to study technologies to meet their own business. Due to the greater seasonal impact, Amazon has also commercialized earlier and launched cloud computing business services in order to save costs and avoid waste. After the emergence of Amazon's cloud AWS, no matter the new Internet company Google, or the traditional software and hardware companies Microsoft and IBM have successively launched their own cloud. At the same time, many new cloud computing companies such as Salesforce and Rackspace have been born. The most interesting one is Salesforce, which is different from AWS, Microsoft, Google, IBM and others focusing on infrastructure cloud services. Salesforce has seen the high cost and relatively few services faced by many small- and medium-sized enterprises when using ERP (Enterprise Resource Planning) and CRM (Customer Relationship Management) software. Salesforce has rapidly launched the cloud computing service based on CRM software, which is a great breakthrough and innovation. In fact, Salesforce started very early. Marc Benioff, the founder of Salesforce who was still working for Oracle, boldly proposed 'no software' and 'software as a service' in 1999. In the 20th anniversary of the establishment of the company, the company's business has run out of a very sexy growth curve, accounting for about 20% of the CRM market share, far ahead of other CRM software company.

After looking at the United States, let's look back to China. Alibaba was also an e-commerce company at the beginning. It had similar business scenarios with Amazon, and the complexity was even greater, because China's double 11, 618, New Year's Day and other large-scale promotion activities were larger and more frequent, and the demand for flexible resources was stronger than other region Internet companies. On the one hand, we can see the development trend of international technology. On the other hand, due to the extremely high cost of traditional software and hardware in the e-commerce industry, Alibaba also started the research on cloud computing in 2009, established Alibaba Cloud, and commercialized it on a large scale in 2014. So far, the business form of cloud computing has undergone tremendous changes, not only the layout of traditional cloud computing business, but also the development of cloud computing. It is further to the smart city, retail cloud, financial cloud and other new areas with strong industry attributes. Subsequently, a group of cloud computing companies focusing on the segmentation industry have sprung up, including UCloud and QingCloud. With the rapid promotion of Alibaba Cloud, AWS, Azure and other cloud computing platforms, we all see that cloud computing has a very bright future and huge market potential. After 2016, Tencent, Huawei, Baidu, Kingsoft, JD and other domestic enterprises have started to rapidly launch their own cloud computing. China's cloud computing has entered the fast lane and entered the spring and autumn and Warring States period. However, with the increasingly obvious Matthew effect, it may quickly enter the Three Kingdoms era in the future.

It is because of the high cost of traditional hardware and software provider, the unfriendliness of architecture for new business, the difficulty of traditional hardware and software provider to choose their own life, the rapid development of the Internet era and other factors that cloud computing was born at the peak of traditional hardware and software business, and developed rapidly. However, compared with the traditional technology architecture, the overall market share of cloud computing is still a relatively small number, and there is still a very broad development space for cloud computing in the future.

1.1.1.2 Four Stages of Cloud Computing Development

Cloud computing has experienced several different development stages since its birth, and it is still in high-speed evolution.

1. Public cloud stage: Public cloud is the earliest form of cloud computing, and it is also the ultimate form that many cloud providers expect to achieve in the future so far. It starts from the earliest elastic computing shared resource rental service. So, what is public cloud? Generally speaking, public cloud is a kind of cloud computing mode in which users or customers connect to a third party through the

Internet to provide public computing, storage, network, security, data and application sharing services. The following chart is a typical public cloud usage scenario on Alibaba Cloud. The emergence of public cloud not only reduces the difficulty and cost of ownership for small- and medium-sized enterprises and individual users to use basic resources such as computing, storage and network, but also improves the convenience, flexibility and timeliness of use, which also takes effect for large- and medium-sized enterprises. At the same time, for large- and medium-sized enterprises, it has also changed their information construction mode of holding proprietary data center, server, network, storage and other equipment with heavy assets in the past, and can further optimize and upgrade the personnel structure, and focus more on their own business development. Of course, the public cloud is not only limited to the infrastructure to provide services, but also takes the platform, software and data as a new mode of public cloud to provide services, which are called IaaS (infrastructure as a service), PaaS (platform as a service), SaaS (software as a service), DaaS (data as a service), etc. In this way, cloud computing is like hydropower. Users are tenants one by one, applying for the required resources and paying the required fees on demand. The rapid arrival of public cloud has brought about qualitative changes in the traditional IT market. On the one hand, as enterprises and ordinary users, they need to build their own data centers, purchase hardware and software equipment as assets, and train a large number of technical personnel to carry out the operation and maintenance of these infrastructures. With the arrival of public cloud, many technical personnel begin to transform to be technical architects who understand the business, and many enterprises also began to change thinking, and migrate the data and applications to the public cloud. On the other hand, with the popularity of PaaS, SaaS and DaaS services on the public cloud, enterprises are not limited to renting infrastructure shared resources, but directly act on the container platform, middleware and big data platform on the public cloud to build their own applications, and even directly rent the software made by the third-party cloud platform for direct use. The most famous is the SaaS service of Salesforce. Of course, there are also a lot of data public services that are becoming more and more popular (Figure 1.7).

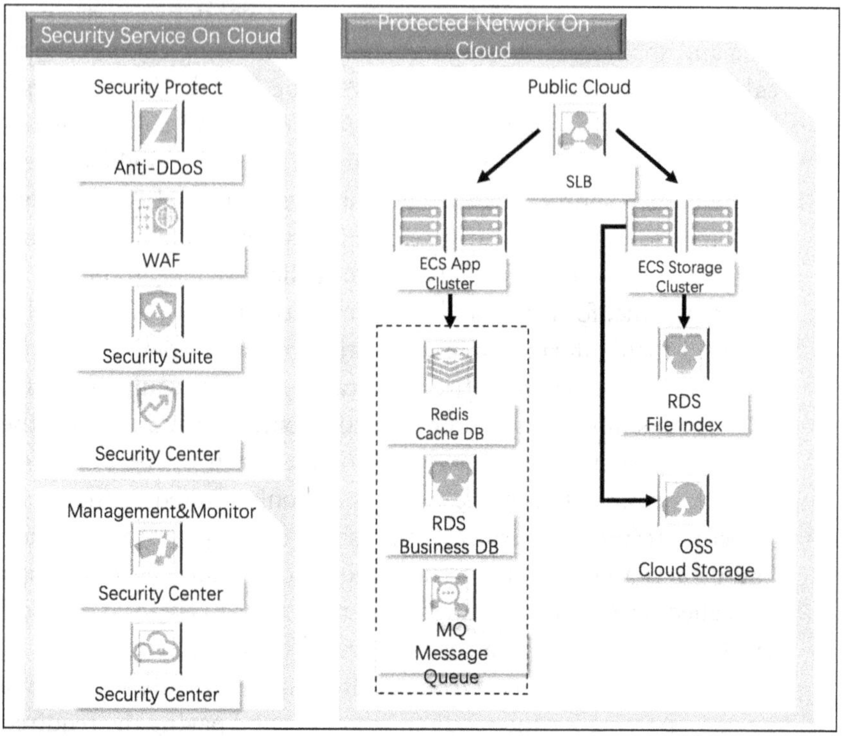

FIGURE 1.7 Typical Architecture on cloud.

2. Private cloud stage: Compared with the public cloud, people's attempt to private cloud appeared earlier, but at that time it was not called private cloud, but in another form, called virtualization. In the early enterprise IT architecture, in order to better save resources and avoid the waste of exclusive hardware resources, virtualization technology of hardware resources appeared, which can achieve resource isolation, scalability and high security attributes. This is not a real cloud, but it is a gratifying attempt, and in a long period of time, even today, there are many enterprises using this traditional virtualization technology. VMware, Hyper-V, Citrix, PowerVM and other virtualization technologies have emerged, which has greatly improved the utilization mode of the exclusive and extensive resources in the past. But virtualization technology does not change the traditional IT usage essentially. We still need to plan a few months in advance, purchase software and hardware, install and

deploy them a few months in advance, and the scale of virtualized equipment is relatively small. Every enterprise or institution is still carrying on all kinds of construction. With the emergence of the public cloud, this concept has changed. In the fifth year of the emergence of public cloud, the private cloud represented by OpenStack began to appear. In China, Alibaba Cloud also began to develop a private cloud platform with the same technology architecture and product code as its own public cloud in 2014. Alibaba Cloud called it dedicated cloud (exclusive public cloud). This platform took a different technical roadmap from OpenStack and chose the unified framework of public and private cloud structures. At the critical time point when the world's leading cloud providers take the public cloud as their ultimate goal and make every effort to promote the public cloud, why is there a large number of demands in the private cloud market? Because the initial cloud appeared in several Internet companies, no matter Google, Amazon or Alibaba, they are all typical Internet companies, and their business scenarios are also based on the Internet, so the customers they initially sought are mainly Internet-type customers. However, with the concept of cloud computing gradually gaining popularity, more and more enterprises in traditional industries are also opening up. Although more and more enterprises are trying cloud computing, they have high concerns about data security, high control of resources and so on, and naturally think that private cloud will be more secure. Therefore, with the rapid development of public cloud in the market, there are a lot of demands for private deployment of cloud. At present, there are many kinds of private cloud architecture models in the market. Alibaba Cloud adopts the same architecture and code of public cloud, which can realize the private cloud deployment of more than 10,000 machines. AWS adopts the localization of infrastructure and operation mode, which is similar to a localization extension node, called outpost. It can connect with the VMware localization deployment environment, or independently deploy AWS software and hardware integrated equipment. Many other cloud computing providers also have different attempts.

Figure 1.8 shows three types of private cloud products of Alibaba Cloud. To analyze and explain the product direction of private cloud, they are enterprise version, agile version and all-in-one version.

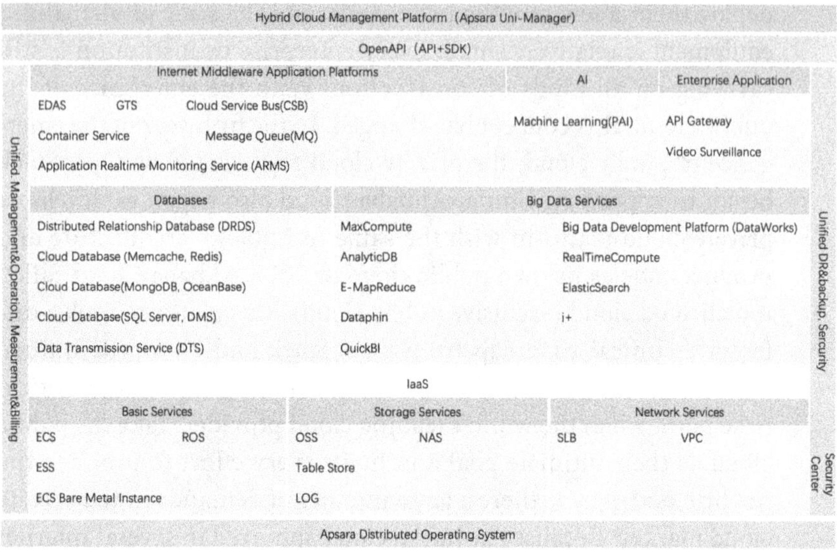

FIGURE 1.8 Enterprise version.

Enterprise version: Because it has almost the same product capabilities as the public cloud, but different in scale, the same cloud operating system, rich IaaS and PaaS products, provide services in the form of OpenAPI (Open Application Programming Interface) and SDK (Software Development Kit), and configure disaster recovery, operation and maintenance management, security management and other management functions for the private cloud, and carry out unified management with a unified management portal. This product form is suitable for enterprises or institutions that want to build a complete cloud of their own needs.

Agile version: Different from enterprise version, agile version mainly focuses on specific product use scenarios or industry scenarios, such as dedicated cloud version focusing on artificial intelligence (AI) field and dedicated cloud version focusing on Internet architecture. Agile core solves the problem of minimum deployment scale of dedicated cloud, which is one order of magnitude smaller than enterprise version. And the agile version supports upgrading to the enterprise version (Figure 1.9).

All-in-one version: Focus on a specific scenario and product, for example, focus on database products, realize out of the box use and do not purchase any software license (Figure 1.10).

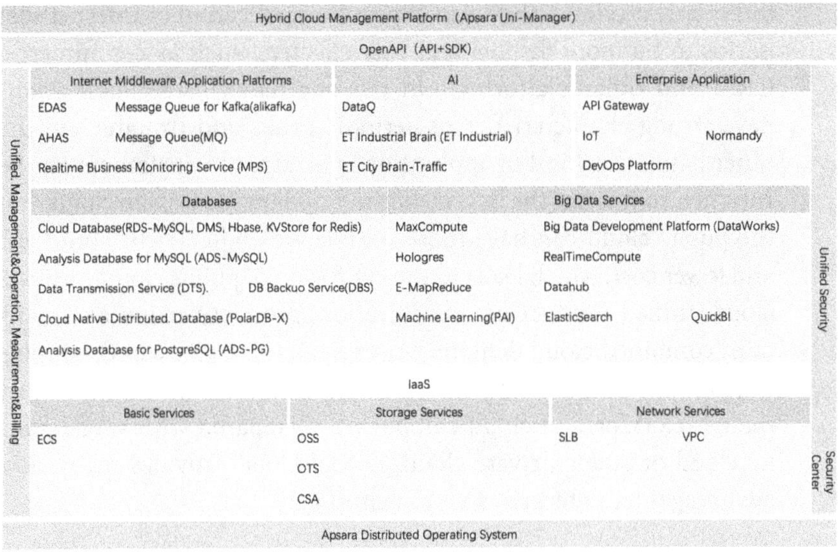

FIGURE 1.9 Agile version.

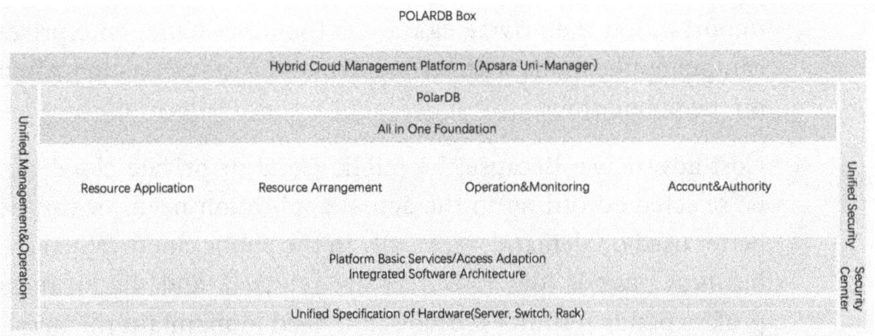

FIGURE 1.10 All-in-one version.

To sum up, we can see that cloud computing providers are trying to provide private cloud solutions for different scenarios and industries based on a unified architecture and platform. As the needs of enterprise customers become clearer, this trend should evolve further.

3. Hybrid cloud: With the advent of public cloud and then private cloud, this is not the end point. More and more enterprises are not satisfied with a single public cloud or a single private cloud. For example, in the face of many large enterprise customers who expect

to use private cloud, they also expect the application of Internet scenarios to be more flexible and cost-effective, such as e-commerce, CRM and other platforms. On the one hand, some applications have strong characteristics of periodic peaks and troughs. On the other hand, this kind of application is oriented to customers on the Internet, so it is not the best choice to build on the private cloud. But the public cloud can have more flexible scalability, richer products and lower cost, which leads to the confusion of public cloud + private cloud unified product use specification, product version and standard combined cloud demand. At present, the major cloud computing providers are making efforts to hybrid cloud, which is the most mainstream form of cloud computing. Compared with single public cloud or single private cloud, hybrid cloud provides many new advantages for enterprises and institutions.

- Perfection: On the one hand, it solves the concerns of enterprises or institutions about data security. Enterprises or institutions can put data security applications and data that they think are very important on the private cloud. On the other hand, enterprises can put applications with strong demand such as elastic scaling on the public cloud.
- Cost advantage: Because the public cloud or private cloud can be selected according to the actual application needs, it can be better used on demand, especially in the public cloud. When the business peak is low, resources are released, and the business peak expands resources. It does not need to maintain the maximum number of resources required by the business peak all the time, thus greatly reducing the cost.
- Expansibility: Private cloud has natural resource constraints. Whether it is data center or hardware resources, it is not very different from traditional IT infrastructure in this respect. It needs to purchase hardware in advance. This cycle is very long. Therefore, private cloud has greatly improved compared with traditional IT infrastructure in terms of scalability, but there is no fundamental change, but public cloud can essentially solve the problem. To solve this problem, it is theoretically possible to expand the capacity infinitely. At present, the scale of public

cloud machines of major cloud computing providers is in the order of millions. Therefore, by using hybrid cloud, we can take advantage of the scalability of public cloud to help enterprises or institutions solve the scalability problem.

- Connectability: The public cloud can reach almost anywhere in the world where there is Internet, which provides good connectivity for enterprises or institutions. This is also the advantage of hybrid cloud.

4. Industry Cloud: The public cloud, private cloud and hybrid cloud we saw above are mostly from the perspective of cloud technology itself. However, for many enterprises or institutions, there is also a strong demand for cloud computing with industry attributes. For example, in order to build digital government governance capability, government institutions need to have cloud computing with industry attributes based on AI and big data. There are a lot of different kinds here, for example, Alibaba Cloud's industrial cloud solutions, such as traffic brain, public security brain and agricultural brain, are different from the previous solutions that are simply pieced together through a variety of basic products. Instead, they actually enter into the business details of the industry, and combine the big data ability and AI ability to realize the Industry Cloud solution products that can effectively solve the industry problems. In this regard, the choice of cloud computing providers is very different from that of IaaS and PaaS in the past. US cloud computing providers are quite different, either focusing on IaaS and PaaS products, or SaaS and DaaS products. However, Chinese cloud computing providers have taken one more step in industrialization by combining various technologies and commercial capabilities of their own business groups, such as Alibaba's retail cloud, logistics cloud and financial cloud. Tencent cloud also provides similar industrial cloud service capabilities with the help of Tencent group's overall commercial and technical capabilities, which is a major watershed in the development path of the head cloud computing providers.

1.1.1.3 Cloud Computing Associated Services

With the continuous maturity and wide application of cloud computing technology, the related services needed by cloud computing are also quickly spawned. At present, the main service forms of cloud computing include digital transformation consulting, cloud migration, cloud native application design and development, big data consulting and implementation, cloud hosting, digital operation service and cloud computing after-sales service. The form and content of services are similar to traditional IT services, and there are many new innovations. For example, consulting, migration, cloud hosting, after-sales and other services are the main force of traditional IT services. However, many new contents have been added, such as cloud native application development, industrial big data servicesand digital operation services, which begin to provide application development based on API (Application Programming Interface) encapsulating cloud computing providers' own business capabilities, and data-driven services based on data and AI analysis results, new service content of operation and operation-driven business. There are also microservices that can be purchased out of the box based on the cloud market. On the one hand, the cloud computing service market is learning from the previous experience; on the other hand, it is also making great strides to explore and move forward.

At present, in the cloud computing service market, there is a pyramid development trend, the scale of service providers presents Matthew effect, the head service providers are further gathering, the traditional service providers are busy with transformation and new cloud service providers are springing up. Service providers that only focused on consulting in the past began to provide more cloud-based customized development, cloud migration, big data implementation and other services, and further expanded their service boundaries. While doing services, new cloud service providers developed their own cloud products in a certain field, hoping to compete in the vertical industry.

Another interesting phenomenon is that cloud computing providers are also showing polarization. Although the ultimate goal of most cloud computing providers is 'integrated' mode, there are great differences in their development paths. In the Chinese market, on the premise of moving toward the 'integrated' mode, cloud computing providers have undertaken a large number of system integration (SI) projects and completed them together with their ecological partners, while the major cloud computing providers in the United States focus on cloud products and cloud services themselves, and their professional services are handed over to their ecological partners. So, what is the reason for this apparent divergence? This divergence probably started from 2017 to 2018. In 2018, China's digital governance was launched on a large scale, and there was a lot of business demand for building a digital government, such as 'run once at most', 'public security brain', 'traffic brain', 'agriculture brain' and 'water conservancy brain'. In 2017, Alibaba Cloud put forward the concept of 'Middle Office strategy', and a large number of enterprises have started the road of digital transformation, such as 'Business Middle Office', 'Data Middle Office', 'management Middle Office' and 'AIoT (Artificial Intelligence and Internet of Things) Middle Office'. Subsequently, China's domestic cloud computing providers have launched their own 'Middle Office' products and services. All kinds of 'brain' and 'Middle Office' are based on the technology base, and need to continue to go deep into the actual business scenarios of customers, and further improve and sublimate in combination with customer scenarios. Therefore, the development direction of 'big integration' first and then 'integrated' gradually appears. In the US market, on the one hand, the IT market itself has been relatively mature, a lot of information infrastructure has been built, and the IT service providers are relatively large and mature, and the maturity of information also hinders the digital transformation. On the other hand, the government and enterprises do not have the strong demand for digital transformation such as China. Lack of a broad market environment for digital transformation, therefore, cloud computing providers pay more attention to the development of basic cloud products themselves, and do not further explore industrial cloud products. This is also the first time in the development process that it has been divided into two obvious branches. As for the future direction, it is likely that it will return to the 'integrated' mode, but the content of cloud services will be very different.

For the current form of cloud services and cloud service providers, make a simple classification:

- Managed service provider: Providers that provide end-to-end cloud consulting, cloud migration, development and big data implementation, cloud hosting and other one-stop services are market leaders.
- Consulting company: Consulting companies focusing on digital transformation and cloud-based consulting.
- Independent software developer: Industry-oriented customized cloud native software developers.

- SI service provider: Provide services such as hybrid cloud platform deployment, cloud migration and cloud product integration.
- Operation service provider (TP): Provide digital operation consulting service and agent operation service.

1.1.2 Cloud Computing 1.0

1.1.2.1 Network That Provides Resource Leasing on Demand

When cloud computing first came into being, it was mainly to provide distributed computing, perform distribution tasks and finally merge distributed computing results. In the early days, Google's MapReduce was a similar application, so it was also called grid computing in the early days. It used thousands of servers to return computing results in a very short time. In the early stage, cloud computing also provided distributed computing, storage and network services according to the demand of computing and other services. Therefore, from a narrow perspective, cloud computing is like a 'network that provides resource leasing on demand'.

1.1.2.2 Classification of Cloud Computing Services

1. Infrastructure as a Service (IaaS)

   IaaS providers provide users with cloud-based IT infrastructure, including processing, storage, network and other basic computing resources. Users can remotely deploy and run any software (including operating system and application program), while suppliers charge service fees according to the amount of storage server, bandwidth, CPU and other resources used by users.

   Public cloud IaaS is an 'asset-focused' service mode, which requires large infrastructure investment and long-term operation technology experience. This business has a strong scale effect. Therefore, once the giant's advantages appear, it will produce Matthew effect and build a wide 'moat' through price, performance and service. With the rapid maturity of technology, cloud computing giants have started a large-scale price reduction, allowing the technology dividend to feed back the market. Since Amazon launched AWS service in 2006, its price has been reduced 42 times by April 2016; from October 2015 to October 2016, Alibaba Cloud has reduced its price as much as 17 times in a year.

The development of IaaS can be roughly divided into germination stage, growth stage and shuffling stage.

- IaaS in its infancy: In 2008, IBM established its first cloud computing center in China, and IaaS officially entered China. In the following year, Shanda and Alibaba Cloud began to develop pilot-related cloud services.
- IaaS growth stage: Around 2013, Microsoft and Amazon's AWS IaaS business officially entered China. In the same year, UCloud, Qingyun and other IaaS start-ups were established and began to provide services. Tencent, Huawei and other giants also joined the cloud service camp.
- Shuffle stage: By 2015, the development of IaaS cloud industry tends to be stable, the industry structure and profit model become increasingly clear, industry leaders begin to appear, and some cloud computing providers gradually begin to make profits. At the same time, the industry competition intensifies, entering the shuffle stage.

In 2020, several giants of cloud computing still maintain a high growth rate and make great strides forward. For example, according to Alibaba Group's financial report, Alibaba Cloud's revenue in the fourth quarter of fiscal year 2020 (from April 1, 2019 to March 31, 2020) reached 40.01 billion RMB, covering a wide range of industries and enterprises such as finance, health care, public transport, energy, manufacturing, government institutions, games and multimedia. The cloud computing industry is gathering to the giants.

2. Platform as a Service (PaaS)

PaaS refers to the platform of software development as a service, and provides to users. Users or enterprises can quickly develop their own applications and products based on PaaS platform. At the same time, the application developed by PaaS platform can better build enterprise application based on SOA (Service Oriented Architecture) architecture. As a complete development service, PaaS provides all the functions of development platform from development tools, middleware to database software.

In 2009, Sina launched SAE Alpha, the first PaaS platform in China. In 2013, Microsoft, Amazon AWS and other overseas public cloud PaaS services entered China; after that, BAT (Baidu, Alibaba, Tencent) launched developer PaaS platforms (Baidu's BAE, Alibaba's ACE, Tencent's Qcloud); vertical PaaS platforms in IM (Instant Message), push and other fields began to provide services (personal push, financial cloud, etc.). In 2015, domestic large-scale cloud service providers began to open more cloud service capabilities for developers (such as Alibaba Baichuan). At the same time, vertical PaaS platform also developed rapidly, such as PaaS platform for logistics network and voice recognition.

At present, the basic development status of PaaS providers is to expand service forms and customer groups based on the original technical functions and advantages. The expansion direction mainly focuses on the following three aspects:

- Expand service function: Expand multiple PaaS services from a single PaaS service to form a function store.
- Expand service forms: On the basis of the original PaaS services, expand similar SaaS services.
- Develop new customers: On the basis of PaaS general module, it provides customized process management for enterprises.

At present, the development trend of PaaS mainly shows in three aspects:

- Business-type homogenization: PaaS providers providing single function will grow into PaaS tool stores horizontally, which may result in a high degree of business coincidence among PaaS providers.
- Developer value-added services become growth point: On the basis of providing functional modules, the platform ecology is formed to provide technical services that meet the whole process of enterprise/application life cycle.
- International business became a new growth point: In the process of international promotion, technology modules are less

restricted by geographical location and usage habits. Breaking through the overseas market will become a new growth point for PaaS providers.

3. Software as a Service (SaaS)

The concept of SaaS service came into being in 2003, which is the earliest form of cloud service introduced into China. SaaS is a mode of providing software through the network. With SaaS, all services are hosted on the cloud. Users no longer need to buy software and maintain the software. Now, a complete enterprise web application can provide an agile and unified enterprise cooperation platform on the cloud. Just like using tap-water, enterprises can rent online software services from SaaS providers according to actual needs. SaaS can help enterprises reduce costs and manage hardware, network and internal IT departments.

The early SaaS service is mainly used for sales management, such as 800 customers and xTools. ASP is the early SaaS service. In 2008, traditional enterprises began to transform, and office software enterprises (such as UFIDA and Kingdee) began to develop SaaS business. SaaS enterprises are concentrated in the field of CRM and ERP process management. In 2013, SaaS services began to be subdivided, tools such as communication, email and Internet disk began to be SaaS based, and SaaS platforms based on mobile terminals began to appear. With the popularity of mobile terminals, SaaS has been growing rapidly since 2014.

At present, the scale of enterprise SaaS service market maintains rapid growth, and continues to be favored by capital. Relevant participants continue to explore new service forms. The basic development status of SaaS providers can be summarized as follows:

- Traditional software enterprises transform into SaaS service providers by expanding service forms.
- Based on the original vertical SaaS services, start-ups expand similar PaaS services to the bottom.
- Internet giants expand platforms, introduce third-party service providers on the basis of SaaS basic services, form SaaS service entry platform and create cloud service ecology.

The future development trend of SaaS mainly shows in three aspects:

- Intelligence promotes process management efficiency: The introduction of AI and other technologies will greatly improve the efficiency of process optimization, improve services and become an important driving force to improve SaaS services.
- Industry vertical SaaS space is large: SaaS enterprises that accumulate certain customers in the vertical industry have the opportunity to explore other business opportunities in the supply chain, such as B2B and value-added services.
- Get through fragmented services: Most service providers focus on a single SaaS process and face the problem of service data interconnection, so there are opportunities to provide SaaS services integrating fragmented data and processes.

4. Data as a Service (DaaS)

DaaS is a new service concept developed after IaaS, PaaS and SaaS in recent years. DaaS, data as a service provides a new way for enterprises and other enterprises to share data through centralized management of data resources and data scene. In the past, enterprise data were either scattered in various teams or departments and could not be provided as an internal service to improve enterprise operation efficiency, or each enterprise regarded data as its own gold mine and would not share it with other enterprises or individuals. In this case, the data gold mine would not produce much value, because it doesn't circulate. Only by cashing out the gold mine and allowing other enterprises to use data resources, can they obtain the materials they want. And with today's big data explosion, no enterprise can collect all the data they need. With DaaS service, they can purchase the data they need with other companies, and enhance the competitiveness of enterprises through division of labor and cooperation.

1.1.3 Cloud Computing 2.0

1.1.3.1 The Emergence of Industry Cloud

In the past 10 years, the general service capability of cloud computing has grown rapidly, but cloud computing is not a panacea. It has greatly solved the problems of flexibility, scalability, cost, timeliness and so on.

At the same time, it is also faced with how these basic technical capabilities can better and more effectively help customers realize the real business value. The previous article (please refer to 1.1.1–3) has mentioned why the Industry Cloud appears, so let's further uncover the mystery of the Industry Cloud.

Industry Cloud first has several unique attributes: full link, industry segmentation, highly customized and closed-loop service.

- Full link: Full link cloud product service capabilities, from IaaS to PaaS, from PaaS to SaaS and DaaS.
- Industry segmentation: Each Industry Cloud has a strong industry attribute, with its own unique laws and regulations, technical specifications, and specific industry business, such as financial cloud and retail cloud.
- Highly customized: Industry Cloud is actually close to customer business. Although many Industry Cloud providers want to focus on the bottom IaaS and PaaS layer to provide services, customers who like Industry Cloud just pay more attention to the business and data service capabilities of Industry Cloud. So, at present, most Industry Cloud providers are moving up from the basic cloud platform service capabilities to business services and data services ability development direction.
- Closed-loop service: Industry Cloud providers will have their own complete service system from consulting, implementation, after-sales, operation and so on.

Next, let's take a concrete look at the general mode and technology architecture of a typical Industry Cloud.

Government affairs cloud: As we all know, the most complex management system in the world should be the government's administrative management system, which involves all aspects of the national economy and people's livelihood. The Industry Cloud of an administrative cloud is also very complex. Let's see what the Industry Cloud architecture of a typical government industry should look like (Figure 1.11).

Typical government cloud includes several core modules: cloud platform, business platform (government application support), data platform

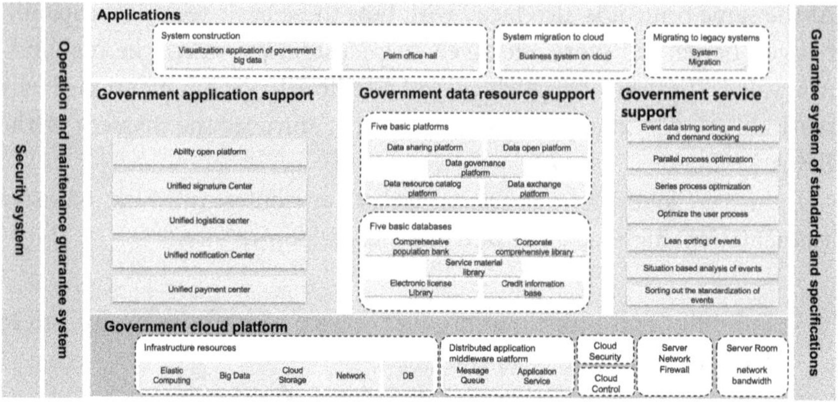

FIGURE 1.11 Cloud Architecture of typical government industry.

(government data resource support), process platform (government service support), operation and maintenance management platform, and customized application.

1.1.3.2 Business Operation System and Cloud Computing

In the process of transformation from IT services to digital services, based on technology, business capabilities and data capabilities are being served. Alibaba put forward the concept of business operating system in 2018, which abstracts the capabilities precipitated from Alibaba's own business operation into services and provides them as an extension of cloud services. Business operating system can be understood as technology as the foundation, data as the blood, shared business module as the skeleton, scene business as the vein, and ultimately form a complete and vivid digital service system that can provide services for many enterprises and institutions, also known as business operating system. Big aspects can be abstracted into financial cloud, such as advertising services, global marketing services and digital supply chain services. Commercial operating system is different from traditional cloud computing. On the one hand, business operating system makes traditional cloud platform transparent, and makes commercial service ability and data service ability stand in the front. On the other hand, business operating system provides commercial ability service, not simple technical service.

The core problem to be solved by business operating system is: Under the background of Internet customer-centered era, facing the rapidly

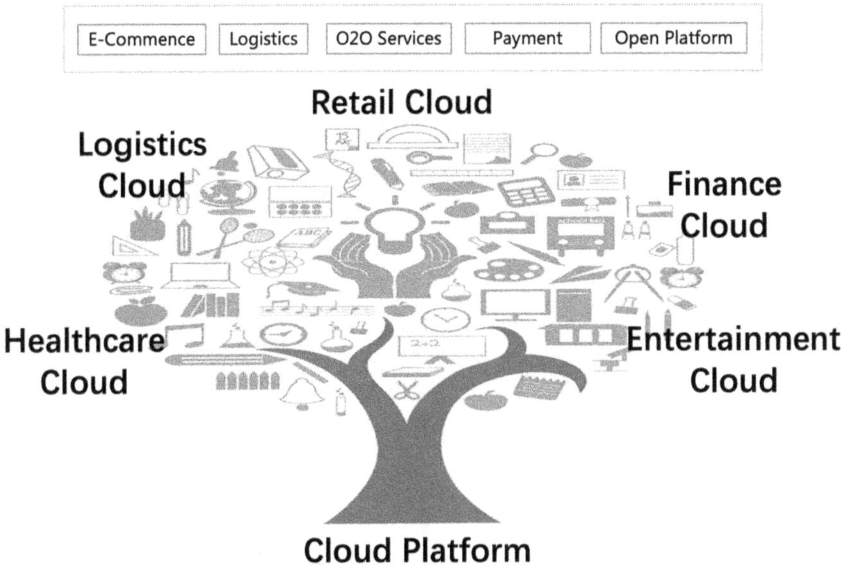

FIGURE 1.12 Evolving cloud computing.

changing market and business, a new design pattern and concept is used to solve the problems of uncertainty boundary and unpredictability. At the same time, it builds a customizable, pluggable, scalable business capability and technology ecosystem, and has a complete development ecosystem (Figures 1.12 and 1.13).

1.1.3.3 IoT Accelerates the Development of Cloud Computing

Human society has gone through several key milestones in the way of linking people to the world. The way of linking is constantly evolving, especially in the past 20 years, great changes have taken place.

The first stage: The age of letters: As a great invention in human history, papermaking not only solved the problem of civilization inheritance, but also made the way of human communication change for the first time. People can communicate with each other through letters. Of course, the disadvantages of the letter are also at a glance, poor timeliness and poor interaction.

The second stage: The age of telephone: Since the invention of the telephone, sound through radio waves, no matter how far away people can have a high timeliness of communication and communication.

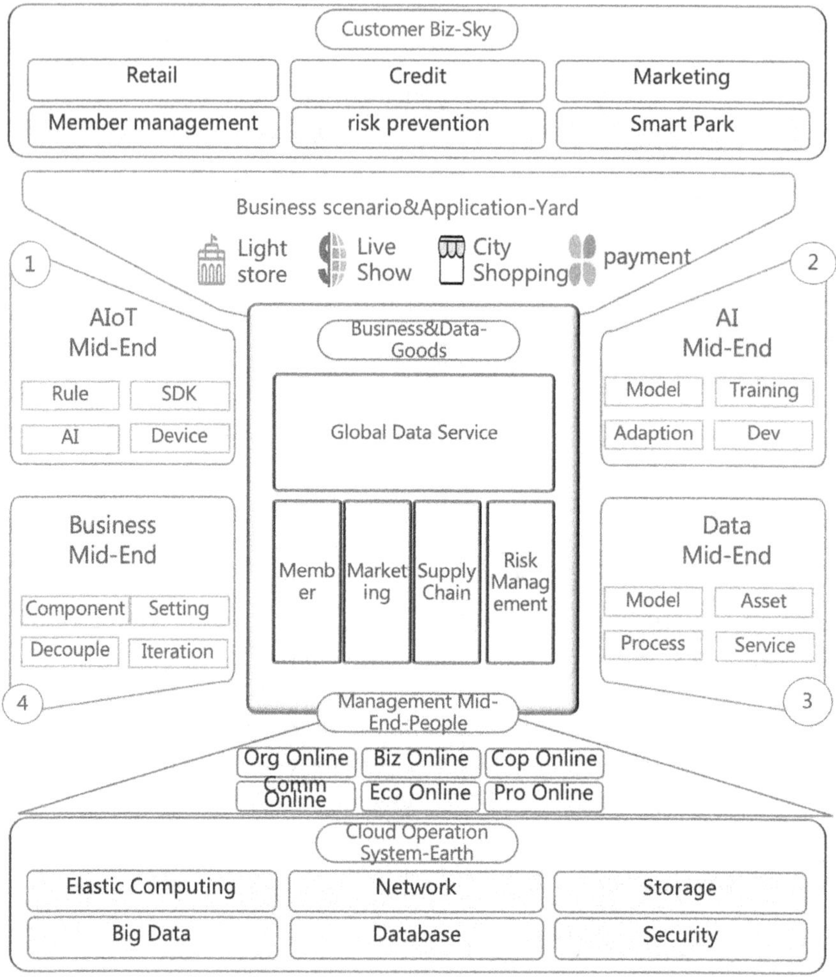

FIGURE 1.13 Sky-earth-people and people-goods-yard.

The third stage: The age of instant messaging: With the advent of instant messaging technology, the one-to-one problem in the telephone era has been solved, which is another great leap. Multi-person interaction and multi-style communication mode appeared. Simple communication evolves into complex scene communication.

The fourth stage: Video age: The desire of human beings is always endless. In the past, the communication methods of words and voice greatly improved the ability of communication and linking, but they still did not meet the demands of people who want to see images. The arrival of the era of video communication has once again improved and enriched the communication methods and channels. At present, live broadcast and short video are the new communication mode.

The fifth stage: The age of interconnection: In the past, communication was between people, but there was no link and interaction between people and things, between things and things. The emergence of the Internet of things has greatly changed the communication mode. Through a variety of Internet of things devices, like mobile terminals, people can feel all kinds of emotions (temperature, humidity, scars, etc.) of things, and can also directly receive people's instructions. Let things have feelings for the first time. Combined with AI and big data, the whole world seems to have become a living organism. People are the brain, data and AI are brain cells, 5G and other communication technologies are tendons, Internet of things devices are nerves, and things become various tissues and organs of the body.

1.1.4 Development Status of Cloud Computing

1.1.4.1 Matthew Effect in Cloud Computing Market

Cloud computing has gone through nearly 15 years, from a sudden rise to a hundred flowers blooming to the rise of giants. Matthew effect has also played an important role in the cloud computing market.

The current cloud computing market highlights several obvious trends:

- Pyramid spire: Both the market share of IaaS and the overall revenue market share of cloud computing have obviously closed to several head cloud providers. The era of giants has come, and the total share of the four head cloud computing providers has further increased compared with 2018.
- Steady growth: With the rapid growth of more than 10 years and the growing scale of cloud computing companies, the growth rate is gradually stabilized.
- The gap narrowed: The first few cloud computing companies are catching up with each other, and the gap between them is narrowing, especially the gap between Microsoft and AWS. The competition among cloud computing companies is further intensified.
- Regional differences are obvious: As you can see from Figure 1.14, there are obvious regional effects in the Asia Pacific region. Alibaba

Worldwide cloud infrastructure spending and annual growth
Canalys estimates, full-year 2019

Cloud service provider	Full-year 2019 (US$ billion)	Full-year 2019 market share	Full-year 2018 (US$ billion)	Full-year 2018 market share	Annual growth
AWS	34.6	32.3%	25.4	32.7%	36.0%
Microsoft Azure	18.1	16.9%	11.0	14.2%	63.9%
Google Cloud	6.2	5.8%	3.3	4.2%	87.8%
Alibaba Cloud	5.2	4.9%	3.2	4.1%	63.8%
Others	43.0	40.1%	34.9	44.8%	23.3%
Total	**107.1**	**100.0%**	**77.8**	**100.0%**	**37.6%**

Note: percentages may not add up to 100% due to rounding
Source: Canalys Cloud Channels Analysis, January 2020

FIGURE 1.14 Changes in market share of cloud computing.

Cloud, which originated in the Asia Pacific region, has a significantly higher market share than other regions. In Europe and the United States, AWS, Microsoft and Google have absolute advantages.

1.1.4.2 Segmentation of Cloud Computing Market

With the advent of the era of giants, many small- and medium-sized and even entrepreneurial cloud computing companies have begun to find a living space in the cracks. It is more and more unrealistic to do large and comprehensive cloud computing at the beginning of the outbreak of cloud computing, so there are many cloud computing-related companies in the industry.

From a technical point of view, many small- and medium-sized cloud providers are currently concentrated in the following popular segmentation areas:

- Container: Through the container technology to simplify the cloud native transformation of customers, at present, it mainly focuses on the management of PaaS layer and DevOps.

- Video: With the advent of the social mode of live and short video, the requirements of video technology are higher and higher, and many cloud providers focus on the field of video technology.
- Big data: It focuses on big data intelligent solutions in some vertical industries.
- Smart park: It provides industrial cloud products and services in intelligent buildings, community life, equipment interconnection, human and vehicle management, business incubation, etc.
- CDN: In order for users to have a better Internet experience, some providers have been focusing on CDN service provision. At present, after combining with cloud computing technology, they further optimize the service experience and quality.

1.1.5 The Future of Cloud Computing

1.1.5.1 Northbound Expansion – Industrial Cloud and Intelligent Cloud

At present, technologies in IaaS and PaaS have become more and more mature, and these technologies are relatively common. However, SaaS services have strong industry attributes and customization requirements. Although there are excellent SaaS cloud service providers like Salesforce, there is still a big gap to meet the market demand. In the future, cloud computing will further expand to the industry. Today, there are many industrial clouds, but they are still coarse. In the future, they will be further subdivided to better meet the accurate needs of customers. For example, the financial cloud may be further subdivided into banking cloud, securities cloud and insurance cloud in the future. The Business Middle Office may also be further subdivided into the automotive industry Business Middle Office, the energy industry Business Middle Office, the banking industry Business Middle Office and so on.

In addition, with the continuous development of cloud native technology and the continuous development and maturity of container, serverless, AIops and other technologies, the future cloud will be further intelligent in several aspects:

- Business configuration: With the emergence of microservice, service grid, business platform, data platform and other concepts and technologies, business addition and reduction will become easier than ever, and business can be flexibly adjusted through plug-in mode.

- Resource transparency: The emergence of serverless has gradually evolved into FaaS (function as a service). At present, there are still several ways to make IaaS resources transparent. However, in the future, it may further expand to abstract and transparent business capabilities, and further develop to provide more powerful serverless programming and choreography capabilities. So as to further optimize the cost of basic resources and application systems, and enhance the competitiveness of enterprises.
- Fault self-healing: Through AIops for monitoring, data analysis, machine learning and so on, it can carry out self-diagnosis and self-repair when the fault occurs, and inform the engineer to carry out manual intervention when the machine cannot complete self-repair.
- Expansion and contraction automation: Cloud computing has a strong capacity to expand and shrink, and the large-scale use of containers further improves this capacity. According to the actual operation of long-term business, setting the corresponding expansion and contraction rules can achieve a certain degree of automatic capacity management, further improve resource utilization and reduce costs.

1.1.5.2 Southward Integration – Integration of Cloud Computing, Internet of Things and Blockchain

As mentioned above, on the one hand, cloud computing expands its business and data services in the north direction, and on the other hand, cloud computing also deepens its integration with the Internet of things, blockchain and other underlying technologies. Especially with the combination of the Internet of things, the cloud extends its tentacles to the end. A large part of the data that originally needed to be calculated on the centralized cloud can be decomposed to the end to do the first-level calculation, instead of transmitting all the data every time. On the one hand, it greatly alleviates the computing burden of the centralized cloud and the bandwidth waste of invalid data transmission. On the other hand, it also greatly improves the autonomy and timeliness of the end.

1.1.5.3 Decentralized Cloud

At present, the mainstream cloud computing is highly centralized cloud, cloud computing providers turn massive servers, storage, networks and other hardware devices into virtual resources through virtualization technology to provide users with on-demand purchase services. This method provides a very powerful computing ability; users can buy the required resources at a relatively low price. However, in addition to this kind of centralized computing resources, there are still more massive computing resources idle around the world, such as personal computers (PCs), mobile phones and other mobile devices. At present, some organizations have begun to use these resources through distributed technology, and BOINC (Berkeley open network computing platform) is a typical example, However, this model also has this natural defect, because the individuals who provide idle computing resources cannot get the corresponding benefits, and the incentive mechanism is not clear, so there are relatively few volunteers. However, the emergence of blockchain brings the dawn to solve this problem. We can provide a proven consensus and traceable trust mechanism through the contribution proof protocol, so as to form an effective incentive mechanism.

Cloud computing has a long-term development goal since its birth, which is 'trusted, reliable and controllable', which is highly consistent with the trust mechanism of blockchain. The combination of blockchain and cloud computing will bring a new service experience. At present, there are some blockchain cloud computing services that try to contribute idle hard disks and mobile phones' idle computing power. For example, we contribute idle hard disks, integrate a large number of idle computing resources into supercomputers/computing pools and rent them to users through the rules of blockchain and in accordance with the service mode of cloud computing. Similar providers include Storj and Thunderbolt.

Of course, blockchain also has its own performance problems to be solved. So, the future is likely to be a cloud era in which many cloud computing giants and many small clouds based on blockchain coexist.

1.2 DIGITAL TRANSFORMATION TO DIGITAL INTELLECTUALIZATION

1.2.1 Information Construction and Digital Transformation

1.2.1.1 Gains and Losses of IT Construction

What is informatization? Informatization is a technology based on modern communication technology, database technology and network technology, which collects all the elements of the research object into the database for specific people's life, work, learning, decision-making and other behaviors closely related to human beings.

Enterprise IT construction has experienced the era of mainframe, PC and minicomputer, and Internet Data Center (IDC). In the first decade of the 21st century, the era of cloud computing is coming. Today, more and more enterprises and information builders begin to embrace or deeply intervene in various innovations and changes in the era of cloud computing, and the arrival of cloud computing has also brought great impetus to social development and change power. With the extensive promotion and application of cloud computing era and Internet era, data from the original scattered to the overall precipitation of data center, and then to the explosive growth of massive data, data management and data value mining have become a problem that many enterprises have to face and think about. Let's review the development of enterprise IT system.

- From the mid-1960s, enterprises began to use mainframe. The threshold and cost of using mainframe is very high, only a few enterprises can use it, and the computing power is occupied by a few enterprises, while ordinary enterprises cannot use this kind of resources.
- Since the 1980s, PCs and minicomputers have appeared. Enterprises acquire computing and storage capacity by purchasing hardware. However, there are some problems, such as inflexible architecture, low resource utilization rate, easy to be locked by providers and uncontrollable.
- Since the mid-1990s, enterprises have begun to host and rent hardware in IDCs, giving up building their own data centers. Wind, fire, water and electricity are provided and guaranteed by the data center, but the computing equipment and computing system still need to be provided by the enterprise itself, and the enterprise's IT operation cost has not decreased significantly. Computing capability is still occupied by a small number of large enterprises, and general enterprises cannot enjoy the convenience of computing due to capital, technology and other factors.

Now it has entered the era of cloud architecture, a service mode provides computing capability, which can be obtained on demand, and the threshold of use is reduced, which reduces the investment of enterprises in capital, technology, time and other aspects, making computing a social public infrastructure like hydropower.

The elasticity of cloud architecture leads to the intensification of resources, and the use on demand, payment on demand and high-quality professional operation and maintenance bring the service of computing. Through the transformation and optimization of IT-centralized architecture to distributed cloud architecture, it can support the requirements of high concurrency and high-performance architecture, enable enterprises to embrace the Internet with confidence and carry out business innovation based on Internet and big data.

Human beings have moved from the IT era to the DT (Data Technology) era. The IT era focuses on self-control and self-management, while the DT era focuses on serving the public and stimulating productivity. It seems that there is a technical difference between the two, but it is actually a difference in ideology.

Information construction helps enterprises and institutions to bring new productivity represented by computer-based intelligent tools, greatly improves enterprise efficiency and reduces enterprise costs. The core of information technology solves the problem of high sharing and application of information resources, and information technology promotes the full exploitation of the potential of social resources.

In the past, the informatization construction of enterprises and institutions mainly focused on business function flow, information flow data storage and application sharing, network connectivity, infrastructure, information security and other aspects. The informatization construction mainly focuses on the 'business function realization', and achieves the extreme in improving efficiency and reducing cost. Although it has done a lot of things based on data analysis, it has done little in how to focus on the value of data mining and promote business innovation. This is also where there were gains and losses in the past informatization construction.

1.2.1.2 Why Are Enterprises Eager for Digital Transformation

Since the emergence of the Internet, information explosion has also happened simultaneously, and with the continuous development and maturity of AI, AI combined with massive data promotes the rapid arrival of digital technology era. But in the face of massive data, it is a happy worry for many enterprises. Happy things can be more valuable through data mining. The trouble is how to store and mine so much data.

Today's enterprises are facing complex and fast changing market demand. How to respond to market changes faster, more accurately and more quickly is a practical problem faced by many enterprises. Although there has been good information construction in the past, facing new problems, especially the impact of emerging industries, enterprises begin to turn their eyes to digital transformation. So, what is digital transformation? Digital transformation is based on the construction of digital transformation and digital upgrading, using a kind of data thinking for enterprise operation, through data mining value, to establish a higher dimensional business model as the core goal.

Previously, we talked about information construction, which is usually called IT. Now we talk about the digital transformation, and we call it DT. Digital transformation is not only a technical and data perspective, but also an organizational and cultural perspective. Today, many enterprises are very excited to hear about the digital transformation. They think that a big data platform, some data models and data mining will realize the digital transformation. In fact, this is putting the cart before the horse. The first step of digital transformation is to make a change in mindset, which needs to change from the previous simple 'using data' to today's 'collecting data, storing data, standardizing data, using data', which is referred to as 'collecting, storing, standardizing and using'. For example, in the past, many enterprises have built CRM systems and sales systems, but there is often an interesting phenomenon, that is, the data of different systems cannot be linked, the same user has different data descriptions in different systems, and the data island phenomenon is very obvious. In the CRM system, customer service may only need the basic information of the customer, and in the sales system, only the order information of the customer is related, so a customer is cut alive. This cannot completely describe the real characteristics of a user, and each system may only have a small part of the user's characteristic data.

In the case that many emerging Internet companies have achieved very good experience in operating enterprises with data thinking through data mining and application, traditional enterprises and government departments have started their own digital transformation strategy.

1.2.1.3 The Relationship and Difference between Informatization and Digital Transformation

Informatization and digital transformation are closely related, informatization construction is the foundation, and digital transformation is based on informatization construction, changing thinking, operating with data thinking and seeking the breakthrough of new business model. So digital transformation is based on information construction, but different from information construction; it is not a simple upgrade, but also needs to change in organization and thinking.

1.2.2 Digital to Digital Intellectualization

1.2.2.1 What Is Digital Intellectualization

In fact, the word 'digital intelligence' has been mentioned in the previous sections. Digitization focuses more on business data, showing business processes and business capabilities in the form of data, and then realizing business execution and flow through data flow. For example, when we look at the reimbursement process of an enterprise, we first fill in the electronic reimbursement form, and then paste the invoice. The first step of electronic insurance policy and invoice information is to convert them into digital form. In the second step, the data flow among different people through the process. Each person completes his own responsibilities in the process, and finally completes a business process.

Digital intellectualization is based on digitalization and further combines AI and Internet of things, etc. to make digital intelligent and reach more places and fields through Internet of things devices. For example, in the past information age, including the digital age, more and more people deal with computers, and the human-computer-edge-machine-edge-computer-person closed loop is difficult to achieve, because there is a lack of edge device as a two-way connector and also there is a lack of intelligent decision-making, so it is difficult to truly achieve more efficient automation execution, and good two-way feedback and control.

In the cloud habitat conference in 2020, Alibaba Cloud intelligence President Zhang Jianfeng announced Alibaba Cloud into the 2.0 era: flying cloud this 'supercomputer', will be installed on a digital native operating system, like Windows let the computer into thousands of households, so that people do not understand the code can also use the ability of the cloud. This will provide a new model for human interaction with cloud computing, making the cloud easier to use and application development easier. Any enterprise and individual do not need to understand the code, can have the ability of cloud, data, intelligence, mobile and IoT. The upgraded cloud makes it easier for people to interact with cloud computing, making the cloud accessible to more businesses and more people. Zhang Jianfeng believes that cloud integration and cloud nail integration

FIGURE 1.15 Alibaba Cloud 2.0 and CEO of Alibaba Cloud: Zhang Jianfeng.

are important parts of the digital native operating system. These two strategies will change the way people use the cloud: change the way applications are developed, and usher in a new form of cloud computing (Figure 1.15).

> 'Low code 2020 is basically the beginning, and 2021 is definitely the buzzword. It's important to let people know that digital is something I can do myself. It's a huge step forward for society.'
>
> – *Ali Yun Xing Dian*

Behind 2020, different enterprise collaborative platform not only embodies the path dependence on natural resources, but also includes the organization for the future development direction of different judgment, and thus increasingly reflect very different product ideas – Lark changed the traditional message-centered design concept, with the documents and meeting coordination experience on the map; after the enterprise WeChat gets through WeChat, it becomes a marketing tool to manage and activate the private traffic. After the integration of 'cloud +DingTalk', DingTalk has gradually evolved into an enterprise-level application development platform, providing users with low code development ability, and making a leap forward in the direction of an operating system that stands on the cloud base.

1.2.2.2 AI Helps Digital Upgrade to Digital Intellectualization

Intelligence has become the trend in the future. AI technology is the core element from digital upgrading to digital intellectualization. Let's see how AI can help digital upgrading to digital intellectualization through several cases.

1. Intelligent customer service: As an important cornerstone of the service industry, customer service plays an important role in customer experience, user stickiness, secondary business opportunities mining and so on. The core elements of typical customer service business generally focus on such key metrics as 'service attitude', 'response speed', 'one-time solution rate' and 'problem solving time'. In the past, there were several prominent problems in the customer service industry, such as 'focusing on human service model', 'accurate matching of customer problems', 'single service means' and 'fault of service process', which ultimately affected customer service cost, customer experience and customer stickiness.

   These headache problems can be solved through intelligent technology. Figure 1.16 is a typical intelligent customer solution. It is divided into four aspects for intelligent construction: 'intelligent coverage', 'intelligent solution', 'intelligent experience' and 'data intelligence'.

   - Intelligent coverage: First of all, sort out the customer's possible service channels and scenarios, and do a good job of scenario embedding, so that once there is a service demand, it can quickly classify and match through the access service channels (telephone, work order, IM, web page, etc.) and access scenarios (product problems, service experience, price problems, etc.). Accurate division for different customer demands.
   - Intelligent solutions: Once customers are formally connected to the service process, intellectualization can do a lot of things. For example, before customers begin to describe their problems, quickly go to the knowledge base to match the things that

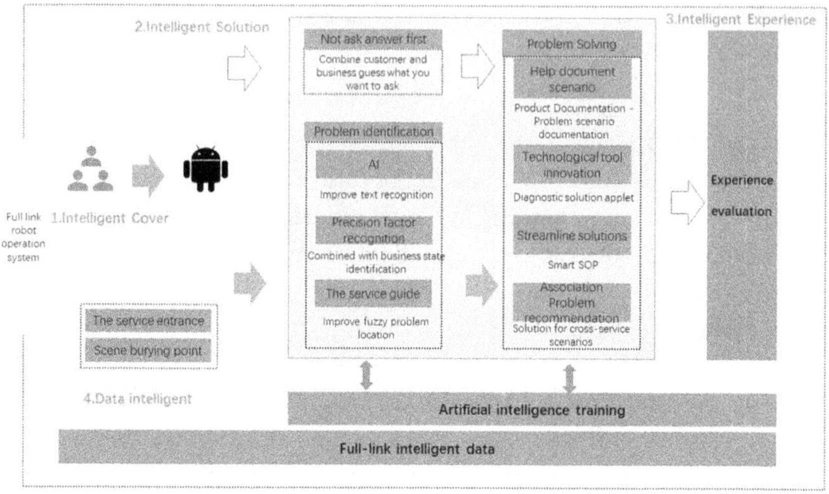

FIGURE 1.16 Intelligent customer service.

customers are most likely to want to consult according to the problems they have encountered before or the access scenarios, and give the questions that customers may want to ask in advance, so as to turn passive into active service and let customers feel comfortable that I really understand him. And all of these are operated by intelligent robots. In the following further Q&A sessions, text recognition, knowledge base comparison and further positioning of customer questions are conducted by robots. This can greatly reduce the cost of customer service.

- Intelligent experience: After positioning the problem, intelligently output the problem resolving solution to the customer through standard SOP and tools to help the customer solve the problem.

- Data intelligence: The first three steps are based on the last step, that is, a large number of full link service data, accurate data model, long-term AI robot training, etc.

2. Intelligent decision engine: Here is an example of intelligent decision-making through AI technology. A common intelligent decision

engine generally consists of 'data center', 'decision factor library', 'decision scenario configuration' and 'my decision scenario'.

'Decision factor library' is the basis of intelligence. It is necessary to design data metrics, data dimensions and atomization scenarios, and build model library and algorithm model.

'Decision scenario configuration' is to better combine with the business, show the calculation results of the business scenario and the underlying data, data model and algorithm through scenario business matching, and give decision suggestions and assist the implementation of decision actions.

'My decision scene' can be friendlier and more personalized to show the settings and preferences of each different user (Figure 1.17).

3. Intelligent manufacturing: There is often a kind of process in manufacturing, that is, 'slicing'. Once there was a case that the 'yield rate' of silicon chips of this manufacturing enterprise had some problems, and we didn't find a particularly good solution. Later, we tried with

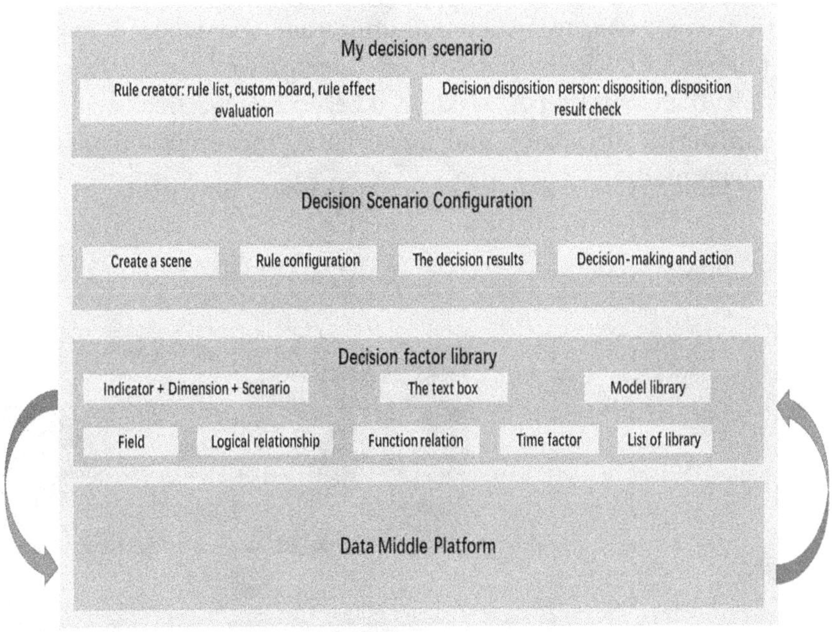

FIGURE 1.17 Intelligent decision engine.

IoT equipment and algorithm, and finally solved the problem. The 'yield rate' increased by 2%, saving a lot of costs. So how is it done and what role does AI play in it?

First of all, IoT equipment is deployed on the equipment used in each link of the process to collect data.

Second, design data model and algorithm model, and build AI platform.

Then, the collected data are cleaned, processed and calculated.

Finally, through the analysis of the noise data (burr part) in the calculated data, it is found that the problem lies in the rotating shaft of the slicer. When the speed of the rotating shaft reaches a certain degree, the vibration phenomenon will occur. Finally, the slicer will produce offset, resulting in the decline of slicing quality. After the analysis, the engineer of the factory was asked to adjust the shaft, and the problem was solved (Figure 1.18).

These three examples are just a small part of the many aspects of AI helping digital upgrade to digital intellectualization. However, through this, we can see that digital solves the problem of 'collection, store, standardization and utilization' of data, but it does not completely solve the problem of 'intelligence' in the 'utilization' link. Therefore, after AI blessing, it becomes 'collection, store, standardization, utilization and intelligence', and evolves into digital intellectualization.

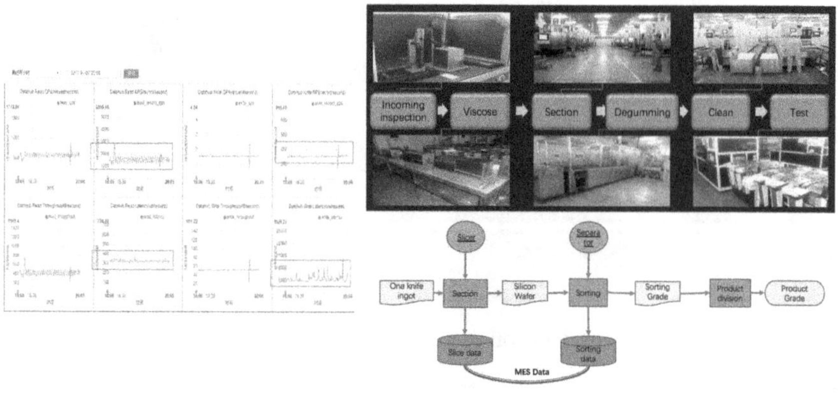

FIGURE 1.18 Intelligent manufacturing.

1.3 DIGITAL TRANSFORMATION OF CLOUD COMPUTING BLESSING

1.3.1 Cloud Computing and Digital Transformation Complement Each Other

1.3.1.1 Major Changes Brought by Cloud Computing

It's not the first day that the idea of improving efficiency, reducing costs and carrying out business innovation through data came into being. This idea has already appeared in the old era of mainframe and minicomputer. At that time, we also successively developed databases, data warehouses, data marts, BI analysis, etc.; voice recognition and AI also appeared very early. But why in those time points, the digital transformation has not really erupted? Computing power and massive data are the most important constraints. In the 21st century, with the large-scale use of Internet technology, distributed technology is becoming more and more popular. Cloud computing has arrived as scheduled. The arrival of cloud computing, on the one hand, brings the possibility of theoretically unlimited computing power and storage capacity expansion. On the other hand, it also makes us realize that massive data were thrown away as garbage on the side in the past due to various constraints. In recent years, some innovative enterprises constantly tap business value through data mining technology, more accurate matching of user service needs, looking for new business models, digital transformation has finally opened the veil of mystery, and began to truly show its value, not only enterprises or institutions are enjoying the dividends of digital transformation, but also every individual is really enjoying these benefits, for example, we can enjoy a more convenient lifestyle brought by smart home, can more accurately get our favorite products or movies recommended. Digital economy has officially stepped onto the stage. Therefore, cloud computing is the cornerstone and booster of digital transformation. The arrival of cloud computing has greatly advanced the arrival of the digitization era and further promoted the rapid arrival of the digital intellectualization era.

1.3.1.2 Cloud Computing Provides Several Ways for Digital Transformation

So, in which aspects does cloud computing facilitate the digital transformation?

- Agility: In the past, the first thing for many enterprises to carry out information construction is to make a long-term plan. One of the most important links is to reserve a long time for data center construction, hardware equipment procurement, software installation and deployment. When they encounter the need to expand capacity, they have to run this process again. These are often a few months to 1 or 2 years of the time cycle, and cloud computing just gives a good new solution, rent on demand, rapid expansion, theoretical unlimited expansion capacity, DevOps and other agile development and operation/maintenance integration management methods, all of them greatly shorten the long construction cycle, from a few months to 1 or 2 years directly reduced to a few minutes or hours. Who doesn't like the time cost and flexible use?

- Cost and labor income: For any enterprise, cost and labor income, which is often referred to as ROI, are one of the most important business metrics. For the convergence of industry competition and the rapid change of market environment, if we can have lower cost and higher ROI than other competitors, it will undoubtedly enhance the market competitiveness of enterprises. Cloud computing uses on-demand leasing. At present, many cloud computing services have come to measure the cost by day, by the number of calls and so on. They are increased when needed, and released when not needed. This is a huge cost advantage. Once a customer clouded the data center as a whole, the cost was reduced by about 50%.
- Security: When it comes to cloud computing, we have to talk about security. This is the first reaction of almost every customer who starts to consider using cloud computing. They always feel that the cloud is not in their own data center, and the security will not be as good as the traditional data center, especially in the aspect of data security. In fact, it's the same as keeping money at home and in the bank. Is it safer to keep money at home or in the bank? Generally speaking, security includes 'information security', 'data security', 'compliance security' and other dimensions. If you want to build a complete set of security system in your own data center, you not only need to buy a large number of security equipment (firewall, anti DDoS attack, behavior monitoring, etc.), but also need to design perfect security management specifications and processes, including complete security operations management. Even if it is built, there will be another problem, that is, it will not have such a large number of security sample data. In fact, it is difficult to really protect well. In the cloud, the security protection system is a security

sample that has been continuously optimized and iterated through a large number of security attacks and protection processes of millions of customers, and the richness is very obvious. Therefore, the protection capability is not of the same order of magnitude at all. In general, the security of cloud is better than that of self-built data center. At that time, there were still many people worried about whether their data were safe on the cloud and whether it would be stolen. In fact, this can be seen from several dimensions. First, the cloud is a multi-replication multi-availability zone architecture, which naturally has disaster tolerance capability. Second, every cloud computing provider has to do business. No one makes fun of customers' data security. If something goes wrong, they have to close down. In other words, cloud providers are actually more afraid of user data problems than users themselves. So let's think about it. Do cloud providers care more about protecting customers' data security? If we solve the problem of data security morally, there will be no problem technically.

- Fast execution: This is one of the advantages of cloud computing. For example, if there is a need to build a virtual machine, in the past, it may take 2–3 months (2 months to purchase hardware equipment, and a few days to install software and configure virtual machine), but now it can be done by clicking the button and waiting a few minutes through the image. I have also seen customers do more automated management. Which customer is expanding or shrinking based on the combination of container technology, cloud virtual host and storage, but at the same time, in order to solve the problem of business peaks and troughs, an intelligent business monitoring system is built to predict the trend of business volume, and the automatic scheduling container platform is used to expand or shrink resources.

Some of the above benefits of cloud computing technology greatly promote the process of digital transformation era.

1.3.2 Cloud Computing and Digital Transformation Are Not Panacea

In the previous chapter, we have seen all kinds of good cloud computing, so is it really perfect and omnipotent? This must not be the case. Any enterprise or institution that wants to implement cloud computing must fully assess and recognize the possible challenges and risks.

The transformation of technology stack: Migration from traditional IDC to cloud, and further transformation of application based on cloud native technology, is accompanied by a large number of changes in technology stack, such as the transformation from minicomputer and virtualization to container technology, the transformation from monitoring and other commercial software to open source technology, and to use new technology stack of microservice and AI technology. All of these will make the enterprise's technology stack encounter a relatively big impact and change, and face great challenges for the reserve of technical personnel, the choice of new technology stack, the transformation of old system, etc. In this way, we need to fully evaluate that after the changes of these technology stacks, which technicians need to transform, which need to introduce and which need to find special service companies to provide support. At the same time, the transformation of the old system is also accompanied by the investment of cost and the impact on the business stability when cutting over. Therefore, after these risks are fully assessed, and the corresponding coping strategies are well done, the risks will be within the controllable range, and the most taboo is to be too optimistic or stagnant.

Function upgrading of IT department: Similarly, with the introduction and implementation of cloud computing and digital transformation, the functions of IT department will face an upgrade, which will evolve from the current cost center to the value center, because cloud computing and digital transformation bring an opportunity to help business innovate through technology and data. Of course, this is an opportunity on the one hand, and also a challenge on the other hand. If it is not done well, the business department will challenge that IT department should just do a good job in technology. Even more technical support work has been transferred to the cloud, which is more automated and intelligent. For IT departments, a lot of work is also facing transfer and upgrading. Therefore, only when IT department grasp the opportunity well, they can have the opportunity to upgrade to a value center. If IT department do not grasp the opportunity well, they may lose our position. For example, through the implementation of Data Middle Office, an enterprise has mined the value of data on a large scale. How to interpret the value of data, how to give suggestions to business departments in combination with business, how to continuously optimize data and how to intelligentize the linkage between data and business, etc., if IT department fails to grasp it well and makes isolated data and BI analysis, and then business department will not be satisfied.

Change of management mindset: The introduction of cloud computing will also pose challenges to management mindsets. For example, the original development and operation/maintenance are quite different, which will evolve into DevOps, the original pure technology orientation of IT department will change to data orientation, the proportion of technical personnel will expand and technology will drive business innovation.

1.3.3 Reference, Practice and Promotion

1.3.3.1 Step Out of Your Own Way

Digital transformation is a process of general methodology and technology+best practice+self-verification. It is impossible to copy other people's experience completely. So how to step out of your own way?

1. Recognize yourself: From the multi-dimensional analysis of the industry, business situation, IT situation, personnel situation and enterprise strategy, give yourself a portrait of digital transformation. Then, consider the situation of digital transformation methodology and best practices, set a future enterprise goal, and analyze the gap, to determine the start time of digital transformation programme, clarify the implementation roadmap, plan human and financial resources, design organizational formation requirements, and identify possible risks and difficulties, etc.. Take the organization as an example, what kind of organization does digital transformation need? Let's take a simple example, if an enterprise wants to develop a food marketing business, they will encounter the following problems, what kind of people are needed in this organization to achieve greater efficiency? Actually, there is a very simple principle. If you choose someone who is particularly interested in or sensitive to food to do this food marketing operation, the effect may be better. Because he has a strong personal interest, he will first go deep into the business and seriously understand the business. He will have stronger interest and motivation to do it, and is more likely to do it better. Digital transformation is the same. In the construction stage, business systems and big data platforms have been established. After the completion of the construction, people who are sensitive to data and have a strong interest and firm idea in driving business change through data and technology are needed to be in the organization. This digital transformation is more likely to be more effective.

2. Mindset first: Before we really start the work related to digital transformation, the most important thing is to let the key people in the organization reach a consensus on the mindset. If everyone has a certain mindset of digital transformation in mind, it will greatly reduce the resistance of the organization in the implementation of the digital transformation campaign. In the early stage, we can seek some external digital transformation professional teams to share and train some digital transformation mindsets, methodologies, best practices, organizational culture design, etc. Then we organized more brainstorming internally, so that everyone could raise questions and reach a consensus.

3. Overall consideration, small step and fast run: Digital transformation needs to focus on high and start from low. In terms of strategy, we must consider the overall situation, try to think further, and have a big blueprint. But the implementation process should not be greedy, and the tasks should be prioritized to make a complete plan. However, it is suggested to start from the most painful and most effective scenes and points to quickly cut in and get results, so that it is easier to verify the digital transformation implementation route suitable for the enterprise itself. After the initial verification method and route are suitable for the enterprise itself, the concurrency can be amplified later. What I fear most is to do it all together at one time. If you want more, you will lose.

4. Both inside and outside: Many enterprises feel that it is a big mistake to ask outside digital transformation experts to help them achieve digital transformation once. Digital transformation needs external support, which is very important, but more importantly, it is essential to cultivate both of internal and external skills. External experts will bring the methodology, technology and best practice of digital transformation, and then take the enterprise to do it in parallel for a period of time. We commonly call it 'supporting the horse' and 'sending a journey'. Within the enterprise, the team can really carry out digital operation independently and smoothly transfer it to the enterprise's own organization. Therefore, the enterprise must establish a suitable organization, find the right people, give the right power and responsibility and appropriately simplify the digital operation-related process.

5. Difficulties drive innovation: The 'crisis transformation' in the digital transformation is a very interesting phenomenon. As mentioned above, some people may follow suit to start the digital transformation. However, I believe that most enterprises decide to start the digital transformation because they feel the pain. If there is any pain, there will be demand. The process of digital transformation is doomed to not be smooth sailing and will definitely encounter difficulties. The author has been working for many years from experience; I have never seen an enterprise's digital transformation is smooth sailing. They are all moving forward in setbacks. The more frustrated, the braver they are, and they often win the final victory. Those who are afraid in setbacks basically give up halfway. Therefore, we should dialectically see the difficulties in the process of digital transformation, and there are potential driving forces in these difficulties to help the digital transformation forward. Let's share a successful case and an unsuccessful case. First, look at the successful cases and see how they do it. This is an enterprise in the food industry, which is in the upper middle position in the market. It started the digital transformation project two and a half years ago. From the start of the project, the enterprise itself and the consulting and implementation service provider jointly formed a joint delivery project team, with the president of the business department and CIO as the project sponsors. Each team has members from both sides; make sure you have me and I have you, and defined everyone's responsibilities. The project designed the complete blueprint stage and the scope of the first phase. There are several key points in this. One is the definition of No.1 position, which is the person in charge who can really mobilize multiple resources in the enterprise, and also has real business pain points. One is that the project is not blindly greedy, and it is spread out in an all-round way. The first phase focuses on the planning and design of the overall blueprint, plus the capacity building of the technical platform and the data analysis and modeling part in the middle and bottom of the data platform. It is the integration of the project teams of the two sides, with equal rights and responsibilities, which has played a great role in moving forward the project. However, despite this, the project still encountered a very big challenge and difficulty in the later stage of implementation, that is, the business department felt that the project did not bring real value to the business. Fortunately,

the blueprint was made in the early stage of the project, and the goal of phase I was to build the basic technology base, and the sponsor of the customer's business line was the president of the business department. He was not satisfied with phase I implement result and has some anxiety, but he understood that digital transformation need time, he insisted on supporting the implementation of digital transformation projects and put forward optimization suggestions, hoping to find business scenarios that can be effective quickly. This difficult point not only did not hinder the digital transformation process of the enterprise, but also accelerated the progress. Due to the expectation of business effect, they started the second phase of digital marketing capacity building on the basis of the original blueprint, and accelerated the construction process and completed this part faster than expected. After completion of the construction, they can see the business effect. Although it is not particularly amazing, it has made a big step forward than before, which has strengthened the determination of the enterprise's digital transformation. Later, they have started three and four phases one after another. Next, let's take a look at a less successful example. This is a sales enterprise. At first, the founder of the customer heard about the digital transformation solutions, and also heard about the attempts and successful cases of other enterprises in the industry, and had a strong interest in the digital transformation, hoping to complete a business innovation and breakthrough, and form a new competitive barrier in the fierce competition. The project is divided into two phases. During the implementation of the project, there are several major problems. One is that the No. 1 of the project is the CIO, not the person in charge of the business department. Second, in the second phase of the delivery process, because the project has not yet been delivered, there are some difficulties such as delay in the implementation process, unable to quickly see the business effect, the CIO was replaced many times in the middle. After the CIO was replaced, they all had their own new ideas and priorities, and they could not go on according to the original plan. At the same time, the human resources of the enterprise participating in the project were also transferred to other things. The follow-up of this project can only be concluded hastily. This kind of thing is not uncommon in the past information construction process and today's digital transformation construction process. The main reason is that when they encounter difficulties, they start to panic or lose the strategic nature, and completely deny the past. As a result, they cannot find a solution to the difficulties, and things can only have one end in the end.

6. Firm in of purpose: The implementation of digital transformation may be a relatively short cycle, but it will be a relatively long process to complete the whole digital transformation. In the process of small step and fast running, it is necessary to have a strong heart after solving all kinds of noise in the process of moving forward. The digital transformation process cycle of different enterprises is different, but it often takes several years to iterate. Therefore, there are strong requirements for strategic qualitative and resilience.

1.3.3.2 Be Honest with Yourself

In a thousand words, digital transformation not only needs to know what it is, but also needs to be down-to-earth to practice, learn from other people's success and failure experience, and really walk out of their own way. This is the most suitable way for their own enterprises in the digital transformation. Therefore, the author thinks that 'Be honest with yourself' is very helpful in the digital transformation. Digital transformation is most afraid of several categories of people, that is, 'talking theorists', 'doing things only', 'impulsive radicals', 'conservatives who are afraid of hands and feet' and 'opportunists who take advantage of the opportunity to seize power'. Digital transformation needs 'calm response', 'macro planning', 'down-to-earth', 'modest and prudent' and 'consistent' people who are honest with themselves.

So, what does the digital transformation talent model of being honest with himself look like? In Wang Yangming's 'Being honest with yourself, ending in the best', the most important thing is 'there is practice in knowledge, and there is knowledge in practice'. This 'knowledge' does not mean simple knowledge, but refers to 'conscience', and 'practice' refers to practice. You will be very curious about the relationship between 'conscience' and digital transformation. Let's give you a detailed analysis. Because digital transformation is a systematic, large-scale, long-term process, involving a lot of people, it needs highly precise coordination. In the process, whether we can adhere to the 'good' in our hearts, consistently follow our inner conscience, and ensure that the practice process is not deformed and distorted is a very important prerequisite for success. There is also a sentence in the book 'heart' written by Kazuo Inamori, a famous entrepreneur who is also proficient in the study of the heart, which is that 'everything begins with

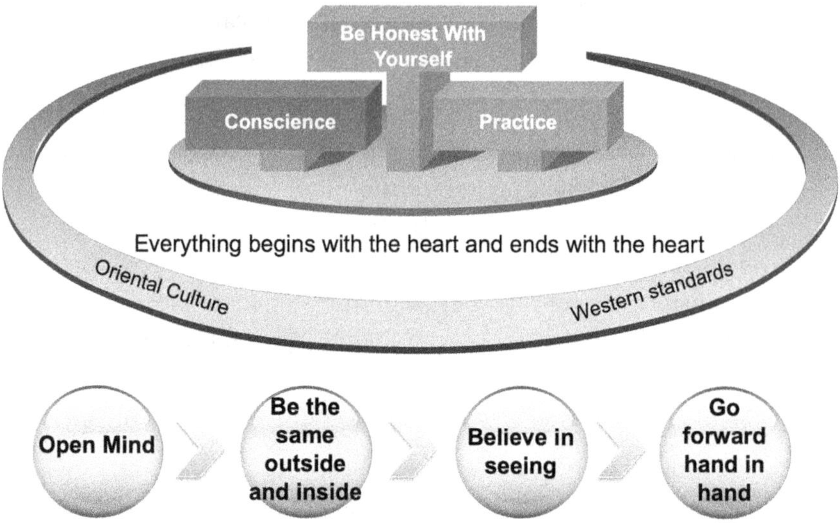

FIGURE 1.19 Be honest with yourself.

the heart and ends with the heart'. Alibaba has also encountered many difficulties in the process of digital transformation, and Alibaba also has some vernaculars related to values such as 'never forget the original intention' and 'simple because of trust', hoping to create a simple, kind and loving cultural system and organization, which will play a very good lubricant role in the process of digital transformation. Values really play an important role in the process of digital transformation (Figure 1.19).

1.4 DIGITAL TRANSFORMATION FROM CONSTRUCTION TO REFINEMENT OPERATION

1.4.1 The Original Intention of Digital Transformation

In the whole process of digital transformation, we always need to ask ourselves what is the original intention of starting the digital transformation strategy. Each enterprise may have its own different understanding, but they all focus on "using digital technology and intelligent technology to promote the transformation of business model, corporate culture and organizational structure". After the digital transformation strategy is launched, many enterprises are very excited and invest a lot of money in the construction of various platforms and systems related to digital technology. Many people expect to lead enterprises to realize digital transformation through the construction of several platforms or systems, but often after the construction of these platforms and systems is completed, they suddenly find that it seems different from their imagination. They always feel that something is missing, and the value is not obvious. Therefore, we need to return to our original intention of digital transformation and see what we expect in the end. Technology can promote the change of business, but only technology, business will not change. Therefore, we need the organic combination and close cooperation of the two wheels of business and technology. At the same time, we must think about who the end users are that we serve from the beginning. Digital operation is the best link between business and digital technology.

What digital operation needs to do is "consumer-centered", "data-driven" and "full link integration service".

- Consumer centered: In other words, what is our original intention? No matter how we change and transform, the ultimate goal is to better serve our customers, so as to keep the enterprise green and lasting. The consumer is our end user. Therefore, when we design the digital transformation strategy, we must focus on how to provide better products and services for end consumers, which is our starting point. "New customer global marketing", "user operation management" and "member operation management" are some operation means focusing on consumers.
- Data driven: The biggest uncertainty and certainty brought by the digital age are "massive data". Why is it both uncertainty and deterministic events? In the face of massive data, some people regard it as treasure and some people abandon it as garbage. People who regard it as a treasure think that they can mine valuable content from a large amount of data and continue to look for how to create more data. People who abandon it like garbage think that a large amount of data takes up too much storage space and wastes too much enterprise IT cost. Here is a fork. People who can make good use of data to mine value often bring new business sales opportunities, new product incubation, cost reduction, etc. to the enterprise, making the enterprise evergreen more certain. People who don't like data may continue with the past business model until one day they suddenly find that consumers don't like to buy. The development momentum that they thought could be sustained and better in the future suddenly stops. This is the

uncertainty caused by ignoring the value of data. "Data label system construction", "data model optimization", "member insight analysis", "marketing insight analysis", "supply chain insight analysis" and so on are all methods of using data to drive business value.

- Full link integration service: What kind of data is most valuable? What kind of problem is the biggest headache for enterprises? Many enterprises exist for the same customer, but there are different information and labels between different business departments. For the same commodity, there is no complete data chain from the beginning of the order to the whole chain of supply chain, manufacturing and external procurement, so it is difficult to analyze and optimize the data on the whole link for the commodity. Therefore, as an important part of digital transformation, digital operation needs to be closely combined with the construction stage to form a full link data integration service. The data integration of the whole link includes such aspects as "activity operation management", "precision marketing management", "supply chain production scheduling analysis", etc (Figure 1.20).

1.4.2 What Is Digital Operation

Facing different industries, digital operation also has different contents, but the implementation path and foundation are the same. In general, digital operation will protect several core parts, "business operation", "data operation" and "technology operation".

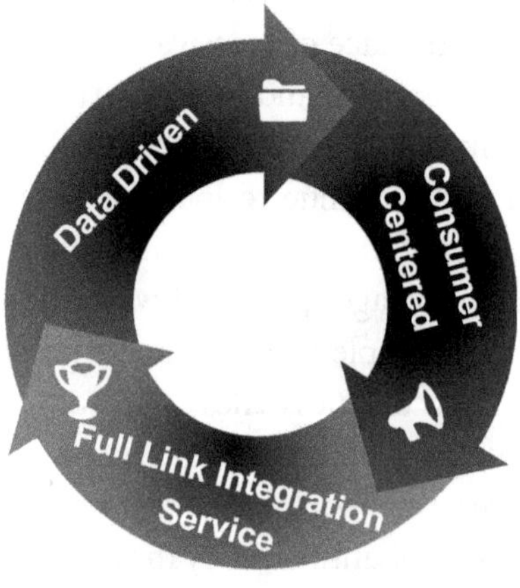

FIGURE 1.20 Three elements of digital operation.

Business operation: Digital operation, as the link between business and data technology, must not be separated from business and do things purely based on data, which will put the cart before the horse. Therefore, business operation is very important. It is necessary to establish relevant operation management systems such as "member operation", "marketing operation", "supply chain operation", "position operation", "content operation", and "risk control operation" in combination with the actual business scenarios, so as to effectively combine "people, goods and fields". In this way, data analysis and analysis results can be targeted and effectively support business innovation and optimization.

Data operation: A core foundation of digital operation is to do things around data and intelligence, so data analysis, algorithm optimization, data label system analysis and optimization are effective starting points and methods. Through effective data analysis results, the business can accurately carry out precision marketing for different consumers, effectively improve the efficiency of the supply chain and reduce production and manufacturing costs.

Technical operation: What I mentioned earlier is the business layer, but no matter what operation we want to do, another very important thing is that we can't leave the underlying technology platform. For this platform and all kinds of products on the platform, we also need to operate. For example, capacity planning and management for the cloud platform, image analysis and design of tenants on the cloud, daily management of measurement and billing, security and compliance management, as well as configuration and use management of big data products on the cloud platform, all belong to the category of computing operation.

Taking the new retail industry as an example, digital operation hopes to achieve three "insights", namely "seeing", "insight" and "innovation".

Seeing: The first thing to do is to set a "Polaris indicator" and some core related indicators for yourself, so that the business process of the enterprise can be digitized, quantified and visualized. Different enterprises have different Polaris indicators in different development stages. For example, for a search engine company, the total number of users may not be his Polaris index, while the number of active users is the real Polaris index, because in this way, we can truly know the current business situation and development trend of the enterprise.

* Polaris indicator is simply understood as the development goal set by the company, and there will be different goals in different stages. It is

called "Polaris" index, which means to guide the company's direction like Polaris. When setting goals, the company generally needs to comply with the smart principle, specifically referring to specific, measurable, achievable, relevant and clear time-bound.

Insight: If you can see the operation of the enterprise, you need to deeply analyze the data you have seen, so this stage is called insight, and you can do a lot of predictive analysis and industrial analysis.

Innovation: The first two stages still do things based on the current data itself, and innovation is essential if you want to compete with other competitors in different dimensions. The innovation stage is to realize the innovation of new businesses and products through a large number of scenario, whole life cycle, product relevance and other data analysis, combined with some marketing means, flow realization, channel optimization, and so on(Figure 1.21).

1.4.3 Organization and Capability Required for Digital Operation

The departments involved in digital operation will be complex and need high-frequency business interaction. Therefore, if digital operation wants to do well, it needs complete link and organization design. It is generally designed as follows:

1. KPI quantitative disassembly: Operation is a strong result-oriented management behavior. Operation without KPI cannot be measured and is not efficient. Therefore, we must first clearly design a

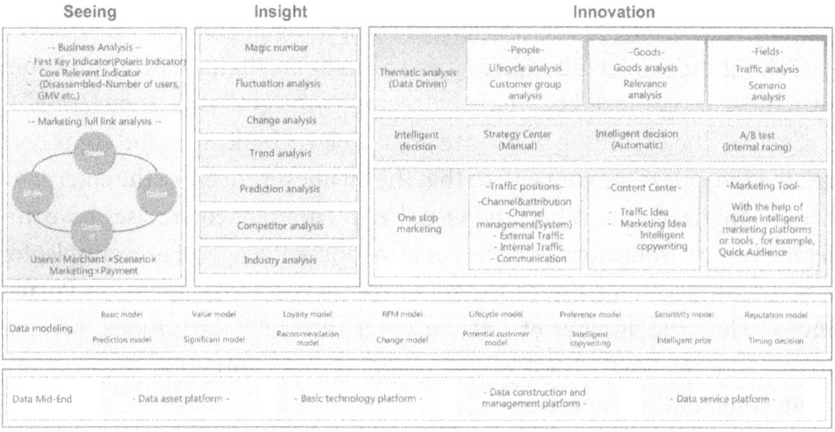

FIGURE 1.21 Three "opinions" on digital operation.

quantifiable and detachable business KPI to guide all relevant business parties to cooperate with the whole operation activities.

2. KPI for business departments: Total business dismantling. First of all, after having the overall KPI, it needs to be further disassembled to each business unit so as to refine the management, enable an operation activity to be carried out accurately, and avoid spending a lot of marketing expenses, but it is used in the wrong place.
3. Sand table for business owner: For the business commander in chief, he needs to know the disassembly and progress of the whole business path, so needs to build a sand table for the business owner, which can be managed overall.
4. Itemized operation:
    - Channel operation-sand table: The most important thing is to sort out a relatively perfect scene layout to ensure that there are no major channels and scenes missing.
    - Product operation-funnel monitoring: The principle of shortest link / least jump out should be used in product operation, so a funnel-shaped monitoring model needs to be set to help the product launch better meet the two demands of consumers' efficiency and experience.
    - Institutional operation: Competitive products and supervision: Operation is not only to know itself but also to know the market competition. Therefore, competition is also an essential part of product analysis and product quality supervision.
    - User operating costs and users: At present, consumers tend to have multi-channel and multi-scenario consumption patterns, so complete user life cycle management is very important. Only in this way can we effectively reduce the cost of user management and improve the effect of user cross drainage. Therefore, it is necessary to establish a full life cycle management mechanism.
    - Data operation and analysis: Data analysis to identify opportunities. This is also the basis of digital operation, so it is necessary to establish data analysis tools and models.

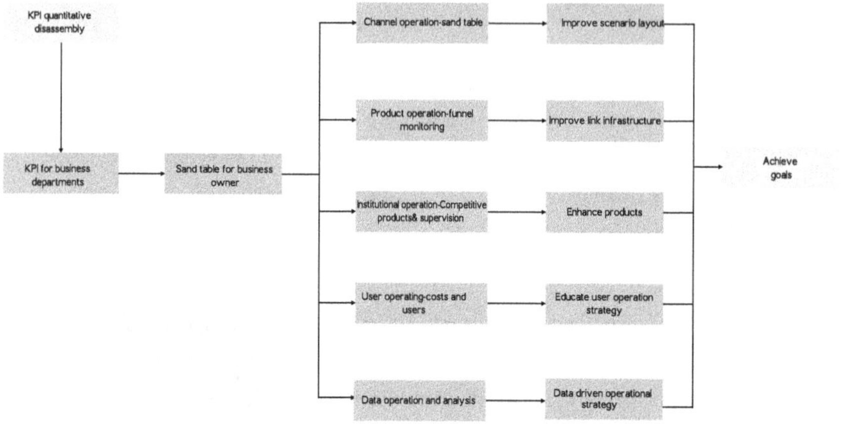

FIGURE1.22 Digital operation chain and organization.

5. Achieve goals: Also the principle of total score and disassembly score. From the initial setting of KPI to the actual results after the final completion of operation activities, a closed loop needs to be done. Whether it meets the standard or not, it needs to make a complete review, learn lessons and optimize the business (Figure 1.22)

With the above methods, an organization still can not do a good job in digital operation. We need to work together from "Tao", "art" and "method". "Tao" is data awareness. If everyone doesn't like doing business through data, there will be no foundation. Everything else can't be achieved. The most important thing is to cultivate members in the organization to develop data thinking. Secondly, we also need to have methods to operate, not simple data analysis, intuition and experience. Finally, a good tool is a powerful assistant for efficient operation, so it is essential to build a good operation tool (Figure 1.23).

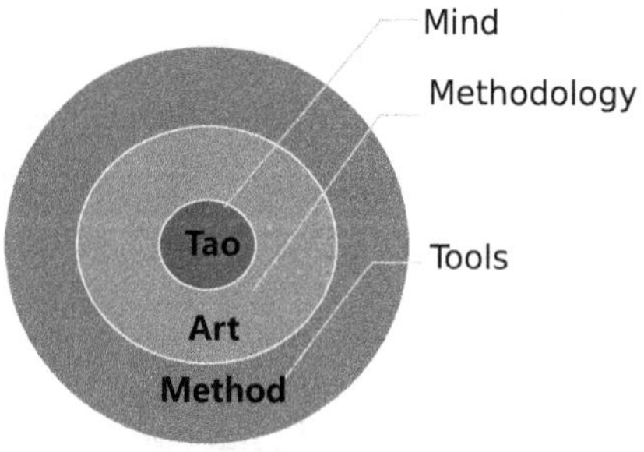

FIGURE 1.23 "Tao", "Art", "Method" of digital operation.

CHAPTER 2

The Top-Level Design of Enterprise Digital Transformation

Zhuang Longsheng, Huang Peidong, and Li Yuchen

2.1 THE TIMES BACKGROUND OF TODAY'S ENTERPRISES

2.1.1 The Digital Economy and Its Characteristics

The world has now come to an unprecedented technology and industrial revolution, based on innovative fusion of new information technologies such as the internet, big data and the artificial intelligence (AI), and has gradually established rapid, efficient and low cost calculation, data processing and storage of the new system. Our understanding and exploration of the objective world has also moved forward into information space from physical space, building channel, hub, and platform to support connecting, accurately mapping, interactive feedback and effective control between virtual and reality, atoms and bits. The context and trend of the global digital transformation has been increasingly clear: reshaping the whole system of production, growing with data as the core factor of production, and forming development outlook, methodology, value judgment and the operating mechanism are now the essential elements and paths for reshaping the core competitiveness and for improving the efficiency and quality of network and intelligence fields.

DOI: 10.1201/9781003272953-2

With digital knowledge and information as key production factors, digital technologies as core motivation, modern information network as important carrier, digital economic is a new economic form which accelerate reconstructing economic development and governance model by deeply combining digital technologies and physical economic and by continuously improving level of digitalization, internalization and intellectualization.

With the advent of the era of digital economy, the world is undergoing digital transformation at an increasingly sonorous pace. Digital economy is not only the product of technology, economy and development in the current era, but also leads the era forward with its remarkable characteristics.

2.1.1.1 Background of Digital Economy Era

(1) Digital economy has become a new growth engine for economic development

The digital economy has become a major driver of global economic growth in the 21st century. According to the 2020 White Paper on the economic development of China's digital economic (2020) released by China's information and communications research institute, in 2019, the added value of digital economy has reached to the scale of 35.8 trillion yuan, which is 36.2% of Chinese GDP, increasing by 1.4pt year-on- year. On comparable basis, the nominal GDP growth rate of China in 2019 was 15.6%, with a year-on-year increase of 7.9pt, which further highlights the position of digital economy in national economy.

China attaches great importance to the great potential of digital economy to social development. At the 13th G20 Summit on November 30, 2018, General Secretary Xi Jinping pointed out that 'The digital transformation of the world economy is the trend of The Times, and the new industrial revolution will profoundly reshape human society. We should not only encourage innovation and promote deeper integration of the digital economy and the real economy, but also pay attention to the risks and challenges brought about by the application of new technologies, strengthen the institutional and legal systems, and pay attention to education and job training. We should not only fully tap the potential for innovation based on our own development, but also open our doors and encourage the spread of new technologies and knowledge so that innovation will benefit more countries and people.'

In the government Work Report for 2020, it is proposed to comprehensively promote the 'Internet Plus' and build new advantages of the digital economy. As the advanced stage of informationization development, digital economy is a new form of social and economic development following agricultural economy and industrial economy.

(2) Steady development of digital industrialization

In recent years, with the further consolidation of information infrastructure and continuous optimization of enterprise internal structure, the digital industry is developing steadily. In terms of scale, the added value of digital industrialization reached 7.1 trillion yuan in 2019, with a year-on-year growth of 11.1%. From the perspective of structure, the digital industrial structure continues to soften, and the proportion of software industry and Internet industry continues to increase slightly (Quote from *The White Paper on the Development of China's Digital Economy (2020)*).

(3) Digital transformation is an active choice to adapt to the development of digital economy

In the era of vigorous development of global digital economy, digital transformation has become the consensus of all walks of life. Digital transformation is changing the operation rules of enterprises and many industries. Not only digital native enterprises, but also traditional enterprises are actively exploring the way of digital transformation.

Digital native enterprises can enhance the competitiveness of products and services through the in-depth application of the new generation of ICT technologies, so as to realize the leapfrog development of enterprises themselves. Start-ups and Internet companies are attacking traditional market rules and boundaries, gaining broader advantages with more intensive capital investment, stronger customer relationships, more agile operating systems and more personalized brands. For example, ride-sharing platforms have brought a disruptive impact on the traditional taxi industry.

Under the trend of macroeconomic slowdown, traditional enterprises are facing more fierce market competition than before. Using the new generation of ICT information technology and implementing the digital transformation with innovation as the core are important paths for the transformation of traditional enterprises.

(4) Digital transformation of the industry into the 'Deep Water Zone'

Digital transformation has become a consensus in the industry, and most enterprises have started digital transformation.

With the in-depth development of digital transformation, the deep problems of digital transformation gradually emerge, such as the lack of overall strategy and roadmap, the lack of consensus reached by senior leaders on digitalization, insufficient representation of business value, unclear responsibilities and power of digital transformation.

Change is the main feature of this era, and digital transformation is innovation in the midst of change. In the future, it will not just be about capital and resources, but the ability to embrace change and make deep transformations.

2.1.1.2 Characteristics of the Digital Economy

(1) A new generation of ICT becomes a new factor of production

Digital transformation is to take the new generation of ICT technology as a new factor of production and add it to the original factor of production, thus leading to the innovation and reconstruction of enterprise business. Therefore, whether the new generation of ICT technologies can be effectively applied and generate significant business value for enterprises is the key characteristic of transformation.

(2) Digital assets become a new source of value creation

Digital transformation is not only the simple application of technology to the production process, but also the continuous accumulation and formation of digital assets in the process of transformation, the construction of digital world competitiveness around digital assets and the continuous creation of value for enterprises.

2.1.2 The Internal/External Opportunities and Challenges Faced by Enterprises

The era of digital economy has arrived. With the Internet subject gradually penetrating into enterprises and the whole industrial chain and life cycle, the Internet has gradually completed the 2C revolution and the 2B revolution is beginning. How to harvest digital dividends and realize upgrading and transformation has become an opportunity and challenge for traditional enterprises facing the new digital economy.

2.1.2.1 The Opportunities for Businesses

The following are the opportunities brought by the digital transformation of enterprises in the era of data economy:

(1) Accelerate application innovation

In order to keep up with the development of market, all kinds of the industries are changing the new products, applications and the way of new release of the products. In traditional models, data collection, design and manufacturing take a long time, and updates, tests and releases are planned in advance, taking months or even years to complete.

More and more companies are moving to agile design, manufacturing and delivery, achieving a better balance between speed and quality, and being able to quickly withdraw unsuccessful new products or services without compromising the continued performance of critical services and systems. To build more agile workflows, organizations must achieve closer team collaboration and seamless system integration, and they need to be able to monitor the effectiveness of collaboration and integration in real time.

(2) Use big data for insight

We all want to arm ourselves with big data, but only if we understand what data mean can we turn information into competitiveness. In fact, every enterprise has a large number of customers, competitors and internal operation data, so it needs to adopt appropriate tools and processes to mine the true meaning of the data, so that it can make informed decisions quickly, promote innovation and make forward-looking development plans.

(3) Provide next-generation workspaces

The concept of a workspace has fundamentally changed as the trend toward technology consumption and the rise of mobile devices have led to a much more mobile work environment for corporate employees than ever before.

Work will no longer be limited by time and place, and in order to attract and retain talented people, companies must create an environment and culture that can adapt to this new way of working. This is where the right digital tools and policies enable employees to respond effectively to the complexities of the workplace.

(4) Security guarantee suitable for business development

While accelerating innovation and shortening product cycle, enterprises are also faced with more security risks and threats. As more applications are connected, hackers can gain access to all connected systems by successfully breaking into one, and the remote access granted to employees and partners leaves companies dealing with the possibility of more backdoors.

From a security perspective, simplifying security processes and continuously refining, testing and upgrading all systems are critical. With automated tools and better protocol configuration, companies can significantly reduce the time it takes to find and fix vulnerabilities, thus minimizing the potential for system intrusion and data loss.

2.1.2.2 The Challenges for Businesses

(1) Clash of cultural concepts

The digital enterprise of the future will operate in a completely different form and manner. The process of digital transformation will greatly break through the 'comfort zone' of traditional enterprises and explore in the unknown areas lacking experience. There will be long-term conflicts between the old and new cultural concepts.

(2) Lack of high-level digital strategies

If the decision-makers of enterprises do not realize the urgency and importance of digital transformation, then enterprise digitization will not succeed. Competition in the digital age requires corporate leaders to be highly sensitive to digital technologies and emerging business models, and to constantly reflect on or adjust corporate strategy. Leadership and accountability in digital transformation mean that digital transformation must be supported and empowered at the highest levels (*White Paper on Digital Transformation of Enterprises in 2020*).

(3) Lack of digital transformation culture

In the process of promoting digital transformation, many enterprises do not endue their corporate culture with new digital connotation, or blindly start digital transformation when there is no unified understanding among all parts of the staff. Such a start without adequate preparation will bring a series of strong subsequent resistance, which will lead to the failure of enterprise digital transformation (*White Paper on Digital Transformation of Enterprises in 2020*).

(4) Lack of digital organizations and talents

In order to effectively promote digital transformation, it is necessary to carry out organizational transformation at the same time. The transformation itself is dynamic, and how to establish and adjust the organizational structure in the process of transformation is an important aspect of the comprehensive challenge of transformation.

Transformation talents are also a major challenge in the process of industry transformation. Digital transformation requires not only new technical talents and business innovation talents, but also cross-field talents who can combine new technology and business. It is an unavoidable issue in transformation to cultivate high-level transformation talents.

(5) Lack of appropriate technology platform

In the digital transformation of enterprises, business needs are fast and changeable, new technologies emerge in endlessly, and the digital system needs stable expansion and smooth evolution. Closed systems or platforms can seriously hinder digital transformation. It is difficult for a heavy and inflexible technology platform to respond quickly and flexibly to the needs of customers in the era of digital economy. In the new round of digital transformation characterized by digitalization, networking and intelligentization, the technology platform of suitable enterprises plays an important role (*White Paper on Digital Transformation of Enterprises in 2020*).

(6) Lack of system design ability

Digital transformation without top-level system design will not be successful. It is believed that digitalization is to build a system, improve a business line and follow pure hardware suppliers or developers without industry experience to engage in digital enterprises.

It is easy to turn digital transformation into informatization (*White Paper on Digital Transformation of Enterprises in 2020*).

(7) Harnessing and integrating new technologies

Digital transformation not only requires enterprises to quickly learn and master new technologies, but also needs to integrate new technologies into a combination of advantages, and find the right combination points in business transformation, so as to make them apply and change existing businesses. Digital transformation poses a great challenge for enterprises to harness new technologies.

2.1.3 Digital Transformation Is an Irresistible Trend

2.1.3.1 Several Stages of Enterprise Informatization Experience

The informatization development of domestic enterprises can be roughly divided into the following four stages (Figure 2.1).

The first stage: The electronic era dominated by stand-alone version system. Electronization is the process in which the heavy daily manual work of an enterprise is transformed into machine work. Paying attention to the individual's work behavior can improve the individual's work efficiency. This stage is represented by financial computerization, manufacturing automation, etc.

The second stage: The information age of business process. Informationization is to solidify business processes through enterprise management restructuring and management innovation, combined with IT advantages, pay attention to the process of the whole organization and

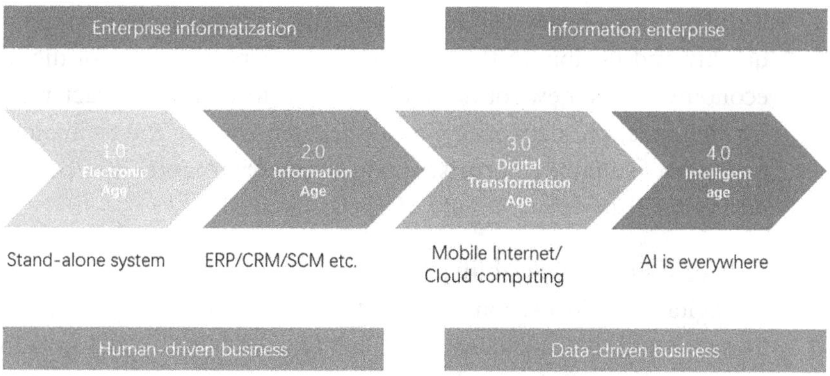

FIGURE 2.1 Four stages of enterprise informatization.

improve the efficiency of the organization. In this stage, process combing and informatization construction were extensively carried out, such as ERP (Enterprise Resource Planning), CRM (Client Relationship Management), SCM (Supply Chain Management); and BOSS (Business & Operation Support System) system construction.

The third stage: Digital transformation era. The core of the digital age is that the enterprise opens up the ecology between the external and the user, which not only effectively connects the customer, but also connects the production- and manufacturing-related back-end equipment. Its core technology is the mobile Internet, cloud computing.

The fourth stage: Intelligent business decision. Intelligent means that on the basis of the existing knowledge of the enterprise, the enterprise can intelligently create and excavate new knowledge, which can be used for business decision-making and daily management of the enterprise, etc., to form a self-organizing, self-learning and self-evolving enterprise management system. In this stage, the advanced ideas of AI and expert system will be applied in the field of enterprise management.

The four stages of enterprise development also reflect the stage of data development.

In the information age, data are used for the statistical analysis of local simple data within the software and system as a byproduct of auxiliary process application.

In the digital age, enterprise-level data analysis needs to appear. The system represented by data warehouse and business intelligence, with data visualization and report analysis as the main service means, provides auxiliary decision-making for the operation of enterprises.

No matter in the information age or in the digital age, data has been used by people and it was still people to make final decisions. However, in the intelligent age, date is no longer only explained and understood by people, but is also used by application system itself. System is now able to communicate with another system, for example: the AI algorithms could get insight from data center to guide operating behavior, and directly drive operating systems.

2.1.3.2 The Relationship and Difference between Digital Transformation and Traditional Informatization

(1) The difference between traditional informatization and digital transformation

Since the 1990s, the high-frequency words associated with most enterprise informatization are not more than OA, EAS, ERP, PDM, CRM, SCM, MES, etc., and BI/Business intelligence.

Informationization pays more attention to the realization of business process system support, emphasizing business efficiency and risk control. And digitization and To C are close, belong to extroversion. Reflected in the enterprise-related processes and customers to achieve direct docking, so as to traction, force the change of internal mode.

The arrival of the mobile internet has differentiated digital age from traditional information age. Even in the Personal Computer (PC) Internet era, people only sit near the computer, they are online. In traditional information age, even with personal computers, people are online only if they were sitting near their computers. The internet was actually internet among computers. The emergence of mobile Internet, marked by the emergence of iPhone and the popularization of 4G, has greatly changed the face of human society. That is, the internet world has moved forward to human-to-human interactions from computer-to-computer interactions. With thousands of applications providing various functions, people could meet most of their daily life needs as long as they have mobile phones in their hand, which changed the internet world tremendously.

By the iron law of the business world, service revolves around the customer. Customers are online now, so our services and products have to be online. Internet enterprises provide a platform to bring traditional offline services to online customers. Even if the services we enjoy are offline, the whole transaction link is online. The core of trading is to realize the exchange between the demand side and the supply side through the market. The core essence of trading is matching information, that is, buyers know where to buy and sellers know where to sell. Obviously, this information exchange market is the digital platform. So, one way to go digital is to join a platform.

Nowadays, many retailers have established their own online channel (mini program, wechat-shop, etc.), and by the same time, they also expanded to other platforms like Tmall or JingDong. Under this circumstance, the original internal inventory system of a retailor needs to be synchronized on all the platform. Behind the inventory is a set of production management, procurement management and so on. When customers' various requirements are transmitted from the C-terminal system, what they see from the perspective of the customer is only the state of availability, lack of availability and express delivery, but the whole internal operation needs to be supported by a huge system group. The connection of these system groups is not only the data communication at the technical level, but also the process communication and business communication behind. Therefore, each department within the enterprise needs to break down barriers and optimize organizational processes to adapt to this scenario of rapid response to C-terminal requirements. Therefore, it can be found here that information (inventory, production management system, etc.) must be the premise of digital (fast to customers). If we do not do a good job of enterprise informatization foundation, digitization is out of the question.

Informatization is the first stage of digitization. In fact, many enterprises have not really completed the process of informatization. Unlike software upgrades, Six Sigma or supply chain improvement projects, enterprise digital transformation is a comprehensive digital transformation of enterprise processes, reconstructing products and services.

Enterprises need to complete the information process first to ensure the smooth development of digital transformation. So here we define informatization and digitization:

Informatization: The online transformation of the core business process belongs to the internal optimization of the supply side (with heavy To B attribute).

Digitalization: The docking of internal business processes and customer demand scenarios belongs to the connection between the demand side and the supply side.

The core of the so-called supply-side reform is that the supply side can quickly identify and respond to changes on the demand side so as to adjust itself. Here it not only needs the system support, but also needs the enterprise internal process optimization transformation.

(2) Digital transformation leads the transformation of traditional informatization

The core of Marxist political economy is a sentence: The productive forces determine the relations of production. The biggest problem with classical economics is that it treats producers and consumers as if they were two different kinds of people. In fact, both producers and consumers have to be specific to individual people and enterprises in the end. They play a dual role in the whole market.

Digital transformation has a very important purpose, to bring customers and final producers closer, to reduce the internal transaction costs of customers and enterprises and ultimately to reduce the transaction costs of this society. The key to digital transformation is 'transformation', including business, organization, ideas and tools. The key of traditional informatization is 'online'. All businesses get through online, and the 'tool' attribute is relatively heavy.

Digital is the need to connect the client demand and enterprise internal production and operation. This requires quick collaboration and flexibility among various departments within the enterprise, which is the most difficult.

Now the core business of Internet enterprises has always been online, with digital genes (production relations, business models) inherent.

Traditional enterprises (manufacturing, retail, real estate, etc.) have different business models. However, as long as they enter a stock market environment (where production is greater than consumption), they need to find ways to quickly reach customers with their products and services. Both e-commerce and live streaming are for the purpose of quickly matching demands and services.

So back to the topic of digital transformation, how to judge whether an enterprise is making digital transformation depends on the following three points:

- Is it possible to quickly reach customers with products and services (e-commerce and self-built online channels are both means);
- Whether the internal processes of the enterprise have been optimized and adjusted (internal production relationship adjustment) to adapt to the rapid response to market demands;
- Whether the processes and data of To C and To B are interoperable;

The biggest difference between digitization and informatization is that it aims at transforming and optimizing production relations. By quickly aligning requirements with services, IT becomes more than just a 'tool'; IT becomes part of the business.

Digital transformation is not only the transformation of productivity but also the transformation of production relations. For enterprises, it is the transformation of business processes, models and even business models, and for individuals, it is the comprehensive upgrading of ideas and skills.

2.1.3.3 Why Are Enterprises Eager to Digital Transformation?

Since the emergence of the Internet, the explosion of information also occurs simultaneously, and with the continuous development of AI and cities, AI combined with massive data promotes the rapid arrival of the era of digital technology. However, in the face of massive data, it is a trouble for many enterprises to be happy. What makes them happy is that they can find more valuable things through data mining. What worries them is how to store and mine so much data.

Nowadays, enterprises are confronted with complex and rapidly changing market demand, How to respond to market changes faster, more accurate and more agile is a realistic problem faced by many enterprises. Although had good informatization construction in the past, but in the face of new problems, especially the impact of emerging industries, companies are turning to the digital transition. So, what is digital transformation? Digital transformation is based on the construction of digital transformation and digital upgrading. It carries out enterprise management with a kind of data thinking, and takes the establishment of a higher dimensional business model as the core goal through data mining value.

The informational construction mentioned previously is often called as IT, and the digital transformation we are going to introduce below is often called as DT.. Digital transformation is not only a technical and data perspective, but also an organizational and cultural perspective. Today, many enterprises are excited to hear about the digital transformation. They think that a big data middle office, some data model and data mining will realize the digital transformation. In fact, it just puts the cart before the horse. First of all, digital transformation requires a change in the concept, from the previous simple 'use data' to today's 'use data, set data, access data, use data' concept change, we referred to as 'adopt, set, access data, use data'. For example, in the past, many enterprises have built CRM systems and sales systems, but there is often an interesting phenomenon that the data of different systems cannot be connected. The same user has different data descriptions in different systems, and the phenomenon of data island is very obvious. In the CRM system, serving customers may only need the basic information related to the user, while in the sales system, only the order information of the customer is related to the customer, and a customer is cut alive. This does not give a complete description of a user's real characteristics, and each system may only have a small part of the user's characteristic data.

After seeing that many emerging Internet companies have achieved good empirical results through data mining and application, traditional enterprises and government departments have begun to launch their own digital transformation strategies.

2.1.3.4 Digital Transformation Is the Inevitable Direction of Enterprise Survival and Development

Today, no country, region, industry or even a company can ignore the need for digital transformation. In the 'Second Machine Revolution', Erik Brynjolfsson, director of the MIT Digital Business Center, argues that digitization is everything today and the only beginning of the future.

According to *The White Paper on the Development of China's Digital Economy (2020)*, the digital economy continues to expand and grow at a high speed, which has become a key tool for China to cope with the economic downturn. At the same time, companies that make a successful digital transition are expected to increase profitability by 26%, valuation by 12% and the revenue-to-assets ratio by 9%, according to the study. Digital transformation will bring disruptive changes to enterprises, and enterprise users will need to rethink corporate culture, strategy, business processes and other aspects, including the cooperation with partners.

Therefore, we can be sure that for enterprises, digital transformation is the only way! This development process cannot be separated from cloud computing. In the first stage, cloud computing should be applied first to solve the risk of physical damage. The second stage is virtual storage, scalable and on-demand distribution; in the third stage, with cloud computing and big data, users can be provided with more personalized services and value-added services.

Not only enterprises, but also governments and individuals are facing such a digital transformation. In the future, in addition to the upstream and downstream, the intermediate links will be 100% digital. In addition to the final physical product, it is not necessary to produce other physical products. In order to realize this physical product, it is not necessary to produce other physical products. The physical products in the intermediate link only play a role of bridge or auxiliary.

Driven by the development of technological upgrading, people's lives have also been greatly changed. Twenty-six years ago (April 20, 1994), China was officially connected to the Internet, ushering in the Internet era. Nowadays, our life is inseparable from the Internet. In addition, the digitalization of our life, especially the digitalization of money, has become increasingly obvious. Our life has been extended, and everyone can have 'clairvoyant eyes' and 'wind ears'. The physical objects that used to be mediators are dying out, and the Internet is disintermediating the medium. The development of technology has promoted the emergence of many new forms of consumption, such as the unmanned stores in the retail industry, opening the container, taking out the goods, automatic settlement and leaving the store. Such a smooth consumption experience shows the so-called convenience, personalization and refinement.

The arrival of these technological revolutions is cultivating and even changing the consumption habits of the audience, which is why technologies such as cloud computing and big data play a very important role in the economic development of China today. This kind of change and transformation is unprecedented in the history of world economic development, and will involve more enterprises and industries in the future. Therefore, enterprises facing digital transformation are not a choice, but a necessary way.

2.2 INTRODUCTION OF ENTERPRISE DIGITAL TRANSFORMATION

2.2.1 Definition of Enterprise Digital Transformation

In today's world, digital transformation has been given different definitions, and countries have different strategic goals as well. With the increasing influence of digital technology, digital transformation is driven by data as the core elements. The only way for China's industrial transformation has become a broad consensus. Digital is promoting the major changes of national governance, government services, and enterprise development, which makes China's digital transformation promotion system accelerating to form.

Enterprise digital transformation refers to the significant change of effectively reconstructing enterprise resources with data as the core driving factor, combined with the new generation of cloud computing, big data, AI and other digital technologies. It is an important strategy to promote the sustainable development of enterprises. For enterprises, digital transformation has become an enterprise-level action. The importance of digital transformation is not only in the IT field, but the leading team is often the company's core executives.

Market pressure and external uncertain factors are the main factors of digital transformation of enterprises. Fierce market competition, regulatory requirements of regulatory authorities and the impact of new epidemic (COVID-19) accelerate the consensus of enterprises on digital transformation. The external uncertain factors are mainly reflected in the following aspects:

- In response to the diversification of market demand, more and more consumers pursue personalized customization (c2m), and pay attention to and respond to the segmented market or even niche market; they can get more benefits from the 'long tail effect'. On September 16, 2020, rhino intelligent manufacturing platform, a new manufacturing platform that has been sneaking around Ali for 3 years, officially appeared. Rhino intelligent manufacturing factory, known as 'No.1 Project', is also officially put into operation in Hangzhou. Rhinoceros intelligent manufacturing takes the clothing industry as the breakthrough point, which can realize production on demand, production based on sales, and rapid delivery, and is used to test the water to cope with the uncertainty of market demand (Figure 2.2).
- In response to the value-added products and services, the market is facing a strong demand for the improvement of product at technology level and service level, and improving the added value of products and services has become an urgent demand of enterprises. Enterprises with service-added value as their reputation have sprung up, and service star enterprises represented by Haidilao have been learned and imitated by more and more brand enterprises on demand (Figure 2.3).

In response to the coordination of the whole link production process, the production and manufacturing are becoming more and more complex. A single enterprise cannot treat the production and manufacturing in isolation, but must cooperate with the upstream

FIGURE 2.2 Rhino intelligent manufacturing factory.

and downstream. The production and marketing coordination bring strong demands of cost reduction and efficiency improvement, and full-link digital monitoring.

- In response to the synergy of industrial ecology, the industrial division of labor is more detailed, leading to the lengthening of the industrial chain, the competition and cooperation relationship between the upstream supplier ecology and the downstream customer ecology is becoming more and more complex, and the ecological management has become an important part of the internal digital transformation of enterprises.

Alibaba Cloud has launched a one-stop, full-link digital intelligence transformation and upgrading 'five steps' tailored for the retail industry, including infrastructure cloud, front-end contact digitization, core business online, operation data and intelligent supply chain, which further provides a detailed roadmap for the digital intelligence transformation of the retail industry.

FIGURE 2.3 Web celebrity Haidilao hot pot migrates to cloud.

The first section of digital transformation would be enabling the infrastructure to be clouded. Clouding is the fastest and the most effective way to reduce costs and to increase the efficiency. Alibaba Cloud as Alibaba all technology and product output platform, operating the world's largest retail platform, so far 80% of Chinese apparel enterprises have reached cooperation with Alibaba Cloud (Figure 2.4).

The second part is the premise of digitization of front-end contacts to gain customers and 360-degree consumer insight. Alibaba Cloud's 'seven-star linkage' solution can connect the offline consumer attribute, behavior and transaction data in series, and then converge with a 'North Star' Tmall, finally breaking the space limit and understanding consumers more stereoscopic through the global data. So as to realize a stock and the whole area marketing, the formation of online and offline customer behavior doubles funnel management.

The third section of the core business online is the more efficient operation of enterprises, better service to consumers. Collecting the core capabilities of the enterprise along with its business development on platforms through digitalizing, and establish an operating system which is

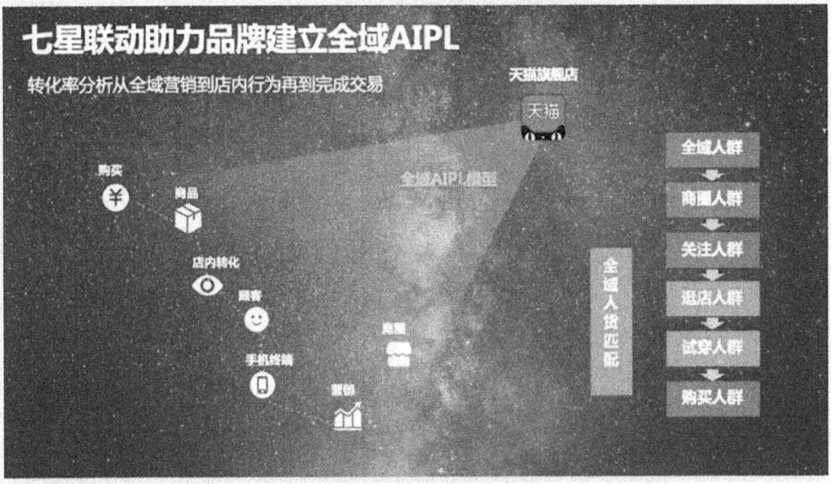

FIGURE 2.4 The whole domain AIPL of brands.

centralized in service, and close-looped managed by business middle platform and data middle office for the enterprise to explore and innovate business more efficiently and build up its own core competence.

In the fourth section of the operation of data, data are necessary condition. Enterprises in the operation of each link will produce data, how to manage the data and to develop and use? Multiple departments have data, how to integrate? Only when the data are in the middle, can we effectively solve this problem.

The fifth section of the supply chain intelligent and the data business is supported by cloud on the Middle Office and the whole station, which forces the reform of the supply chain. The front-end contacts are digitized and the core business is online. These key data are integrated between the data middle office and business middle office. Through the computing power support after the cloud on the whole station, enterprises will be forced to carry out the reform of the digital supply chain and finally realize the end-to-end intelligence.

Although enterprises have reached a consensus on the importance of digital transformation, they have different expectations for the business value brought by digital transformation, and even have strong demands for short-term benefits. This often brings great challenges to the leaders of enterprise digital transformation. Specifically, the digital transformation of enterprises faces many practical difficulties:

- The path of enterprise digital transformation is vague, lacking of medium- and long-term planning, unable to produce a clear digital strategy and implementation path, and most enterprises have the will but no method. Enterprise digital transformation project is usually considered as the cost center, and ROI is difficult to prove by data, which is also an important reason why enterprises cannot make up their mind.
- The barriers of enterprise organization system and mechanism are serious. New customers, new products, new organizations, new technologies, new cultures and new models challenge and impact the existing traditional organization mechanism. Enterprises break the barriers of mechanism from the inside out and burst out new opportunities.
- The enterprise data operation ability is weak, the digital transformation technology ability and application ability are still weak, the digital talent shortage and loss are serious, and there are effective means in data application, data governance, data development, data quality and so on.
- Enterprise digital transformation is not only technical solutions, organizational and behavioral elements also play important roles in the process of digital transformation, digital transformation leaders with a broad organizational vision and wisdom of leadership is the catalyst; but the complex enterprise organizational level, fragmented KPI assessment, conservative internal culture, deep-rooted ideas have become obstacles to digital transformation The stumbling block of the model.

During the period of new epidemic (COVID-19), enterprises' response to the epidemic reflects new opportunities for digital transformation. For epidemic prevention and control and resumption of work, enterprises need to operate digital means to solve practical problems. On the one hand, there is a consensus that the willingness of enterprises to transform has been significantly improved, and digital transformation has been the survival strategy of enterprises; on the other hand, enterprises' demand for digital transformation has been rapidly released during the epidemic period, and business scenarios of rapid trial and error through digital means have emerged in large numbers.

2.2.2 The Essence of Enterprise Digital Transformation

Digitization is to connect people's real world and virtual digital, so as to seek a new business model. Digital transformation is based on the emergence and development of new digital technology to help enterprises combine the original traditional business with digital technology to solve the practical problems in the process of enterprise development. At the same time, it also helps enterprises to innovate rapidly to cope with uncertainty and realize the requirements of enterprise performance growth and sustainable development.

Digital transformation of enterprises is not for the purpose of digitalization, but for the purpose of business growth. Enterprise digitalization is not only limited to the IT field, but also includes customer service, operation, sales, marketing, finance, senior management, supply chain, personnel, logistics, customer relationship, ecology, legal affairs and other kinds of enterprise affairs, which is the unremitting pursuit of enterprise for business growth.

There are two characteristics in the development of enterprise digital field:

- With the development of enterprise information technology, enterprise informatization gradually covers all departments within the enterprise, but there are differences in the development levels of internal information flow, capital flow and logistics. The level is mainly divided into single business unit information integration (such as a separate R & D management system), cross unit information integration (such as ERP system) and full value chain cycle integration (such as the integration system of production, supply, marketing and service). Then the development of enterprise digital field is to solve the information seamless docking and linkage of all units in the enterprise internal value chain.
- In the past, the development of enterprise information technology is more concentrated in the enterprise, but with the transformation of customer needs and internal and external environment, the enterprise internal information alone cannot meet the needs of enterprise transformation. The driving factors of enterprise growth are from internal information integration to industrial chain information integration, from single enterprise resources to industrial chain resources integration.

Enterprise digital transformation needs to coordinate enterprise strategy, not the tactics of pursuing immediate benefits. Its essence is to reconstruct enterprise resources, cut into enterprise business flow through data technology, form enterprise data intelligent application closed loop and make the whole process of enterprise production and operation measurable, traceable, predictable, evaluable and optimized.

The reconstructed enterprise resources bring new value chain digitization tools to enterprises, and help them transform into digital enterprises:

- Overall collaboration: Business Middle Office, Data Middle Office, AIoT Middle Office and Organization Middle Office.
- R & D design: Trend prediction, design crowdfunding, simulation testing and new product trial marketing.
- Manufacturing: Real-time data analysis, production data analysis and optimal state.
- Product service: Product life cycle design, customer journey service, remote diagnosis, intelligent and agile operation and maintenance.
- Organization and management: Group and regional unified and separate management and control, regional authority decentralization, independent business unit closed-loop, Middle Office architecture.

2.2.3 The Path/Stage of Enterprise Digital Transformation

Based on the theory and practice of digital transformation, Alibaba can make a clear decision on the nature of the digital transformation strategy. Specific reference can be made to 'Alibaba Cloud's overall digital transformation methodology'. According to their own actual situation, enterprises can formulate the path of digital transformation (Figure 2.5).

With Alibaba Group's suggestions on the digital transformation of enterprises:

(1) Cloud-based IT infrastructure (hybrid cloud + appropriate application)

- Build a hybrid cloud based on the existing IT architecture to make the Internet, business applications and internal systems on the cloud, and improve the utilization rate of IT resources, operation and maintenance efficiency and user experience.

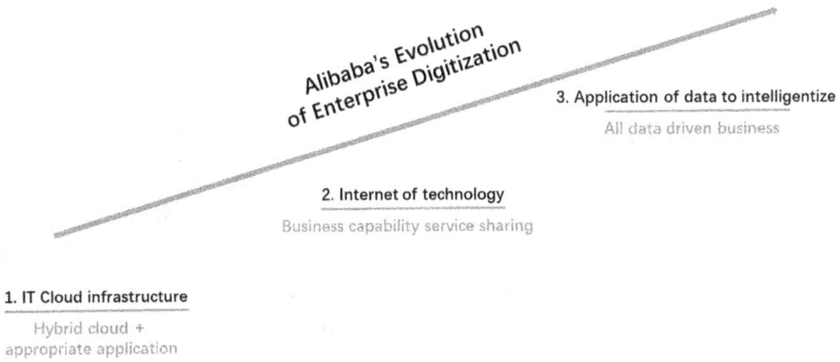

FIGURE 2.5 The path of Alibaba's digital evolution.

- Reasonable security strategy is adopted to ensure information security.

(2) Internet technology

- Using enterprise-level Internet technology, build enterprise business middle platform, break the information island, precipitate and share business modules, release enterprise IT agility and promote business innovation and development.

(3) Application data and intelligence

- The construction of enterprise Data Middle Office through the extraction and connection of business data to form a business-based data system, through the provision of data services to help business development.
- Through machine learning, AI and big data and other means, we can deeply tap the value of data, solve the problems that cannot be solved by human in the past and expand the cognitive boundary.

At the enterprise level, we are also actively exploring a new mode of digital transformation for enterprise development. Individual enterprises consciously strengthen their transformation ability:

- Improve the ability of digital transformation, use it to set up cloud technology and speed up the cloud deployment of core equipment and business systems.

- Promote the transformation of traditional organizational structure, build a platform organization (similar to Alibaba Zhongtai) and build an appropriate organizational management system.
- Strengthen the introduction and training of digital talents, introduce digital transformation leaders, cultivate digital transformation professionals, shape enterprise innovative culture, develop short, smooth and fast business organization circulation, and establish and improve enterprise innovation management, all staff co-governance and internal entrepreneurship mechanism.

2.2.4 Challenges Faced by Digital Transformation of Enterprises

In the process of promoting digital transformation, traditional enterprises need to face up to the current situation of the enterprise, including the concept within the enterprise, the innovation degree of the business team, the technical level of the IT technology team, the organizational structure within the enterprise and the competing action plan. Therefore, whether the enterprise can define the problems to be solved in the digital transformation of the enterprise is particularly important, with the appropriate rhythm, agile and rapid implementation to ensure the normal progress of the transformation process.

Although digital transformation is facing many challenges, some enterprises show great resilience in dealing with external uncertainties:

- Alibaba Group's short-term and rapid implementation of health code project with digital ability during the new epidemic in 2020 has become a typical application of digital ability to assist urban governance, and has played a very important role in controlling the spread of the national epidemic (COVID-19), which is inseparable from Alibaba's good digital ability foundation.
- During the period before the new epidemic (COVID-19) in 2020, Alibaba Cloud DingTalk has used digital capabilities to enable enterprises and families to work online, hold online meetings and conduct online education based on DingTalk in a short time. With the cooperation of Alibaba Cloud DingTalk, it has rapidly supported the online office and online education scenes of hundreds of millions of users across the country. Digital capabilities once again show great ability to cope with uncertainty.

Of course, there are many challenges faced by enterprises in digital transformation. Facing up to the current situation, defining problems, transformation path and small steps are all issues that enterprises need to consider. We summarize some technical and non-technical challenges faced by enterprises in digital transformation, as follows:

Is it necessary to consider the transformation of IOE to SaaS application? Is it better to reconstruct traditional IT system or to make add-ons to the original system? Enterprises with the purpose of restructuring should pay attention to the new business model,comprehensively build the roadmap of enterprise strategic transformation, and implement it in batches and stages; while enterprises with the purpose of increment should pay attention to solving the current practical problems of enterprises, and use the opportunity of cloud to quickly adapt the new formats brought about by the transformation, so as to spawn a new local cloud IT architecture.

- With digital transformation tools, can the transformation be successful?

  The results of a large number of digital tools delivery projects of traditional enterprises confirm that digital tools cannot cure all diseases, and the superstition on tools may repeat the mistake that many traditional enterprises find it difficult to soft land large software of ERP enterprises. Successful transformation needs to ensure that digital technology and tools are applied in the right business and product remodeling and improvement points, so as to promote the whole transformation from point to area.

- How to ensure the security of Open Cloud Architecture?

  Cloud computing has become the foundation and base of digital transformation. The enterprise-level Internet architecture deployed on the cloud can support the changeable business needs of enterprise digital ecology and global data governance. The IT architecture deployed on the cloud provides the driving force for the sustainable growth of enterprises and becomes an important weapon for enterprises to drive business innovation.

  Security has become a major factor restricting enterprises to go to the cloud. Reasonable security strategy, public cloud or hybrid cloud deployment, hierarchical rights management and audit become the major technical challenges of enterprise cloud.

- The leadership of mobilizing the masses to solve problems is very important. No leadership, no digital transformation?

  In the process of enterprise resource reconstruction, business model reconstruction is particularly significant for enterprises. Because the leadership that can drive the overall resources of the enterprise is the key to the digital transformation. The basic measures to ensure the digital transformation are to establish a departmental transformation group and a resource adaptation mechanism for the purpose of transformation. Some enterprises often mix this kind of digital transformation project with enterprise informatization project, and the CIO of information and process department will lead the whole enterprise digital transformation. It is generally recommended that the leading senior executives or CEOs within the enterprise work together to ensure the full integration and interaction of business functions, IT functions and digital functions at the organizational level.

  Based on some cases of Ali, we can see that many traditional enterprises are in the transition to new retail, and there are considerable difficulties in resource coordination among e-commerce, traditional offline, new retail and other business departments, as well as the overall resource coordination at the company level. Once again, organizational and human factors play an important role in the process of digital transformation, which needs 'digital leadership'.

- Is the agile organization of big, middle and small front desk the organizational guarantee of digital transformation?

  Based on Alibaba's practical experience of 'large Middle Office, small front office', and on the basis of common technical ability, an agile organization is formed to cope with the rapid change of business, which is a suitable choice in highly uncertain business environment. Traditional enterprises pay attention to the process and specification. In the digital form, the business chain is greatly shortened, because the flat and agile small business unit is a good choice to adjust at any time with the change of the market.

- Is the data-centric decision-making process an effective means of digital transformation?

  Although the enterprise has a relatively perfect internal IT system and even ERP system, the problems of internal value chain data fragmentation, limited data insight scenarios and lack of intelligent decision scenarios are still very serious. Data in the upstream and downstream flow, often faced with data inconsistency, labor efficiency bottleneck and other cost risks, data-centric decision-making process, can help enterprises improve the efficiency of data analysis, and solve the bottleneck.

2.2.5 A Multi-dimensional Framework for Enterprise Digital Transformation

When talking about digitalization, most people think that AI and data could make it happen, including many of my friends. I was shocked when hearing the CFO of one of my clients saying that "The enterprise can complete digital transformation as long as we could build up the data middle office" when we chatted. The author listened to was very surprised. People may have encountered with so many lectures or reports filled with technological explanations and data that they think technology and data are all needed. The figure below is a relatively complete picture of enterprise digital transformation in the author's opinion. We position digital transformation in a four-dimensional business framework, namely, organization, business, technology and ecology, if can organically combine the power of the four-dimensional digital transformation, we could create a 'panoramic digital perception', 'whole link research insight', 'wise policy decision' and 'automation business execution', 'speech' adaptive closed-loop digital management system, then we further detail the framework and multi-dimensional model (Figure 2.6).

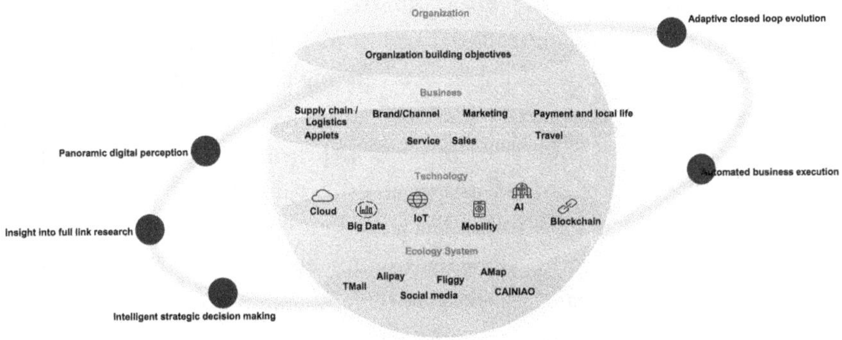

FIGURE 2.6 Multi-dimensional model of enterprise digital transformation.

Just like multi-dimensional space in mathematics, enterprises and organizations in the real world can use multi-dimensional space to design their digital models.

(1) Zero Dimension

The individual skilled person, which is the most basic element of the society, is also the basic element on which every enterprise and institution depends for survival. Every individual technologist uses his or her skills to do things. This is the zero dimension in the multi-dimensional model of digital transformation.

(2) One Dimension

Technology is like a line that consists of many different functions, like points, which would eventually connect together. This is the one-dimensional space in the multi-dimensional model of digital transformation, including the technology platform, AI, data and blockchain.

(3) Two-Dimensional

Business is a two-dimension plane in the multi-dimensional model of digital transformation, composed by multiple technologies which we mentioned as lines in the last paragraph.

(4) Three-Dimensional

A complete organization have multiple businesses, with each business as a two-dimension plane, and finally form as a three-dimension organization after organization design and management. At this point, a business model and framework based on a traditional architecture emerge. But here is a very big problem; people are more around the enterprise or the organization itself to operate. Although it will also cooperate with other enterprises or organizations, it is still carried out in order to improve its own architecture, which is the so-called informatization in the past. Everything is built around the enterprise itself, rather than really connecting with the surrounding, and there is no real data to drive the associated ecology together, so as to truly form a digital society.

(5) Four Dimensions

Ecology. Out of the enterprise itself, each enterprise is in the environment of a large number of enterprises and institutions, their development and operation are not completely determined by the individual enterprise. Like time, Like time, the whole ecology never stops.Enterprises only real complying with the trend of the ecological development, in which to find the right orientation and target, can be more effective in the four dimensions in the same dimension in the development of better, and to the positive feedback affects the development of the digital transformation of the four-dimensional space. In the digital transformation, by linking the enterprise itself with the ecology, and leveraging the environment in the ecology effectively with AI, blockchain and other technologies, enterprises can find the key to success more efficiently and easily than other enterprises.

(6) N Dimension

The golden key to higher order dimensional iteration – five elements of digital transformation iteration. From the zero dimension to the four dimensions, after the enterprise completes the model design and construction of digital transformation, there may also be a situation, that is, put down the wrong choice, the wrong process, then is there any way to help the enterprise to succeed again? The author's point of view is that, just like many great enterprises in history, they have gone through many near-death experiences, but successfully took off, such as Big Blue IBM, General Electric, Nissan, Nokia, Apple and other great enterprises. Investigate its reason, it is innovation, break old pattern, rebuild brilliant next. In the digital transformation model, we hope to be able to define a technique which could save companies from transformation failure, and help them try again. I think, we can construct the five-key ability, forms the enterprise digital transformation for days, they are 'panoramic digital perception', 'whole link research insight', 'wise policy decision' and 'automation business execution', 'the adaptive evolution of closed loop'.

Panoramic digital perception: just like eyes, ears and nose. The key number metrics can be displayed clearly, just like the eyes and ears. Problems can be found through the key number metrics displayed, so as to provide decision input for final decision.

I. Full-link research insight: Meridians. Each key business node is labeled, just like the meridian of a person. Through the input of I, it can quickly locate the specific key business node that has gone wrong, so as to make accurate decisions, adjust the problematic business and take global consideration of other related businesses.

II. Intelligent strategic decision-making: The brain. We don't want every decision to be set and executed by human beings, just like the Stone Age and the agricultural age. We want to make intelligent judgments and give key suggestions through AI, so as to assist decision-makers to make optimization or reform strategies.

III. Automated business execution: Limbs. Strategy formulation has been completed in III, so a set of intelligent and highly automated tools are also needed to assist in implementation. And that's a lot of tools and platforms.

IV. Adaptive closed-loop evolution: The heart. I–IV mentioned above is a complete set of executive path, but as of now, we could only have one chance to mend, we are continuously trying to find paths which allow more chances to mend, that will require us to have the ability of self-revolution. To keep the enterprise active with effective organization and culture, so that we could self-revolution when facing problems, could be an effective way.

2.2.6 Enterprise Digital Transformation Is a Long Process

Going back to what the author mentioned about the CFO of the client, you can definitely understand her perception of a period of digital transformation, which should be relatively short term. Which should be relatively short term, maybe less than a year. However, this is a very big cognitive mistake. Digital transformation is definitely not a process of building a big data platform, which can be achieved by doing some data analysis. Moreover, it requires long-term operation and operation and continuous iterative presentation process. For example, the digital transformation can be divided into two stages. The first stage is like building a house. If a house is built, it will definitely not be comfortable to live in without property management. Digital transformation is like building a house. Both phases are very important. In addition, organizational security and concept need to evolve along with it. If you really want to make a decision on digital transformation, you should be psychologically prepared from the first day. It will be a long process, but you can decompose the business goal you most want to achieve in different stages of the process and gradually realize it. At present, some enterprises that have done well in the digital transformation of the industry insist on continuous evolution and iteration for a long time.

2.2.7 Exploration of Enterprise Digital Transformation

After a period of trial, digital transformation has had some interesting exploration and effects in many fields. For example, the novel Coronavirus epidemic trial played a very important role in the epidemic prevention and control. Moreover, an interesting new term 'strong information person' appeared. What does this mean? That is to say, in the past information construction, people more is to foster information goods such as tag, but the outbreak period, in order to better control the outbreak, around the 'people' built a set of information system, including the basic attribute of people, travel, health attribute, and then further with the 'family', 'affiliate' attribute to combine to form a digital information network, so if anyone infected with the virus, can quickly to troubleshoot a baseline, ultimately all involved to find out, and effective to prevent further spread of the virus. Again, for example, the milk from the cow, the milk collection, production, packaging, transportation and sales process on the label information reoccupy blockchain management, so that it can be the advantage of food quality management, any link has a problem can quickly find the problem point and precise action. Still has a lot of interesting attempt, such as is now in the research and development and promotion of Cloud Computer, turning and change the traditional mode of the use of PC, can provide not only the traditional PC some function, but also provide a number of traditional PC cannot provide things, such as large-scale complex calculations, huge amounts of data mining, AI visual rendering and so on. So, the age of cloud and digital transformation is one of constant disruption and surprise.

2.2.8 The Common Misunderstanding of the Enterprise Digital Transformation

2.2.8.1 Obsess Over Other Enterprise's Success Stories

In recent years, more and more manufacturers and digital transformation experts have stood on the platform and shared a lot of digital transformation methodology, products, successful cases, etc. You may feel excited and determined to carry out digital transformation in your own enterprises, thinking that digital transformation is a good medicine for the world. For example, the construction of big data platform helps an enterprise to develop new business and reduce the cost. Through the establishment of intelligence control center achieves the management of the whole link and so on. These cases all sound very nice. So, everybody is excited and going digital transformation.

First of all, let's see if we can make the digital transition at all. There is no doubt that digital transformation is a must. But the success of other people's home is really so simple on the success, so easy to replicate it? The answer is definitely not.

Second, how can we distinguish the success of digital transformation in other people's stories from what we can learn from and what we must learn from. Sharing on the stage is usually very short, and most of the time it's busy showing the good side, but in most cases, it doesn't share much background of the story. Therefore, we must have an urgent heart for the hot digital transformation, but we also need to keep a calm and cool head. Before you decide to make a digital transformation, you must do three things: 'Self-transformation feasibility assessment', 'Digital transformation SWOT analysis' and 'No. 1 Project'.

Self-transformation feasibility assessment: The author suggests that some statistical methods can be used for analysis, such as the Analytic Hierarchy Process (AHP), the feasibility of digital transformation can be divided into 'target layer', 'constraint layer' and 'scheme layer'. The three layers are interrelated, with the upper layer dominating the lower layer, and the factors in the same layer are independent of each other. For example, the 'target layer' is dominant to the 'constraint layer', and the 'constraint layer' can only be defined after the 'target layer' is defined. It is also advisable to have no more than nine or less than five factors per layer, because according to psychological research, human judgments of things beyond a certain range are distorted. It is neither too much nor too few.

Goal layer: Focus on the goal of solving the problem. In this layer, such factors as design can be considered. Of course, different enterprises have different goals and requirements, so they need to design according to their own characteristics: 'Organizational change', 'GMV growth', 'new members', 'Traffic realization', 'ROI optimization', 'Cost reduction', 'Process optimization', etc.

Constraint layer: Focuses on various measures, criteria and constraints that need to be taken to achieve the overall goal.

Solution layer: Various solutions and measures that focus on solving problems.

For AHP, it is necessary to design and construct a judgment matrix. The factors that have influence on the target are compared in pairs, and the comparison results form a matrix. Further consistency check is also required for the constructed matrix, and the comparison matrix is adjusted continuously according to the consistency check calculation results until a certain range of consistency calculation results is reached. Because the AHP is more complex, it is not the key content of the book, interested readers can query special information for learning (Table 2.1).

SWOT analysis of digital transformation: In addition to the feasibility analysis mentioned above, it is also suggested to make a SWOT analysis on own ability, industry environment, and maturity of digital transformation technology, and make some judgments on multiple dimensions.

The No. 1 Project: It is important on the before points, but the most important one is probably the last one, that is, the digital transformation must be positioned as the 'No. 1 Project' of enterprises or organizations. If it cannot be presided over by the first one himself, he will often face problems such as giving up halfway or great organizational resistance.

Finally, once the decision is made, it is necessary to have the determination not to let go, organizational change, persistence, focus on long-term benefits and so on to keep up.

2.2.8.2 There Is No 'Silver Bullet' in Digital Transformation

Digital transformation can be help enterprises to take off again, but also be sure to prepare for the digital transformation, it is not a 'silver bullet',

TABLE 2.1 Example of Analytic Hierarchy Process Judgment Matrix

Priority	Scale	Level	Rules of judge proof	Factor 1	Factor 2	Factor 3	Factor 4	Factor 5
1	5	Rule 1	1	3	5	3	3	5
1	5		1	1	3	3	5	3
2	3	Rule 2	1/3	1	3	1	1	1
2	3		1/3	1	1	3	3	1
2	3		1/3	1	1	1	1	1
2	3		1/3	1	3	3	1	1
3	1	Rule 3	1/5	1/3	1/3	1	1	1
3	1		1/5	1	1/3	1	1	1
3	1		1/5	1/3	1/3	1/3	1/3	1

can't solve all the problems enterprises and provide digital transformation is a set of concept, a set of technology, a set of best practices, but these are the external cause, external cause can make enterprises or institutions to look into the mirror, find the key, but willing to look in the mirror, find the key, this need internal cause, if the organization or institution holding the manner which give it a try, or blind worship of the digital transformation, feel is all-purpose adhesive, which tend to end up a miserable. Therefore, there is no 'silver bullet'. Only by taking solid steps step by step, can digital transformation be truly successful.

2.2.8.3 Are You Ready Mentally, Intellectually and Physically?

Digital transformation is a long run, in which the final winner is often not the fast sprinter who starts first, but the runner who keeps steady state and distributes energy, maintains a relatively steady rhythm, and even follows for a long time. The 'mentality', 'intellectuality' and 'physical strength' of an athlete are measured during long runs. It is same as true of digital transformation, which also tests the 'mentally', 'intellectually' and 'physically' of managers and participants.

- Mentality: Ideal, passion, understanding, care, etc., to put one's heart and soul into doing one thing or undertaking with cooperators.
- Intellectuality: Logical analysis ability, abstract ability, summary ability, structural ability, etc., is a person's professional performance.
- Physical strength: Execution, toughness, etc.

So, what is the order in which these three are used? 'Mentality' must come first, because you need to empower yourself so that everyone can do things with passion, understanding and a shared vision. When hearts are together, things are easy to do. The second is to use 'intellectuality', with their own professional abstract problems, to find solutions to ideas. The last thing is to use the 'physical strength'. If we use up the 'physical strength' at the beginning, then things may not go as expected, because people's physiology is exhausted. How can we have the energy to talk about ideals and rationally design and solve ideas?

If one of the three is not ready, it is recommended to prepare again, and then make a concerted effort to complete the digital transformation and transformation in a planned and rhythmic long distance.

2.2.8.4 Organization and Concept Are Key to Success

It is an interesting phenomenon that many enterprises have built a big data platform and built a relatively complete data analysis platform and statements, but customers find that they cannot read the data. This is not an isolated case, but a relatively common phenomenon. There are also many enterprises that are confident to make digital transformation, but in the process, they always find it difficult to move to achieve the expected results.

Therefore, in the digital transformation and transformation, it is very important to ensure the stable organization and let the data penetrate into the daily business philosophy. So, what kind of organization is suitable for digital transformation? How do you really get the idea of digital into the head?

Organization: Through a general survey of the better organizational forms of enterprises in digital transformation, there are some common things that can be used for references. For references: flexible process control, weak concept of rank, flat management matrix, young people in key positions, and long-term strategic goal. In order to carry out digital transformation and transformation, if everything is done strictly according to the process, the organization will lack flexibility and restrict the innovation of employees. However, the success of digital transformation does not depend entirely on the 'external brain', but more on the 'internal brain' to constantly evolve and innovate in the process. The external brain is the trigger. The internal brain is the key. Moreover, if the concept of rank is relatively heavy, it is difficult for people to break the boundary between their respective businesses. Any change is inherently destructive. If proper trial and error and destruction are not tolerated, or if there is no soil for survival, then the change cannot be started. Similarly, since it is a change, it needs more new blood, new thinking and new attempts, and young people are the best representatives of this aspect. However, there is also a great risk of failure due to the fact that many organizations lack a clear strategic positioning and strategic nature. As they do so, they will quit halfway if they encounter difficulties instead of stepping forward.

Digital business philosophy: The so-called digital business philosophy, I think the most need to do is: (1) know what is the digital business philosophy; (2) according to the digital business philosophy, develop a daily reading of data and make judgments and decisions based on data analysis; and (3) we should neither blindly believe in nor ignore data. First of all, let's take a look at the concept of digital business. To sum up, there should be three words: 'Digitization of all businesses', 'Digitization of all data businesses' and 'Intelligentization of business data'. In the first place defining own businesses clearly and make them digitalized, which would be the foundation, and then use these data for modeling and analyzing, for digging stories behind data, for combining data with business scenarios, and eventually provide insights to support decision-makings, so that data could bring far more values. Moreover, we also are trying to explore the data with AI, to find more values beyond artificial abilities.. In a word, digital business should be digitized, data business should be digitized, and data should be intelligent. Then, the habit of looking at the data every day and constantly optimizing and iterating the data should be formed. Meanwhile, the business potential or innovation points should be deeply explored through the data.

2.3 TOP-LEVEL DESIGN OF ENTERPRISE DIGITAL TRANSFORMATION

Section 2.3 will introduce from the value, scope and three principles of the top-level design of enterprise digital transformation, as well as detailed introduction of how enterprises carry out the top-level design of digital transformation to elaborate its importance in the enterprise digital transformation.

2.3.1 The Value of Top-Level Design in Digital Transformation

Digital transformation is the revolution and innovation of enterprise development mode, the inevitable choice for enterprises to move from industrial economy to digital economy era, and the most concerned topic for most enterprises at present. If enterprises want to develop their digital economy, they must provide the support of digital combat capability, and 'top-level design' is the way to go. Magnates always have strategic vision, broad pattern and large minds. The informatization and partial digitization of enterprise technology often fail to solve the problems of current reality and future long-term development of the transformation. Therefore, it is particularly important to have a set of top-level design with a panoramic view of God.

In the process of realizing enterprise digital transformation, it involves not only solutions and products, but also organizational structure changes, and even the adjustment of the overall strategy and culture of the enterprise. All of these require a set of correct top-level design to indicate the direction of the enterprise.

Top-level design is a kind of structure that guides and controls various relationships and processes of enterprises. It can guide the development of various work of enterprise informatization, improve the enterprise architecture, make it add value and help enterprises achieve their goals by means of information. Second, in the process of development, enterprises also need the delivery and empowerment of products and services, as well as the division of labor and collaboration between various business departments, with the application system as the support. However, with the rapid development of enterprise business, it will be found that the original information system no longer meets the needs of business, data are not unified, and the trend of information island is becoming more and more serious. One of the key elements in the digital transformation of an enterprise is the empowerment of data, turning it into assets and capital. Then how to ensure the input, output and output of data, the enterprise needs a variety of applications to work. In the enterprise top-level design, the enterprise needs business architecture, application architecture, data architecture and technology architecture, and security architecture for support. Finally, all the strategies, transformations and iterative updates of the enterprise need to be carried out around the vision. Digital transformation is the beginning of a new strategy for organizations facing the future. Technical difficulties are only a part of the difficulties. Enterprises need to regard transformation as a vision: viewing the present from the perspective of the future and creating infinite possibilities for future businesses with technology. Layers are designed to give the enterprise direction.

Therefore, top-level design is the only way to digital transformation of enterprises and is of great significance. The significance of top-level design lies in the overall vision and future perspective to lead the digital innovation of business. A set of top-level design that fully understands the business operation and digital technology can promote the digital transformation of enterprises, stimulate the impetus of all employees and improve the probability of successful digital transformation.

2.3.2 The Scope of Top-Level Design in Digital Transformation

Without a panoramic perspective and strategy, digitalization from the technical level may result in the separation of production data and data, as well as the separation of digitalization and operational reality. It is equivalent to using local thinking to plan the whole, which will inevitably fail to achieve successful transformation results. The plan of digital transformation must have strategic height, business breadth and practice span, and integrate the complicated business and system in the enterprise through strategic thinking and structured way.

The top-level design of digital transformation is not only a complete system, but also a comprehensive design of planning objectives, implementation means and methods, and resource guarantee. As for the scope of top-level design in digital transformation, the author believes that it can be described from the following six levels:

(1) The business layers

The top-level design of the business layer is to sort out the core business logic of the company, and the most important point to realize the digital transformation of the enterprise is to do a good job in the business transformation. In the process of combing through the core business, it combines the current 'Platformization' design ideas of most enterprise business systems. So as to achieve from the internal management and operation chain through, integrate service resources and break through service chain externally.

(2) The technology layers

The top-level design of the technology layer is to build the technology architecture that can realize the core business, with the goal of realizing the future business development enabled by technology. Design the technology architecture from the five layers of infrastructure, data, services, applications and portals to ensure that each layer has its own core capabilities.

(3) The organizational layers

The top-level design of the organizational layer is the organizational form used to ensure the landing of the design of the business layer and the technical layer, and the implementation of the planning and design is carried out through the form of special planning and design communication meeting, so as to establish the information management organizational guarantee of the company layer and continuously optimize the management mechanism.

(4) The operation layers

The top-level design of the operation layer is to realize the user as the origin of the operation value, so as to build a more perfect operation system and maximize the operation benefit. In the process of digital transformation and upgrading of enterprises, the construction of operation capacity is particularly important. If the construction of operation system is not carried out simultaneously, the investment in software system and resources in the early stage will be basically wasted.

(5) The staff layers

The top-level design of staff layer is to realize the final landing of digital transformation and promote the digitization process of enterprises. Once the digitalization strategy is formulated at the company level, professional talents are required to follow up and participate in it, and talent excavation should also transform from development and maintenance to product design, process reengineering, architecture design, AI development and data value mining.

(6) The cultural layers

The top-level design of cultural layer is to realize the agility and mobility of enterprise digital transformation. The change of corporate culture drives the process of digital transformation. Only by changing the traditional culture can an enterprise realize agility and lean and help promote the behavioral change of the whole company.

2.3.3 Three Principles of Top-Level Design in Digital Transformation

How to carry out top-level design in digital transformation, enterprises should follow the following three principles:

(1) Business Drive principle

Through the top-level design of the business layer, the core business architecture and logic of the company are sorted out to achieve the reconstruction of information business flow. (1) A clear set of business boundaries, clear boundaries on responsibilities to avoid the phenomenon of cross-duplication of work and process; (2) a set of correct business process connection, avoiding fragmentation by combing the core business process, so as to make reasonable connection between different departments and processes; (3) the synchronization of innovation can avoid the imbalance of innovation degree among businesses in different enterprises or the synchronization of the same business among different departments, so as to realize the synchronization of innovation and smooth operation of process, which is conducive to the optimization of business process.

(2) Unified architecture principles

Only by following consistent standards, consistent experiences, modular microservices and mastering core capabilities can the top-level architecture achieve the unification of platforms and businesses, and build a set of technology architectures that can support businesses and rapidly improve enterprise efficiency and profitability.

Consistent standards: Achieve integration and expansion by unifying the standards of key technical nodes such as data, development, interface and security, so as to avoid multiple losses of time, manpower and material resources caused by incompatibilities caused by large differences in standards after parallel development of multiple systems.

- Experience the same: Front-end design should be simple with user's perspective, to reduce learning-cost so that the enterprise could promote their system easily and quickly, and take advantage of digital transformation immediately. Establish confidence, let enterprise digital go far and long, also avoid because of the new system is too complicated, errors and risks for the enterprise operation, rather than the icing on the cake, but worse.

- Modular microservice: The establishment of a general modular microservice, standardized module assembly to improve the reuse rate, not only can greatly improve the efficiency of development, but also can reduce security risks, and at the same time meet the above-mentioned principles of consistent standards and consistent experience, to facilitate the smooth docking between systems and users quickly get started.

- Grasp core ability: In the process of digital transformation, the enterprise should pay attention to grasp its own core ability, and deeply understand the logic behind, in order to ensure the transition process and architectural design are controllable, and notice chances and risks quickly and concisely, so that enterprise could adjust accordingly to reach the best result.

(3) Principle of division of labor and coordination

The principle of coordination requires a unified goal, unified thought, and unified design, with clear business information planning and integration committee layer, technology, organization layer three top-level design goals for unified goal, with special design communication will as the carrier, in accordance with the unified information requirements in every platform the overall design idea, strictly follow the company's technical standards, a unified architecture, data, interface standards, through planning and integration of the committee for examination and approval of technical solution, achieve rapid iteration, methodically digital transformation.

2.3.4 How Do Enterprises Carry Out Digital Transformation Top-Level Design

2.3.4.1 Top Design Methodology for Enterprise Digital Transformation

In the process of digital transformation, reengineering business process optimization ability is still mainly relying on various business departments (individual companies have the professional team of process combing with the IT team), using the built-in technical ability of the IT team, to control the whole system construction link to system integration, in order to avoid the technical ability or absence caused not integrated control (Figure 2.7).

The main models are as follows:

- Process combing of business departments + independent development

  Business departments put forward requirements, IT department independent development and construction;

- Process combing of business department + independent design + outsourcing development

  Business departments put forward requirements, IT departments independently control the design, code, property rights, outsourcing construction;

- Process reengineering center + system construction

  Business departments put forward requirements, process reengineering center reengineering process, system construction can adopt the first two ways of construction.

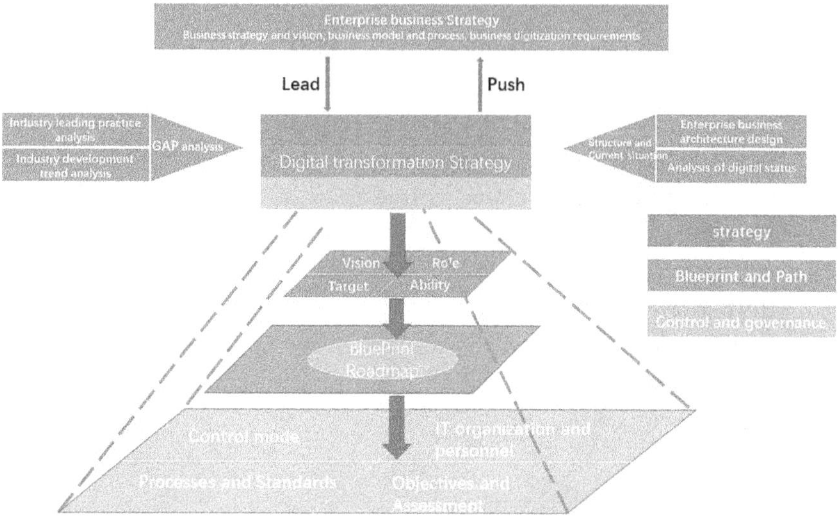

FIGURE 2.7 Top-level design methodology for enterprise digital transformation.

As we can see, the first and the second model emphasizes the mastery of technology integration, technology architecture design, code and property, to avoid technical barriers from outsourcing, and the main difference between the two models is team size. The third kind of pattern is one of the few enterprises to adopt, or team, in the center of the IT department set up process reengineering emphasizes the process optimization of the whole process reengineering, is a group of technical personnel, research process with advanced information technology thinking innovation enterprise processes.

All of the above models require self-design ability, which is the basis of the digital transformation. With a professional team, combining with its own situation enterprise could tailor its own information planning and architecture design, control-process design, and data house design, so that to avoid unnecessary repeated works and inconsistences.

2.3.4.2 *The Working Steps of Enterprises Digital Transformation*

(1) Requirements gathering

In 2011, E-Works launched the bi-integration maturity assessment system, with six industrial versions. Later, it launched the intelligent manufacturing and intelligent factory assessment system,

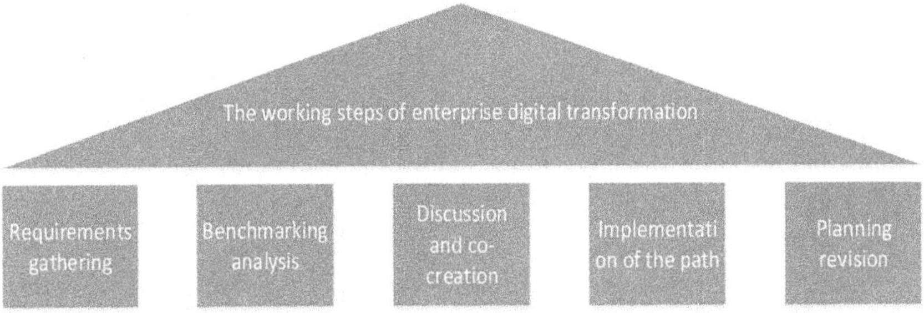

FIGURE 2.8 Work steps of enterprise digital transformation.

and relevant research institutions of the Ministry of Industry and Information Technology also launched the relevant assessment system. Manufacturing enterprise could understand the depth and width and also effectiveness of digitalization applied in each chain, also to see some breakpoints through transformation evaluation. The enterprise could also compare itself with other competitors or industrial pioneer (Figure 2.8).

(2) Benchmarking analysis

Through extensive internal research and industry best practice analysis, combined with industry standards, norms and compliance requirements, and based on the enterprise's development strategy, the needs of enterprises to promote digital transformation are sorted out, and the breakthrough point for enterprises to promote digital transformation is determined according to the importance and feasibility.

(3) Discussion and co-creation

This paper analyzes how the business process of an enterprise should be optimized in the process of digital transformation, determines the key assessment metrics for the enterprise's digital transformation, formulates the overall framework of the digital system and defines the overall plan for the enterprise's digital transformation in the next 3–5 years.

(4) Implementation path

Define the specific function, deployment mode and integration scheme of each digital system; determine data acquisition, device networking, IT and OT integration scheme; develop an annual investment plan for digital transformation; define the organizational system to promote digital transformation; analyze the investment returns of digital transformation; and predict the possible risks and avoidance strategies in the digital transformation process.

(5) Planning revision

The plan for digital transformation should also be a 3-year plan and a rolling 1-year plan. Enterprises shall conduct annual inspection of the status of digital transformation, and revise their plans for digital transformation in the light of the changes in the actual situation of the enterprises and the development of emerging technologies. At the same time, enterprises should attach great importance to the core team construction of digital transformation, combine IT department, automation department, planning department and lean department, and employ external expert consultants, so as to ensure the digital transformation process of enterprises step by step and achieve real benefits.

To promote digital transformation, enterprises are the main body. To truly realize the digital transformation, it needs the determination and perseverance of the senior management of enterprises, to truly understand the connotation of digital transformation and to lead the process of digital transformation.

Digital transformation should not be formalistic or go through the formalities, nor should it be big and complete. It should be customized according to an enterprise's own needs, its position in the industrial chain, its strength and development vision. It can be predicted that more and more manufacturing industries will put forward the demand for digital transformation consulting services, and at the same time, the consulting institutions that can provide such consulting services also need to have experienced consulting service teams after long-term practice.

2.3.4.3 Maturity Assessment Model of Enterprise Digital Transformation

After enterprise clarifies its goal of transformation, and completes its top-level design according to the above steps, the enterprise could evaluate its own strengths and weaknesses according to "Enterprise digital transformation maturity" published by Alibaba Intelligence, and make the best decisions when deploying.

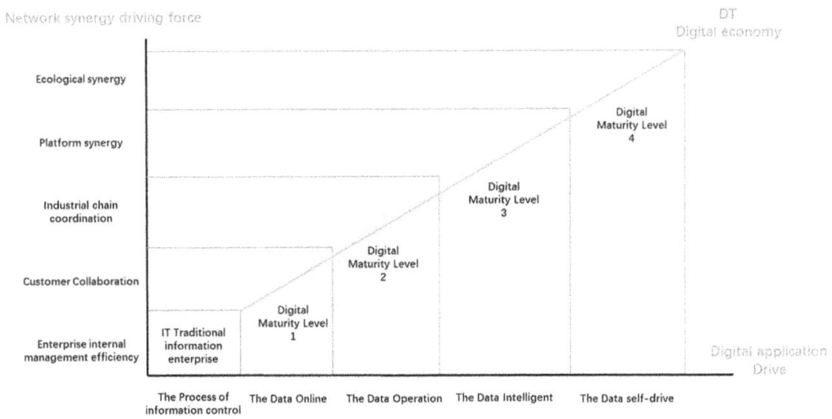

FIGURE 2.9 *Alibaba Cloud Research Center – New Digital Transformation White Paper (2019)* – Alibaba Cloud digital transformation maturity model.

As described in the second chapter of the *White Paper of Alibaba Cloud Research Center – New Digital Transformation (2019)*, with the arrival of AI era and the realization of big data and microparticle cognition, 'digital transformation' has been introduced into a new connotation: data have replaced traditional physical materials and become an important asset; the 'digital outfield' compresses time and space and changes production relations. Therefore, we should start from the two dimensions of 'network collaboration' driving force and 'data application' driving force to help enterprises make a new judgment on their position of digital maturity and the roadmap of transformation and upgrading from the perspective of only business. It should be noted that the 'digital outfield' may give the enterprise a non-linear basis when quantized strategic view is used to measure the enterprise's digital maturity, so the maturity model essentially allows jumps and superpositions (Figure 2.9).

Based on the model shown in Figure 2.9, enterprises can answer a series of research questions, evaluate the digital interval they are in and find the digitalized transformation plan suitable for their own industry according to comprehensive factors such as industry and enterprise strategic objectives and different maturity metrics.

CHAPTER 3

Application Cloudification and Cloud Native

Li Yichao, Wang Keqiang, Cheng Zheqiao, Zhang Yuchen, Liu Xiao, and Liu Xun

After analyzing the difficulties of traditional IT architecture, this chapter discusses the requirements of application Cloudification. Then it analyzes the concept of cloud-native architecture in detail, and finally introduces the methodology, practice and case of digital transformation.

3.1 CURRENT SITUATION AND PROBLEM ANALYSIS OF TRADITIONAL IT ARCHITECTURE

3.1.1 Difficulties and Challenges in IDC (Internet Data Center) Era

At present, enterprise innovation, especially the innovation of small and microenterprises and start-ups, is facing barriers in IT technology and IT cost. In addition, traditional IT enterprises are facing 'the pain of security', 'the confusion of elasticity' and 'the fatigue of technology', and then facing 'the suffering of innovation'. We might imagine the following scenarios:

- With the continuous expansion of business, an IT company with a small scale is increasingly unable to cope with IT…

DOI: 10.1201/9781003272953-3

- 'The original eight O & M, three DBA, two network engineers, tired of working hard, living a life of' 007, 'do you need to recruit more people? But the labor cost budget is limited…'
- 'The information department has more than 80 engineers, how to change from the cost center to the profit center…'
- 'Last year, one IT company spent 1 million yuan on the servers, and the business has not been online yet, but the price of servers has dropped by 30% this year. How to avoid it in the future?'
- 'We bought SAP Hybris, but it doesn't seem to play a big role. What should we do?'
- 'We bought SAP HANA, the price is expensive, is there a higher price-performance ratio plan?'
- 'For a device with the storage capacity of 2 PB, the business flood peak has increased by 400 TB, which is difficult to support for a long time. What should I do? Do I need to buy a new device?'
- 'With more and more business data, disaster recovery is very important. What kind of solution has a higher price-performance ratio?'
- 'The operation team made a promotion and was dumped in an instant without any reason…'
- 'What should I do if the website is maliciously attacked by hackers and the customer data are leaked?'
- '618 sales promotion is approaching, 100 new machines are needed according to the expected business traffic, but they are not used on a daily basis. Do you want to buy or not?'
- 'A batch of machines are now under warranty and some of them are faulty. How to smoothly migrate business and data?'
- 'What should I do if the database concurrency increases and frequently hangs? DBA maybe lose sleep again….'
- 'In order to launch the business quickly, a lot of open-source software has been installed. How can bugs be solved continuously?'
- 'How should the Business requirements Department support it?'
- 'How to integrate the large and small chimney business systems?'

- 'How to quickly build a large screen like Alibaba Singles' Day so that the management can make decisions?'
- 'How to make big data work for the business, how to conduct accurate marketing, and how can user portrait be quickly built?'

Yes, the middle and senior managers of the enterprise are facing many problems that need to be solved, which can be summarized as the following five aspects:

- How to reduce the TCO (Total Cost of Ownership) of enterprises while maintaining rapid business growth?
- How to make the enterprise's infrastructure more flexible to cope with the peak of traffic during the big promotion period and the waste of resources during the business downturn?
- How to make the IT facilities of enterprises safer and more reliable and maintain the business continuity?
- How to keep the IT technology stack of an enterprise at the forefront of the industry, improve employee efficiency and maintain competitiveness?
- How to enjoy the technological dividends brought by the development of science and technology conveniently, feedback business innovation and bring greater profits to enterprises?

Generally speaking, it can be summarized into six words: cost, flexibility, stability, security, efficiency and innovation. This is the problem that enterprises strive to solve in the traditional IDC era, and it is also the goal that enterprises pursue.

3.1.2 Opportunities for Enterprises Shifting to Cloud

3.1.2.1 The New Technological Revolution

At present, a mighty wave of scientific and technological revolution is sweeping the world, and ICT (Information and Communications Technology) technology is the real protagonist. Technologies, such as cloud computing, big data, artificial intelligence, IOT (Internet Of Things), and blockchain are rapidly entering our work and life. In particular, the novel coronavirus, which is raging all over the world, will advance the digitization process of the whole society by 3–5 years. As a result, new technologies and new applications will permeate into every corner of our production and life, making people connect with people, things and things, enterprises and enterprises, industries and industries more and more closely, boundaries are becoming increasingly blurred. The new technological revolution will definitely change our whole world.

Among these new technologies, cloud computing, as an infrastructure, will fully carry all kinds of new technologies and applications, thus supporting various technological innovations, application innovations and model innovations, and then defining new pattern.

As a public computing resource and service, cloud computing has the following advantages compared with computing resources in the traditional IT era: first, the intensification of hardware; second, the intensification of talents; the third is the intensification of security; the fourth is the inclusiveness of services; the fifth is the business innovation

Cloud computing is an important support for 'smart' and 'intelligence'. Intelligence has two big supports: network and big data. Various networks, including the Internet, mobile internet and internet of things, are responsible for collecting and sharing data. Big data, as a 'raw material', is the foundation of various smart applications. Cloud computing is a platform that supports network and big data. Therefore, almost all smart applications cannot work without cloud computing.

Cloud computing is a powerful guarantee for enterprises to enjoy IT application and innovation environment equally. At present, enterprise innovation is facing barriers between IT technology and IT cost. The emergence of cloud computing has broken this barrier. IT has become a low-hanging basic resource. Enterprises do not need to focus on IT support and implementation, but can focus more on their areas for business innovation; this will play a vital role in improving the informatization level of the whole industry and stimulating the enthusiasm for innovation and entrepreneurship.

For our country, the strategy of developing cloud computing industry is of great significance. Cloud computing is no longer just an 'IT infrastructure'. IT will become a 'national infrastructure' like power grids, mobile communication networks, the Internet and transportation networks. Fully serve the implementation of many major national strategies.

3.1.2.2 The Development of Cloud Technology

At the International Information Processing Conference in June 1959, a paper titled computer time-sharing application by Christopher was recognized as the earliest discussion of virtualization technology in the world. In the late 1990s, after the virtualization technology and distributed technology matured, the prototype of cloud computing initially formed. At the beginning of the 21st century, Salesforce took an initial attempt of SaaS (Software as a Service), and cloud computing really entered people's vision. Since 2005, in order to solve the problem of idle resources, Internet enterprises such as AWS, Alibaba and Google have accumulated distributed computing in the e-commerce industry for a long time, security and other technologies and idle resources have been sublimated and utilized. For the first time, a large-scale IaaS (Infrastructure as a Service) has been formed and services have been formally provided to external customers. Since then, the era of cloud computing has really begun. Then, many traditional software enterprises, IT service enterprises and cloud computing start-ups have devoted themselves to the world of cloud computing, each one showing special prowess.

The demand for private cloud is strong, and the integrated architecture cloud platform is developing rapidly. Around 2014, the concepts of cloud computing and big data have been accepted by enterprises in China. No one doubts whether cloud computing is needed. People are more concerned about how to build cloud computing and how to integrate cloud computing with enterprises. During the same period, public cloud vendors based on the native architecture were busy improving the off-premise service and had no time to take into account the strong demand for On-premise. OpenStack, which focusing On the On-premise market, has developed rapidly. IBM, HP, Huawei and other well-known enterprises have launched their own OpenStack cloud platform solutions. These solutions meet the urgent needs of enterprises for cloud computing. Many enterprises (such as Shenzhen Stock Exchange and Industrial and Commercial Bank of China) have embarked on the road of using OpenStack.

The native cloud platform of Internet companies began to deploy On-premise. With the in-depth use of cloud platforms by enterprises, users find that the integrated architecture represented by OpenStack requires enterprises to have strong integration capabilities and high operating costs. As a result, the popularity of integrated architecture begins to decline. Enterprises like IBM and HP have withdrawn from this market. At the same time, native cloud vendors represented by Alibaba began to dabble in the On-premise cloud market to provide On-premise cloud computing solutions. In 2015, Alibaba released its first private cloud version Apsara Stack 1.0.

Open source does not mean open up. 'Cloud users are not concerned whether you use open source to implement cloud technologies, but they concerned whether you can provide open interfaces for customers to use and provide sufficient services'. However, most cloud platforms with integrated architectures emphasize IaaS rather than PaaS (Platform as a Service), and PaaS is deficient.

3.1.2.3 Technical Route and Market Pattern

From the perspective of cloud computing industry chain, it mainly includes infrastructure layer, platform and software layer, operation support layer and application service layer. The infrastructure layer is mainly based on hardware resources such as underlying components and cloud infrastructure devices, which guarantees the stability of the entire service. The platform and software layer is based on the basic layer and provides tool software and application development platforms; it is the source of entrepreneurship and innovation in the industry. The operation support layer is located in the middle of the industry chain and plays an auxiliary role in planning, consulting, integration and security. The application service layer includes cloud terminals and cloud application services; it is the driving force for the continuous growth of the industry.

Cloud computing can be divided into IaaS, PaaS and SaaS according to different service modes. In addition, it can be divided into public cloud, private cloud and hybrid cloud according to different deployment.

The public cloud market, represented by IaaS, PaaS and SaaS, reached US $136.3 billion in 2018, with a growth rate of 23.01%. It is expected to exceed US $270 billion in 2022, with a compound annual growth rate of 19.00%. Among them, IaaS market grew rapidly, with a growth rate of 28.46% in 2018 and a market size of US $32.5 billion, which is expected to grow to US $81.5 billion in 2022. PaaS market grew steadily, with a market size of US dollars in 2018, up to 22.79%; with the development of big data applications, the demand for database management systems has increased significantly. SaaS is still the largest component of the global public cloud market, with a market size of US $87.1 billion, far exceeding the sum of IaaS and PaaS, the growth rate of 21.14% slowed down slightly, among which CRM (Customer Relationship Management), ERP (Enterprise Resource Planning) and office suite accounted for 75% of the market share, while services such as business intelligence application and project portfolio management grew rapidly, although the scale was small (Figure 3.1).

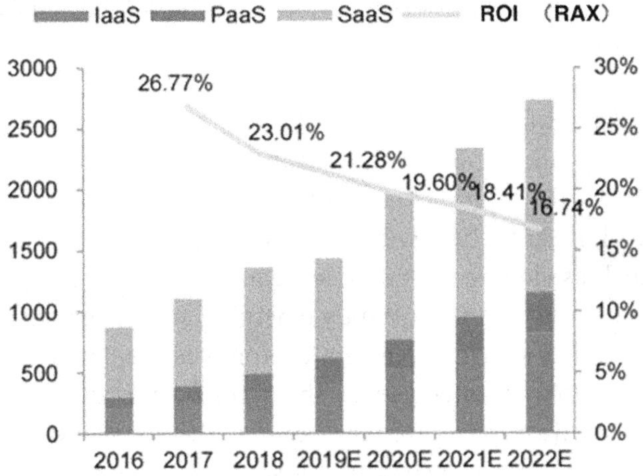

FIGURE 3.1 Global cloud computing market scale.

3.1.3 Cloud Technology Is Mature

As can be seen from the scale and growth rate of the cloud computing market, the global market scale has shown a stable growth trend, and the Chinese market scale has maintained a high growth rate. It is expected that it will continue to maintain stable growth in the next few years with promising prospects.

3.1.3.1 Wide Acceptance

According to the data of 'China cloud Trust report' jointly released by Alibaba Cloud and IEEE China, Cloud has gradually become the mainstream IT technology and business model accepted by enterprises in China. More than 60% of enterprises use cloud to support their main business websites, and more than half of enterprises use cloud to support their core business systems. In the group of start-ups, more than 59% of all businesses go to the cloud. The trust of Chinese enterprises in the cloud is higher than expected. More than 74% of enterprises already believe that the cloud is trustworthy, while half of them believe that the cloud is safer. Enterprises that have already gone to the cloud, the degree of trust in cloud computing is 52% higher than that of non-cloud enterprises. Cloud computing defines an on-demand and pay-as-you-go resource utilization mode. In addition, in recent years, the policy system has been improved, the society has been trying to accept and the technology is mature. Cloud computing has already passed the hype period, and has become the basic productivity tool of the whole society and the new infrastructure.

3.1.3.2 Ecology Is Becoming More and More Mature

At present, all walks of life are talking about ecology. The author believes that for cloud computing, cloud ecology mainly includes two levels. At the first level, from the perspective of technology integration, new technologies such as cloud computing, big data, Internet of things, machine learning and mobility are constantly emerging and integrating with each other to build a brand-new ecology; it promotes the development of the whole society toward intelligence. The second level, from the perspective of industrial cooperation, whether it is the ecological development of the upstream and downstream industrial chains of cloud computing with cloud service providers as the main body, or in line with the national 'the Belt and Road' strategy, expands the scope of industrial integration beyond the country. Cloud computing ecosystem must be the general trend and become more mature.

3.1.3.3 Cloudization Is an Inevitable Choice for Enterprises

Cloud is needed where computing demands exist. Cloud, as an infrastructure, is moistening things silently and everywhere, changing our economic activities and life. Cloud migration is a must-do multiple-choice question for global enterprises. They will either shift to cloud or eventually drown in the wave of cloud development. In a word, Cloud is an inevitable choice for enterprises.

3.1.4 The Dilemma of Enterprise Cloudification

Having said so much about the current situation and opportunities, facing the originally outdated system, how companies can start to complete the system cloudification step by step, smoothly and safely, has become the biggest problem when it truly comes to the realization of the cloudification strategy. It is mainly reflected in two types of problems: one is the difficulty of decision-making, and the other is the difficulty of technic.

The key factors that make decision-making difficult are as follows:

- Weak digital technology: Many companies still use office software to collect and use data manually, which leads to low level of informatization and low efficiency in data acquisition.

- Insufficient cost input: Enterprise digital transformation is a complex system project with a long-time cycle and large investment. From software and hardware purchase to system operation and maintenance, from equipment upgrading to human resource training, continuous capital investment is required.
- Shortage of talents reserve: Gartner estimates that 30% of technical positions will be vacant in 2020 due to the shortage of digital talents. At present, the shortage of talents has become the main bottleneck for the development of digital transformation no matter for government departments or traditional enterprises.
- Lack of organizational system: A new flat and platform-based organizational structure is more expected to be built based on more miniaturized, autonomous and flexible decision-making unit on the cloud. However, traditional enterprises are prone to an organization model with complex levels, multiple leaders and slow responses.

It can be summarized as follows: First, the lack of digital transformation capabilities leads to 'won't transfer'; second, the high cost of digital transformation and insufficient capital reserves cause 'can't transfer'; third, the lack of corporate digital talent reserves makes 'not dare to transfer'; fourth, the unclear digital transformation strategy at the decision-making level of enterprises leads to 'bad transfer' and fifth, the ineffective multi-layer organizational model of enterprises leads to 'unwilling to transfer'.

The key factors leading to technical difficulties are as follows:

- Solidified IT architecture: There are a large number of enterprises with digital experience in traditional IDC. The extensive IT infrastructure includes VMware, OpenStack, x86 physical machine, centralized storage, load balancer and other traditional devices. Therefore, how to smoothly migrate applications and make use of some unprotected hardware is the dual trade-off between architecture evolution and cost control.
- Adaptation and reconstruction: In order to adapt to the architecture on cloud, enterprises often need to transform a single application into an application cluster based on microservice architecture. The migration of a local data center to a multi-sit deployment requires

proper reconstruction of the original application. Key parts of the migrating system often result in downtime, and new deployments still cannot guarantee success in the first run.

- Mass data transmission: Relational data often depend on related storage systems. Whether it is a database cluster with a relational database as the core or a big data platform based on Hadoop, the technology stacks on and off the cloud are often different. Related data transmission also needs to overcome the compatibility problem of the storage underlying system.
- Continuous iteration on the cloud: All modern continuous integration and continuous delivery (CI/CD) platforms support multiple cloud providers, but the automation is usually associated with vendor-specific APIs (Application Programming Interface) that require to configurate, monitor and disassemble the server. However, these APIs are not suitable for small local data centers unless the engineering team spends time building it on their own.

To sum it up, the engineers consider the actual scenarios in combination with the constraints of technology and cost. How to balance the cost and benefits of architecture transformation, how to reasonably use the advantages of cloud architecture, how to select data transmission schemes and how to open up CI/CD on the cloud and off the cloud will all require deep thought in technology.

3.1.5 The Important Technique of Enterprise Cloudification

In view of the difference between traditional IDC technology and the current mainstream cloud technology, there will be many key technologies in enterprise cloudification to supplement the gap between cloud scenarios and traditional scenarios, and meet the requirements of enterprises in the cloudification process. The technologies are divided into three major categories according to the computer architecture and cloudification process, which is basic technology, relocation technology and transformation auxiliary technology. With reference to the current Alibaba Cloud ecosystem, examples are given in turn.

3.1.5.1 VPN and Intelligent Gateway

To achieve a smooth cloudification process, the first thing to solve is the network interconnection between the offline IDC intranet and the cloud site. When we talk about the interconnection between two network instances, our first reaction may be VPN (Virtual Private Network) technology.

- The full English name of VPN is Virtual Private Network. It can establish a temporary and secure connection through a public Internet network, which is a safe and stable tunnel through a chaotic public network. The purpose of using the Internet safely can be achieved by encrypting data several times with this tunnel. When two independent network instances need to be interconnected with insufficient cost to build a physical network link from them, the need for an efficient VPN technology comes into being. Among the widely used corporate offices, the VPN can also be an extension of the corporate intranet. The VPN can help remote users, company branches, business partners and suppliers establish a reliable and secure connection with the company's intranet to connect to the secure extranet VPN of business partners and users cost-effectively. Therefore, many office workers also need to establish VPN connections in their computers to facilitate remote work and so on.
- Alibaba Cloud VPN: Alibaba Cloud provides VPN gateway, which realizes the secure and reliable connection between enterprise data center, enterprise office network or Internet terminal and Alibaba Cloud proprietary network (VPC) through encrypted channel based on Internet network connection service. It supports IPsec-VPN connection and SSL-VPN connection. In addition to providing the functions of traditional VPN, Alibaba Cloud VPN gateway has the following advantages: Security: Uses IKE and IPsec protocols to encrypt transmitted data to ensure data security and reliability; High availability (HA): Dual-system hot standby architecture achieves switching in seconds when failures occur, which ensure that the session is not interrupted and the business is not aware; Simple configuration: Real-time configuration and the deployment is completed quickly.
- Smart access gateway: In addition to the basic VPN gateway, Alibaba Cloud also launched a smart access gateway based on the cloud-native SD-WAN architecture. The hardware and software versions are provided to help enterprises achieve one-stop access to the cloud and obtain a smarter, more reliable and safer cloud experience. In

the mode of intelligent access gateway: ZTP install, deploy and centralize management and maintenance. The control plane is based on the software defined of the Alibaba Cloud platform. It can manage all client-side intelligent access gateway CPE devices through the console, API, and cloud monitoring platform just like managing resources such as VPC and Elastic Compute Service (ECS) on the cloud. Hybrid network access: The forwarding layer is an enterprise private wide area network (WAN) based on a cloud platform, supporting dedicated line, broadband and 4G hybrid network access, increasing the utilization of dedicated lines and improving network reliability. Cloud network integration: Realize the integrated architecture of cloud, network and end, cloud on-cloud VPN protocol auto-negotiation, cloud next-key access to cloud services and cloud on-cloud unified end-to-end security policy control.

3.1.5.2 Express Connect and Cloud Enterprise Network

- Express Connect and Cloud Enterprise Network: In addition to virtualization, the WAN realizes the intercommunication of network instances. But using dedicated lines is a more secure and stable solution.

- Alibaba Cloud Express Connect: Alibaba Cloud Express Connect can establish high-speed, stable, and secure private network communications between local data centers and private networks on the cloud. The high-speed dedicated line connection bypasses the Internet service provider in your network path, which can avoid the problem of unstable network quality and avoid the risk of data being stolen during transmission at the same time. The high-speed channel connects your local internal network to the access point of Alibaba Cloud through a dedicated line. One end of the dedicated line is connected to the gateway device of your local data center, and the other end is connected to the border router of the high-speed channel. This connection is more secure, reliable, faster, and has lower latency. After the border router and the Alibaba Cloud private network to be accessed are added to the same cloud enterprise network, the local data center can access all resources in the Alibaba Cloud private network, including cloud servers, containers, load balancing and cloud databases.

- Cloud Enterprise Network: Alibaba Cloud Enterprise Network is a highly available network carried on a private global network with high performance and low latency provided by Alibaba Cloud. Cloud Enterprise Network can help you build private network communication channels between VPCs in different regions, between VPCs and local data centers. Through automatic routing distribution and learning, it can improve the rapid network convergence and cross-network quality and security of communication, and realize network-wide resources. The intercommunication helps you build an Internet network with enterprise-level scale and communication capabilities.

3.1.5.3 Elastic Compute Service

After solving network interoperability, choosing the carrier of enterprise applications on the cloud has become a top priority. To ensure the same performance off and on the cloud, and even to get better services on the cloud, you need the carrier of the cloud application with higher stability and security. And as an important consideration for enterprises to go on the cloud, operability and elastic scalability will also be key considerations.

- ECS (Elastic Compute Service): Alibaba Cloud cloud server is a basic product of Alibaba Cloud's IaaS layer. Compared with ordinary IDC computer rooms and server vendors, Alibaba Cloud cloud server has stricter IDC standards, server access standards and operation and maintenance standards, which are used to ensure the HA of cloud computing infrastructure, data reliability and HA of cloud servers. And benefit from the precipitation of Alibaba Cloud in the security field for many years, it has very strict requirements for the privacy of user data, the privacy of user information and the protection of user privacy, and it has passed a variety of international security standard certifications, including ISO27001 and MTCS. When it comes to elasticity, Alibaba Cloud has the ability to create IT resources required by a medium-sized Internet company within a few minutes, ensuring that most businesses built on the cloud can withstand huge business pressure. And the flexibility of ECS is not only reflected in the flexibility of calculation. The elasticity of storage, the elasticity of the network and the elasticity of your business architecture re-planning can allow customers to combine services in any ways.

- Elastic bare metal server (Shenlong ECS Bare Metal Instance): A new computing server product based on the next-generation virtualization technology completely independently developed by Alibaba Cloud combines the flexibility of virtual machines with the performance and functional characteristics of physical machines. Compared with the previous generation of virtualization technology, the next generation of virtualization technology not only retains the flexible experience of ordinary cloud servers, but also retains the performance and characteristics of physical machines, and fully supports nested virtualization technology. Elastic bare metal servers combine the advantages of physical machines and cloud servers to achieve super-strong and ultra-stable computing capabilities. With the virtualization 2.0 technology independently developed by Alibaba Cloud, your business applications can directly access the processor and memory of the elastic bare metal server without any virtualization overhead. The elastic bare metal server has complete processor features at the physical machine level (such as Intel VT-x), supports ARM and other instruction set processors at zero cost, and can also achieve the advantage of physical machine-level resource isolation, which is especially suitable for cloud deployment. 3.1.5.4 Server Migration Technology

The Server Migration Center (SMC) is a migration platform developed by Alibaba Cloud independently. Compared with traditional P2V and V2V tools, Alibaba Cloud SMC is fully compatible with Alibaba Cloud ECS and can migrate your single or multiple migration sources to Alibaba Cloud. The migration source (or source server) refers to your IDC server, virtual machine, cloud host of other cloud platforms or other types of servers to be migrated. It supports the synchronization of incremental data generated by the source server system to Alibaba Cloud without business suspension. And it can realize batch migration, multi-threaded acceleration transmission, and provide rich API for automatic expansion. In addition, at SMC, IDC servers can also be migrated to container mirroring services to further assist customers in cloud-native transformation.

3.1.5.4 Data Migration Technology

- Data dump (DUMP): Whether it is a file object, a relational database, or non-relational data/text. Most storage components provide their own file export and import functions for the reuse of this function, such as packaging batch, progress control, status control and other functions can effectively adapt to data migration between various components. However, due to the limitations of DUMP technology, only snapshot migration at DUMP time can be performed. This solution is not suitable for frequently modified data. In addition, identifying data increments will become particularly important. If there is no suitable method to record the time point of the DUMP snapshot, the storage medium itself cannot perform incremental snapshots at the time point. The only way to ensure that the snapshot of DUMP is consistent with the real business data can be ensured by temporarily suspending business writing at the source. This again has clearer requirements for business characteristics. To sum up, DUMP itself has general and simple characteristics, but is limited by its static data relocation characteristics, and it is necessary to pay special attention to the business increment during the DUMP process and after the DUMP.

- Data transmission server: In addition to the static DUMP series of technologies, Alibaba Cloud also provides streaming data migration technology DTS. The full name of DTS (Data Transmission Service) supports data transmission between RDBMS, NoSQL, OLAP and other data sources. It provides multiple data transmission methods such as data migration, real-time data subscription and real-time data synchronization. The launch of DTS just made up for the gap in static data migration technology led by DUMP, and supports the collection and synchronization of real-time changes in data. Users do not need to consider complex information such as increments, modifications and migration time points during use.

3.1.5.5 Application Discovery Technology

Application Discovery Service (APDS) is a cloud migration assessment tool for assessment, planning, construction and migration needs of enterprises to go on the cloud. It is used to help companies that are going to the cloud to automatically discover and organize offline IT assets, analyze and identify host and process information, resource usage levels, and dependencies among applications and components. APDS installs collectors and probes in the business system under the cloud, records the mutual access of applications under the cloud to a local file, and automatically recognizes and analyzes it in Alibaba Cloud through offline upload mode to discover/inventory IT resources under the cloud, analyze and identify the host and process information, the level of resource usage, and the dependencies between applications and components. APDS, which supports Linux, Red Hat, Debian, CentOS, Ubuntu and other types of operating systems, non-intrusive collection, does not affect the performance of online business, and supports data encryption collection and storage. It automatically marks and clusters processes, identifies the three-party system components used by users, and further simplifies the system architecture. It is a powerful tool for the migration assessment needs of large- and medium-sized enterprises and an assisted cloud architecture.

3.2 THE CONCEPT, STRATEGY AND PRACTICE OF CLOUD NATIVE

3.2.1 Confusion and Limitations after Cloudification

3.2.1.1 Confusion and Limitations after Cloudification

Enterprise applications or enterprise systems, we also use words like 'information system'. They are huge and complex systems often, including salary systems, medical records, leaving systems, CRM, inventory management, insurance, logistics control and email. The first feature of these enterprise systems is that they are large and provide great value to the enterprise, although they don't have millions of users than Internet applications. There is no doubt that in the modern days we can find business opportunities in big data, find out the solution of reducing cost, or improve production efficiency; these practices must rely on mature and stable enterprise systems. And cloud computing has accelerated this process. It supplies a better operation environment, almost no limit storage, different databases, and variable cloud products. The amount of systems modern enterprises are operating has far exceeded before as the result.

We also know that during the rise of cloud computing, various types of service are appearing in markets, like IaaS, PaaS and SaaS; the top vendors are the combination of these cloud services, from the infrastructure layer to sell software to the tenant directly. However, from the perspective of enterprise applications, the IaaS could achieve better revenue; it sealed the initial virtualization decoupling the operating environment and the physical machine so that the enterprise applications will not bind to the hardware, and it will be easier to plan, purchase, upgrade and maintain.

It doesn't require more knowledge about the cloud for the enterprise employee; it's close and similar to the server.

Many companies are not very aggressive in adopting cloud computing; at the beginning they want to reduce IT costs, and increase efficiency or just reserve some elastic resources; they don't realize that these new cloud products could impact our business. For example, the distributed RDS allows us to use relational databases to process data almost in unlimited; without designing a specified data structure and complicated query logic, we don't need to care about the servers, the database engine, security and upgrade major version. But if we still use 'virtual machines' to design and deploy enterprise applications, our systems and business will be limited in the past.

We have a profile of a traditional enterprise that has just finished cloudification like:

- Cloudization: The systems have been deployed on clouds, and using cloud services is more traditional, mainly based on three major components similar to ECS, RDS and OSS, and have certain cloud operation experience.
- Management: The management method has improved a lot since the IDC age, adopting cloud monitoring and creating a lot of dashboards and alarms, but granularity is instance level. No DevOps practice.
- Application: Most of them are using monolithic architecture, and the tech stacks are not popular than Internet applications. The applications are huge and complex to deployment, migration, or adding new features; the development cost is high; and the cycle from inspection to release is long, without agile methods.
- Data: Although RDS is adopting, other types of database warehouse and storage are rarely used. It's not common to select persistence technology according to different business scenarios.
- Resources: Hard to improve resource utilization, elastic and implement the high availability because use machine as the operation facade for the applications.

Obviously, the benefits of cloud computing to these companies are not enough to make people applaud. One of our customers mentioned that cloud computing is just an enhanced virtual machine platform with a good-looking dashboard and an automatic backup function is most valuable for them, by the way, planning the network architecture on the cloud is also good – by clicking mouse buttons. I admit that in this scenario there is not too much value for them, just like after the jet fighter was invented, we still used the way to drive the piston fighter to fly the new machines and no results of course.

In fact, these old problems like 'improve resource utilization', 'follow new technology trends' and 'inspect value from data' are not that these companies are not aware of it, but do not have enough capacity to spike and adopting. A motorcycle producer is certainly not as good as an Internet company in using container technology; traditional logistics company doesn't have enough resource in R&D department to customize systems for itself. In the age of cloud computing, enterprises do not need to worry about infrastructure, operating environment, data persistence. The other application dependencies have standards or common solutions and should be provided by the cloud vendor. The enterprise just needs to pay attention to their own business in the application layer and that's the intention of cloud native.

In the early days, enterprises applied the classic 'lift and shift' model in cloud migration, but this was just a cloud migration; the outdated applications could not flexibly take the technical profits brought by cloud computing. The traditional application model is more dangerous; it will limit the thoughts, organization and work processes for the people. Before CD is implemented, many people can't even imagine that software features can be iterated so quickly, and in the cloud-native age, the cycle time is few hours.

There is no free lunch in cloud native. We need to use new patterns to create or refactor 'native' cloud applications, the architecture design, tech stack selection, development, and delivery process may be different from before, and the new apps will be accompanied by microservices, containers, DevOps and other latest concepts. As Kubernetes and CNCF have become the de facto standard in this domain, the risk of adopting cloud native has been greatly reduced, and there are popular solutions in many use cases, and most of them are open source. Cloud native also has a deep impact on our applications. When designing programs, we emphasize statelessness, open to extension, disposal, prefer asynchronous and elastic. The goal is to unlock the value of cloud computing, respond rapidly to the business creating possibilities that were impossible before.

3.2.1.2 Cloud Native Is Next Step after Cloudization

Native is an adjective in vocabulary, has a meaning of original, local, born in a particular place. Perhaps at the beginning, the cloud-native concept just describing application is suitable to the environment on the cloud, the cloud environment is the prerequisite when building a native application.

Let's start with the classic ECS usage scenario. As a basic virtualization product, ECS provides advanced virtual machine usage features, it's the cornerstone of the cloud age. We can use ECS to install different software, deploy our codes and running services such as websites, FTP, databases and middleware. ECS is almost omnipotent because it's just oriented to universal computing and everything is computing generally. But gradually, we found that FTP servers are often not very sensitive to the number of the CPU cores. The bottleneck is disk IO or capacity, or network bandwidth, and the cost of maintaining an FTP server is also very high. We need to handle the backup, security, file verification, etc. So, in this case, computing is not the core business, so we will consider using OSS for file storage. If your application is designed directly using OSS, we can call this program a cloud-native application because it is naturally oriented toward cloud products.

For the company internal usage website which is deploying on ECS, during the peak time (such as 9 a.m.), the application needs to perform a lot of calculations, such as task assignment, message delivery and arrange calendars. At this time, we prefer to use big instances maybe 16 CPU cores. But during idle hours, this kind of internal website doesn't have much traffic, a t2 instance is enough. This typical tidal application often uses ASG (Auto Scaling Group) for resource allocation, but there are not too many scenarios where companies use ASG to improve resource efficiency because many applications cannot support elasticity well (such as sticky session or state retention), and ASG effective time is too slow (minute level) so it's hard to respond to workload rapidly, that's why ASG is not popular in traditional enterprises than Internet companies.

In addition, many traditional systems, or extremely important dedicated production systems have specific dependencies and configurations. They often require specific operating systems and runtime dependencies, or due to authorization reasons, once these systems are deployed to ECS. It's difficult to take it out, or even to redeploy it, these systems are static, and bind to ECS instance tight. Enterprise applications should be decoupled from cloud products such as ECS, because ECS is not an asset to them. In the cloud age, the most important asset is not enterprise application programs, but data. Many companies are unable to modify or update the legacy system, but they are reluctant to deploy it to the cloud platform. So that there is not much benefit after cloudification.

Our vision is clear. We want to run websites and applications during the day, and perform batch processing of big data calculations at night. We hope that each subsystem adjusts resources according to its pressure, we don't want A giant Oracle instance with 1,000+ tables, we don't want the entire enterprise to share a self-built, super-large Redis cluster, we hope that new technologies can be adopted immediately, and could quickly create values, we hope that the dashboard system could fully display everything, from infrastructure to applications, we don't need to recruit senior, top-level operation people with a big heart and cool head, we just want more people to pay attention to the business, infrastructure should not be a blocker to the business.

Therefore, since the concept of cloud native has been proposed to the present, whether the representative technology is container or serverless, the core value is to solve these pain points after Cloudification. The value advocated by cloud native is consistent with the value of cloudification, but there is more clarity. Of course, methodologies and practices are constantly evolving, and they are also changing with the improvement of technology, the best practices and products are extremely dynamic. For example, the book of Kubernetes from last year and we found many example cases that can no longer run in the current version.

3.2.1.3 Evolution of Cloud-Native Concepts

According to the definition of CNCF, cloud native is a method in software research and development.

> Cloud native technologies empower organizations to build and run scalable applications in modern, dynamic environments such as public, private, and hybrid clouds. Containers, service meshes, microservices, immutable infrastructure, and declarative APIs exemplify this approach.
>
> These techniques enable loosely coupled systems that are resilient, manageable, and observable. Combined with robust automation, they allow engineers to make high-impact changes frequently and predictably with minimal toil.
>
> *https://github.com/cncf/foundation/blob/master/charter.md*

These designs take into account that loosely coupled technologies can improve the flexibility, observability of our system, combined with powerful automatic technologies, and these technologies allow us to efficiently and stably support business requirements.

Most companies that practice cloud native have similar technology selections: use a large number of docker containers to run micro-service-style applications. These containers are orchestrated by an orchestration tool such as Kubernetes and using tools such as log center, service management, adopting DevOps and CICD for multiple deployments.

Indeed, with the accumulation of practice, the cloud-native landing approach advocated by CNCF has become the consensus of the industry, and tremendous progress has been made in enterprises with strong technical capabilities. The market size is also gradually increasing, the Gartner report pointed out that in 2022, 75% of global enterprises will use cloud-native technology in production. According to the 2019 Container Adoption Survey report, 87 of the 501 IT projects sampled are using container technology, compared to 55% in 2017.

Of course, not only CNCF has a definition of cloud native, but the definition from VMware-Pivotal is also popular enough and can be considered to be inherited and developed from the PaaS age. Therefore, many concepts and practices of cloud native come from these PaaS ladders: Cloud Foundry and Heroku. For the PaaS's user, we don't want to pay attention to the operating system, network and underlying dependencies, we want the platform to handle them all. This leads to application standards such as 12-factor app and the interactive mode of using APIs to integrate. This is the original intention of many companies to develop PaaS based on Kubernetes. The latest definition from Pivotal is that:

Cloud native is an approach to building and running applications that exploits the advantages of the cloud computing delivery model. When companies build and operate applications using a cloud-native architecture, they bring new ideas to market faster and respond sooner to customer demands.

Similar to CNCF, the definition focuses on building and running an application program. Through the decoupling of infrastructure, the program can be deployed to any cloud environment. Pivotal technically converges it into four aspects: microservices, containerize, DevOps and CD, and in this age, agile methodology is the most suitable for cloud applications, because it can respond quickly to changes.

3.2.2 Methodology in the Cloud Native

3.2.2.1 Balance Evolution and Revolution

As enthusiasts who love new technologies, our passion for learning and adopting new technologies is unstoppable, especially for the technologies that have proven their capabilities in certain scenarios, it is difficult for us to control ourselves not to adopt them. With the arrival of the cloud-native age, how should we deal with these new technologies? With so many successful experiences, can we all-in our tech stack to cloud native? These questions can't be answered at this time, but we can start with a simple mantra: balance evolution and revolution.

It is very tempting to use brand-new technology to create new products, everyone hopes that they will stand at the front of the industry, use the latest technology to prove themselves and show their business value. It's time to sweep trash in old days and recreate a new world. But the real world is cruel. First of all, all technologies in the cloud native are developing rapidly, and we cannot anchor a stable technology stack. For example, in container technology, the orchestration of containers starts from docker-compose, swarm to Kubernetes, and microservices are also moving to service mesh. The technology stack you determined may not be suitable for next year. At this time, the question is hard to answer: continue grudgingly or change the direction. There is no business benefit when the enterprise application is refactoring or rewriting, and a lot of knowledge will be lost or cannot be verified because enterprise applications are accumulated over the years. I have seen business logic based on database sorting, and it is really easy to miss when we refactor that.

Adopting Spring Cloud is a very valuable example, and many companies use Spring Cloud as a microservice platform or framework, especially in China after 2016. But Spring Cloud is not suitable with containers at the beginning (it does not mean that it cannot be containerized). The registration discovery of Eureka works well in the environment of virtual machines because the infrastructure changes rarely, which is aligning with Netflix's infrastructure. But the other companies are not Netflix, they can't consider the consequences in depth. With the popularity of containers and Kubernetes, more and more people have found that putting Spring Cloud applications on Kubernetes is not very straightforward, but the application is coupled with Spring Cloud too much, so the cost of modification is unpredictable.

From the systems I have developed, there are core web apps that use Spring MVC, a large number of RESTful applications based on Spring Boot, running in containers, new services are using serverless as the operating environment. In the continuous improvement process, we've replaced the old module in the core website by using a separated microservice one by one and at last offline to the old site. During the landing of cloud native, evolution and continuity can ensure that we can go to the end and guide the next step of practice through the experience accumulated at each step.

3.2.2.2 Open for Extension

Although this methodology comes from half of OO's Open/Closed Principle, the expansion refers to the expansion of the function, and the expansion capability means the responsiveness to the business.

Traditional technology and R&D models are difficult to open to extensions. The first thing is that code is difficult to modify, so it's hard to have new features or fix bugs. Once chosen, the programming language and framework will be used for many years, and with the replacement of maintainers, we will find that it will be more and more difficult to add new features because it is impossible to determine whether the change is safe, and the larger the application, the more difficult the test. At the end of the life cycle, we can't avoid the fate to rewrite.

This kind of 'Tar Pit' application is not to be difficult to modify at the beginning, the boundary between the two modules is very clear and independent, but due to the lack of isolation and changing rules, new features added in the future may cause coupling, which cannot be solved in the end. In the cloud-native scope, container and microservice technology provide better isolation than before. Container forces you to design the application in stateless style, and microservice forces you to use advanced protocols to replace the method invoking, that because this architecture is distributed and decoupled, and we know it will reduce the performance, but no free lunch in the distribution world.

In the migration of enterprise applications to the cloud, we often face huge Oracle, which has thousands of tables. Indeed, there are many reasons for this situation. The data structure provided by relational databases cannot match the requirement of all applications. Sometimes we need to store files and pictures, sometimes we need to store graphs, DAG and trees, and sometimes we need simple, high-performance data structures, we want to have K-V store and OLAP, but the reality is that we only have a huge database instance, so we must put all data into.

In the traditional cloud environment, the database has the highest importance, and the maintenance of the database is also the most expensive, but after going to the cloud, different database types can be used to replace old, huge instances. The cloud platform significantly reduces the maintenance cost of the multi-type database instances, and we can let the application select the persistence, for example, we can use a K-V store for user's profile data, and choose an ADB to support BA analysis sales data in real time. Choose different databases for different services and hide the details are also very important points of the microservice practice. And it also reduces our dependence on a specified database, helps us retire a huge, centralized Oracle.

In the cloud-native age, the interaction between systems is various. The communication method upgrades from function call to multiple methods: HTTP or RPC, or use middleware implement pub/sub, or adopt queues and topics. The data publishing is also more improved, we can use Kafka to implement a new type of enterprise bus, establish an Event-Driven system and use compensation events to implement transactions in a distribution system.

Since the invocations between enterprise applications are protocol-based, we can also handle the direct boundary with the legacy system, the difference is sending HTTP requests in code to replace method invoke. A clear boundary and protocol-oriented system are friendly to develop and test. We can easily use mock to simulate upstream and downstream and cooperate with the local minikube to open a fully equivalent environment immediately, improving the efficiency of developers.

Container supplies a standard execution environment for us, so we don't need to consider the infrastructure too much, microservices define the relationship between systems and internal structure, observability methods describe the system state so that we can catch every detail inside the running applications, and adopting DevOps toolchain to implement CD ensures that the system is always online and rolling. If we have a cloud-native service called Order Service, and after a new version is deployed we found an online issue, the operation will roll back this version immediately after receiving the alert, thanks to the containers and orchestration platform, this rollback can be very fest in seconds even. The developer could collect error information from the log center, use containers to reproduce the issue locally, then fix that. Finally, after the merge request of the code changes is merged, and all checks, tests are passed, the new round of deployment will start and deployment in canary, by the way. It's common for cloud-native practice, from code changes to production, will reduce from 1 week to several minutes, automatic and safe.

3.2.2.3 Prefer the Application Layer

We all agreed that the calculation will have higher efficiency when closer to the data. Using JDBC Driver to read RDBMS and then performing calculations in Java code is less efficient than using DB functions or stored procedures in databases. Exclude the I/O consumption, stored procedure with a dedicated language for the database is also better than an advanced programming language. Of course, many enterprise-level applications will use a large number of database features such as custom functions, stored procedures containing business logic to develop applications, we will hire DBA to create those complex SQL scripts, each script can accurately and efficiently complete business requirements. However, the consequence of this is that the stored procedure is not suitable for writing complex business logic, and its language design is too early, often difficult to read, understand and modify, and only experts can operate it. Once people change, it's very difficult to understand the business logic inside. It could be worse than stored procedures are bound to the database, and you cannot efficiently use automated methods for testing. Once a system does not have enough automated testing, it will lose maintainability. Database features are two-sided, and it's difficult to have both high efficiency and easy extension.

Therefore, adopting advanced programming languages will greatly ease this process, because the readers of the code are not machines, but humans. For engineers who use high-level languages, it is not difficult to read a snippet of C# code, as opposed to system programming languages (C++, Rust, Go), advanced languages encapsulate too many low-level details, you don't need to manage memory, operate the network directly, and most of the time, it's fast enough. It's simple to read, write and run, easy to modify and maintain, it's soft, and represents the value of software.

In a cloud-native world, the weight of controllability and easy maintenance are greater than efficiency. Of course, this is not to say that system efficiency is not important. As the person in charge of enterprise applications, there will always be trade-offs and choices. We can use NGINX as a reverse proxy to implement the most basic routing function of API Gateway, which is efficient and easy to verify, but for the advanced features, like throttling, authentication and access rule restriction, we have to use the application layer. Of course, if the delay requirements are not accurate to the millisecond level (most applications are like this), we can use Zuul to write Java code to replace NGINX's lua.

Another important feature of cloud-native applications is out-of-process governance. For enterprise applications, such as TLS connection, logs, breaker, access control, throttling, performance monitoring, registration discovery, etc., they are often irrelevant to the business logic. Since these capabilities have nothing to do with the business, can we move them out of the main process? The answer is yes, the sidecar pattern has also become popular in recent years and has spawned an advanced microservice pattern called service mesh. We hope that engineers will focus on the business than before, the rest functions are just provided by the outside, from the platform or sidecar.

3.2.2.4 Be Aware of the Piper

We believe that CHANGE IS THE ONLY CONSTANT. Industries that rely on software technology are full of changes. Let's use the term cloud native as an example, the concept is changing always and depends on the leading companies and organizations, we need to be aware of this and remember no 'silver bullet', and no free lunch in the world. Many companies adopting Spring Cloud as the microservice framework, but do not produce enough value, they use Spring Cloud because it's popular, they don't realize the gap of use case, and require technical capabilities. In the war of programming languages, every year new challenges are claiming to replace Java, such as Kotlin and Scala. Of course, they did not succeed in the end. We welcome new technologies to be invented and promoted. This is extremely important to the industry and represents the vitality, but we also need to realize that advocates the omnipotence of new technologies is very dangerous, even evil.

Well, check our experiences in cloud native, we started to use docker-compose for the orchestration of containers in the production environment at first, docker-compose is very simple and easy to use, but it lacks support for flexibility and handling, and it is difficult to rolling-out update. To support CD, we have to invent many tools to do that. Containers are just processes, they are no value without orchestration. Then, Swarm and Mesos became popular in the market as orchestration tools, and cloud computing giant AWS also began to launch AWS Elastic Container Service. They all claimed to be the future, but the cruel thing is that you have to choose only one and avoid all-in. Fortunately, our application strictly follows 12 factors: stateless, shared-nothing, and has no dependencies outside the container, so we can easily change the orchestration tool and the operating environment. In the end, we know that the winner is Kubernetes, but who can guarantee that Kubernetes is a technology that can be used for the next 10 years? Serverless with greater potential has appeared on the horizon! We didn't spend too much during the transformation, because we believe that the underlying orchestration platform is uncontrollable, so we should keep the distance appropriately.

3.2.3 Cloud-Native Strategic Planning

3.2.3.1 Strategic Perspective and Planning

In the previous two sections, we analyzed the concept of cloud native, discussed the technologies in it and sorted out the commonly used methodologies. Now we should have a vague feeling in our minds. This is a brand-new technology, and mode is the embodiment of the value of cloud computing. There are countless new technologies provided by companies and communities such as CNCF, Alibaba Cloud, Pivotal, AWS and many companies' practical stories before our eyes. Now it seems that it's time to adopt them. Then, the implementation of cloud native must have a top-down strategy, and this section will begin to discuss this aspect.

(1) Technology choice: The difficulty of cloud-native landing lies in the inability to anchor the core technology. Even if CNCF promotes Kubernetes, Kubernetes is a good choice as a winner in the coordinated world, but it may decline. There is dimensional vibration. Therefore, when choosing a technology, the principle of the reservation should be open to that technology, not just relying on it. Especially for the bottom layer of the business process platform, even if the container fails or is forbidden to use, we must consider how to run the container. The cloud-native era is open, infeasible and loosely coupled. Various programming languages and frameworks are welcome, so you should use 'controllable' as the standard. Once you find that technology is out of control, you can replace it at a lower price in that state. This is a reasonable technology choice logic. It's time to

say goodbye to the entire company using only one Oracle database. The Java version of all companies is 1.6 era.

(2) Evolution route: We know that technologies related to cloud native are developing very rapidly and are the end of misconduct. Therefore, migrating and converting existing enterprise applications to the native cloud are like standing at sea, melting an iceberg and jumping to another huge ice block. Both the starting point and the ending point are unstable. In the previous chapter, we mentioned that this is an evolutionary process, not a process of building a house at the destination and then moving in. The entire evolutionary process should be uninterrupted, and the business of the enterprise should not be affected. Therefore, this requires extremely high engineering control capabilities. Lean thinking (lean) can play an important role at this stage. We can first rent a small number of cloud products for proof of concept, and then perform the conversion and migration of edge applications. Once you find a problem, please iterate the updated operation immediately. The efficiency of slow migration will increase. At the same time, this is also a good opportunity to organize existing enterprise applications. Of course, the destination is constantly changing, but if observability and service governance are successful, we don't need to worry. For large enterprises, their applications will run in various places, such as ECS, containers, serverless, browsers and mobile terminals, but will not lose control.

(3) Formulate rules: Change application technology, change the operating environment, release new services every week, update and deploy services every day. These different services use different databases and open protocols to communicate and help each other, use different programming languages, and even developers are not familiar with it. What is certain or stable in this environment? In the cloud-native era, unified constraints are far greater than shared resources. We should impose complete constraints throughout the life cycle of the application. You can use the classic 12-factor app or Open Application Model (OAM). Leave enough space for enterprise applications to limit external performance and dependencies, rather than internal implementation. Sharing resources is terrible. If your service relies on the registry, the registry will always exist on your system until the last system using the registry is offline.

(4) Change the organizational structure: The IT budget model of many companies has greatly affected the value of cloud computing. If the budget is executed for half a year, we will not be able to predict IT investment within half a year, which is more consistent with online investment. The provision of cloud computing is the opposite. We hope that when performing the native conversion, we hope that the system will gradually evolve. Initially, we did not need too much IT investment, nor did we need to rent too much infrastructure and SaaS. Then, this will be consistent with the determined budget, which will lead to conflicts, which will result in the need to waste resources to match the budget, or the progress will accelerate and eventually lead to development failure. At the same time, the internal organization also needs adjustment. In the cloud-native era, we emphasize that applications should be completely business-oriented, and developers are usually directly responsible for applications. This is a 'who build, who run' rule, so the use of functional group division is very important and unreasonable. According to the outdated Conway law, this old-fashioned organizational structure will cause the service after the split to be unable to operate, operate and develop. In the end, everyone still has a huge database. Because there is a team dedicated to operating and maintaining the database, the interesting thing is that the data should belong to the application, just as the goods belong to the buyer rather than the warehouse.

3.2.3.2 Alibaba Cloud's Cloud-Native Layout

Alibaba Cloud is a top member of CNCF, and a standard maker and promoter: Alibaba Cloud has started the containerization process for a full 10 years. As one of the earliest companies to deploy cloud-native technologies, Alibaba Cloud Kubernetes (ACK) has become a real technical protagonist in the top Internet-scale scenario of Alibaba Singles' Day. Standing at the beginning of the new decade, Alibaba Cloud predicts that ACK will become a new interface for users and cloud computing. In the future, more and more application loads will be deployed on ACK, including databases, big data, AI intelligence and Innovative applications that have become the cornerstone of cloud-native computing. At the first KubeCon 2020 conference, Alibaba will bring many cloud-native technologies to the show, focusing on the three major areas of ACK scale operation, microservice system, and serverless ecology, and in-depth discussions on AI, edge containers, Serverless, Service Mesh, Spring Cloud, Knative, Nacos, operation and maintenance, scheduling and other cloud-native hot technology topics.

Alibaba Cloud has world-class experience in containers and microservices: For containers, Alibaba Cloud launched container products as early as 2015. After constant iterations, the current container ecosystem with ACK as the core is reported. In Gartner's report, Alibaba Cloud has the most comprehensive containerized layout and is the only one in China. In the microservice system, Ali has a relatively deep accumulation, through some open-source projects, such as Dubbo, Nacos, HSF and SOFAStack, to export the experience and practice in the Ali microservice system. Also, Alibaba Cloud is integrating the technology of the microservice system and open-source system with the cloud, so that customers who use the cloud can directly use open-source products.

Alibaba Cloud's attitude toward cloud nativeness, standards and development: The ecosystem of cloud-native products must be standardized and open. Services and infrastructure between different vendors can all be interoperable. User-developed applications can run on Alibaba Cloud containers or containers from other vendors. The things that applications depend on can essentially be open interfaces. This has always been the direction Alibaba Cloud is working on. From the perspective of cloud-native development, standards are open and integrated with the community ecology to reduce user costs and concerns.

3.2.4 Application Architecture in Cloud Native

3.2.4.1 Changes in Code Structure

We know that cloud-native architecture is a set of architectural principles and design patterns based on cloud-native technology. It is a new architecture design pattern that is influenced by cloud-native products, which has caused tremendous changes in the programming model of developers.

Maybe 10 years ago, you might still be writing your I/O implementation for local file reading and writing. Now, this low-level, general-purpose scenario has been replaced by a large number of class libraries or services, you don't need to write your own JSON Parser. Many programming languages have the basic tools of files, networks, storage and concurrency, and these basic tools have spawned frameworks to solve network invokes, HA, distributed storage problems, etc.

But in the cloud native, these common problems are often solved by services and other out-of-process modules.

Consequently, our code will clearly distinguish between business and non-business domains. In the business domain, we hope that the code and functions are clear and simple, which precisely represents the requirement. This part of the code has nothing to do with the framework, database, network communication, and does not depend on JPA or MyBatis. At the same time, this part of the code can also best reflect the business value.

Compared with the business code, the non-business code becomes a kind of detail. For business code, we don't care to use which database to store user's annual income information, and this happens in the non-business part, and it's a detail.

Developers are more concerned about technological progress and innovation, and we often check the performance metrics of a certain database or evaluate the new framework is simple to use. Discuss how to design the business code is rare in the community, and many people think that this is something that can be done by if-else.

In the cloud-native age, we hope that applications can focus on business, use 'design patterns' or other methodologies to create flexible, agile and market-responsive code, and eliminate the dependence on business code and other irrelevant details. You can check the import statements to evaluate the dependency, and think about how long will it take to change a low-level framework for the application.

3.2.4.2 Delegate Non-Functional Features

Strictly speaking, there is no standard definition of software architecture. The term software architecture often describes the composition of software. Some people often use architectural terms to balance software architecture. The structure of the house is similar to the software architecture, but the problem is software architecture doesn't determine the functions.

Let's review the graduation design program you wrote when you finished university education; it is not a good practice, full of unreasonable design, coupling, incorrect abstraction, and no clear boundary between modules, but it works. It doesn't need to be a Toyota solid that could drive 100,000 km, but it could help you to graduate.

For example, it is very difficult to build distributed storage. Assuming that we need to provide storage space for storing uploaded files, using single-node storage cannot meet 'non-functional' requirements, a single node is not enough if there are too many uploading files, and the system must pause for capacity scaling, or space is enough, but I/O could be the bottleneck again. Of course, you can use your favorite programming language to write your implementation, you can use technology such as HDFS to achieve, but we should avoid it in the cloud-native age.

The distributed storage problem mentioned above can be solved simply by using OSS or other technologies. The representation in the code may be to delete the 'distributed file system storage module' written by yourself, then introduce the OSS SDK, and then use three methods to wrapper upload, download and access meta, which finish the complex storage problem almost. To be perfect, we also need to control the scope of using the OSS SDK. Generally, we write wrapper encapsulation or use adapter pattern, so from the business code, we don't need to know whether OSS exists. In the future, if OSS is replaced with other technologies, the scope of changes can also be controlled within the wrapper.

We can follow this idea to traverse the non-functional requirements in the existing systems, and try to think about which requirements can be extracted, such as web access log, application log collection, application configuration, service discovery, circuit breaker and health check. When you move these non-functional requirements out of the code, the application will become very clean, so that developers will no longer pay extra attention to the details.

3.2.4.3 Fully Automated: From Testing to Deployment

Software testing is for proving wrong rather than prove right, which is irrelevant with automated. In many cases, software testing can only guarantee the basic quality of the product, but cannot guarantee that there are no problems. The testing pyramid theory points out that we should put more investment at the lowest level because the cost is lower but results will be significant, finding out errors early will be easier to fix. Second, the efficiency of the low-level test is also very high. Generally, unit tests only rely on code, and do not require dependencies such as databases and web servers. They run quickly and can give results immediately.

Automated testing is very important, at least, the code of unit testing is as important as business code. According to the low-level unit test from the test pyramid theory, the investment is low, the value is high, and the dependency is less. We can run the automated tests frequently and automatically from the laptop to CICD servers. On the other hand, during the development, we need to encourage to write the business code and unit test code synchronized, the unit test code could prove the business code meets the requirements.

Of course, landing TDD: Test-Driven Development is not an easy task. It requires the developers to transform their thought and the business code which driven by test code is counter-intuitive, and to adopt depends on the team's situation. We want to establish a tunnel of quick automated feedback: whenever there is a code modification (before merge to the release branch), automated testing is triggered, and a large number of test cases will make sure we don't break anything and give us the confidence to deploy. In our practice, if the unit test coverage can reach 'branch 100%', it means that the system has very good quality because branches often represent business cases.

In the cloud-native age, we want the team to achieve 'fast, fast' and rapid response to changes. However, in a project team, there is often another voice 'stable, and stable', many operations and even project leaders do not want to release more frequently. They often think: The system needs to be online and needs to be released. Since the release is so troublesome, it is necessary to shut down to update the code. The time is also accompanied by database changes, can we do less to reduce risks? The automated release is more dangerous, and manual release is not very stable; let alone automation? It does seem to be the case, especially every time a new version is launched, a long release plan is required, and a rollback plan needs to be specified. The more you do, the more possibility of error.

The scenario just mentioned is the problem that the DevOps culture wants to solve, the problem is 'granularity' control. If the granularity is under control, you can know the difference between the current version and the future version, and you can determine whether it can be released. The emergence of automated deployment tools has returned publishing capabilities to developers, and at the same time forced developers to decouple application dependencies and achieve independent publishing.

Therefore, in the future cloud-native environment, it is not when to release, but every moment, there is a node, service and module that is releasing. Due to the small granularity, rollout and rollback are very fast, and operations could understand the content of the change.

3.2.5 Principles of Cloud-Native Architecture

3.2.5.1 Service Principle

When the code size exceeds the scope of cooperation of small teams, it is necessary to split into services, including splitting into microservice architecture and mini service (Mini Service) architecture, and separate modules of different life cycles through service architecture. Carry out business iterations separately to avoid frequent iteration modules being slowed down by slow modules, thereby speeding up the overall progress and stability. At the same time, the service-oriented architecture uses interface-oriented programming, and the functions within the service are highly cohesive. The extraction of common functional modules between modules increases the degree of software reuse. In a distributed environment, current limiting and downgrading, fusing compartments, grayscale, back pressure, zero-trust security, etc. are essentially based on the control strategy of service traffic [not (network traffic)], so the cloud-native architecture emphasizes the use of service. The purpose is to abstract the relationship between business modules from the architectural level and standardize the transmission of service traffic, thereby helping business modules to carry out policy control and governance based on service traffic, no matter what language these services are based on.

3.2.5.2 The Principle of Flexibility

Most of the system deployment and online need to prepare a certain scale of machines based on the estimation of the business volume. From the purchase application to the supplier negotiation, the machine deployment and powering, the software deployment, and the performance pressure test, it often takes months or even a year; and if the business changes during this period, it is very difficult to readjust. Flexibility means that the deployment scale of the system can be automatically scaled with changes in business volume, without the need to prepare fixed hardware and software resources according to a prior capacity plan. Good flexibility not only shortens the time from procurement to launch, but also saves the enterprise from worrying about the cost of additional software and hardware resources (idle costs), and reduces the enterprise's IT costs. More importantly, when the business scale is facing massive and sudden expansion, at that time, the company no longer 'said no' because of the usual insufficient reserves of software and hardware resources, which guaranteed the company's profits.

3.2.5.3 Principle of Observability

Today, the software scale of most enterprises is constantly growing. The original single machine can do all the debugging of the application, but in a distributed environment, it is necessary to correlate the information on multiple hosts to answer clearly why the service is down and which services Violation of its defined SLO, which users are affected by the current failure, which service metrics have been affected by this recent change, etc., all require the system to have stronger observability. Observability is different from the capabilities provided by systems such as monitoring, business probing and APM. The former is in a distributed system such as the cloud, actively using logging, link tracking and measurement to allow multiple services behind one APP click. The call time, return value and parameters are clearly visible, and you can even drill down to every third-party software call, SQL request, node topology, network response, etc., such capabilities can enable operation and maintenance, development and business personnel to grasp the software operation in real time. It combines multiple dimensions of data metrics to obtain unprecedented correlation analysis capabilities, and continues to digitally measure and continuously optimize business health and user experience.

3.2.5.4 The Principle of Resilience

When the business goes online, the most unacceptable thing is that the business is unavailable, which prevents users from using the software normally, which affects experience and revenue. Resilience represents the ability of software to resist when various abnormalities occur in the software and hardware components on which the software depends. These abnormalities usually include hardware failures, hardware resource bottlenecks (such as CPU/network card bandwidth exhaustion) and business traffic exceeding the software design capabilities. Failures and disasters affect the system, including software bugs, hacker attacks, and other factors , and have a fatal impact on business unavailability. Resilience interprets the ability of software to continuously provide business services from multiple dimensions. The core goal is to reduce the software's MTBF (Mean Time Between Failure). In terms of architecture design, resilience includes service asynchronous capability, retry/current limit/downgrade/fuse/backpressure, master-slave mode, cluster mode, HA in AZ, unitization, cross-region disaster tolerance and multiple active capacities in different places, etc.

3.2.5.5 Principles of all Process Automation

Technology is often a 'double-edged sword'. The use of containers, microservices, DevOps and a large number of third-party components reduces distributed complexity and speeds up iterations, as it increases the overall complexity and components of the software technology stack. Scale, so it inevitably brings the complexity of software delivery. If the control is improper here, the application will not be able to appreciate the advantages of cloud-native technology. Through the practice of IaC (Infrastructure as Code), GitOps, OAM, Kubernetes operator and a large number of automated delivery tools in the CI/CD pipeline, on the one hand, standardize the internal software delivery process of the enterprise, and on the other hand, based on standardization. Through the self-description of configuration data and the final state-oriented delivery process, automation tools can understand the delivery goals and environmental differences, and realize the automation of the entire software delivery and operation and maintenance.

3.2.5.6 The Principle of Zero Trust

Zero-Trust Security has re-evaluated and reviewed the traditional border security architecture ideas and gave new suggestions for the security architecture ideas. The core idea is that no person/device/system inside or outside the network should be trusted by default, and the trust basis for access control needs to be reconstructed based on authentication and authorization, such as IP address, host, geographic location and network. It cannot be used as a credible certificate. Zero Trust has overturned the paradigm of access control, leading the security system architecture from 'network centralization' to 'identity centralization', and its essential appeal is identity-centric access control.

The first core issue of Zero Trust is Identity, which gives different Entities different identities to solve the problem of who accesses a specific resource in what environment. In the R&D, testing, and operation and maintenance of microservice scenarios, Identity and its related strategies are not only the basis of security, but also the basis of numerous (resources, services and environment) isolation mechanisms; in the scenario where employees access internal applications in the enterprise, Identity and its related strategies provide a flexible mechanism to provide access services anytime, anywhere.

3.2.5.7 The Principle of the Continuous Evolution of Architecture

Today's technology and business are evolving very fast. It is rare that the architecture is clearly defined from the beginning and is applicable throughout the software life cycle. On the contrary, it is often necessary to reconstruct the architecture within a certain range. Therefore, the cloud-native architecture itself should and must be architecture with continuous evolution capabilities, not a closed architecture. In addition to factors such as incremental iteration and target selection, it is also necessary to consider structural governance and risk control at the organization (such as the structure control committee) level, especially in the context of high-speed business iterations in terms of structure, business and balance. Cloud-native architecture is relatively easy to choose architecture control strategy for new applications (usually choosing the dimensions of flexibility, agility and cost), but for the migration of existing applications to a cloud-native architecture, the migration cost of legacy applications needs to be considered architecturally/risks and migration costs/risks to the cloud, and technically fine-grained control through microservices/application gateways, application integration, adapters, service grids, data migration, online grayscale, and other applications and traffic.

3.2.6 Cloud-Native Architecture Deep Dive

3.2.6.1 Separate the Computing and Persistence

As we mentioned before, architecture often affects non-functional requirements. At this age, cloud native has given us greater freedom and feasibility in architecture. The most important resource in the world is not oil or nuclear fuel, but data. When we start to learn programming, we were told that the closer to the data, the higher the efficiency the program will have. That's maybe a reason for a lot of business logic in many traditional systems is implemented in 'database stored procedures', because the database is often used as the single point of the system and represents the consistency, is the source of truth. The system built in this way is very hard to refactor, migrate and modify, because stored procedures are too difficult to read, write, and test.

Not only the database, in the big data world the system with Hadoop as the backbone also has similar problems. This leads to a conclusion that the coupling of computing and persistence will lose the ability to expand and evolve, which also wastes resources. Then centralized storage and persistence have to face these problems below:

(1) Resource waste: Computing and persistence will reach the bottleneck during the processing, but the timing is different often. So, if we want to increase the persistence, we must get a new instance, and the calculation resource is naturally wasting.

(2) Hard to scale: If you want to make a horizon scaling, prepare a new instance is the only option, for the IDC age we will have a lot of types of servers in the cluster, which is hard to manage and maintain.

(3) Hard to evolve: The biggest problem is that it's difficult to evolve the whole system, the coupling of computing and persistence will make the code hard to modify, it's bound by the low-level framework often.

Systems around a centered database have become an outdated pattern, it's time to separate the computing and persistence in modern architecture, we should consider storing the state in different places, whether is a database, NoSQL, NAS or OSS, and the computing should be stateless and purely functional as much as possible. During this process, we should get rid of sticky sessions, shared memory, and use the temporary file is reasonable, for the data with high-frequency access adopt Redis, Memcached to deal. When the system no longer depends on the local state, development and deployment will be simple, and it is also easy to find the bottleneck: often the computing part that can be scaled horizontally will not be the bottleneck.

3.2.6.2 Service-Oriented

Compared with monolithic applications, microservices encourage that each module is an independent service, using network protocols to call each other, each service will iterate separately. Because the modules are distributed, the big business requirement could be split into different microservices, and each service will update the code, test and deploy separately. The key point is, the individual service modification must follow the backward compatibility principle, to ensure every minor step is safe and won't break the existing business. The following issues from monolithic applications are the microservices want to solve:

(1) Becomes a big junkyard: If a code repository has more than 10 contributors, the code is difficult to maintain unless the engineers have good capabilities and habits. Monolithic applications are not

complicated at the beginning, they often have beautiful architecture design, clear boundaries and standardized readme. But after people changes and increment of requirement, we have to add features in place which is not very appropriate, that makes the repo complicated and become 'Tar Pit' finally.

(2) Difficult to deploy: The monolithic application still has the difficulty of deployment, the process is manual often, and is bound to a virtual machine (many applications cannot leave the instance to redeploy even). The rollback is difficult too and eventually made the operations reject or postpone the change, and they don't like deployment.

(3) Learning curve and knowledge fragment: The cost of learning codes for monolithic applications makes every newcomer intimidated. With years of iteration, many people's knowledge and designs have been accumulated in the codebase. It does not mean that the code is unreadable, but without the certainty of boundaries, it's hard to understand the context.

However, microservices are not simple and easy to implement. Once we use multiple services and combine them together for the business, the communication between the multiple processes has lower efficiency and instability than the function invokes inside the monolithic application. In the world of microservices, services are deployed in different places. The constraints or agreements between them are changed from method signatures to higher level protocols, such as RESTful and PRC. In this case, calling a module requires the network, we must know the network address and port of the target end and also need to know the protocol, and then we can write code such as using HttpClient to invoke.

Even the popular service mesh will help us solve the invocation problem outside of the process, there are more problems to face: service downgrade, throttling, security, consistency, service interface changes, cascade errors, etc. These are common in the cloud-native age.

3.2.6.3 On Provisioning

On provisioning may be the most important function of the cloud. We often think that cloud computing resources are expensive than local computing resources, if you compare the prices of virtual machines or instances, this conclusion is partially correct. However, the operation and maintenance cost of cloud computing is less than local, because the ability of management and control tools and resource utilization are enough to fill the cost of a virtual machine.

An RDS instance with 32G memory may cost about 1,000 dollars a year, but it's not enough for half a month's salary cost of an operation people. The platform capabilities and automation capabilities provided by cloud computing can improve the efficiency of operation and maintenance. Many maintenance tasks will be delegated to the platform, reducing a large amount of people resource.

Enterprise IT often has to face the explosive growth of traffic and workloads such as promotions and emergencies. Whether to use basic elastic scaling group, or the container technology and serverless, the scaling is in an extremely high priority. This benefit allows us to don't need to allocate many resources but by demand.

3.2.6.4 Disposability

The fast start and graceful shutdown and maximum robustness are the pursuits of modern applications. Why Serverless is so hot, and everyone thinks it is the value of the future, the reason maybe it provides a very standardized programming model: simple and direct function calls. This kind of input–output logic in serverless is easier to understand nature and more stable, it's consistent with the functional programming. Creating a fast start, flexible and scalable application, and rapidly deploying changed codes and configurations, must depend on a solid infrastructure. At the same time, applications with disposability can also improve the resilience of the system. Generally speaking, resilience means that the system can handle properly after a failure and then return to normal.

Container technology is welcome, the one reason is it supports fast startup, you don't need to worry about dependencies, start the process with a simple command 'docker run'. The disposability of the application means graceful shutdown, after receiving SIGNTERM, the application will complete the last job before exiting. You can use Java annotations or the other hooks to implement that, to avoid dirty data, unfinished transactions or exceptions during the frequency deployment.

3.2.6.5 Event Driven and Event Sourcing

Event Driven and Event Sourcing are two different design patterns based on the event, and they are widely used in cloud-native applications; for event driven, we are not stranger. In many cases, use middleware to aggregate system by messages is a kind of event driven, which help us implement a pub-sub architecture easily. First of all, event driven is one of the best solutions for decoupling. These events often represent 'things happened', like Order Submitted, User Created and Job Started; it shows the state for the system in some time. But if your events in the system have imperative verbs, it's not an event but a command, such as Publish Sub, Delete Item and Refresh Order.

To practice this pattern, you must define the event right and in the system which part will fire the event and who will be observer, the receiver is irrelevant to the publisher, and that will significantly reduce the difficulty of each service. When a user registers, the mail service will send an email for verifying, and the promotion service will set a discount for this specified customer, these two services don't need to know each other, so they can update separately. But everything has two sides, this pattern is a bit hard to see the full picture, and it has weak consistency, implement a transaction is another challenge.

The event sourcing pattern is another design pattern based on events. The thought is that events are the only source of truth, it stores entire state of the system, and the state is stored as a sequence of events. Our bank account is an example, from you open the account, the current balance is the result of sum of all the financial operations, and because these operations happened, you can't modify the past. The most interesting part is the value of these events depends on the purpose, we calculate the balance from all events, and bank staff who review your loan request only focus on the income, he just cares about your salary to ensure you can afford the loan. And all events are here, we won't miss a thing. Event sourcing can establish the unique state of the system, the observers cannot impact the eventual consistency state of the system.

3.2.6.6 Backend for Frontend – BFF

BFF and API Gateway are two concepts that are very easy to confuse. Many microservice architecture designs like to draw API Gateway on the diagram. If the diagram is describing the low-level components, we can have the API Gateway, but if the diagram just describes the relations between the business modules, it's undesirable. In many cases, API Gateway should stay away from business: routing, authentication, throttling, access log, etc. But BFF is different, it's a business concept, which represents the creation of different backends for different frontends, although some API Gateways also can aggregate services, just merge the responses. Initially, our application was to serve the desktop web UI, but with the growth of mobile users, we need to create a dedicated service to mobile users, for example, in the product detail page, we don't need to return all details as the desktop, because the mobile screen is small.

These significant differences will make mobile applications and desktop applications have different requirements for backend APIs, especially services with plenty of information and complicated interactions. For this situation, we often use two different interfaces, because the cost of using one interface to support two clients is very difficult. However, these two sets of interfaces have not separated in code, it's impossible to follow the workload in different clients. Therefore, create BFFs for each client, and we can tune each BFFs to have the best performance and suitable interface, to match each client's requirement, and won't break the other ends. But the core business services are still the same, there's no reason to change.

Of course, BFF is not perfect, and the cost of implementing an independent service is high. Although container technology and DevOps methodology can reduce the startup cost of new services, once the system is created, it is difficult to offline in the future.

3.2.7 Create a Cloud-Native Application

3.2.7.1 Boundary

After so many discussions, we have a full understanding of cloud-native architecture. Compared with traditional enterprise applications, we hope to build a new application that takes the advantages of cloud computing, and helps the business more agile, and reduces the cost of development and maintenance, it must be flexible and scalable. Regardless of the application architecture, the business code is the core of the application, and the other parts are just accessories.

In the cloud-native age, our applications are naturally distributed, which forces us to clarify the boundaries between businesses. This is like when we write Java code, if your class has many public methods, it must be a very unclear implementation, and the other people don't know what it does, and it will eventually become a junkyard. After clearly dividing the business boundaries, you can select the communication methods between the boundaries, whether it is synchronous or asynchronous, one-to-one or one-to-many.

3.2.7.2 Tech Stack Selection

When selecting tech stack in the cloud-native age, we never had so many choices.

General computing, or operating environment, persistence, network, observability, container or serverless, security components, automatic platform, DevOps toolchain, programming language, framework, deployment tool and log center are very good candidates, most of which are open source, so are there guidelines to help us choose?

Automation is a good aspect to improve efficiency. The easiest or most valuable automation scenarios are frequent and daily tasks, followed by those that could not be automated before. We can use automation to solve many problems, such as prepare a test environment, in the past, it was necessary to create a MySQL instance and prepare test data separately, and everybody will share it. Now you can have a shell script to start a container locally and then start the service in another. Allocate resources on the cloud, deploy the application, scale up the containers, all these cases can be done automatically, even on CICD. For infrastructure, use Terraform to manage the resources as code, or use helm to quickly deploy your applications.

Make sure the technology is controllable, most cloud-native technologies are open source and easily decoupled from the IaaS layer. The popularity of docker technology is because it balances the flexibility and openness required for development and the automation of operation and maintenance. We can use it in the development environment, automated testing and application distribution. The open-source Kubernetes fully demonstrates the value of container orchestration and become a de facto standard. Most importantly, Kubernetes sealed the differences in the IaaS layer and enables our applications to run in different environments, avoids us being locked by different cloud vendors and finally becomes the most important infrastructure in the cloud-native age.

3.2.7.3 DevOps and Continuous Delivery

Whether it is DevOps, CD and automated operation and maintenance, there is an enlightening principle: do things that make you painful early and frequently. When you hate every release, then try to add enough tests before submitting the code, from unit testing to functional testing, and constantly improving automated tests before release to increase coverage; when you are worried about scaling, especially manually, you need to put the relevant things in the code base and make frequent modifications until the completion, and CICD tool will run it for you; if other people's code always conflicts with you, you should integrate as soon as possible, use the merge and rebase to clean up the code at the beginning.

Software delivery is the responsibility of everyone, not just for the developer, business and operations, it's a continuous improvement process. The developer has completed the requirements provided by the BA, it is impossible to say the work is 'DONE', even if the latest code has been deployed to the production environment. Whether the task is completed or not, in the cloud-native age, in this distributed environment, it cannot be controlled by a single one. It requires the entire team to handle. This is the DevOps spirit we advocate to break the barriers, sit together and share responsibility. Finally, in this cloud-native age, good luck to you.

3.3 DESIGN AND PRACTICE OF CLOUD DISASTER RECOVERY

3.3.1 Introduction

This section first introduces the relationship between cloud native and cloud disaster recovery. Then presents the relevant background, concepts and principles of disaster recovery. Next, it takes the development process and technical requirements of disaster recovery of banking institutions as the breakthrough point, and describes the experience of disaster recovery based on cloud native in Alibaba Cloud. Combined with the characteristics of commerce scene and general industry from Alibaba Cloud, we share how Alibaba Cloud abstracts its internal unique disaster recovery ability into general ability to help users quickly build a disaster recovery architecture based on cloud native. Finally, it makes a simple analysis of the development trend of cloud disaster recovery and the practice of Alibaba Cloud.

3.3.2 Cloud Disaster Recovery Is an Advanced Usage Scenario of Cloud Native

The previous chapter introduces the architecture concept of cloud native in detail. We think the next field of cloud-native subversion may be in the traditional disaster recovery field. Through the cloud-native technology, how to build a disaster recovery scheme for application system is a very interesting topic.

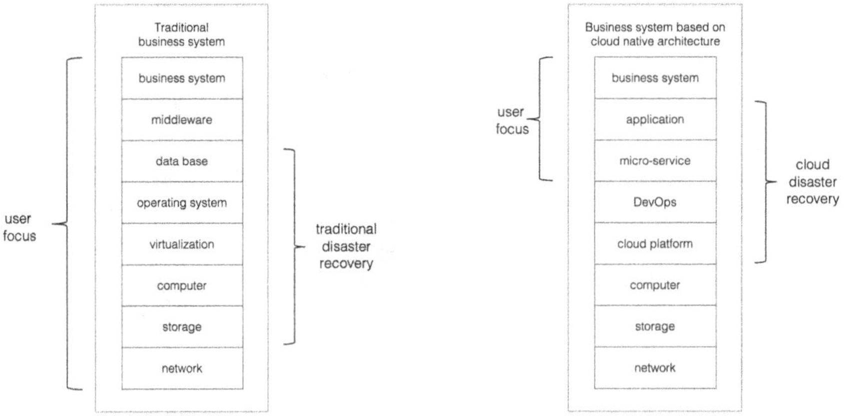

FIGURE 3.2 Comparison of traditional disaster recovery and cloud disaster recovery.

Although the rapid development trend of cloud native, the traditional migration and disaster recovery still stay at the data migration level, the vision of cloud computing is to make cloud resources usage on demand like water and electricity. So, disaster recovery based on cloud native should also conform to such historical trend. Why traditional disaster recovery mode cannot meet the original needs of cloud native? In short, the core of these two theories focus is different. The core of traditional disaster recovery is storage. But there is no effective scheduling method for computing, storage and network infrastructure layers, which cannot realize highly automated scheduling. Storage and database have become the cloud-native service itself. When business systems are all in cloud, users do not have the absolute controller of the underlying storage and database, and no longer needs to pay attention to the underlying disaster recovery architecture when coding. Therefore, the traditional disaster recovery mode needs some changes (Figure 3.2).

The author believes that in the construction of cloud disaster recovery solutions, we should think about the construction method with business as the core focus, and use the arrangement ability of cloud-native service to realize the continuity of business system. We will start from the basic knowledge of traditional disaster recovery. Gradually in-depth, to see how cloud disaster recovery based on cloud native technology concept brings more business values.

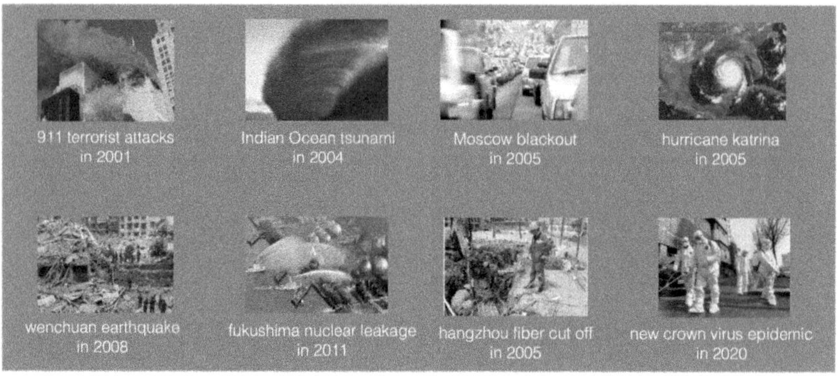

FIGURE 3.3 Major disasters in the past 20 years.

3.3.3 Basic Knowledge of Data Backup and Disaster Recovery

3.3.3.1 Definition of Disaster Events

We often define disaster as a sudden, unplanned, serious unfortunate event that can lead to major injuries or losses. Disasters can happen anytime. Look at some major disasters in recent 20 years, which are the most painful memory (Figure 3.3).

3.3.3.2 Disaster Dimension of Scene

From the perspective of disaster recovery, how to define disaster events? We can focus on the following three aspects:

- The disruption time of critical business functions or processes caused by emergencies exceeds the maximum tolerated by enterprises.
- Usually the recovery time object value (RTO value) is used as the basis for determining whether it is a disaster.
- By evaluation, when the expected outage time of key business functions exceeds the predetermined RTO value, it is considered as a disaster, and the corresponding plan and plan should be initiated.

Let's look at a set of data. In the 20 years before 2000, the disaster events caused by natural disasters such as power outages, storms, snow and floods were at the forefront. According to Swiss Re, an international authority,

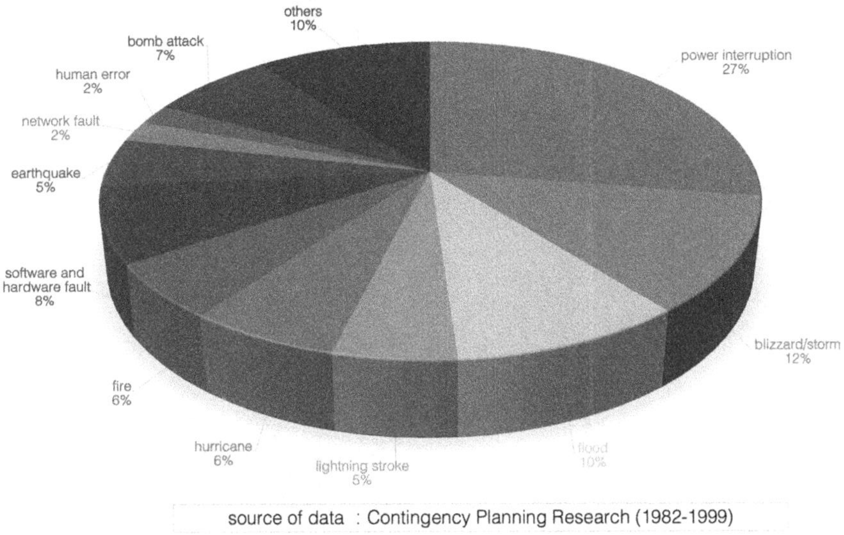

FIGURE 3.4 Disaster scene classification.

the world's statistical incidence of natural disasters increased threefold and economic losses ninefold from the 1960s to the 1990s. In the 21st century, this number is still increasing rapidly! (Figure 3.4).

3.3.3.3 *Concept of Disaster Recovery and Backup*

The purpose of disaster recovery system is to ensure that the data and services of the system are 'linear'. When the system fails, it still can provide data and services normally. In order to prevent disasters, establish the same IT system in two or more places, synchronize with each other, and then switch over at any time when disaster happens.

The purpose of backup is different from disaster recovery. Backup is the process of transferring online data to offline data, which aims to deal with logic errors and historical data preservation. Backup is the cornerstone and the last line of defense for HA of data. Its purpose is to restore data when system data' collapse. After building a backup system, whether do not need disaster recovery system? The answer is negative. The premise is the RTO/RPO metrics by business department's expectations. Backup can only meet the purpose of data recovery when data were lost or broken, but cannot provide real-time business failover functions. Therefore, disaster recovery system is essential for some key servers.

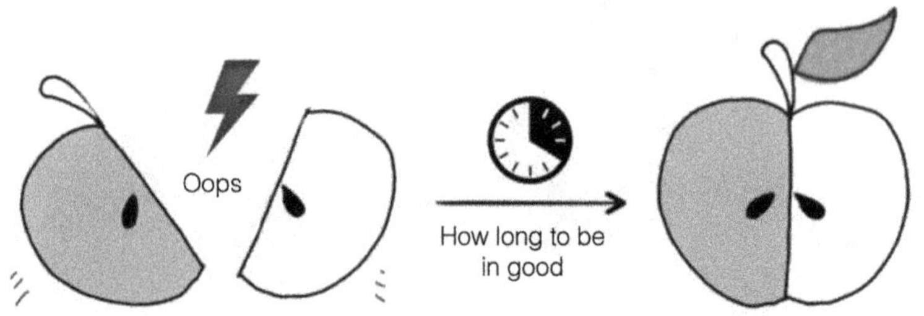

FIGURE 3.5 What is RTO?

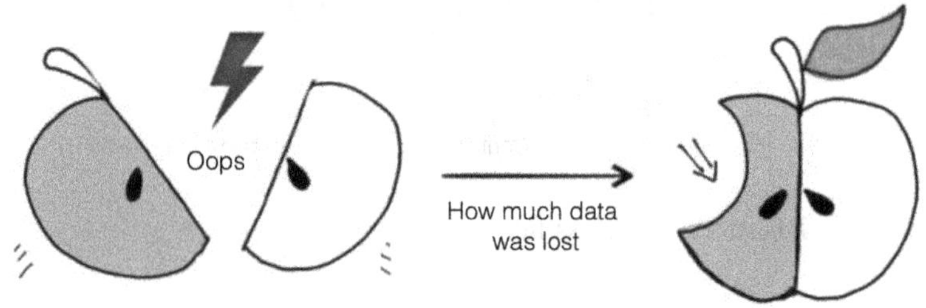

FIGURE 3.6 What is RPO?

(1) What is RTO?

Recovery time objective (RTO): The period of time within which IT systems and services must be restored after an outage. RTO indicates the timeliness of service recovery, that is, the maximum recovery time for IT systems that enterprises can tolerate. A smaller RTO indicates a higher disaster recovery capability, but requires a higher enterprise investment (Figure 3.5).

(2) What is RPO?

Recovery point objective (RPO): The point in time to which data are restored by the disaster recovery system after an outage. RPO indicates the amount of data loss, that is, the maximum amount of data loss that enterprises can tolerate. A smaller RPO indicates less data loss and less harm to the enterprise (Figure 3.6).

3.3.3.4 Disaster Resilience Assessment System

According to the National Standard of the People's Republic of China GB/T 20988-2007 Information Security Technology – Disaster Recovery

Specifications for Information Systems,[1] the protection levels are determined as follows (Table 3.1).

3.3.3.5 Classification of Data Backup

According to the amount of data, backup solutions can be divided into full backup, incremental backup and differential backup.

- Full Backup

  It refers to a complete copy of all data or applications on a certain time point. The advantage of this backup method is that the backup process and recovery process are simple. The disadvantage is that there is a large number of duplicate data in each full backup, due to the large amount of data, and backup and recovery time is long. The ideal way for full backup is to create snapshot of the data and applications, and then use the snapshot to perform full backup. However, whether to support snapshot creation and full data backup based on snapshots depends heavily on the ability of the backup system. If the backup system does not support snapshot capabilities, it means full backup needs to be in a specific time window without any new data production and storage. If the data source changes during the backup period, the full data backup may be actually incomplete. When a disaster occurs, using these incomplete data to recover the business system, it is likely to cause a secondary disaster. Because it is very hard or need to invest very large costs to determine which part of the data is missing. For some core systems, such as mobile banking APP, it requires 7 * 24 hours of continuous operation, while data will be continuously generated. If the backup system cannot support creating a snapshot first and then a full backup, such core systems need to apply for a shutdown window once to ensure that no new data are generated during the backup process. Obviously, this way is difficult to be accepted by the business department, and the operation and maintenance department will also be challenged by insufficient professional ability.

- Incremental Backup

  It means that after a full backup or the last incremental backup, only the changed data need backup, including new addition and

---

[1] See: http://www.djbh.net/webdev/file/webFiles/File/cpzg/20122616046.pdf?spm=a2c4g.11186623.2.11.1869f6795rMbM3&file=20122616046.pdf.

TABLE 3.1 Disaster Recovery Protection Levels

Protection Level	Data Backup	Measure	Preventable Risk	RTO	RPO
Level 1: Basic support	• All the data are backed up once a week • The backup media are stored offsite	–	The service data are damaged	2 days or more	1–7 days
Level 2: Secondary site support	• All the data are backed up once a day • The backup media are stored offsite • The data are regularly synchronized in batches several times each day	Backups are called after an outage	• The service data are damaged • The service processing site is not available	24 hours or more	1–7 days
Level 3: Electronic transmission and partial device support	• All the data are backed up once a day • The backup media are stored offsite • The data are regularly synchronized in batches several times each day	Backups are provided for some data processing devices	• The service data are damaged • The service processing site is not available • Some devices or networks fail	12 hours or more	Several hour to 1 day

(*Continued*)

TABLE 3.1 (*Continued*) Disaster Recovery Protection Levels

Protection Level	Data Backup	Measure	Preventable Risk	RTO	RPO
Level 4: Electronic transmission and full device support	• All the data are backed up once a day • The backup media are stored offsite • The data are regularly synchronized in batches several times each day	Backups are provided for all devices in the available status (cold site)	• The service data are damaged • The service processing site is not available • All the backup devices or networks fail	Several hours to 2 days	Several hours to 1 day
Level 5: Real-time data transmission and full device support	• All the data are backed up once a day • The backup media are stored offsite • The data are replicated in real time	Backups are provided for all devices in the ready or running status (warm site)	• The service data are damaged • The service processing site is not available • All the backup devices or networks fail	Several minutes to 2 days	0–30 minutes
Level 6: Zero data loss and remote cluster support	• All the data are backed up once a day • The backup media are stored offsite • The data are synchronized and backed up in real time to ensure zero data loss	• The disaster recovery site and the production site have the same processing capability and are compatible with each other • Software clusters are used to implement seamless failover and failback • Real-time monitoring and automatic failover to the remote cluster system are supported (hot-active)	• The service data are damaged • The service processing site is not available • All the backup devices or networks fail	Several minutes	0

modification. The most significant advantage of this method is that there is no duplicate backup data. So, the amount of backup data is small and the backup time is short. The disadvantage is that the recovery process is more troublesome, and more time-consuming than full backup recovery process. The recovery process must have the last full backup and all incremental backup data, and must be restored one by one along the time sequence from full backup to increasing backup, which greatly increases the recovery time. For example, a system completes a full backup at 00:00 on Sunday, and then performs incremental backup at 00:00 every morning. If the system fails on Thursday morning, resulting in a large number of dirty data leading to inaccurate problems, it is now necessary to restore the system to Wednesday night. Then the operation and maintenance personnel need to recover the full data of Sunday, and then recover the incremental data of Monday, Tuesday and Wednesday morning at 00:00. In this backup technology, the relationship between the full data and the incremental data is like the chain of bicycles. One chain is buckled, and any problem in one chain will lead to the disconnection of the whole chain.

- Differential Backup

  During the differential backup process, only the selected files and folders marked are backed up. It does not clear the tag. After the backup is not marked as a backup file, it does not clear the archive attributes. Differential backup refers to the backup of files added or modified during the period from one full backup to differential backup. In recovery, we only need to recover the first full backup and the last differential backup. Differential backup not only avoids the disadvantage of the other two backup strategies, but also has their own advantages. First, it has the advantages of short incremental backup time and less disk space; second, it has the characteristics of less storage and short recovery time required for full backup recovery. The system administrator only needs two disks of tape, the full backup tape and the differential backup tape the day before the disaster, to restore the system.

Comparison of three backup methods (Figure 3.7):

- Sorting by the amount of backup data: full backup > differential backup > incremental backup

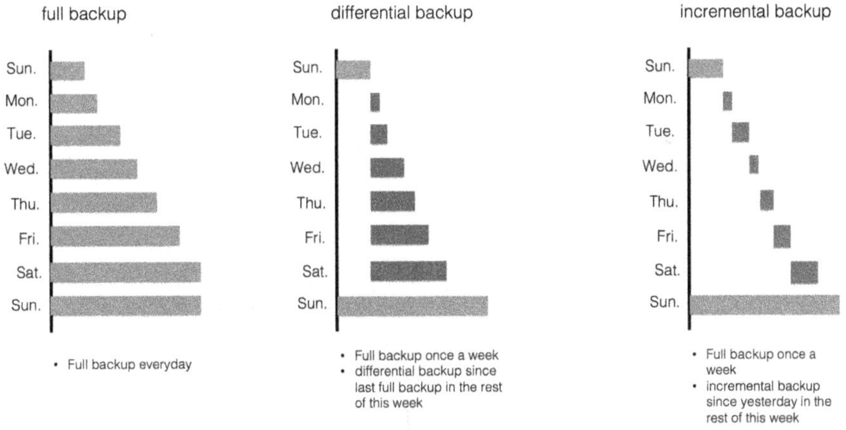

FIGURE 3.7 Classification of data backup.

- Sorting by data recovery speed: full backup>differential backup>incremental backup

3.3.4 Construction Demand of Cloud Disaster Recovery

3.3.4.1 Deployment of Cloud Disaster Recovery

In order to facilitate readers and authors to have a unified understanding of cloud disaster recovery and its services, the definition of 'cloud disaster recovery' refers to the realization of the disaster recovery purpose of the business system through the cloud. The forms of cloud disaster recovery include cloud backup and cloud disaster recovery. The cloud forms include public cloud, proprietary cloud and hybrid cloud.

Cloud backup refers to the usage of data transmission technology, which directly copies the production data from the production center to the cloud disaster backup center. When a disaster occurs and recovers in the production center, the backup data from the cloud disaster recovery center can be transmitted back to the production center, and then the production center can resume business access. When a disaster occurs in the production center, the business flow can be quickly switched to the cloud disaster recovery center to ensure business continuity.

The form of cloud disaster recovery center can be public cloud, proprietary cloud and hybrid cloud:

- Public cloud refers to making cloud services available to the general public on the Internet. Users use cloud services on demand and pay on demand, with greater flexibility.

- Private cloud refers to the cloud environment created exclusively for a user or group and usually runs within the firewall of the user or group.
- Hybrid cloud is the best intermediate scheme to absorb the dual advantages of public cloud and private cloud. Hybrid cloud has the maturity of public cloud level in terms of stability, ease of use, and takes into account the safety and controllability of private cloud. It can better meet the specific performance, application and safety compliance of government and enterprises.

There is a simple example to approximately describe the difference of the three cloud forms.

Tom, Andy and Bill are neighbors. And there is a restaurant not far from their houses. Tom has used to eat at the restaurant, and pays on demand. Restaurant likes a public cloud. Andy has used to buy and cook food at home. The kitchen likes a private cloud. Bill's family suddenly has some visitors, including young and old. The old people do not want to have lunch outside, they cook at home. And the young people go to the restaurant to have their lunch. Home and restaurant are like the hybrid cloud.

At present, there are two mainstream deployment modes of cloud disaster recovery:

One of the mainstream deployment modes is that the cloud disaster backup solution built on the cloud platform, including cloud backup and disaster recovery of databases, files, middleware.

The user's business system is deployed based on the cloud platform. After the corresponding adaptation transformation, the business system has backup and disaster recovery capabilities.

Figure 3.8 is one-stop disaster recovery architecture based on Alibaba Cloud.

Another mainstream deployment model is to provide customized disaster recovery technology solutions based on numerous third-party commercial software and disaster recovery technology services, without changing or minimizing the original business system architecture. The typical implementation path is to build the disaster recovery of cloud business layer based on the IOE (IBM Power Series, Oracle database software, EMC storage array) dedicated architecture.

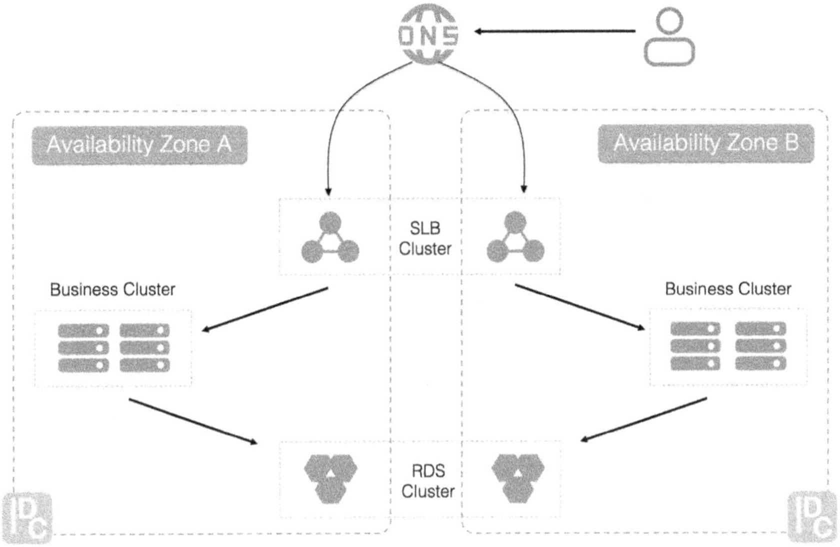

FIGURE 3.8 One-stop disaster recovery architecture based on Alibaba Cloud.

Figure 3.9 is about the traditional disaster recovery architecture. The owner needs to purchase, manage and maintain multiple commercial software and hardware products. Like a self-built house, buy your own furniture.

No matter what kind of deployment mode, the technology architecture level can be divided into storage layer, calculation layer, application layer, network layer and transport layer. From the perspective of cloud backup, timing or real-time backup can be used to select according to the actual needs and costs of users. From the perspective of cloud disaster recovery, including local to cloud, cloud to cloud, single cloud product fault, network fault and application system fault switching dimension. Disaster recovery switching mode can be divided into manual switching and automatic switching. No matter what kind of mode, it is necessary to consider the RTO and RPO of disaster recovery switching, and deal with the client user experience and data consistency.

3.3.4.2 The Value of Cloud Disaster Recovery

Cloud native has distributed, flexible, rich middleware and database products. Cloud disaster recovery mode based on the characteristics of cloud native has many advantages over traditional disaster recovery mode.

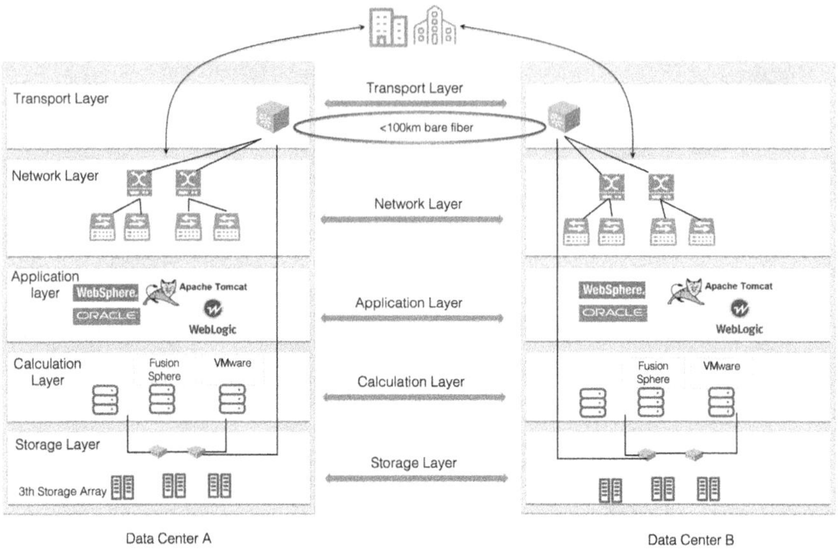

FIGURE 3.9 Disaster recovery architecture based on multi-commercial products.

(1) Reduce IT costs

The traditional disaster recovery mode highly relies on the data replication ability of special storage hardware and commercial database software, such as the IOE architecture that we often hear. In this way, the data and computing ability are concentrated on a few storage and hosts devices. This mode is called the centralized architecture. This centralized architecture requires high performance and reliability of hardware and commercial databases software, so component replacement and maintenance costs are high.

The distributed architecture based on cloud native is designed to solve the problems of difficult to expansion and high price of centralized architecture. Cloud storage and cloud database and other cloud products, mainly using a large number of cheap ordinary hosts, are using distributed collaborative software. Data are stored in a number of general storage servers, and through multiple copies and consistency algorithm for integrated management, providing a unified logical storage space for end users. Compared with dedicated servers and storage devices, these cheap servers have a higher probability of failure. Therefore, based on the HA and disaster recovery mechanism of cloud storage and cloud database, the design and development of cloud platform architecture must be considered. This design idea is called failure-oriented design.

(2) Safe backup and quick recovery

When a disaster occurs, different industries and different users have different requirements for RPO and RTO. The requirements are usually implemented in accordance with the norms of superior regulatory units.

For storage, databases, middleware, load balancing, flexible cloud hosts and many cloud vendors have multi-copy cluster technology based on distributed architecture.

Compared with the stand-alone mode with centralized architecture, cluster mode has faster recovery speed in stand-alone fault. At the same time, some excellent cloud vendors can realize cross-regional data automatic backup at the platform, and provides a unified logic data entry for users. Users do not even need to care about the logic implementation of underlying data replication, and only need to pay attention to their own business logic implementation. This cross-regional data automatic backup technology can help improve the backup security level, and automatically complete the switching of physical layer data access path when disasters occur. At the same time, the logical entry is guaranteed to be constant. It is a very good support way to meet the disaster recovery requirements of superior supervision units.

(3) Professional maintenance

The core of cloud-native technology is to realize the HA of corresponding database/storage/middleware products by distributed concept and failure-oriented design. Compared with the traditional centralized architecture, new requirements are put forward for the breadth/depth/thinking mode of operation and maintenance personnel. In the disaster recovery system on the public cloud, there is a cloud manufacturer professional 7 * 24 hours operation and maintenance team. They cover datacenter operation and maintenance, network, computing, storage, security, database, middleware and other complete professional talents, which greatly simplifies the user's operation and maintenance. If the disaster recovery system of private cloud is built by users, the above fields need to be equipped with corresponding professional operation and maintenance personnel.

Otherwise, it is difficult to achieve the expected disaster recovery effect when the disaster occurs.

3.3.4.3 Attention to the Implementation of Cloud Disaster Recovery

Cloud disaster recovery implementation refers to the process of building disaster recovery center in the cloud. This section mainly discusses the deployment and implementation process of IaaS, PaaS and SaaS layers. Disaster recovery drills and recovery are discussed in the subsequent chapters. Cloud disaster recovery construction is a huge investment project for each user involved, so it is necessary to do a full assessment in the project demand analysis stage. The following points as a general reference direction for the needs analysis phase reference.

(1) Professionalism

Choosing a professional cloud manufacturer is the most critical factor in the implementation of disaster recovery projects. There are several reference bases for evaluating the professionalism of a manufacturer, including whether the construction of cloud disaster recovery has been completed within the manufacturer, whether there are successful cases and peer customer evaluation, whether there is a professional consulting/delivery/maintenance system, the peer ranking of cloud manufacturers, and the cloud ecological construction.

(2) Timeliness

Timeliness is divided into two kinds: one is RTO, and the other is RPO. The implementation time of disaster recovery depends on the complexity of the user's IT architecture, the scheme constructed by the user, and the business system involved. If the hardware and software equipment at both datacenters of the disaster recovery are isomorphic, and only involves data, files, databases and system backup, the implementation time will not be long; if there are more heterogeneous IT architectures at both datacenters, and a variety of business disaster recovery needs to be built, the implementation time will be relatively long.

Data recovery and business takeover are also suitable for the above situation. Users should clarify the types of disaster recovery business, data volume, cycle, recovery time, network bandwidth and so on. RTO and RPO are used as quantitative standards for cloud disaster recovery construction. Under the condition of meeting the regulatory requirements, the faster the better.

(3) Security

Security can be divided into several levels: First, cloud infrastructure and network security, mainly to prevent the emergence of system bug and network operation and maintenance problems. Second, during the process of migration, backup and disaster recovery construction, system downtime and outage may be caused. Third, it is about autonomous control. Different from the traditional IOE architecture, cloud platforms are safe and controllable in geopolitical terms. Attention should be paid to the problem of cloud platform, which may cause the loss of business data. The risks come from software bugs, equipment failure, hacker attacks (such as DDoS), thief, operation error prevention work and so on. The solutions can be encryption and multi-copy backup. Do not put all eggs in one basket.

(4) Compliance

Compliance is a regulatory requirement. It should be highly valued by participants, because long-term backup data need to face archiving problems. Although the relationship between data archiving and business continuity is gradually deteriorating, archived data are still subject to mandatory constraints of compliance.

3.3.5 Demand Analysis of National Content Disaster Industry

In the process of disaster recovery construction of China's data center, bank institutions are the earliest and most mature. Therefore, through the disaster recovery process of the bank institutions, it can reflect the level of disaster recovery construction and the future development trend in China by a certain extent. The following focuses on the analysis of bank institutions.

3.3.5.1 Disaster Recovery Business Requirements

Taking bank institutions as an example, the dynamics of disaster recovery construction is very strong, and the level and requirements of disaster recovery construction are very high. Banks have strong financial strength, while the business itself involves the national economy and people's livelihood has very strong social responsibility.

In summary, there are several requirements:

- Support failover: Support intra-plan failover drill, and extra-plan failover with detailed DRP (disaster recovery plan).
- Courage to switch: Build a higher level of data RPO, and ensure data consistency. When the real data center fails, user can dare to execute the DRP.
- Regularly switch: By combing the switching process and often performing switching drills, finally helping maintainers to improve the proficiency of DRP.
- Rational cost: Through more reasonable capacity ratio, controlling the overall TCO and getting the best overall cost performance.
- Online: Application supports cross datacenter level gray release, to flexibly schedule business traffic.

Through the investigation of some large banks and banks in developed areas, it is found that the disaster recovery construction of large- and medium-sized banks has several stages:

- Phase 1: From 2000 to 2010, banks mainly completed the construction of datacenters in the same city, including single center and double center, and tried to switch business flow from one center to another center.
- Phase 2: From 2010 to 2015, banks mainly built three datacenters in two cities, and some of them tried to some service as active – active mode by two datacenters in the same city.
- Phase 3: Since 2015, banks have been continuously optimizing disaster recovery technology architecture and processes (Figure 3.10).

3.3.5.2 Technical Requirements for Disaster Recovery

Through research and analysis, it is found that in the process of disaster recovery construction, each bank will basically encounter several core technical matters.

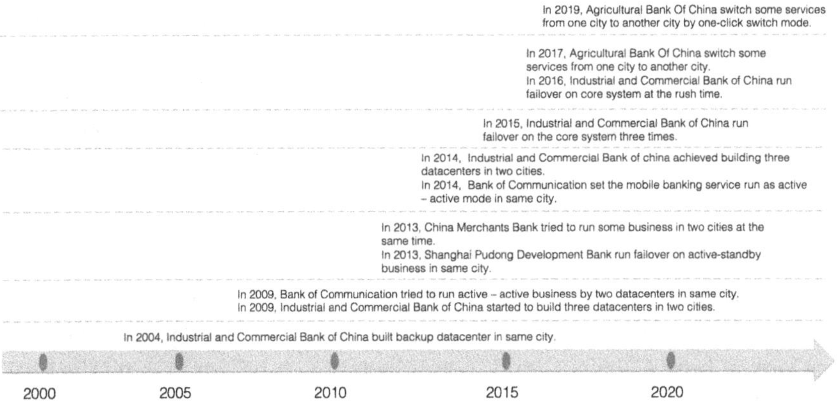

In 2019, Agricultural Bank Of China switch some services from one city to another city by one-click switch mode.
In 2017, Agricultural Bank Of China switch some services from one city to another city.
In 2016, Industrial and Commercial Bank of China run failover on core system at the rush time.
In 2015, Industrial and Commercial Bank of China run failover on the core system three times.
In 2014, Industrial and Commercial Bank of china achieved building three datacenters in two cities.
In 2014, Bank of Communication set the mobile banking service run as active - active mode in same city.
In 2013, China Merchants Bank tried to run some business in two cities at the same time.
In 2013, Shanghai Pudong Development Bank run failover on active-standby business in same city.
In 2009, Bank of Communication tried to run active – active business by two datacenters in same city.
In 2009, Industrial and Commercial Bank of China started to build three datacenters in two cities.
In 2004, Industrial and Commercial Bank of China built backup datacenter in same city.
2000
2005
2010
2015
2020

FIGURE 3.10 Disaster recovery construction process of some banks.

(1) How to design disaster recovery architecture and backup data between two data centers?

The core factor of backup strategy depends on the data synchronization ability. It is hard to guarantee the real-time performance of long-distance data transmission. There is some delay in the data between the two data centers. How to back up data between two data centers, by the business system or by the basic platform, is a question?

(2) How does the business system adapt to the DRP based the basic platform?

When developing and upgrading the bank's business system, developers pay more attention to the realization of business logic. They believe that the multi-center deployment and disaster recovery switching logic should be shielded by the cloud platform. So, what should developers notice when developing a business system?

(3) How to build disaster recovery platform?

In the process of data center disaster recovery drills, a unified scheduling platform is needed. Scheduling platform can manage and monitor the running state of multiple data centers, while covering the IaaS/PaaS/SaaS layer objects. But how to conduct unified scheduling for platforms and systems while involving multiple manufacturers?

(4) How to conduct disaster recovery drills?

After the construction of disaster recovery data center is completed, how to select the order of disaster recovery drill scenes?

Computer room power down, network interruption, some business system failure, database failure and so on? The business impact involved in each exercise scenario, the difficulty of disaster recovery switching and the recovery time are different, and the selection of the scenario is directly related to the exercise results.

3.3.6 Alibaba Cloud Disaster Recovery Architecture and Practice

The above business and technical requirements for disaster recovery are addressed by most agencies carrying out disaster recovery projects. Here we look at how Alibaba Cloud to handle them. After comparing the business characteristics inside and outside, Alibaba Cloud abstracts a generic disaster recovery architecture.

3.3.6.1 Application High Availability Service (AHAS)

One of the purposes of the Alibaba High Available Architecture is to address the following common disaster scenarios:

- Human error: The main scenarios include configuration errors and irregularities...
- Hardware failure: The main scenarios include switch failure and server failure, and cause application services affected.
- Network attack: DDoS and other network attacks.
- Disconnection: Optical cable is cut off.
- Power off: Power failure or voltage instability in data center.
- Natural disasters: Lightning causes power failure in the computer room.

To ensure HA of applications and business continuity, enterprises that use traditional application HA solutions need to make significant changes in their IT architecture. The long go-online duration and difficult construction and maintenance also drive many companies away. Moreover, traditional HA solutions used for physical machines and virtual clusters cannot meet the HA requirements of applications in distributed architectures. To address this issue, Alibaba Cloud has developed the Application High Availability Service (AHAS) as a new solution to improve the availability of applications in distributed architectures (Figure 3.11).

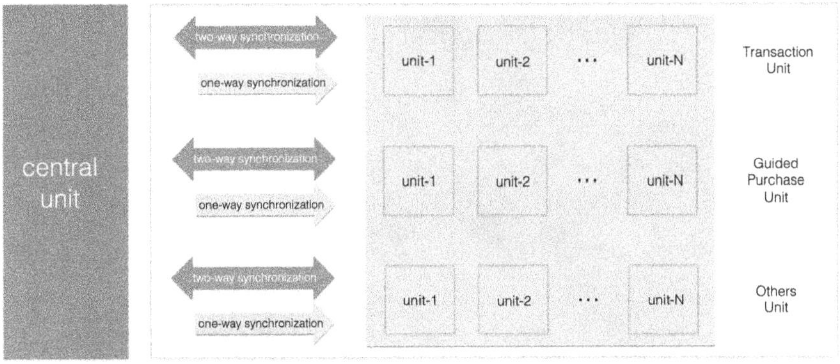

FIGURE 3.11 The architecture of AHAS.

However, AHAS is a very complex architecture and cannot be directly applied to most users. Therefore, a generic disaster recovery architecture is needed to help users with disaster recovery needs.

3.3.6.2 Common Disaster Recovery Architecture from Alibaba Cloud Best Practice

Alibaba Cloud generic disaster recovery architecture includes urban dual-center and remote dual-center. The two architectures have some common architectural principles, mainly related to:

- Flow control: GSLB (global service load balancing/Global Server Load Balancer) is used to control the proportion of traffic diversion to multiple data centers. GSLB monitors the state of the centers, and automatically or manually switches the flow from one center to another. Through the disaster recovery switching drill data of many banks, it is found that GSLB switching technology is widely used.
- Network connectivity: The centers are connected through optical fiber lines, and there are usually two lines for data synchronization between the two centers.
- Application deployment: Each application is deployed in two centers, offering different VIP addresses. When a datacenter fault occurs, another datacenter can provide full services.
- Local DC first: Usually complete all business logic processing in the local data center. The SLB (Server Load Balancer) of the business

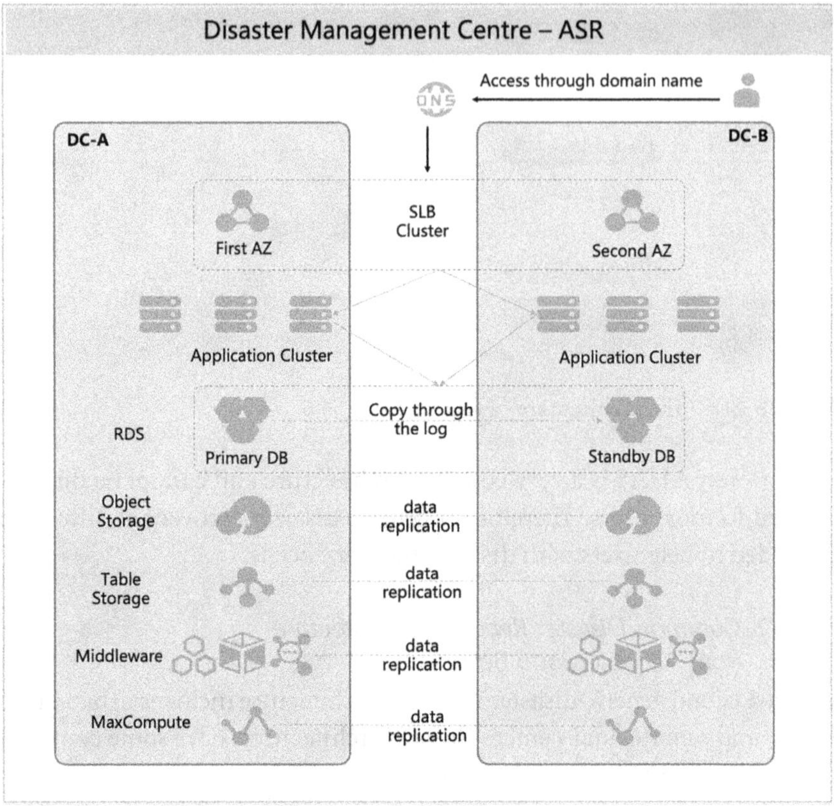

FIGURE 3.12 Disaster recovery architecture in the Same City.

entry mounts only the business host (ECS or container) of the local datacenter.

- Data access: Databases and storage products provide access domain names for users, and parse domain names from DNS within the cloud platform to the actual main instance.
- Backup: Inside the cloud platform, backup data are synchronized from the active center to the standby center asynchronously and periodically. Database access at the business level through domain names. Business flow does not need to process data synchronization logic in two centers (Figures 3.12 and 3.13).

Now, let's have a look at the difference between the urban and remote disaster recovery architecture through some metrics.

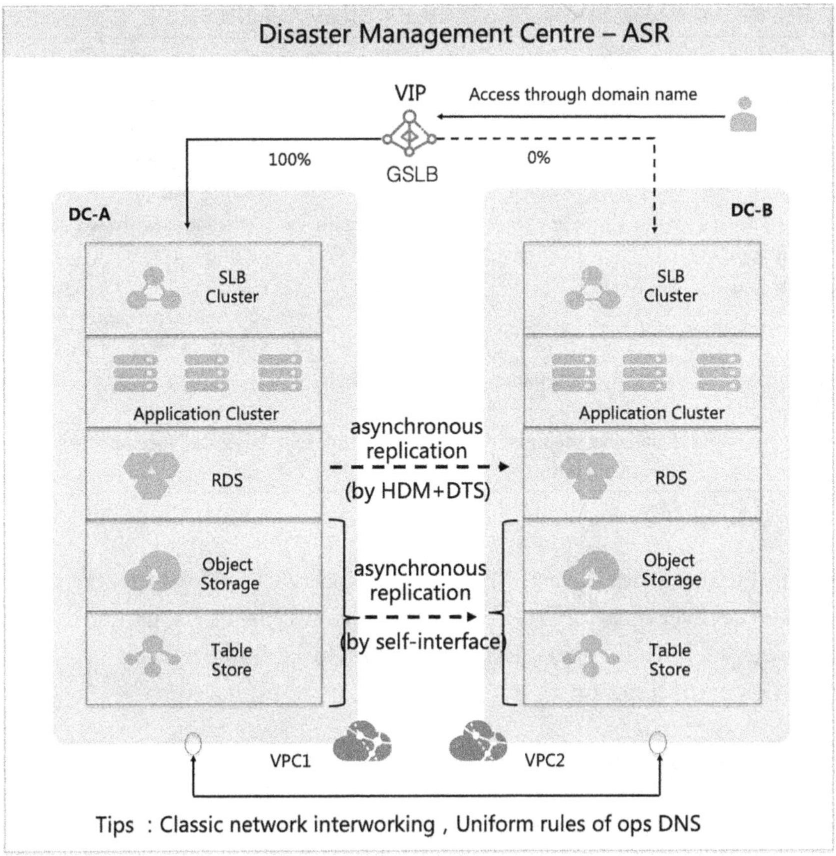

FIGURE 3.13 Disaster recovery architecture in a Different City.

The core factor causing the difference is geographical distance. Urban disaster recovery is to establish two datacenters in the same city or similar areas (≤100 km), and the distance between the remote disaster recovery data centers is longer (>100 km).

Due to the influence of long network bandwidth delay and stability in different places, the data synchronization timeliness of the two datacenters is worse than that of the same city, resulting in architecture differences. The following is a simple comparison between urban and remote disaster recovery architectures (Table 3.2).

When the business system completes the disaster recovery adaptation of the cloud platform, the business system can make full use of the disaster recovery ability of the cloud platform itself and reduce the complexity

TABLE 3.2 Difference between Two Disaster Recovery Architectures

Items	In Same City	In a Different City	Notes
Architecture	One cloud with two available zones	Two clouds with two available zones	One cloud needs low time delay
Failover and fallback	GSLB or SLB	Only GSLB	SLB cannot be deployed across two cities
Ways to access applications	Same domain name	Different domain name	Two clouds with two domain names
Active–active	Support	Non-support	Database doesn't support active–active mode in a different city
Ways to access data	Same domain name	Different domain name	Two clouds with two domain names
RPO	Database: seconds Storage: minutes	Database: minutes Storage: minutes	Long-distance network time delay between two cities

of the business-side disaster recovery. Alibaba Cloud proprietary cloud disaster recovery platform currently includes two products, named as ASR (Apsara Stack Resilience) and HAS (High Availability Service).

ASR's main functions:

- Supports the failover of cloud platform IaaS layer
- RPO monitor
- Supports the failover of datacenters level
- Five Built-in Disaster Scenarios which support one-click failover

HAS's main functions:

- Supports one-click failover of ANT FINANCIAL SERVICES GROUP's products
- Customizable DRP
- Supports scheduling process and calling for custom scripts
- Supports simulate fault and one-click execute DRP

From the architecture of the above two disaster recovery platforms, it can be seen that the disaster recovery of the basic products (including

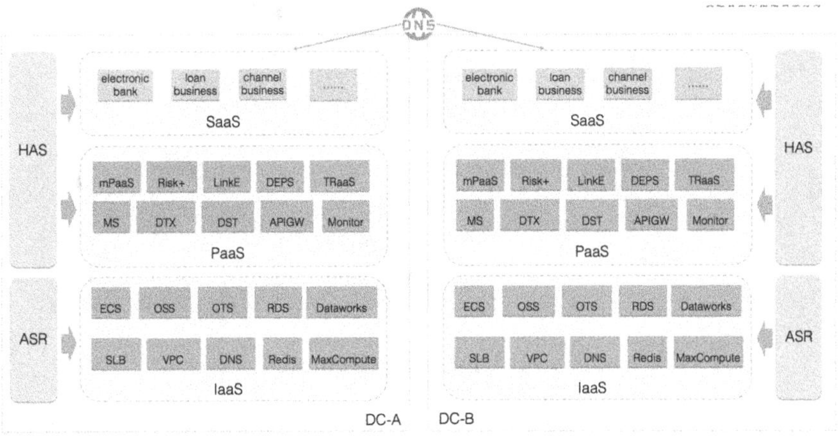

FIGURE 3.14 Disaster recovery architecture in financial industry.

computing, storage, network and database) of the cloud platform IaaS layer can be completed by ASR products, and the middleware product PaaS and the business system SaaS layer of the bank can be completed by HAS products. HAS can support disaster recovery switching capabilities at the business level. The core of HAS is to support calling user-defined scripts and calling process scheduling. Disaster recovery switching at business level often includes several aspects: system initialization, system state checking, system state change, system restart and host restart. These operations can be implemented by script, and then provided to HAS product calls.

From the architecture of the above two disaster recovery platforms, ASR supports the failover of the basic products (including computing, storage, network and database) at the IaaS layer. HAS supports the failover of middleware products on PaaS layer and banking business system at SaaS layer. Business layer disaster recovery often includes several aspects: system initialization, system state check, system state change, system restart and so on. These operations can be implemented by script, and can be called by HAS (Figure 3.14).

3.3.6.3 Application of Disaster Recovery Delivery Methodology

There are many business systems in banking institutions, which are generally developed by different third-party business system vendors. Developers mainly focus on the realization of business logic. In the process of disaster recovery scheme evaluation, we found that there are some

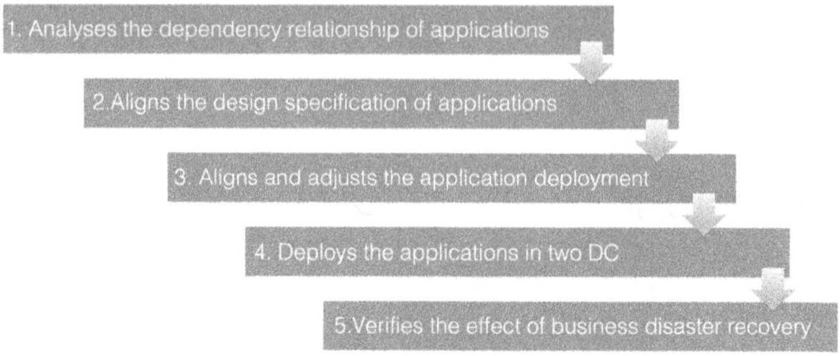

FIGURE 3.15 Adaptation process for business disaster recovery.

common rules that need to be followed by the business system to play the disaster recovery ability of the cloud platform. The business adaptation process is as follows (Figure 3.15):

- Disaster recovery is a systematic project, which needs to analyze the dependencies between applications, and all links should be considered into the list of disaster recovery.
- First confirms which business needs to do what kind of transformation, and then completes the transformation.
- First confirms the deployment mode, and then completes the deployment in the datacenters.
- Completes the business deployment in the datacenters.
- Performs and verifies the effect of disaster recovery architecture by drills.

Here are some general guidelines for implementing cloud disaster recovery based on cloud-native architecture.

(1) Guidelines for Application Development

- Stateless application: A stateless service does not rely on other requests for processing a single request, and the service itself does not store any information. Any business data that need to be persisted are stored in disaster recovery products such as RDS, Redis or OSS. It is not recommended to save business data (including logs) in VM memory or VM disk. It is recommended to export streaming data to persistent products (such as SLS), and do not recommend writing at local files.

- In distributed systems, design for failure is recommended. Applications need to have the ability to detect, capture and deal with network connection errors to cope with service switching in disaster recovery scenarios. Typical cases such as RDS database, when the switch between RDS primary instance and standby instance will have second connection interruption, application access to RDS will be abnormal. Moreover, when RDS completes the failover, the application may still be unable to access the database. The main reason is that the invalid connections held by the connection pool are not updated, resulting in the application error. Therefore, the application needs to have the function of reconnecting and retrying effectively to deal with network or service switching problems.
- Services can be retried: When calling happens in services, due to various failure reasons, the callers may retry. Therefore, the target system needs to support repeated processing of the same tasks and ensure the correctness of the results.
- Using cloud services: In principle, cloud products with disaster recovery are preferred, such as RDS, Redis and OSS. If cloud products fail to meet the needs, it is necessary to design and implement a new disaster recovery architecture. Disaster recovery architectures consider not only data replication, but also service switching and unified service interfaces.
- Local DC first: When the application system is deployed in the dual DC (datacenters), it tries to avoid cross DC calls to reduce the network delay. In microservice scenarios, calls between services are frequent. Therefore, the principle of local DC first is very meaningful to improve the system.
- Processing expired cache: Cache services (such as Redis) usually cache database query results to reduce database compression and response time. When encountering scenarios such as full storage space, cluster failure and cluster switching, data in the cache may

expire. Therefore, applications need to handle scenarios where the cache expires.

(2) Guidelines for Application Deployment

- Deploy key applications in all data centers. Deploy the key applications and dependent modules in all datacenters to ensure that applications in every datacenter provide complete services.
- Service other applications through SLB. SLB has automatic failover function, and can effectively resist the urban datacenters level and product level fault. After disaster recovery, the SLB instance can maintain the VIP of the service. It is not recommended to directly use ECS for external exposure services, because in the fault scenario, the ECS instance itself does not have the ability to provide services across datacenters.
- Access cloud products by domain names. Most cloud products (such as databases and storage) provide domain names. Applications need to use domain names to access these products rather than IP/VIP, and there is no need to worry about the cost of domain name parsing. Cloud products can keep domain name unchanged after change, upgrade or disaster recovery switching, but IP/VIP of some products will change. Therefore, domain name is more suitable for disaster recovery scenarios without application modification.

3.3.6.4 Selection of Disaster Recovery Exercise Scene

Through the above analysis, we can find that the common disaster recovery scenarios include: Switching a single cloud product or business system from active DC to standby DC. Switching all cloud product and business system from active DC to standby DC, and after a period of time in the standby DC, the flow will be switched back to the active DC.

There are many actual disaster recovery scenarios, and the disaster recovery strategy is not exactly the same. Common disaster recovery drill scenes include:

(1) Datacenter power off: Including production center power off, urban standby center power off and remote standby center power off. The power off scenario will involve business flow switching, data center

power recovery, all system restarts and flow back-off. The operation is complex and time-consuming, so few users will perform this drill.

(2) Data center access layer disconnection: Including production center disconnection, urban standby center disconnection and remote standby center disconnection. Since the disaster recovery datacenter is in operation, the business flow can be switched recovery of business. So, disaster recovery drills in the industry are mainly about this scenario.

(3) Disconnection between datacenters: Data backup, whether real-time or timing, depends on dedicated network connectivity between data centers. Long-term network interruption may lead to too much data difference and data loss, namely RPO>0. Therefore, the industry will carefully choose this exercise scenario.

(4) Faults of single platform or single business system: Mainly refers to the fault of a platform or business system in the active datacenter, which needs to be switched to the disaster recovery data center separately. This kind of drill scene is common. Through the intelligent DNS service of the access layer, the domain names of the fault platform and the business system are switched to the standby DC. At this time, the database and storage layer can be accessed normally without data loss, RPO=0.

3.3.6.5 Prospects for the Development Trend of Disaster Recovery Architecture

As the level of disaster recovery construction of banking institutions is very representative, we take the disaster recovery construction of banking institutions as an example and think that there are two obvious trends:

(1) Large and some medium banks: Most of them have completed the disaster recovery architecture of urban and remote. In the future, there will be 'active–active mode normalization', 'disaster recovery drills normalization' and 'core systems on distributed architecture'.

(2) Some medium and small banks: Most of them have completed the dual-center disaster recovery architecture, and some business will be run on a two-site three-center architecture in the future. They attempt to replace a centralized architecture with a distributed

architecture during new system development and old system transformation. Build and improve a unified disaster recovery management platform, and regularly carry out disaster recovery drills.

Based on the above analysis, Alibaba Cloud will continue to improve the disaster recovery of Alibaba Cloud products combined with user needs and technical development trends.

3.3.6.6 Alibaba Cloud Disaster Recovery Service Case

Since 2017, Alibaba Cloud has helped dozens of banking institutions to conduct disaster recovery exercises, including urban and remote disaster recovery. In 2020, several banking institutions carried out fault drills on Alibaba Cloud, including product failures (including ECS, RDS, OSS, Redis, MQ and DataWorks), network interruption and datacenter interruption power interruption. These exercises verify the availability of Alibaba Cloud disaster recovery architecture in banking business scenarios.

CHAPTER 4

Business Middle Office on Cloud

Qi Gaopan, Cao Lin, He Wei, Shi Jinliang,
Sun Zhongpeng, Zheng Fa, Yu Kang,
Song Wenlong, and Yuan Tieqiang

4.1 THE HISTORY AND BACKGROUND OF BUSINESS MIDDLE OFFICE

4.1.1 Definition of Business Middle Office

With the increasingly fierce digital wave, simply building the application cannot allow the enterprises to stay competitive with the ever-changing market. The traditional application construction basically follows the process: business department raises the requirements beforehand. Then the R & D department understands the requirements and completes the system design and construction. Finally, the IT department deploys the system to the production environment and works really hard to keep the system running as expected, to meet the business operation requirements. If you look from the surface, the traditional system construction has no problem at all, but if you dig deeper, below problems emerge:

- The R & D department provides relatively independent solutions for the always same requirements from different business departments.
- Since the traditional systems construction always built independent application for different departments, which in turn created the data silo and finally the data for each system are not unified.

DOI: 10.1201/9781003272953-4

- Hard to maintain.
- Poor scalability, unable to meet the flexibility of Internet business.
- Poor reusability, reinventing the wheel and really high costs.

Therefore, the enterprise application building requirement has shifted from simple system-level reuse to reuse of core capabilities. How to improve the flexibility of enterprise-level core capabilities, the reusability is the core demand of digital transformation and system construction to face Internet business challenges. To help enterprises address this issue, Alibaba has its best practice which is based on the advanced application infrastructure: Business Middle Office.

4.1.2 Alibaba's Digital Transformation and the Development of Business Middle Office

The concept of Business Middle Office came from Alibaba, which is a digital transformation methodology specifically for enterprise. Business Middle Office aims to improve business capability reusability and standardizes enterprise application from system design to business capability reuse, how to connect consumers, business innovation, and to design the enterprise-level digital platform to realize the data-driven business. Taking Alibaba as an example, Alibaba at very beginning centralized Taobao as the platform to link business and consumers. From that time, Taobao gradually built various reusable e-commerce capabilities, such as pre-sales, sales and after-sales capabilities. With the development of Alibaba e-commerce business, the capabilities gradually get abstracted and turned to reusable business capabilities, supporting the rapid development of Taobao e-commerce and Tmall thereafter (Figure 4.1).

Time to Market (TTM) and flexibility are two very important performance metrics for Business Middle Office platform construction. They can be understood from below aspects:

- Deployment flexibility
- Instant extendibility and scalability
- Resilience
- Accelerated iteration cycle

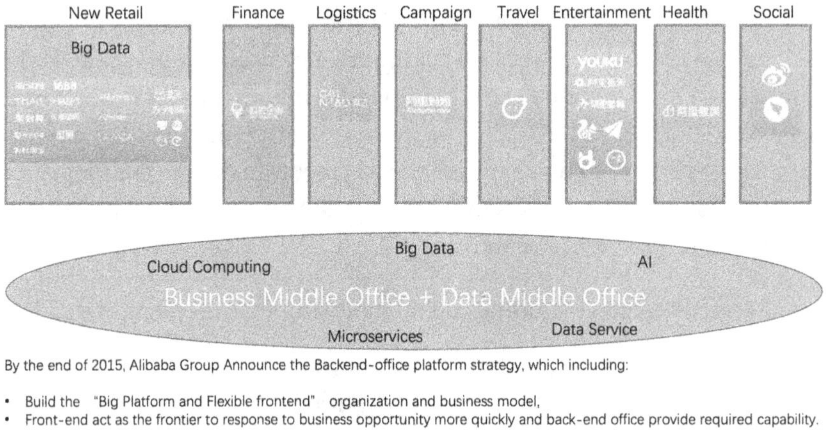
New Retail
Big Data
Finance
Logistics
Campaign
Travel
Entertainment
Health
Social
Big Data
Cloud Computing
AI
Business Middle Office + Data Middle Office
Microservices
Data Service
By the end of 2015, Alibaba Group Announce the Backend-office platform strategy, which including:
• Build the "Big Platform and Flexible frontend" organization and business model,
• Front-end act as the frontier to response to business opportunity more quickly and back-end office provide required capability.

FIGURE 4.1 Strategic schematic diagram of Alibaba Group Middle Office.

The microservice architecture basically decouples applications into isolated, well-designed microservices, and completely addressed the scalability and extendibility problems traditional monolith application struggle with. It is one of the best choices for building a Business Middle Office application among many other options. Apart from the benefits of microservice architecture, such as scalability, auto scaling and agile development of small-scale teams, microservice architecture does bring a lot of arguments and puzzles: should I put my function into microservice A or B? What is the difference between microservices and service-oriented architecture (SOA)? Is there a standard method for microservice boundary splitting? In a summary, the problem encountered by many teams is: how to define the business boundary and application boundary in terms of microservice architecture.

The Middle Office architecture is more focused on how to construct the business structure and determine the business model. The process of iterative defined the Business Middle Office platform is actually a process of continuous analysis and refinement of the business model. To make the microservices architecture really meet the enterprise objectives, it requires clear service boundaries and business boundaries. This gap can be bridged by domain-driven design (DDD) methodology. DDD, as a methodology, can help the business team to define the business model and finally define microservice boundary, which in turn the basic of Business Middle Office platform. These three concepts are perfectly combined and complement each other under the Business Middle Office umbrella.

Alibaba's Business Middle Office platform was built around trading business area, which includes all trading-related scenario associated with transactions. In this model, commodities are input and sold to customers through the orders. The trading process needs the functionality support from marketing, such as coupons, the payment, inventory and logistics. Therefore, a typical Middle Office application in retail industry usually contains multiple service centers.

- Member center, including the membership management, rights, etc.
- Commodity center, including the commodity management capability, like product-related information: product brands, attributes, categories, etc.
- Trading center, primarily includes shopping cart and order capability.
- Marketing center, including coupons and sales quota management.
- Inventory center, including inventory management, the primary capability like inventory lock, deduct, etc.
- Login center, including username and password, login capability, etc.

A typical retail Business Middle Office platform included business center and capability list as in Figure 4.2.

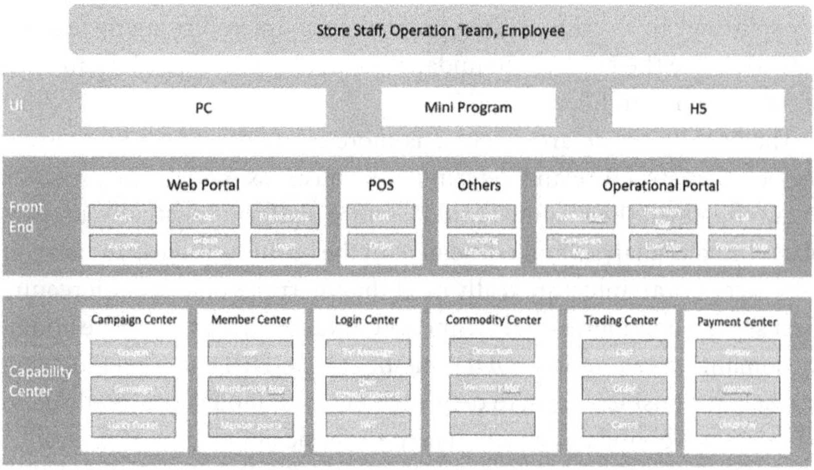

FIGURE 4.2 Big picture of retail service centers.

The member center serves the whole life cycle of customer and provides the customer lifecycle management and right management capability. The enterprise can build customer profiles around the member center and provide differentiated services for different users based on their profiles. The capabilities of the member center include:

- Membership joining
- Membership management
- Membership-level management
- Membership right management
- Membership points management

The trading center is the core of the e-commerce system. The trading center provides full lifecycle management capabilities of transaction orders, it start from adding products into shopping carts to generate actual orders, many e-commerce systems have the ability to split order into suborders, and the order and suborder can be paid, shipped, canceled, searched and finally completed. The capabilities of the transaction center mainly include:

- Shopping cart, including anonymous shopping cart capability
- Order creating
- Order searching
- Cancel order
- Order refund
- Order after sales, etc.

The commodity center provides the ability to manage the core data of commodities. The commodity center builds the ability to manage commodities and related data, including basic commodity information, commodity pictures, commodity categories, items and product attributes. The following are the capabilities of the product center:

- Brand management, including the items and attributes management of product

- Product master data management, including product CRUD capability
- Price management

The inventory center provides the whole life cycle of inventory management, including inventory query, inventory deduction, virtual and channel inventory management, physical inventory management, inventory reconciliation and other capabilities. The main capabilities of the inventory center include:

- Inventory deduction capability
- Inventory management
- Channel inventory and quota management
- Inventory history management
- Inventory reconciliation management

The marketing center provides full-blown management capability of coupons and promotions, including basic marketing-related capabilities such as coupon creation, verification, full discount, red envelopes and bonus pools. The main capabilities of the marketing center include:

- Coupon management, primary focus on coupon CRUD capability promotion rule management
- Quota management
- Red envelopes capability

The payment center provides standardize payment services, including payment pages, payment channels, and the integration capability from payment channel such as WeChat, Alipay, UnionPay and reconciliation capabilities. The capabilities of the payment center include:

- Payment capability
- Payment routing
- Payment status queries

From the above description, we can see that through the construction of the Middle Office system, we can summarize that multiple e-commerce systems run on a set of shared services and capability, because the Middle Office platform builds the core business capabilities of enterprises, multiple e-commerce systems can have higher reusability running on same inventory, payment and commodities services. Even the marketing and membership center, which usually needs to specialize capability, then can have at least 70% of the reusability, and it saves huge system construction and operation costs for enterprises, especially in Internet business area. Through the construction of Business Middle Office platform, it not only reduces the extra cost caused by repeated system construction, but also reduces the system silo and opening up data in various other business scenarios, reduces system operating costs, makes customer portraits more accurate due to data connectivity, increases traffic and promotes business growth.

Let's make a hypothesis. If we build a system through a traditional SOA, the final result is likely to be to a list of service systems and then integrate them together and exposed through by enterprise service bus (ESB), the composition of each system may have functional duplication (because it is not designed by business domain), and the scalability of the system is greatly limited because the existence of the service bus becomes a single point bottleneck. Therefore, the system constructed in this way is not Middle Office, but it is only the integration of all business capabilities and exposed them through interfaces.

Although the concept of Middle Office platform has not been emerged for very long, the Middle Office has already taken its own development path in a few years. Based on the experience of technology architecture evolution, the author believes that Middle Office architecture will continue to evolve and deepen in both breadth and depth.

First of all, from the perspective of breadth, because the core idea of Middle Office is to reuse the core business capabilities of enterprises, the type of Middle Office will appear in the area related to reusable capabilities; for example, from the very popular Business Middle Office to Data Middle Office. On the one hand, AI Middle Office, Mobile Middle Office, Technology Middle Office, Research and Development Middle Office, and even Cloud Middle Office with cross-cloud capability integration will evolve from a technical perspective; Organization Middle Office, Government Middle Office and Management Middle Office will quickly take steps as well. On the other hand, with the development of the enterprise and the further abstraction of the business, the sharing capability within a specific scope of Middle Office will develop as well, from the well-known user center, marketing center, transaction center, gradually expand to ticket center (replacing traditional production support system), content center (replacing traditional CMS), training center, etc. That is to say, if the company finds that the business in a certain direction is developing well, the relevant business can be integrated into the Business Middle Office-end platform, and other teams of the same company can reuse the business capability exposed through restful API, accelerate enterprise innovation and shorten the cycle of new product go to market.

From a deep point of view, the business capabilities accumulated by Middle Office do not match really well with the capabilities required by the front-desk business. That is to say, we cannot build a large and complete Middle Office, but we can deepen the capabilities built on Middle Office. As an enterprise sharing the capability, Middle Office platform needs to further abstract, model, solve common problems, and better serve different businesses. With the development of front-desk business and the use of shared services, it will reverse nourish the abstract business capabilities and promote a higher degree of business abstraction. The business scope of the Middle Office will become wider and wider. Because the better the abstraction is, the more flexible the development of business applications will be. Therefore, the deepening of Business Middle Office-end is another evolution direction for Middle Office to retain its value in enterprises.

What Middle Office platform to solve is the problem of how enterprises deal with changes, it's the core problem that enterprises must solve in the ever-changing market competition environment. Specifically, from the perspective of Business Middle Office-end platform, the market is ever changing, competitors are changing and business forms are changing, especially in digital scenarios, the business of enterprises is mostly reflected in the form of digital services, and the business traffic is also changing. Therefore, how to adapt to these changes better and quickly respond to the new emerging business opportunities, the core is to improve business capability reusability. Business capability reuse is the core of Business Middle Office-end construction. Therefore, the construction of Business Middle Office is essential to solve the problem of enterprises coping with changes.

From the perspective of Data Middle Office, enterprises need to deal with external changes and make decisions. The data needed for decision-making have never been well centralized, and some are always scattered in the enterprise; therefore, in the market, the customers change and the competitors change, and how these changes be provided to the decision makers of the enterprise is the key metric of Data Middle Office platform building. The core of the construction of enterprise Data Middle Office is to reflect the data changes scattered throughout the enterprise with unified enterprise core metrics, helping enterprises to solve the problems of changes. Therefore, the construction of Data Middle Office is essential to solve the problem of enterprises coping with changes.

At present, enterprises are in a changing market environment. Data Middle Office provides better decisions to cope with changes. Business Middle Office deals with changes through business reuse. In other words, Data Middle Office provides effective data metrics to help enterprises identify changes in a timely manner, and then they can effectively respond to these business opportunities. Business Middle Office provides tools to help businesses quickly respond to changes. One is to tell enterprises where to deal with changes, and the other is to provide enterprises with means to deal with changes.

The construction of Middle Office is a huge and systematic project, which cannot be accomplished overnight. It needs effective methodology and process support, and cannot be completed through traditional system construction methodology. Before entering the introduction of the construction process of Business Middle Office, let's answer the core question: Why Business Middle Office can bring value to enterprises?

4.2 THE VALUE OF BUSINESS MIDDLE OFFICE PLATFORM

Compared with the traditional application development, the construction of Business Middle Office applications requires much higher complexity and technical capability. I believe that most enterprises have heard of the Business Middle Office, but not all enterprises will soon enter the Business Middle Office construction cycle, because it is very important to clarify that whether the Business Middle Office application is suitable for the current problems faced by enterprises; whether building a Business Middle Office costs is cost-effective? Think clearly why building Middle Office platform is far more important than how to build Business Middle Office to the IT management team who are in enterprise. Only when the

IT management team has a clear understanding of the value that Middle Office brings to the enterprise, the core team could keep the strong will and determination even the building process is really hard, also could encounter many difficulties, they can overcome all difficulties, stick to the right direction, and gradually realize the value that the Business Middle Office platform brings to the enterprise.

Before introducing the value of Business Middle Office, let's take a look at digital transformation. Business Middle Office is an enterprise resource allocation methodology for digital transformation. By fully understanding the problems encountered in the digital transformation of enterprises, then we can fully understand why the construction of Business Middle Office is important to the digital transformation and why the construction of Business Middle Office is a necessary way for the digital transformation of enterprises.

4.2.1 View Digital Transformation from Middle Office

Let's first take a look of the definition of digital transformation. Digital transformation is based on Digitalization, and further touches on the core business of the company, which is a high-level transformation aimed at creating a new business model. Digital transformation is a process of developing digital technology and supporting capabilities to create a dynamic digital business model.

We are in an era of accelerating the integration of new technologies and breeding changes for industries. The new generation of science and technology driven by cloud computing, big data, artificial intelligence and the Internet of Things is developing really quickly, and it has become a powerful driving force to promote the development of economy, society, politics, people's livelihood and culture. The maturity of new technologies has also promoted several changes in traditional enterprises.

The first change is due to the booming development of the consumer Internet in the past 20 years, which has created a business model of Internet cloud computing, such as Alibaba, Tencent, Baidu, Byte-dance Toutiao, Meituan, and consumer Internet giants such as Didi have also created countless wealth deification, setting off a wave of digital transformation and development for enterprises to adopt new technologies and new businesses.

The second change is cloud computing. Cloud computing uses virtualization technology to achieve elastic scaling of resources, helping enterprises integrate resources and improve resource utilization. IT can flexibly expand/shrink resources to meet business needs according to business needs. In traditional IT platforms, data are distributed across business servers, and there may be a single point of security vulnerabilities. After the cloud system is deployed, all data sets are stored and maintained in the system memory.

The third change will take place in the industrial Internet era in the next 20 years. The consumer Internet market has become stable and matured. After experiencing barbaric growth, Internet companies use their huge consumption data and unique business models to cut into the industrial value chain, drive the transformation of the back-end supply side and form the consumer Internet (demand side) to drag the characteristics of industrial Internet (supply side). In the era of industrial Internet, enterprises must use cloud technology to build sharing platforms, realize online data and intelligent applications, and use data capabilities to improve service capabilities. Only in this way can they better serve customers and employees. Relying on digital transformation, the industrial Internet era has come to realize the deep integration of traditional enterprises and the Internet.

4.2.2 Characteristics of Enterprise Digital Transformation

The digitization of the enterprise takes the actual development stage and core business of the enterprise as the foothold, combinations of the core competitiveness for the enterprise, such as the enterprise's products, channel coverage, user range, as well as the development direction of the industry, the interconnection of the industry, relying on the advantages of the enterprise itself, capture the digital nature of the enterprise itself. What is the essence of enterprise digitization and how should we understand enterprise digitization? Based on the past experience in digital transformation and the experience of participating in Alibaba's digital transformation in the past few years, the author summarizes the essence of digital transformation as follows:

The first feature of enterprise digital transformation is data. In the era of digital economy which is fully based on the Internet, each of us is constantly generating data every day. These data can be generated in a large quantities of numbers, words, pictures, images via personal devices. We classify, analyze, summarize and apply data with the help of big data technology and algorithm models, let it become the most important asset in the digital economy era, namely digital assets. Transforming data into digital assets is the first feature of enterprise digital transformation.

The second feature of enterprise digital transformation: connection. If we observe the traditional business model and carefully observe how enterprises sell their products to customers, we will find several factors in the traditional business model: customers, commodities, enterprises and channels need to be connected in some way to complete the sales work of the traditional business model. For example, enterprises need to connect products and users through TV advertisements and telephone sales; enterprises need to connect products and channels through offline promotion meetings and expositions; sales channels need to connect channels with users by purchasing TV advertising periods and outdoor advertisements. Therefore, the core of business model is connection. For digital transformation, the nature of connection has not changed, but the means that enterprises can use are more abundant, and with the help of big data technology, through customer portraits, enterprises can accurately connect products with customers and give full play to the efficiency of marketing. In addition, through the interconnection of enterprise marketing, service, market, brand, design, research and development, production and other links, enterprises can optimize resource allocation and form an efficient enterprise value chain. Interconnection is not only the connection between different businesses of an enterprise, but also the internal interconnection of the enterprise itself, the interconnection between employers and employees, and the interconnection between employees and enterprises to improve the efficiency of enterprise operation. Finally, interconnection can be reflected as the interconnection of people and devices, such as wearable devices and monitoring of human health metric.

The third feature of enterprise digital transformation: wisdom. For enterprises, digital transformation continuously precipitates data through systems and continuously analyzes data through big data platforms, forming commercial insights and reports, the final solution is the ability to respond to changes in unknown business areas, including the intelligent operation capabilities of commodities and users within the entire ecosystem.

The essence of enterprise digital transformation is summarized in Figure 4.3.

Now all industries are integrating digital resources together to form a larger digital economy platform, providing more convenient, efficient, excellent connection service. Therefore, every enterprise should integrate its own digital resources and share them if possible, so as to obtain greater

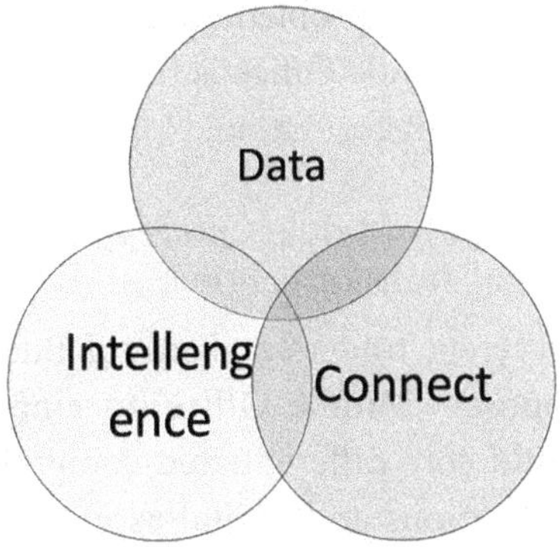

FIGURE 4.3 The essence of enterprise digital transformation.

value returns after connecting users through resource integration. For enterprises, digital transformation is not a question of whether to do it or not, but when to do it.

China's economic development has entered the digital fast lane, leading the innovation of the global digital economy. Every enterprise wants to take this digital transformation trend to win the competitive advantage on the road of innovation. Digital transformation is no longer indispensable. Every enterprise should realize that this is a survival problem related to the future development space of the enterprise.

Digital transformation involves the transformation of industry, ecology, enterprise and other management in the commercial environment. Thinking, innovation and business model determine the digital path of each enterprise. Enterprises will eventually move toward digital Capitalization. By then, the valuation of enterprises will not only be based on metrics such as income scale, but also on user data, commodity data, and, commercial value generated by transaction data, etc. Introduce digital technology to improve the business relationship of employees, customers, channel distributors, business partners, terminal stores or merchants, suppliers, different business partners and related digital interest chains in the industrial ecological chain.

In short, digital transformation is to integrate the application of digital technology into the internal management field of the enterprise and the external changing business environment, thus making decisive changes to the business value chain. Enterprise reform should be constantly innovated, and the original business processes should be digitized to explore new operation strategies. These problems are all the goals of the construction of Middle Office, so Middle Office is the only way for enterprises to make a successful digital transformation.

4.2.3 The Value of Business Middle Office Platform in Digital Transformation

Everyone has very different understandings of Business Middle Office. At first, Alibaba proposed 'Middle Office' to empower and output its own practices and build core differentiated competitiveness in the form of scenario-based applications and digital assets. Middle Office has just achieved the goal of 'enabling enterprises to operate in the business field by digital technology, supporting the rapid change and innovation of front-end business, and conforming to the business operation mode and value creation path in the digital economy era'. Middle Office is the foundation and guarantee of enterprise digital transformation.

After understanding the importance of Middle Office to the digital transformation of enterprises, finally let's talk about the value that Middle Office brings to enterprises. Figure 4.4 shows the value of Middle Office to different levels of the enterprise.

The existence value of Business Middle Office is to serve customers. There are several types of customers that Middle Office can serve.

1. To Management Team

   For enterprise managers, the Middle Office brings the improvement of core competitiveness to enterprises. Enterprise capability

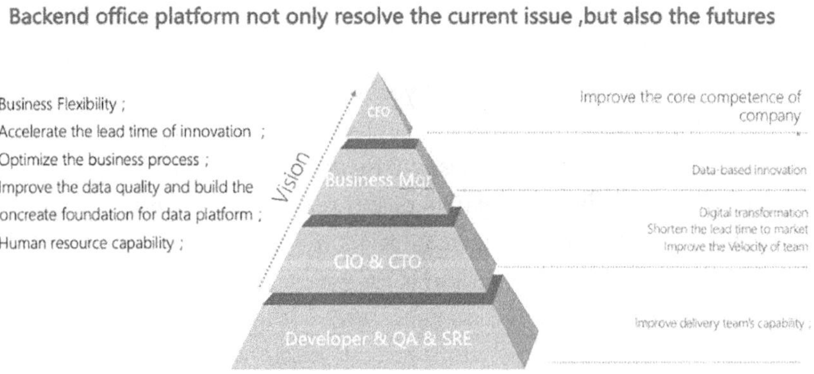
Backend office platform not only resolve the current issue ,but also the futures
✓ Business Flexibility ;
✓ Accelerate the lead time of innovation ;
✓ Optimize the business process ;
✓ Improve the data quality and build the concreate foundation for data platform ;
✓ Human resource capability ;
Vision
CEO
CIO & CTO
Developer & QA & SRE
Improve the core competence of company
Data-based innovation
Digital transformation
Shorten the lead time to market
Improve the Velocity of team
Improve delivery team's capability ;

FIGURE 4.4 The value of the Middle Office to the enterprise at all levels.

service will make enterprise business more agile and efficient. Especially for traditional enterprises in the business transformation period, Middle Office brings not only the service of core business capabilities, but also lower cost business exploration to complete business transformation for enterprises, find the star products and strategic direction of the enterprise's development in the next 10 years. In addition, the Business Middle Office has a complete design. Data can be precipitated with higher quality, laying a solid foundation for the construction of data-driven business and even Data Middle Office.

2. To Employee

For the employees of the enterprise, the construction of Middle Office requires compound innovative talents, and comprehensive talents who have in-depth understanding of business, technology, commerce, data capabilities, cloud computing, architecture design and coding, the construction process of Middle Office is also an excellent opportunity to participate in the development of individual employees in all aspects above, which has trained enterprises to cope with the rapid changes of business in DT era, the booming development of cloud computing and big data has helped the entire stack of business talents.

Enterprises have been shouting for many years to improve quality, transform, reduce costs and increase efficiency, from the enterprise core competitiveness, talent training and enterprise strategic development direction, combined with cloud computing, big data, artificial intelligence and other technologies, Middle Office construction gives the optimal solution to this problem. Enterprises need digital transformation to enhance their competitiveness in the DT era. The construction of Middle Office platform is the only way for successful digital transformation of enterprises.

3. To End Customer

For profit purposes, enterprises can significantly improve the operation and maintenance efficiency of enterprise IT systems through the construction of Business Middle Office platform. Especially in the DT era, Business Middle Office can effectively alleviate the uncertainty brought by Internet applications, which is mainly reflected in:

First, the traffic is uncertain. Uncertain traffic causes two problems. Enterprises invest huge amounts of resources to cope with the sudden surge of traffic, but the traffic is not high during the promotion, resulting in waste of resources; Internet viral marketing causes a large number of users to rush into the system instantly, causing urgent system resources, downtime causes service interruption and affects the end-user experience.

Second, the uncertainty of the market requires that the products and services of enterprises have enough TTM to seize the fleeting business opportunities. For example, a competitor withdraws from a certain product portfolio, which is an important competitive product of the enterprise. The marketing and marketing departments quickly follow up and give a business promotion plan, and IT and digital departments are required to provide support applications immediately to assist in marketing and sales to serve loyal customers of enterprises.

The uncertainties described above can be solved through the construction of Business Middle Office. Because Business Middle Office aims at improving the business capability of enterprises, enterprises can cope with these uncertainties with ease, provide users with a stable, secure and consistent service experience.

After introducing the value that Middle Office platform brings to the enterprise, as a reader of this book, you are probably unable to resist the impulse of start the building right away. We are eager to know how to build Middle Office platform. Don't worry. Let's talk about the history of Middle Office evolution from the perspective of technical architecture development, or let's answer how to move to the Middle Office model step by step.

4.3 BUSINESS MIDDLE OFFICE EVOLUTION APPROACH

Middle Office originates from the platform, but Middle Office is different from the platform; from the perspective of enterprise IT strategy, Middle Office is higher than the platform. Platform construction has been practiced in many enterprises for many years, and there are also many mature implementation experiences. But I believe that when many people first come into contact with the concept of Middle Office, there must be various doubts in their heads: Isn't this another platform construction plan? Come again to collect the IQ tax of the enterprise, change the soup without changing the medicine? Before learning more about the construction of Business Middle Office, let's talk about where Business Middle Office comes from and where it will go.

4.3.1 Alibaba Business Middle Office Strategy

The concept of Business Middle Office originates from Alibaba's Middle Office strategy. On 2015, Alibaba Group announced the launch of the Middle Office strategy. The core points of the Middle Office strategy are as follows:

- Construct DT era of more innovative, flexible 'big Middle Office platform, small front platform' Organization mechanism and operational mechanisms.
- Front desk as the front-line business is more agile and faster to adapt to the market. Middle Office will integrate the digital operation capability and product technical capability of the whole group to form a strong support for front desk of each business.

The Middle Office strategy has provided solid support for Alibaba's development in the following years, integrating the digital capabilities of the entire group and accelerating business innovation. Of course, Alibaba is not building a core system in the direction of Middle Office from the beginning. It is the result of continuous exploration and attempts by Alibaba employees in the systematic construction over the past decade or so. Figure 4.5 shows the development path of Alibaba Middle Office.

As shown in Figure 4.5, Alibaba's system construction is similar to most of the enterprises. Starting with a simple and fast LAMP architecture, the database uses MySQL, the read/write separation feature of the database is adopted to improve the performance of the database. With the continuous development of Alibaba's business, especially the increasing number of users of Taobao business, single applications built by PHP can no longer meet the needs of business development.

Around 2004, Alibaba introduced the Oracle database suite to improve data reliability, and replaced PHP in LAMP architecture with Java to improve the flexibility and robustness of system development, in addition, the search module is provided separately to improve the overall scalability

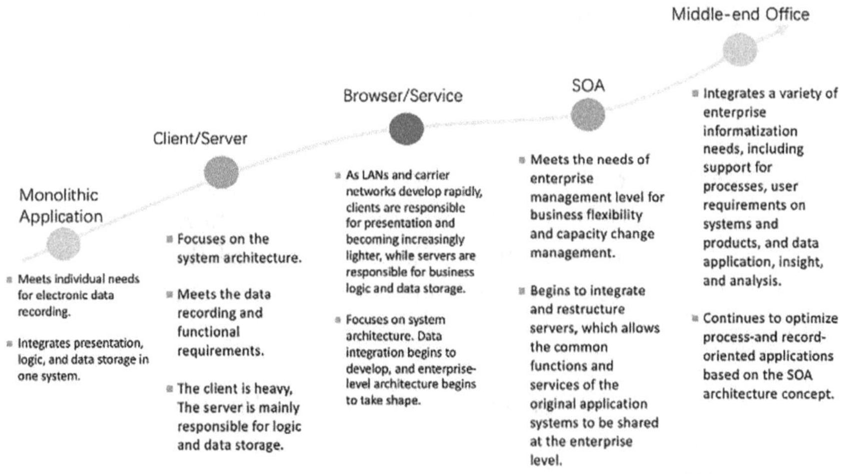

Middle-end Office
Integrates a variety of enterprise informatization needs, including support for processes, user requirements on systems and products, and data application, insight, and analysis.
Continues to optimize process-and record-oriented applications based on the SOA architecture concept.
SOA
Meets the needs of enterprise management level for business flexibility and capacity change management.
Begins to integrate and restructure servers, which allows the common functions and services of the original application systems to be shared at the enterprise level.
Browser/Service
As LANs and carrier networks develop rapidly, clients are responsible for presentation and becoming increasingly lighter, while servers are responsible for business logic and data storage.
Focuses on system architecture. Data integration begins to develop, and enterprise-level architecture begins to take shape.
Client/Server
Focuses on the system architecture.
Meets the data recording and functional requirements.
The client is heavy, The server is mainly responsible for logic and data storage.
Monolithic Application
Meets individual needs for electronic data recording.
Integrates presentation, logic, and data storage in one system.

FIGURE 4.5 The development path of Alibaba Middle Office.

of the system. With the further expansion of Alibaba's size, service, platform and unit have gradually become the main direction of Alibaba's system construction. In the process, Alibaba has gradually developed the distributed application platform HSF, distributed data access layer, infrastructure platform to provide flexible infrastructure to various services, distributed database Ocean Base to meet the requirements of distributed systems for databases, as well as multi-data center disaster recovery, gradually upgrade Alibaba's core system from a single application to a distributed system application that supports multi-data center disaster recovery deployment.

In 2015, Alibaba formally proposed the Middle Office strategy to build a more innovative and flexible 'big Middle Office, small front platform' organizational mechanism and business mechanism in line with the digital age, the front-line business as the front desk will be more agile and adapt to the ever-changing capital market more quickly, while the Middle Office will gather the product technical capabilities and operational data capabilities of the whole group, strong support for front-end business. Let's take a look at Alibaba's definition of Middle Office: Middle Office is a basic concept and architecture. All basic services are built with the idea of Middle Office and connected the business side of the public support enterprise. Therefore, the essence of middle-end is to refine the general requirements of each business version of the enterprise, abstract the business and system and form a general reusable business model.

4.3.2 Evolution of Business Middle Office

Based on the introduction of Alibaba's Middle Office development path, we can roughly see the evolution phase of Business Middle Office platform, which is mainly divided into the following four phases: monolith application, SOA, microservice architecture and Middle Office strategy.

4.3.2.1 Monolith Application

A monolith application is to package all components involved in a system into an integrated structure for deployment and operation. In the field of Java EE, the integrated structure is often embodied as a WAR package, where various application servers represented by Tomcat or WebLogic are deployed and run. For a typical e-commerce application, the following architecture diagram outlines the details.

As Figure 4.6 states, in a monolith architecture, users access applications through browsers. When users log in, they call the service of the

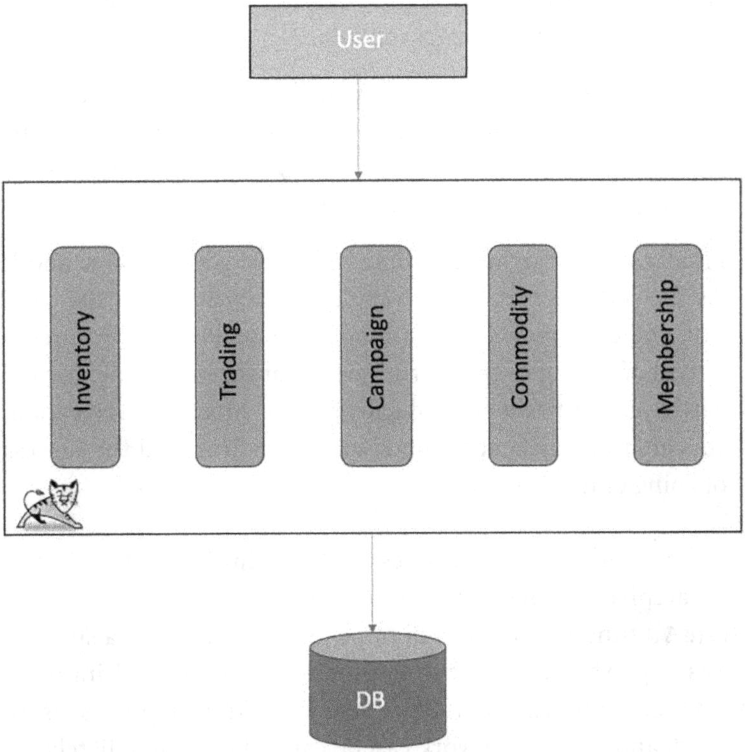

FIGURE 4.6 Monolith application architecture.

member center, and the member center accesses the database layer, the database layer performs operations on data. When a user accesses the transaction center, the service of the transaction center then accesses the inventory center. The inventory center accesses the database to obtain the deductible inventory and deduct the inventory.

This is a typical single application architecture, and I believe everyone is familiar with the problems it brings:

- First of all, the database has a single point. All business databases are stored on the same instance. When a database instance fails, the entire application is unavailable and the availability is very low.
- Second, the scalability is limited. All functions and services of the entire single application are coupled to the same application. When one of the applications needs to be updated to a new version, the entire application needs to be re-released, unable to scale according to the needs of the business module.
- Third, the complexity is high. A single application project contains a lot of modules. The boundary module of the module has unclear dependencies, which affect the whole body. The code quality is uneven, and the maintainability is poor.
- Finally, the deployment evaluation rate is getting lower and lower. As the code increases, the complexity increases, and the construction and deployment time is long. Each deployment requires a long period of testing to ensure the quality, at the same time, because the entire project needs to be deployed each time, the verification process after deployment also takes a longer time, and the success rate of going online is not high.

As usual, everything has two sides, and the single structure is not useless. An archive package contains all functions, which are easy to share and test. All functions are in the same application. Once a single application is deployed, all services or functions can be used immediately, because there is no external dependency, the testing process is greatly simplified, and the testing work can be carried out immediately. All the code of a single application is in the same application project. This mode can effectively improve the efficiency of development, testing, deployment and operation and maintenance in the early stage of the project. A small number of developers can complete end-to-end support for the entire project. However, with the expansion of business scale, the improvement of business responsibility and the continuous expansion of data volume, the shortcomings mentioned above for single applications are becoming more and more obvious. To solve the problem that business growth and single system capabilities cannot match, a SOA has emerged.

4.3.2.2 Servitization and SOA

SOA associates different common functional modules in applications through service-defined interfaces and contracts, that is, by splitting services and databases, improve application scalability to support growing business scale. Interfaces are defined in a neutral manner, independent of the hardware, software and programming language implemented by each interface. In this way, services built in various systems can interact in a common way. Figure 4.7 shows a typical SOA.

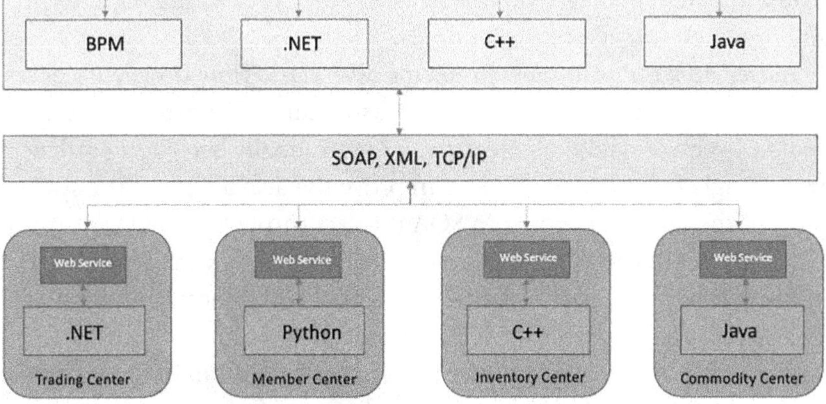

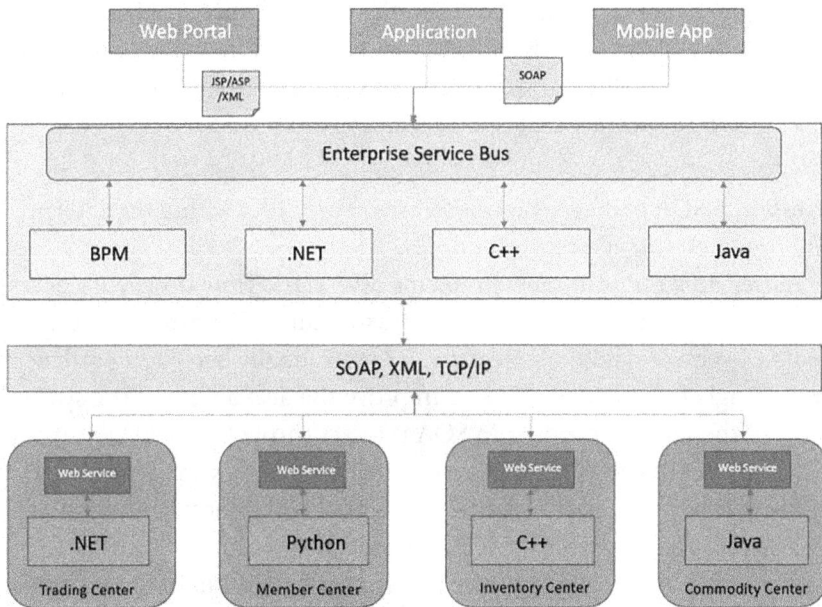

FIGURE 4.7 SOA application architecture.

The SOA has the following advantages:

- Service-oriented system design: The SOA splits huge business systems into highly cohesive service units, and each unit provides independent service capabilities, services and services achieve business value through collaboration.
- Loose coupling: The SOA can apply a variety of technologies. For example, the transaction center and commodity center in the preceding figure use different technology stacks, respectively.
- Clearly defined interfaces and stateless service design: The reliability of the system depends on the characteristics of the external network, and the service is designed as an independent and self-contained functional module.
- Cross-platform, cross-language, service layer can be implemented in any language to make the best use of everything. For example, Java is good at core business processing, C++ is good at high concurrency, DOTNET is good at mathematical computing and Python is good at big data processing. However, the interfaces provided are unified. For the caller, it does not need to be related to the specific implementation technical details of the server, but only cares about the interface contract that can be accessed by the ESB.

In essence, SOA manages various service capabilities within the enterprise and provides unified access standards.

As services are split, new problems also arise. How do client's access services expose through the service bus (secure)? With more and more complex services and more visits, ESB has gradually become a bottleneck and a single failure point. How to improve the availability of the system? Most of the services exposed in SOA interact through the SOAP protocol (SOA protocol). How to better support the mobile terminal of core traffic in the Internet era (note: The SOAP protocol is very heavy, generally, the mobile terminal supports the REST protocol)?

The purpose of SOA is to make it easier for all systems of the enterprise to integrate together and form a unified service for consumers. In other words, SOA focuses more on poor services, instead of the vertical service itself. In the SOA design scheme, the service exposed to the caller through the ESB is the integration of multiple levels and autonomous services at the underlying level. The ESB is more concerned about whether the exposed service contract is reasonable, whether it can meet the business requirements of the caller, rather than the underlying service itself. In the service layer, when designing SOA, we prefer to layer services. For example, each service has a service layer, a business logic layer and a data access layer. After the designer designs the contract, developers develop according to the contract and finally perform integration tests. Generally, this process takes less than half a year and more than 1–2 years to go online. It is a typical top-down development model.

With the vigorous development of the Internet industry, enterprises are striving for their own development and transformation under the Internet environment overnight. In 2015, the State Council issued the Guiding Opinions on actively promoting Internet+actions. Internet application has officially become a major strategic decision in China's economic field. So, let's discuss whether the SOA is suitable for supporting Internet scenarios?

Back to the SOA diagram I gave above, each call to the service exposed on the ESB needs to be routed through the central ESB. For example, when the portal website calls the create order service, each website calls the ESB (once), the ESB calls the inventory service (once), the ESB calls the order service (once), and eventually three service calls occur, all of which are driven by the ESB. From the call logic, the access and computing pressure of the service bus will be very, very large. Therefore, all ESBs must be deployed in clusters to support business peaks. ESB generally contains many functions, such as service discovery, registration, routing and interface listening. It requires high performance server and requires purchase of servers which would result in large infrastructure cost, deployment, installation and deployment, low resource input and output ratio, low operational efficiency of the entire IT industry, especially in the face of uneven access traffic in the Internet industry, there may be a blowout at any time, the SOA seems to be powerless.

The service system based on SOA has become the core hub of the entire enterprise service scheduling. During the construction, it mainly supports system calls within enterprises. However, in the Internet era, these services (such as the order service shown in the preceding figure) need to be open to Internet users, and more traffic will enter the services exposed on the ESB, and more service calls bring more service call pressure to the ESB. With the change of the external environment, the development of internal services will also increase the traffic volume of services. It is difficult to estimate the maximum peak of traffic when the two are combined. For example, a medium-sized retail enterprise estimates that it needs to support up to 10,000 concurrent users and needs 10 machines to distribute routes to the underlying services. During peak traffic hours, the resource usage of each machine reaches 80%. However, when access traffic comes, a server may cause service exceptions due to hardware reasons, resulting in one of the 10 machines being unable to provide services. The problem is, when all the traffic reaches the remaining nine machines during peak hours, the water level of the load will soon exceed 90%. What's worse, if one of the nine machines has another problem, in an instant, the remaining eight machines were completely full, and the entire ESB was washed down by traffic. This is the typical avalanche effect; because one server had a problem, the entire platform is down.

When such problems occur on the ESB, the service recovery time and cost are very high. The traditional server restart cannot completely fix the issue, because once one server is started, the server would be crashed instantly due to large pending requests, you need to cut off front-end access (the entire system is announced), start all 10 servers, and then test that there is no problem and develop traffic. Most of the time, the root cause of server downtime cannot be quickly located. Such a system is fragile and may avalanche again.

Based on the above two points, the scalability of the ESB SOA cannot be linearized, which makes it difficult for applications deployed centrally to meet the scalability, stability and robustness requirements of Internet services.

4.3.2.3 Microservice Architecture

Microservice (Microservice architecture) is an architecture model that advocates dividing applications in a single architecture into a group of small services. Services coordinate and cooperate with each other to provide users with ultimate value. Services are usually built around business capabilities and can be automatically deployed and released through automatic deployment tools such as cloud effect and Jenkins. The microservice architecture model can also be called an architecture concept. Martin Fowler put forward the microservice model as well as the design principles of microservices, and it provides technical guidance for system designers to implement the concept.

The implementation of microservices requires the following items:

- Computing and storage resources be allocated quickly.
- Whether the application is capable of rapid deployment. Applications are usually composed of multiple microservices. Services are independently developed and deployed. Therefore, whether in a test environment or a production environment, both of them need to be capable of rapid deployment to meet the requirements of rapid iteration and independent deployment of multiple services.
- Monitoring requires networks, resources, access links, etc.
- Standardized RPC.

Below is a typical microservice based architecture diagram:

As shown in Figure 4.8, services communicate with each other through REST API. Each service has a clear boundary and the client can access microservices through gateways. The microservice architecture solves the problem that software cannot quickly respond to requirements and business changes under single architecture and SOA. Especially in the Internet era, services need scalability and elastic scaling, with the features of agile development of small teams, microservices can be said to be born at the right time.

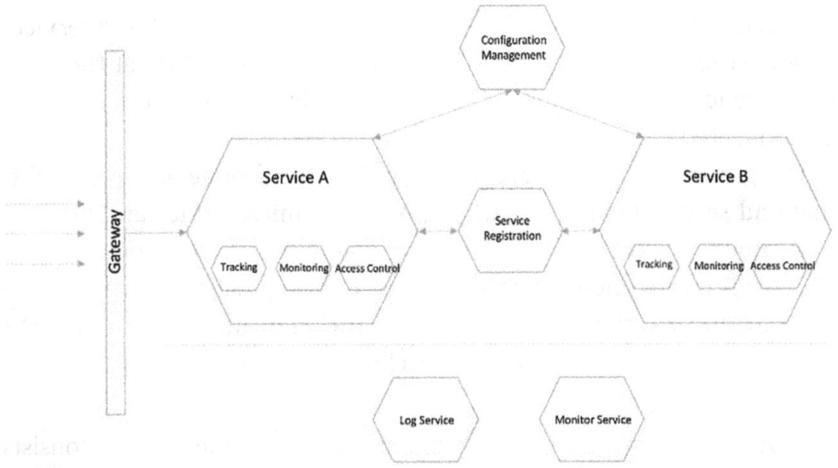

FIGURE 4.8 Typical microservice architecture diagram.

Everything has two sides. While enjoying the benefits brought by microservices, many enterprises also encounter many problems in the actual implementation of microservices: how microservices are? How to design the boundary change of microservices? What dimensions should microservices be split? In the microservice architecture, how to track the access link to improve the efficiency of the O & M team in solving problems? It can be said that for a long time, these problems have no good theory and practice. Because of this, many people in the industry have some misunderstandings about microservices:

- Some people think that Microservices are the upgraded version of servitization, which is to further split each service in the SOA, making the service smaller and easy to develop and manage separately.
- Also, some people think that microservices replace the original single architecture with the technical framework based on microservice architecture, and the smaller the service is, the better.

Apparently, we can find many loopholes to refute the above two points. For the microservice architecture segregation, even ThoughtWorks, the chief scientist of Martin Flol`wler, the proposer, did not tell us how to split it. In my opinion, in essence, Microservices are an evolution of SOA, which is not essentially different from the service-oriented idea of SOA. In other words, Microservices are the product of SOA development, it is a modern and fine-grained SOA implementation method. Microservices no longer emphasize the heavy ESB in the traditional SOA, at the same time, the idea of SOA enters into a single business system to realize real componentization.

It is precisely because Martin Fowler did not elaborate on service splitting and service monitoring when proposing microservice architecture. Practitioners of each microservice architecture have their own unique understanding of microservices, in the process of practicing the microservice architecture, the author agrees with the following description of the typical features of the microservice architecture Martin Fowler:

- A system consisting of distributed services: The system consists of multiple microservices with clear boundaries between services. Each service and its combination constitute the external functional features of the system, instead of building services based on the 'centralized' ESB in the traditional SOA.

- Divide and organize services by business rather than technology: The microservice architecture splits a single business system into multiple services that can be independently developed, designed, run and maintained. The services focus more on capabilities, performance and security, but don't care much about the specific underlying implementation technology.
- Independent deployment and flexible expansion: The traditional single architecture is deployed in units of the entire system, while Microservices are deployed in units of each independent component (such as order service and commodity service).
- Automated operation and system fault tolerance: Applications under the microservice architecture need to achieve high platform availability and stability. Automated O & M can automatically expand resources to cope with system traffic blowouts. When a service of the system fails, the traffic scheduled to the response node can be automatically cut off, which is transparent to the user side and does not affect the user experience.
- Fast service evolution: A good architecture is not designed, but evolved. This principle also applies to microservices. No one can design a system that will never go wrong from the beginning. Admitting the limitation of cognition is the first principle of microservice design. After the service goes online, due to cognitive limitations, the service needs to evolve and gradually move from semi-stable to relatively stable (the service does not need to be stable, but needs to be constantly nourished, different scenarios are required for verification and precipitation, so it is only a relatively stable state).

Microservices are not free lunches. There are intricate call relationships between services and a large number of services. How can we provide stable system services in scenario of sales promotion, to support high volume system pressure and hardware failure, this challenge needs to be considered not only in terms of technical operation and maintenance, but also in terms of service design, microservices should be selected at the appropriate granularity. Alibaba's transformation from traditional application architecture to today's shared service system architecture is essentially a microservice architecture construction process. This is not only the evolution of technology architecture, but also the result of continuous business evolution. If an enterprise wants to build a microservice system architecture, it should not rely on one or two projects to achieve results immediately. As the saying goes, Rome was not built in 1 day. It requires the cooperation of enterprises from business to technology and continuous accumulation, only by depositing the core business of an enterprise into microservices one by one can an enterprise truly see the value of microservices, which is the core of the strategy of Middle Office.

4.3.2.4 Middle Office Strategy

At the end of 2015, when most enterprises were busy making annual summaries, Alibaba announced that it would fully launch Alibaba Group Middle Office strategy in 2018 to build a more innovative and flexible 'big Middle Office' in line with the DT era, small front platform 'organizational mechanism and business mechanism'. As a front-desk business, it will be more agile and capture rapidly changing market opportunities. As a back-end of Middle Office, it will gather the business and data capabilities, and provide more powerful support for front-end services.

To understand the strategy of Middle Office, we should start with the practical challenges faced by enterprises. Informatization gives enterprises a taste of the benefits of system construction. Many companies' information systems, such as the bamboo shoots after the rain version, break through the ground. According to their own pace, the informatization department receives requirements, analyzes requirements, designs software, development and testing, online maintenance, this process is timely for most of the time, but when the business volume is heavy, due to resources and scheduling, it cannot meet the urgent requirements of the business for the system, therefore, a system called 'shadow IT' appeared. Business departments themselves built Information Management applications such as user management and sales management through tools in their hands. Because these applications were not completely designed at the time of development, the enterprise was filled with multiple systems with similar functions after many years, we call this type of system construction 'chimney' system construction mode (Figure 4.9).

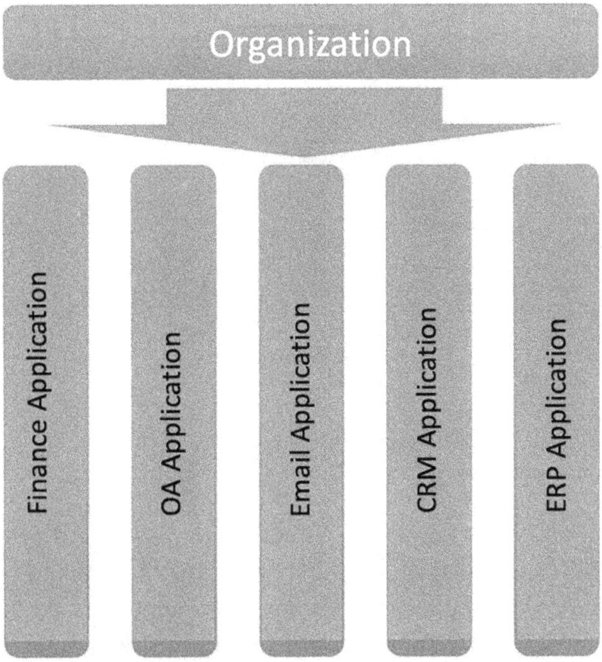

FIGURE 4.9 'Chimney style' system construction.

The construction of chimney system has a specific historical background. Building the system in this way will bring the following disadvantages to enterprises:

- Duplicate investment caused by repeated function construction and maintenance
- The cost of breaking the isolated chimney system is high
- Poor user experience due to inconsistent customer data
- Unsustainable business development

Business supporting has always been the core function of enterprise informatization construction. It is the top priority for enterprises to form differentiated competition to precipitate the core competitiveness of enterprises in a new and comprehensive form with the continuous development of business. The strategy of Middle Office uses scenario-based application

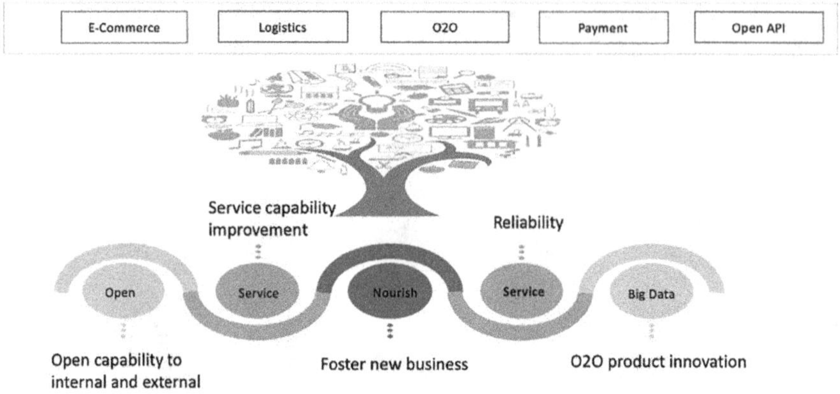

FIGURE 4.10 The comprehensive support and enablement of the Middle Office strategy to the enterprise.

and digital assets to build core differentiated competitiveness, which exactly achieves the goal of 'enabling enterprises to expand and operate their business with digital technology, and supporting front-end businesses to respond to markets and innovations quickly, meet the requirements of the DT era for commercial operation mode and value creation' (Figure 4.10).

Therefore, enterprises no longer need to consider whether to build Middle Office, but when and how to build it. Middle Office is the foundation and guarantee of business operation in DT era.

4.3.3 The Relationship among Business Middle Office Platform, Microservice and DDD

Before introducing the business intermediate construction, I would like to talk about the relationship between microservices, DDD and Middle Office. Microservices are technologies and frameworks at the implementation level of Middle Office. The Middle Office architecture is more inclined to how to construct the business, or how to determine the business model. The process of traveling the Business Middle Office is actually a process of continuous analysis and refinement of the business field. The construction of microservices requires clear service boundaries and business boundaries. This gap can be bridged by DDD methodology. DDD, as a methodology, can guide business modeling and microservice construction of Business Middle Office at the same time. The three methods are perfectly combined and complement each other.

Finally, Middle Office is an evolutionary form of SOA, which is more oriented to the needs of digital transformation of enterprises and focuses on processes, user interaction and Organization Effectiveness Department. Microservices are a way to implement the Middle Office architecture.

4.4 CONSTRUCTION METHODS AND DIVISION IDEAS OF BUSINESS MIDDLE OFFICE

4.4.1 General Introduction

The construction of Business Middle Office is an important link in the digital transformation of enterprises. The transformation of Middle Office can help enterprises design the architecture of the current information system, define the organization, and implement technology and plan Middle Office operations, with the input of current business needs as the business development direction, and support flexible business development as the goal. As a practice of software engineering, the requirements for the complexity, invisibility, compliance, and change of software itself also need to be planned and responded to in Business Middle Office projects.

The construction cycle of Business Middle Office projects generally varies from half a year to a few years, involving large-scale personnel and resource investment of enterprises. Business Middle Office is the construction method directly affects the scalability of the final system. The author summarizes the Business Middle Office planning ideas through several successful practices in medium-sized projects and further verifies and improves them in the projects. Get good results. This chapter briefly introduces the methods applied in the construction process with specific cases (Figure 4.11).

The overall planning and design of Business Middle Office are divided into four levels:

The first level is the product strategic planning. Product strategy is formulated according to the customer's needs and the development trends of competitors, including the portfolio of investment products and the investment process of investment personnel. Product strategy determines the combination of products, and the combination of products also determines the collection of use cases.

Starting from the overall product strategy, we can refine the needs of end customers and the dynamics of competitors' products, and finally form a product portfolio plan. Describe the collection of use cases in

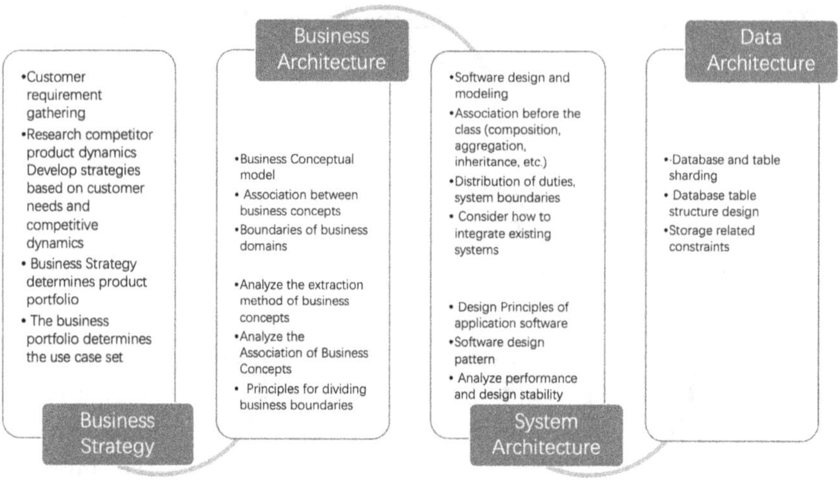

FIGURE 4.11 Four phases of overall planning and design of Business Middle Office.

formatted language, and output business flow diagrams or use-case diagrams to better understand the use case scenarios.

The second phase is business architecture. Based on the previous product strategy, product portfolio planning and customer needs, we can define the business concept model. Domain modeling is performed on business concepts. This involves more conceptual models and the associations between concepts. At the same time, we need to further refine the business boundaries to help us divide the entire system. Experts in the whole field have relatively sufficient communication to define and clearly identify conceptual extraction methods, including considering the correlation of business concepts. Define specific business domain functions by defining domains, subdomains and restriction contexts.

The third phase is technology architecture, which is based on the division of business domains in the early stage. Therefore, the definition of software classes should be considered more at the implementation level, and the design pattern of the implementation should be considered, including the association between classes, and implemented through combination aggregation or inheritance. The boundaries of the responsibility definition of a class, etc. Also, we need to consider how to integrate or extend the existing system. At this stage, we need to consider some software design principles, including design patterns such as software, as well as non-functional requirements such as performance and stability.

The fourth phase is the design of the data architecture, so it will involve how to carry some requirements of data, functional requirements and non-functional requirements, sub-library and sub-table, and the definition of line database or non-linear database. Including design, store, and integration with existing data. Then, we need to consider the performance and stability of the entire system. We will use an example to illustrate the overall process.

4.4.2 Product Portfolio Planning

On the basis of its own business development, Alibaba has gone from a smaller e-commerce website to an existing large-scale system platform across multiple business fields. In this process, the Middle Office architecture design is very critical. With the concepts of 'large Middle Office' and 'small front office', Alibaba can build a business operating system that can support global transactions. Therefore, this set of methodology and Alibaba Cloud's technology platform can help some existing software platforms to carry out Middle Office transformation and provide technical support for the sustained and rapid growth of enterprises. The Middle Office project is also a type of development project. On the basis of the application software systems of traditional enterprises, by using methods and theories such as microservices and domain modeling, the software platform can meet the current business needs while minimizing the workload of reconstruction for subsequent business growth (Figure 4.12).

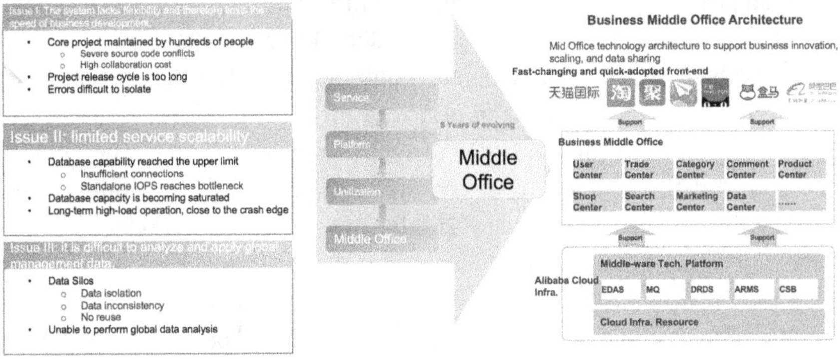

FIGURE 4.12 Middle office technology architecture supports enterprise digital strategy.

4.4.3 Domain-Driven Design

Business Middle Office approach to business modeling, domain modeling is commonly used. Domain modeling is used to unify cognition, define concepts and clarify the relationships between concepts. Based on the analysis of the current business situation and development trend, and based on the domain modeling method, the knowledge of business experts and technical experts is combined to define the business domain capability that is optimal for enterprise business development. Combined with technical practices, the capability division can be refined into the central capability division of the Middle Office. Domain definition as input can guide the design and implementation of code engineering.

Business Middle Office architecture design approach: including business scenario analysis, estimation of the overall effort, definition and design of the capabilities for the entire service center. Through specific business scenarios, analysis of entity objects, precipitation of service capabilities, to define the specific capabilities of each center is finally formed technically.

Field is a general term for the commercial activities carried out by an enterprise. An Enterprise provides services in a specific industry or market. Business scope is what we call 'field'. In the process of starting Business Middle Office construction, enterprises are aimed at specific fields to meet their business strategies and development goals. After the construction starts, the corresponding field boundaries are determined. A system is built to meet the business requirements of a domain. In Business Middle Office, common shared capabilities need to be differentiated by subdomain capabilities.

What is a 'domain model' – a domain model is a software model about a specific business domain. Generally, domain models are implemented through object models. These objects contain both data and behaviors and express the business meaning of preparation.

In single application, in order to realize the functions of the field, it is completed through a single, cohesive and full-function model. DDD can divide a domain into several subdomains. A domain model develops data in a domain-restricted context. This domain model is the scope of the business domain in Business Middle Office projects. By implementing functions in one business domain, the function scope can be split into multiple development centers to reduce risks.

The domain model is obtained mainly based on the collection of use-case sets and the processing of methodologies (Figure 4.13).

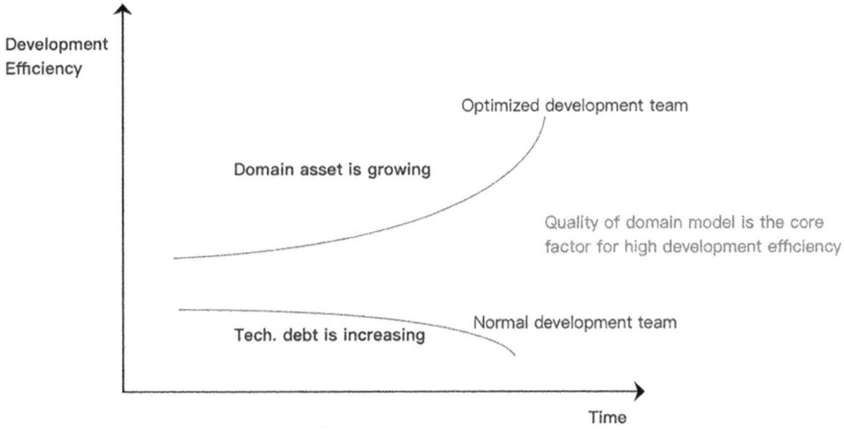

FIGURE 4.13 Relationship between domain model quality and R & D efficiency.

The domain model reflects the business problems in the problem domain, which can be used to describe the current business situation in a concise manner. The domain model has the following features:

- Each domain model is a tool that ensures high consistency between business requirements and technical implementations.
- Well-developed domain models can support rapid business development.
- Currently, domain models are optimized through progressive iteration.
- Each domain model is the core of an R & D organization. It determines whether technical liabilities or technical assets exist.

To perform domain modeling and analysis, follow these steps (Figure 4.14).

4.4.3.1 Example of Use-Case Analysis

Taking the omnichannel commodity management business scenario as an example, commodities need to support the following business scenarios:

Through the analysis of business scenarios, we can structurally define use cases and analyze the entities and associations in the business domain, which can help us better understand the relationships of domain concepts (Figure 4.15).

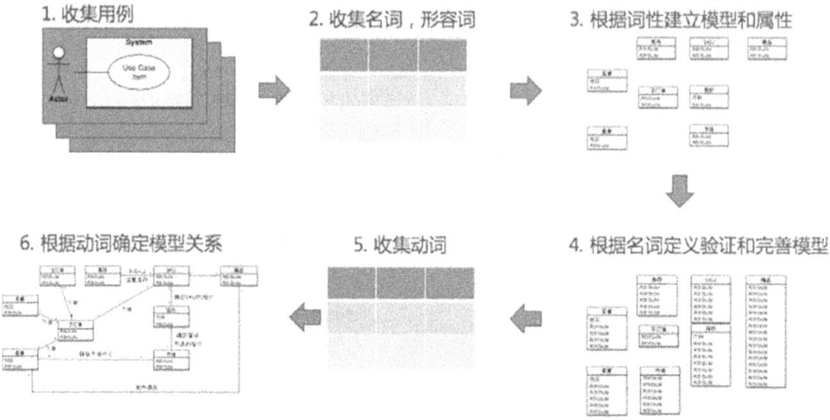

FIGURE 4.14 Domain modeling steps.

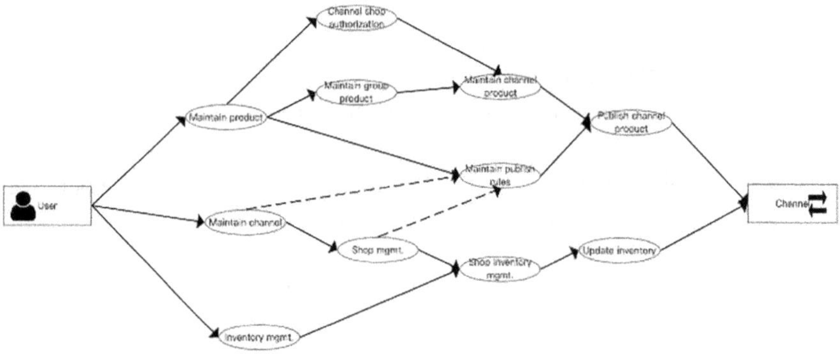

FIGURE 4.15 Business scenario analysis example.

4.4.3.2 Use-Case Analysis

The process of use-case analysis can be analyzed by event storm, which is a team activity. Domain experts and project teams list all domain events in the domain through brainstorming, and form the final domain event collection after integration, and then mark the order that caused the event for each event. Set the role of the command initiator for each event. Can be user-initiated, third-party system calls or timer-triggered execution. Finally, this command classifies events and generates entities, aggregates, aggregate roots and bounded contexts. It can quickly analyze and decompose complex business domains and complete domain modeling. The most important thing is to identify entities, attributes and associations:

Use cases are used to define what roles need to perform in a business scenario. The use-case diagram in the preceding figure mainly defines the tasks that roles need to perform. For specific scenarios, semantic analysis is used to analyze and identify entities and attributes. Analyze the sentence according to the editing of nouns, and find nouns, adjectives and verbs from 'predicate', 'adverbial', 'attribute', 'subject' and 'object':

[Shanghai] [operations 2] [for Tmall platform] [release] [red specifications] [product]

- [Shanghai]: attribute
- [operations 2]: Subject: The subject is a noun, and the noun can try to abstract into the domain model.
- [for Tmall platform]: adverbial: noun: The association between the noun and the object in the adverbial can be determined.
- Publish: This predicate is a verb, which determines the association between the subject and the object.
- 'red': attribute: Contains nouns and adjectives that can be abstracted into objects.

Association models (nouns have multiple values) or attributes (nouns have single values and are text or numbers). The attribute also contains attribute values or associated object values. [Commodity] object: a noun that can be abstracted into a domain model.

Example of entity name (Figure 4.16 and Table 4.1):

4.4.3.3 Association Analysis

Identify attributes and entity relationships, and collect nouns and adjectives from business scenarios to generate an entity instance as follows:

1. Entity and Value Object

During domain modeling, find out the entity that generated the command based on the relationship between the domain objects generated in the scenario analysis, such as commands and events. Analyze the dependencies between entities to aggregate and build the dependencies between models, and delimit the context.

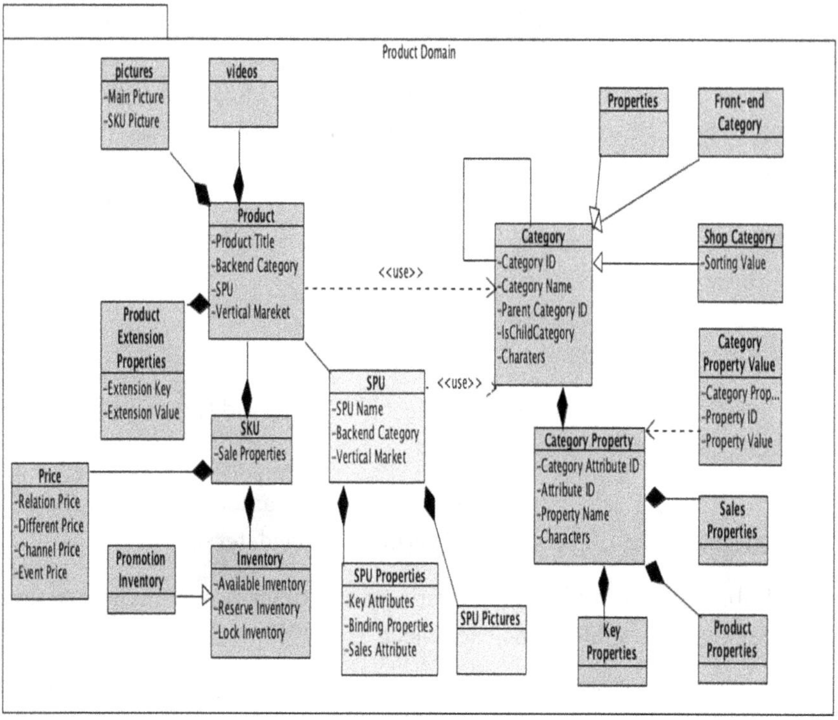

FIGURE 4.16 Product-related ER logical model.

TABLE 4.1 Product Entity Relationship Table

Entity Name	Explanation and Instruction	Example
SPU	SPU (Standard Product Unit) standardized Product Unit is a description of the common characteristic attributes of a certain type of Standard products. It is an extraction of common attributes of commodity information. SPU is a concept between leaf categories and commodities. It refines categories and is the basis for Taobao's standardized and standardized operation. SPU, or product, is a mechanism for identifying products. It is composed of background categories + a set of key background category attributes + a set of constrained background category attributes. These attributes are the key attributes in this category. 'category + key attributes' uniquely identifies a product (SPU).	For example, you can determine an SPU based on the brand and model of the mobile phone. For example, you can use the P40 brand to Huawei and model to determine the P40.

(*Continued*)

TABLE 4.1 (*Continued*) Product Entity Relationship Table

Entity Name	Explanation and Instruction	Example
Product	In particular, items are related to merchants. Each item has a merchant code, and each item has multiple colors, styles and SKUs.	For example, the iPhone is a single item, but when Taobao is fooled by many businesses selling this product at the same time, the iPhone is a commodity.
SKU	Stock keeping unit (unit of inventory): SKU is the unit of measurement for incoming and outgoing inventory, which can be in pieces, boxes, pallets, etc. It is most widely used in clothing and footwear.	For example, the pink S code in a women's wear is a SKU, the M code is a SKU, and the L code is also a SKU. Therefore, generally a women's dress has S, M, L, XL, XXL, XXXL6 SKU.
Key attributes	The unique attribute of an SPU.	For example, brand+model (such as Huawei+P40).
Sales attributes	A special property that defines the available property of the SKU that a category has.	For example, the color and size of clothing category, mobile phone category package and color.
Product attributes	The attribute represents the unique characteristics of the item and cannot be used as an SPU attribute.	

2. Method of Identifying Entities:
    - Specifies whether the bucket has a unique identifier.
    - Whether or not the system can remain unchanged after the state change.
    - These features include business attributes and business behaviors.
    - And can be persisted.
3. Value Object Identifying Method:
    - A temporary value object is invariant and cannot be changed after it is created. It contains all its attributes.

- A group of associated attributes is created, and each attribute is an indispensable component.
- Containers are replaceable. When a referenced value object needs to change the state, it must be replaced with a new one.

4.4.3.4 Division of Domains

To classify all identified entity objects (Figure 4.17).

Entity objects can be classified by domain. For example, objects such as primary orders and suborders can be classified to the transaction domain, while other objects such as sellers and buyers can be classified to the member domain. A domain is a collection of information about one or more entity objects. It manages the lifecycle of entity objects in the domain.

Relationship between the domain and the entity object:

- Allows one domain to manage one or more entity objects.
- Only one entity object can be managed by a domain.
- However, if one entity object is managed by multiple domains, domain responsibilities conflict, resulting in coupling and mutual influence.

A bounded context is used to encapsulate domain objects, provide context environments, and divide boundaries, so that business-related objects have specific meanings. This boundary defines the scope of application of

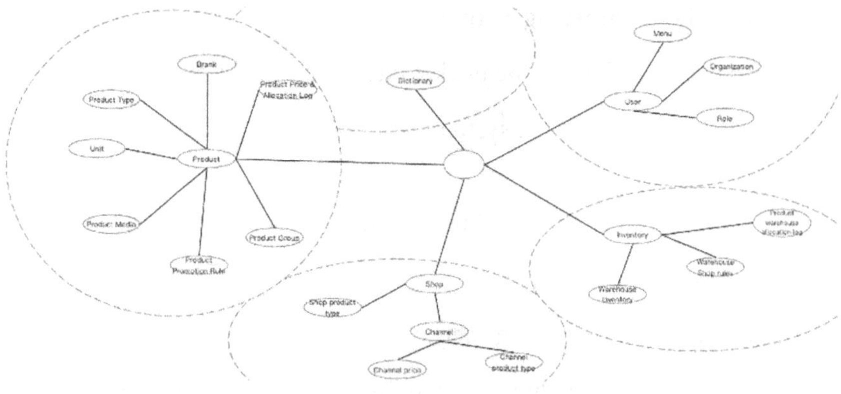

FIGURE 4.17 Domain division of domain model.

the model. The boundary of a domain is defined in a bounded context. The boundary context is the main basis for microservice design and splitting.

- This module classifies aggregations based on the context semantics in a domain to delimit the context.
- It also divides strong dependencies into the same bounded context based on the dependency between aggregates.
- Checks whether the aggregation is in the core part of the subdomain.

Microservices can be formed only when the boundaries defined by the context are finally bounded (Table 4.2).

4.4.3.5 Examples of Business Domain Division

TABLE 4.2 Example of Business Domain Division

Member domain	1. Unified member ID, support multi-channel and product line Member operations, full coverage (PC, H5, WeChat public number, etc.). 2. Including: Registration (enterprise registration, personal registration), login, session, SSO, Diversified user identity verification function and account information maintenance. 3. Account security: Password modification, password retrieval, email address modification, email verification, mobile phone verification and security issues. ...
Product domain	It provides multi-form (physical, service, virtual) Commodity Services (commodity Publishing & Editing) for multi-channel, multi-industry, multi-terminal (PC, wireless, WeChat public number), including 1. Manage and search for category information, product (SPU) information, brand information and commodity rule information. ...
Transaction data domain	The transaction is the exchange of money and products or services between buyers and sellers through the form of contract on the platform of computer network. 1. Link perspective: The transaction link connects products, marketing, payment, logistics and other systems, enabling consumers to purchase physical objects/services from businesses and ensure the successful performance of transactions. ...
Payment domain	In the capital domain, issues in order fulfillment and settlement are resolved. 1. Payment: Mainly solves the problems in the order payment process, including Payer: Who will pay?...

(Continued)

TABLE 4.2 (*Continued*) Example of Business Domain Division

Settlement domain	Settlement: 1. Settle the sub-account, that is, when the order is completed, the money received by the seller will be distributed to other accounts according to a certain proportion. The sub-account needs the following two points of information. 2. Account allocation description information: Mainly records the account allocation business of the order, so as to facilitate Huijin system or Alipay to make decisions on how to allocate accounts. 3. Account allocation rules. Alipay's account allocation rules are written into orders, so this information will be written into the vertical table of orders. ...
Marketing domain	It provides multi-channel, multi-industry, multi-terminal (PC, wireless, WeChat public number) and various marketing solutions, providing a variety of marketing games, which will run through the commodity harmony, shopping cart, preferential display and enjoyment in the transaction process. Link. It includes the following functions 1. Preferential management: Providing various preferential methods such as price reduction, discounts and gifts to empower merchants, and providing interfaces for creating, displaying, querying and using preferential policies. ...

4.4.3.6 *Division of Centers*

From the planning and definition of business domains, we can analyze and plan specific capabilities of business domains. Taking the product domain capability as an example, the planning business architecture is shown in Figure 4.18.

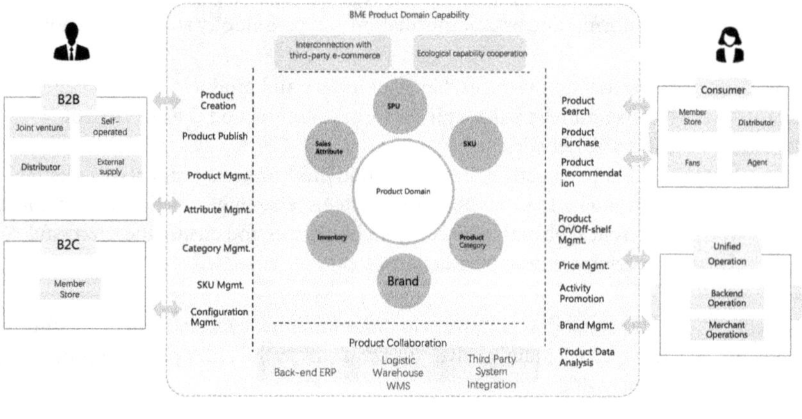

FIGURE 4.18 Product domain business architecture diagram.

Business Middle Office consists of service centers that encapsulate data and capabilities based on domains. The service center provides API objects as capabilities, and Business Middle Office outputs and presents the objects to upper-layer applications in a unified manner. Upper-layer applications orchestrate the capabilities provided by the Middle Office and select the capabilities provided by the center as needed or in part, to rapidly develop and deploy, so as to adapt to the rapid changes of enterprises' business.

Shared service centers and application systems are designed in a microservice manner. Microservice is an advanced architectural design idea, which has been successfully applied in many large Internet companies in China and abroad. The core of microservice is to simplify complex design, break up into parts and break down applications into small service modules for independent development. This feature of microservices makes it easy to deploy to containers, which has a revolutionary impact on the entire development, testing, and operation and maintenance process. It strongly supports DevOps development, facilitates agile development and automated testing, is conducive to independent deployment, maintenance and upgrade and fault handling, and improves efficiency and quality.

Business Middle Office and microservices complement each other and are closely linked (Figure 4.19).

Service Group	Service Capability
Product Search	Obtain product list
	Query product information
	Get promotion information
Product Lifecycle Management	Obtain product list
	Obtain product details
	Obtain the SKU information of a product
	New product
	Product editing
	Save history when editing items
	Delete product
	Delete commodities in batches
	New product type
	Edit product type
	Delete product type
Product On/Off-shelf Management	Query the product loading and unloading list
	Products on the shelves
	Product off the shelf
	Bulk shelves
	Batch off-shelf products

Service Group	Service Capability
Product Template	Obtains a list of product templates.
	Obtain product Template details
	Create a product Template
	Edit a product Template
	Delete a product Template
	Delete product templates in batches
	Commodity template status modification
Product Combination	Obtain a Package List
	Queries package details.
	New plans
	Edit package
	Delete plans
	Batch delete plans
	Modify service status
	Set the product to be associated with the plan
	Obtains the package loading and unloading list.
	Package loading /unloading
Front-end Category	Add front desk category
	Modify foreground category information
	Delete foreground category
	Delete foreground categories in batches
	Queries the category list on the tree table.
	Obtains the details about the tree category.

Service Group	Service Capability
Back-end Category	Add backend category
	Modify backend category information
	Delete background Category
	Delete background categories in batches
	Obtains a tree background category list.
	Obtains the information about the tree category.
Brand	Query brands
	Add a brand
	Modify brand information
	Delete a brand
	Delete multiple brands
Product Attribute Group	Obtain the list of product property groups (loan plan, oil parameter, and auto insurance)
	Queries product property group details.
	Add a property group
	Modify a property group
	Delete a property group
	Bulk delete attribute group
Product Attribute	Obtain the list of product attributes (number of down payment periods, weight, and commercial risks)
	Obtain product attribute details
	Add attribute
	Modify properties
	Delete attribute
	Delete attributes

FIGURE 4.19 Business Middle Office microservice splitting.

Microservice is a technical way to implement Business Middle Office. Therefore, splitting services is crucial to system development and implementation in the future.

If services are split too carefully, the number of applications increases and the maintenance workload becomes more complicated. Moreover, the splitting will cause the number of content and operational tasks that O & M personnel need to pay attention to multiply. If these problems are not properly solved, it will surely become a nightmare for O & M personnel. In addition, interactions between services will become complicated, which may lead to service call timeout and system crashes.

Although the design of microservices is a subjective work that relies on the architect's personal experience and business understanding, there are still some rules to follow. Table 4.3 lists some referential principles for dividing microservices in general. The most important principle is to divide microservices by business domain, which is DDD.

The principles for dividing microservices are described in the following (Figure 4.20).

In response to the above principles, there are the following special instructions:

Microservices are characterized by the ability to extract different services based on the business. The system is split and divided into business systems and shared service centers based on different functions. Each subservice system calls multiple shared service centers to complete functions, and the shared service center calls multiple middleware frameworks at the data layer. In general, the business service system can only access its own database. For data in other databases, it is completed by calling the interfaces provided by its services.

The biggest difference between the idea of breaking up into parts and the traditional software development and design method is that it is not to develop a huge single application, but to put all the services and capabilities into this application, instead, applications are divided into small and interconnected microservices. A microservice generally performs a specific function.

Taking the digital transformation project of a tourism hotel enterprise that the author participated in as an example, the pilot application of service-oriented design is comprehensively considered according to the technical characteristics and universality of microservices:

TABLE 4.3 Principles of Microservices

Rule Name	Rule Explanation
Creator	1. Question: Who is responsible for generating instances of classes? 2. Solution: If one or more of the following conditions are met, you can assign the responsibility of creating an instance of Class A to Class B • Account B contains A • B aggregation A • Account B has the data to initialize A and passes the data to Class A when creating an instance of Class A • Account B records an instance of account A • Account A is frequently used by account B
Information expert	1. Definition: If a class has all the information needed to complete a duty, then the duty should be assigned to the class to implement. At this time, this class is the information expert relative to this responsibility. 2. Solution: Assign duties to classes (domains) that have information necessary to perform a duty.
Low coupling	1. Question: How to support low dependency, reduce the impact of changes and improve reusability? 2. Solution: Assign a responsibility to keep the coupling degree low (depending on the measurement of strength of other domains or models).
High cohesion	1. Question: How to make the complexity controllable? 2. Solution: Assign a responsibility to maintain high cohesion (degree of relevance and concentration of responsibilities). From the perspective of domain boundary division: The close relationship between models is reflected in the interior of domain classes, rather than between domains.
Pure fabrication	1. Problem: How to deal with it when you don't want to destroy the design principles of high cohesion and low coupling, but some responsibilities have no place. 2. Solution: Assign a set of high cohesion responsibilities to a fictitious or convenient class, which is not a concept of the problem domain, but a fictitious concept, in order to support high cohesion and low coupling. The purpose of reuse.
Indirection	1. Question: How to allocate responsibilities to avoid direct coupling between two things? 2. Solution: When we don't know which model to assign responsibilities to, we can see if we can assign responsibilities to intermediary models.

The core is the eight shared service centers that make up the Business Middle Office. The public services that the entire pilot application system relies on are integrated into these eight service centers. With the gradual expansion of the system, the modules of the Business Middle Office will become richer and more functional. In this way, when a new business

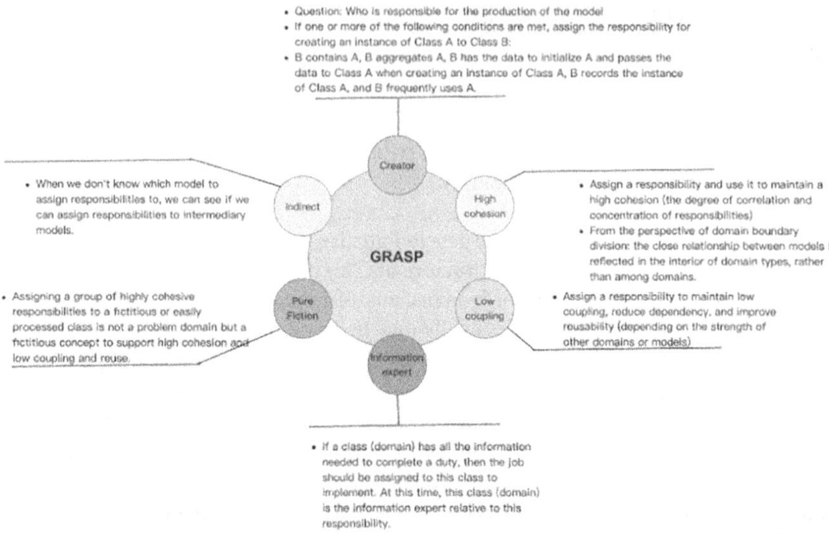

FIGURE 4.20 Principles of microservices.

system is going online, you can directly reuse the existing shared service center.

The entire system is designed with a microservice framework, which includes the five most important functions of microservices: service governance, service gateway, service fault tolerance, service link tracking and service monitoring.

The Business Middle Office construction section will explain some specific benefits of architecture design.

4.5 BUSINESS MIDDLE OFFICE CONSTRUCTION PROCESS

4.5.1 Introduction of Middle Office Building Process

Through the previous chapter introduction, we have a comprehensive understanding of the construction ideas and design of Business Middle Office. What is particularly important is that we have analyzed the problem domains we are facing through DDD methodology, the center division and the capability list of each center are obtained. Based on the output of the design phase of the Business Middle Office, I believe that if readers have a development background, they should have been anxious and eager to know when we will write the code? Before writing code, we still have a lot of work to do, such as technology selection, architecture design and detailed design. In addition to these specific technical-related work, Business Middle Office eventually belongs to the field of software development, which requires engineering construction process to ensure quality.

4.5.1.1 What Is Software Engineering

Let's first look at the definition of software engineering. Software engineering is to study and apply how to develop and maintain software with systematic, standardized and quantitative process methods, and how to combine the management technology proved to be correct after time test with the best technical method currently available in engineering. Software engineering involves programming languages, databases, software development tools, system platforms, standards, design patterns, cloud computing, security, performance, software development patterns, etc.

The goal of software engineering is to develop applicability, effectiveness, modifiability, reliability, intelligibility, maintainability, reusability and portability of software products that can be traced, interoperable, and meet user needs. Pursuing these goals helps to improve the quality and development efficiency of software products and reduce maintenance difficulties.

Software development model is an important research field of software engineering. Software development model is used to describe and represent a complex software development process. When mentioning the software development model, the first thing that comes to mind is probably the famous waterfall model, which is composed by W.W. Royce initially in 1970. Waterfall model defines the software life cycle into planning, requirement analysis, software design, program compilation, six basic activities, such as software testing and operation and maintenance, and stipulate their fixed order of top-down and mutual connection, like waterfall and flowing water falling step by step. In the waterfall model, all activities of software development are carried out strictly in a linear manner. The current activity accepts the work results of the previous activity and implements the required work content. The work result of the current activity needs to be verified. If the verification is passed, the result is used as the input for the next activity and the next activity is continued. Otherwise, the modification is returned. In addition to the waterfall mode, there are the following common software development modes:

- Iterative development mode, which is also called iterative incremental development or iterative evolutionary development, is a software development process opposite to traditional waterfall development, it makes up for some weaknesses in traditional development methods and has higher success rate and productivity.
- Rapid prototype mode: The first step of the rapid prototype model is to build a rapid prototype to realize the interaction between customers or future users and the system. Users or customers evaluate the prototype, further refine the requirements of the software to be developed. By gradually adjusting the prototype to meet the requirements of customers, developers can determine what the real needs of customers are; the second step is to develop software products that customers are satisfied with on the basis of the first step.
- Agile mode, agile development is a human-centered, iterative and step-by-step development method. In agile development, the construction of software projects is divided into multiple sub-projects. The results of each sub-project have been tested and have the characteristics of integration and operability. In other words, a large project is divided into multiple interrelated but independent small iteration projects and completed separately. In this process, the software is always available.

The software development mode needs to be solved in essence: To provide an effective software development process to guide software development to meet the functional and non-functional requirements of software. To help you better understand the software development model, let's focus on the waterfall model and agile model.

4.5.1.2 Waterfall Model

Waterfall model divides the software life cycle into six basic activities, including planning, requirement analysis, software design, program writing, software testing, operation and maintenance, and stipulates that they are top-down, the fixed order of mutual connection is like a waterfall flowing water falling down step by step. In the waterfall model, all activities of software development are carried out strictly in a linear manner. The current activity accepts the work results of the previous activity and implements the required work content. The work result of the current activity

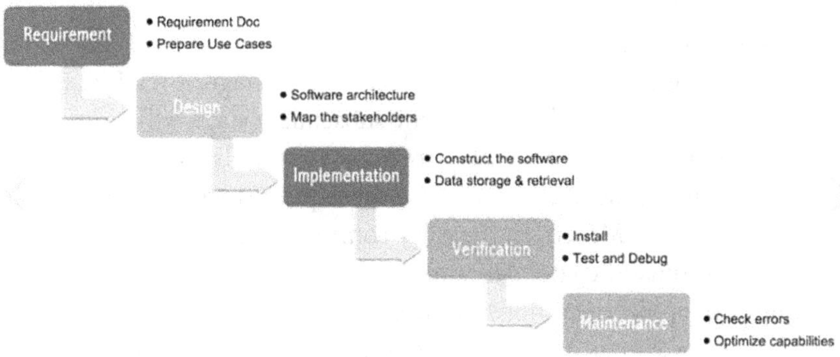

FIGURE 4.21 Waterfall pattern development process.

needs to be verified. If the verification is passed, the result is used as the input for the next activity and the next activity is continued. Otherwise, the modification is returned.

The waterfall model has the advantage of strictly following the pre-planned sequence of steps, and everything is more rigorous step by step. The waterfall model emphasizes the role of the document and requires careful verification at each stage. Figure 4.21 shows a typical waterfall model.

The problem with waterfall mode is that the linear process is too idealized and is no longer suitable for the modern software development mode. It is almost abandoned by the industry. Its main problem lies in:

- The division of each stage is completely fixed, and a large number of documents are generated between stages, which greatly increases the workload.
- Because the development model is linear, users can only see the development results at the end of the whole process, thus increasing the risk of development.
- Early errors may not be discovered until the testing stage in the later stage of development, thus causing serious consequences.
- It takes a long time to connect each software life cycle, and the communication Cost of team members is high.
- The waterfall method is basically infeasible when the requirements are unknown and may change during the process of the project.

4.5.1.3 Agile Model

There is a strong dependency between each stage of waterfall mode. The previous stage is regarded as the input of the latter stage. If the input quality is not high, the output quality of the subsequent stage will be seriously affected. At the same time, if the previous stage fails to meet the standard, it will also cause stagnation in the subsequent stage and lengthen the development cycle. Moreover, the early commitment of the project makes it difficult to adjust the later demand changes, which is costly.

Agile development was born in the context of such a question. In 2001, 17 software developers gathered in Snowbird, Utah, to discuss their ideas about work and various software development methods, looking for similarities among them, finally, the famous 'Agile Manifesto' and 12 principles were released together, officially announcing the beginning of the agile development movement.

Data show that 70% of software development projects adopting waterfall development methods have failed. The reason is that the market demand changes rapidly, and it is difficult to realize the clear and complete collection of product demand. At the same time, the development of technology is also changing with each passing day, the realizability of defined functions also faces multiple uncertain factors. Therefore, when the work of demand collection and product definition cannot be completed well, the waterfall development method naturally cannot get rid of the fate of high failure rate.

The transformation of a core thinking mode of agile development is: from 'Fix Scope, Flex time' (fixed range, flexible time) represented by waterfall development to 'Fix time, Flex Scope' – fixed time, the elastic range.

Under the background of market changes and technological changes, since the 'scope' represented by market demand and product definition cannot be fixed from time to time, it is impossible to target a confirmed target how much resources should be invested to complete it, we might as well fix the existing resources and take resources as constraints to maximize the 'scope'. Therefore, it changed from 'plan-driven' to 'value-driven' (Figure 4.22).

After the ideas of agile development was introduced, 'individuals and interactions are better than processes and tools', 'software that can work is better than all-round documents', 'Customer Collaboration is better than contract negotiation' and 'response to changes is better than follow the

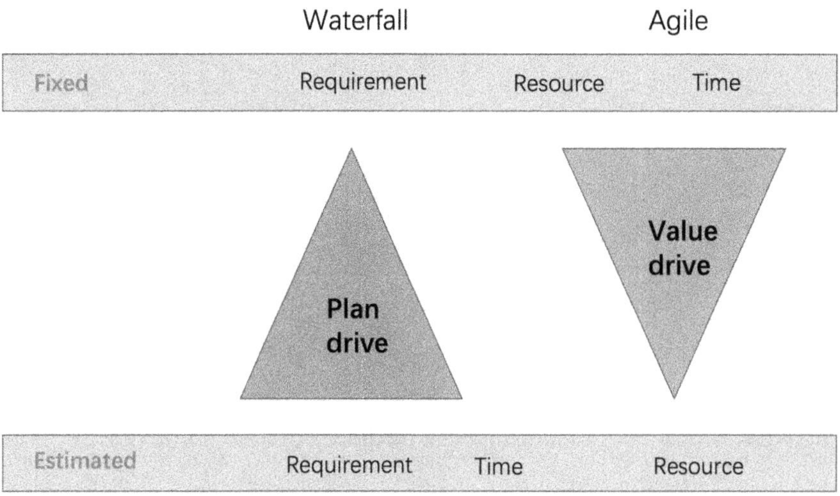

FIGURE 4.22 Waterfall mode vs. agile mode.

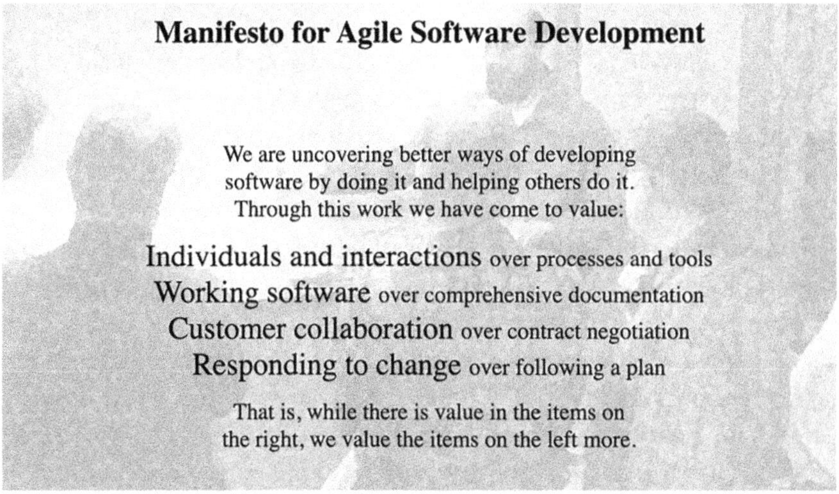

FIGURE 4.23 Manifesto for agile software development.

plan', which represent agile values, attach full importance to the value of people versus merely code writers in the software development process (Figure 4.23).

At the same time, under the guidance of the Agile Manifesto, a variety of agile development methods have emerged, such as Kanban and iterative 'agility' methods; furthermore, the agile values represented by the

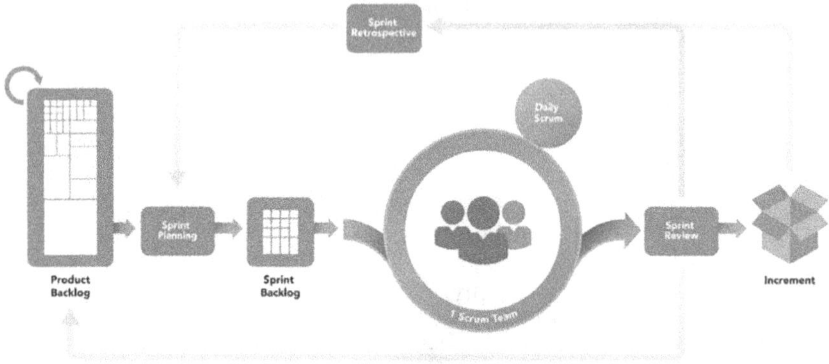

FIGURE 4.24 Scrum agile development method.

'Agile Manifesto' are displayed through specific implementation means (Figure 4.24).

It can be seen from 'agile development' that agility is not only an adjective, but also represents a method. So, what is 'agility' on Earth?

It takes 'complex system' as the background, and 'agile typical' takes 'complex system' as the background. As a method, it is eventually adopted and implemented by 'people'. However, people's cognition and understanding of the world are always moving toward the two directions of reducing Unknown 'Unknown' and uncertain 'Uncertainty'. They need to gradually understand the Unknown (Understandable), while for Uncertainty, it is usually predicted in advance and judgment (Predictable) is obtained through feedback. Therefore, comprehensiveness and predictability have also become two dimensions of people's cognition of the world. However, no matter how scientific and technological progress is, many things in the world have reached the point of being understandable and predictable, but there are still many things that cannot be understood or predicted. In particular, everyone's cognitive ability is also different. The same thing is understandable and predictable for some people with sufficient cognitive ability, but if the cognitive ability is insufficient, there will be 'Chaotic systems' (Chaotic) that are neither understandable nor predictable. The same is true for Complex systems (Complex). When compared with human cognition, it puts forward certain high requirements for comprehensibility and predictability, and then it presents the characteristics of Complex systems. Like the background of agile development, the market is changing rapidly, and the demand becomes unpredictable; with the rapid development of technology, the technology realizability of certain requirements becomes more and more difficult to understand. However, this kind of unintelligibility and unpredictability is not far beyond the scope of people's cognitive potential and has not reached the level of complete confusion. At the same time, through continuous feedback and learning in the process, the unknown and uncertainty can also be gradually eliminated. Therefore, for such a complex system, using agile methods will better obtain the understanding and prediction of the system.

Driven by people as the core, for the application of agile methods, its ultimate goal is to understand the complex system and stimulate the energy of the complex system. More emphasis is placed on the value of 'system' to 'human', rather than simply recognizing its 'complex' characteristics. At the same time, complex system is a relative concept, which is relative to human cognitive ability. However, for Complex systems, the cognitive process will still develop along the two directions of 'comprehensible' and 'predictable', in which 'human' will play a major role, we need to fully tap the potential of 'people'. No matter for 'purpose' or 'process', when using 'agility' method, 'human' is the core driving force in the process of cognition and operation of 'complex system'.

For empirical process control with adaptive ability, the third characteristic of 'agility' is that 'agility' is actually an empirical process control method. As a method, it usually has a certain purpose. In order to achieve the goal, it is necessary to implement certain process control to improve the probability of achieving the goal. Under the background of 'complex system', the Predefined Process Control represented by 'waterfall development' is no longer suitable, but the empirical process control is driven by people as the core. It will have higher adaptability and flexibility, and at the same time, it can give full play to the potential and value of 'human'. Human beings are always facing various unknown and uncertainty process of evolution, cognition and transformation of the world, therefore, human history is naturally an 'agile' process.

To sum up, 'agile' represents a method under the background of 'Human-Driven' complex systems, it's an adaptive empirical process control method.

4.5.1.4 Business Middle Office Building Process Introduction

After introducing the two construction modes of waterfall and agility, let's talk about the construction process of Business Middle Office. As the author begins to introduce in this chapter, business intermediate

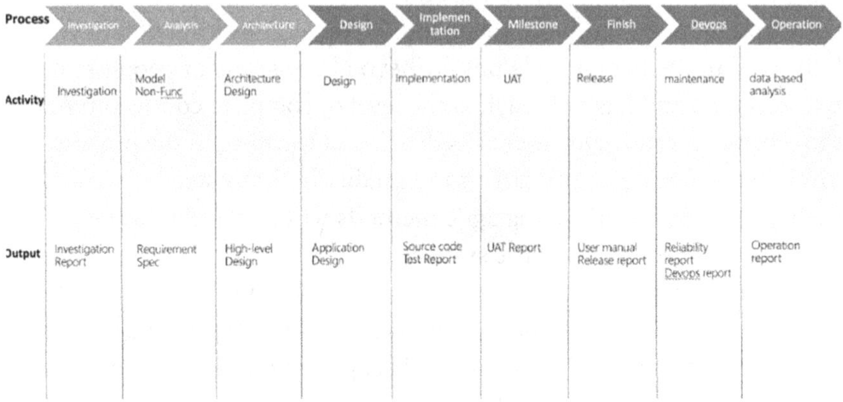

FIGURE 4.25 Construction process of Business Middle Office.

development essentially belongs to the category of software development, so all the development modes of software engineering can be used as the development modes of Business Middle Office construction. Readers may ask, do I choose the traditional waterfall development mode or the agile development mode that is popular now? To answer this question, we need to see what work is needed in the whole construction process of the Business Middle Office and what is the sequence of these works. Figure 4.25 is a summary of the author's years of experience in the construction of Business Middle Office.

As shown in the preceding figure, the assumption of Business Middle Office has a total of nine steps, and each step has clear input and output. This process is exactly the same as the waterfall development mode, does it mean that the Business Middle Office can only use the waterfall model?

4.5.1.5 Business Middle Office Construction Pattern

There is no need to guess the answer to this question. The core idea of agility is to divide large pieces of work into small pieces of work that can be independently verified for iterative construction. After carefully observing the construction process of the Business Middle Office, we find that most of the steps can be completed iteratively and step by step with the help of agile thinking and small steps. In particular, we can use agile methods to define milestones and complete iterative construction in the Middle Office construction section of the business. Figure 4.26 shows the construction

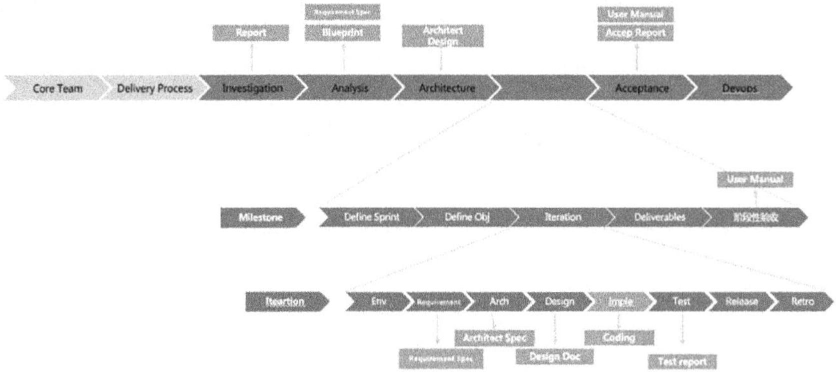

FIGURE 4.26 Combination of Business Middle Office and agile development.

process flow chart that adopts agile mode in the Middle Office development and construction section of the business.

The construction of Business Middle Office platform cannot be achieved overnight, and needs to be gradually improved through iteration. Each step in the nine major steps of Business Middle Office platform requires the enterprise to invest enough resources to build it. Next, let's take a look at the details of each stage of Business Middle Office platform construction, this chapter introduces what the Middle Office should do in each step, what good practices it has and what the output is.

4.5.2 Requirement Gathering

The purpose of the demand survey is to find out the business demand points of the enterprise and lay a foundation for introducing the Business Middle Office to solve problems later. Therefore, it is essential to understand the customer's business, understand the business process and sort out important business scenarios to design a shared service center for the Business Middle Office (Figure 4.27).

4.5.2.1 Preparations before Demand Survey

Before conducting a demand survey, the following preparations are required:

1. Understand the basic standards and professional terms of the industry to which the customer belongs, and at this time, do a preliminary work with the customer's docking personnel communication.

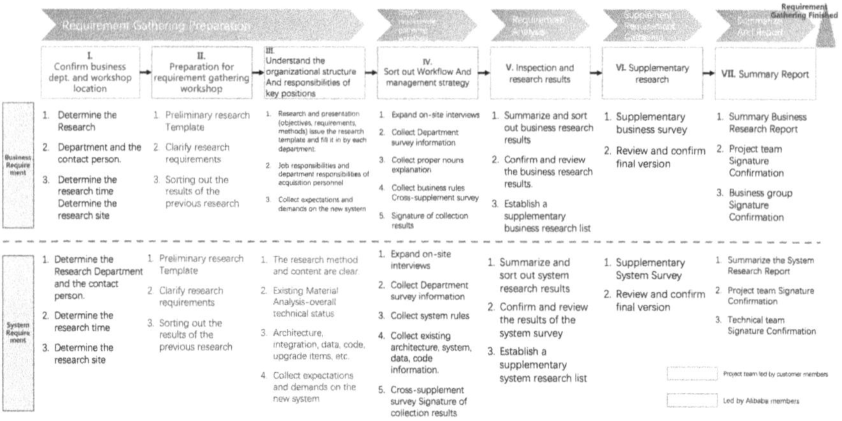

Requirement Gathering Preparation
Requirement Gathering Finished
I. Confirm business dept. and workshop location
II. Preparation for requirement gathering workshop
III. Understand the organizational structure And responsibilities of key positions
IV. Sort out Workflow And management strategy
V. Inspection and research results
VI. Supplementary research
VII. Summary Report
Business Requirement
1. Determine the Research
2. Department and the contact person.
3. Determine the research time Determine the research site
1. Preliminary research Template
2. Clarify research requirements
3. Sorting out the results of the previous research
1. Research and presentation (objectives, requirements, methods) issue the research template and fill it in by each department.
2. Job responsibilities and department responsibilities of acquisition personnel
3. Collect expectations and demands on the new system
1. Expand on-site interviews
2. Collect Department survey information
3. Collect proper nouns explanation
4. Collect business rules Cross-supplement survey
5. Signature of collection results
1. Summarize and sort out business research results
2. Confirm and review the business research results.
3. Establish a supplementary business research list
1. Supplementary business survey
2. Review and confirm final version
1. Summary Business Research Report
2. Project team Signature Confirmation
3. Business group Signature Confirmation
System Requirement
1. Determine the Research Department and the contact person.
2. Determine the research time
3. Determine the research site
1. Preliminary research Template
2. Clarify research requirements
3. Sorting out the results of the previous research
1. The research method and content are clear.
2. Existing Material Analysis-overall technical status
3. Architecture, integration, data, code, upgrade items, etc.
4. Collect expectations and demands on the new system
1. Expand on-site interviews
2. Collect Department survey information
3. Collect system rules
4. Collect existing architecture, system, data, code information.
5. Cross-supplement survey Signature of collection results
1. Summarize and sort out system research results
2. Confirm and review the results of the system survey
3. Establish a supplementary system research list
1. Supplementary System Survey
2. Review and confirm final version
1. Summarize the System Research Report
2. Project team Signature Confirmation
3. Technical team Signature Confirmation
Project team led by customer members
Led by Alibaba members

FIGURE 4.27 Business Middle Office requirement survey process.

2. Understand the customer's personnel organizational structure, rules and regulations, etc.
3. Take the main business scenarios obtained from customers as input and outline the important business requirements to be investigated.
4. Formulate a detailed survey schedule.
5. Obtain the person in charge of the application system to be investigated and contact information.
6. Prepare the interview record table of the survey and the list of core issues of the survey.
7. Prepare the questionnaire.

4.5.2.2 Requirement Gathering Process

Through the process of demand survey, we can conduct questionnaires, separate interviews, meeting discussions and other forms. The general survey process is as follows.

The comparison of survey methods is as follows: In conclusion, the three main methods have their own advantages and disadvantages, because they can be used in combination with the three methods according to the actual situation. Survey Methods advantages is it could conduct a large number of surveys and collect feedback, while the question is limited and cannot fully display the current business process. A single interview allows you to have a deeper understanding of the individual's understanding of the current business and the business processes that can be improved. An individual may not have a comprehensive understanding of the business. A multi-person meeting can fully discuss the business understanding of a department. Perhaps due to the large number of people, some more complicated topics cannot be expressed clearly. At the same time, a large number of people will lead to divergence of views and reduce the efficiency of research. The above research process may also be carried out by specialized business architects in the project, but technical architects must also learn and be familiar with the output of business research in detail, it is convenient to carry out subsequent technology architecture design work.

The survey includes the following aspects: overall objectives, background, system role definition, overall business processes, business scenarios, functional requirements and interaction with external systems. Research is to better understand the customer's business needs; therefore, in this process, we should focus on the following aspects:

1. Overall objectives and business processes

   The scope of demand survey should be determined here, and the overall goal should be consistent with the customer's goal without deviation. The business process is the core basis for dividing the Business Middle Office capability center.

2. Business scenarios

   Generally, business scenarios can be classified according to system role definitions. Taking the demand survey of the digital transformation project of hotel tourism industry that I participated in as an example, the business scenarios are divided into scenarios applicable to consumers, enterprise employees and headquarters operators. By business scenario classification, you can identify entity objects, which are important components of the Business Middle Office capability center.

3. Technical requirement gathering

   Based on the scope of the overall business planning in the early stage, in order to ensure the smooth follow-up research and development process, the surrounding systems that need to be integrated also need to be investigated accordingly. The main content is as follows:

I. Existing IT organization and R & D architecture

The construction of Middle Office involves the adaptation of R & D organizations. Therefore, we recommend that you analyze the current R & D system in the early stage. The technical research phase mainly analyzes the current technical organization and research and development status, from the construction of product design and research team to the system situation of infrastructure operation and maintenance.

Key research contents include:

- The first is the composition of the current technical team. The construction and operation of traditional application systems are mainly divided into the composition of front-end teams, business analysts of back-end implementations, R & D and O & M teams, the corresponding responsible team of the database. For the implementation scenario of the Business Middle Office, the agile development team will focus on one or more technical R & D and operation teams, and the entire team will be built around the center, from the definition and decomposition of requirements, development and implementation, and online version control, etc.

- The second is the current R & D process: From the source of requirements to the overall launch of the current process for detailed research, design of each stage, output products and other definitions. Analyze the process of publishing cycle from waterfall requirements/problems to development and testing. In the future development process of Business Middle Office, we will analyze how to implement faster iteration. We need to define the development process in detail (Figure 4.28).

II. Investigation of existing system conditions

The Middle Office transformation is a process of improving and optimizing the IT architecture of enterprises and continuous evolution. First, we need to clearly define which application systems need to be replaced or optimized. In terms of application systems, based on different industry conditions, there are generally e-commerce platform systems, CRM, CDP systems, and typical back-end systems such as ERP, manufacturing execution and financial management.

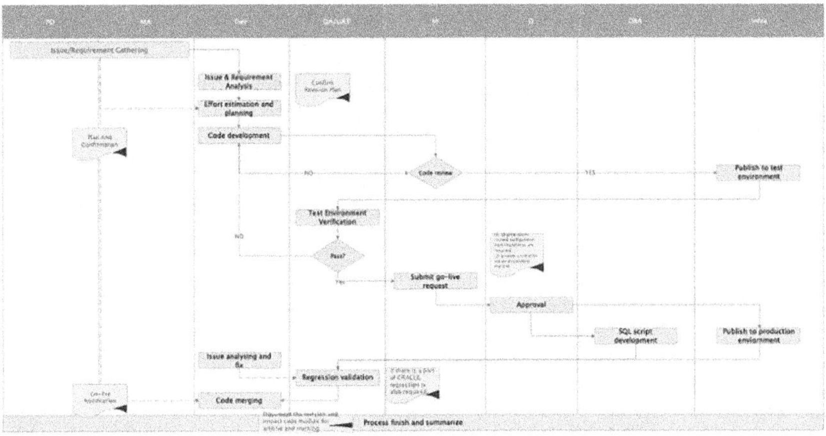

FIGURE 4.28 Problem/demand management swimlane diagram.

For the current situation of existing systems, we need to do a full survey first to understand the system data volume and technical implementation. Through this survey process, we can use it to better understand the system boundaries that need to be replaced or implemented during the construction of Business Middle Office. During the construction of the Middle Office system, the technology architecture is supported based on the collected network sites, integration and data (Table 4.4 and Figure 4.29).

III. Network Status

IV. Safety status

Cyber Security:

- Anti-DDoS: Uses anti-DDoS cloud-native-basic
- WAF: firewall application
- Apsara Stack security (database audit)

Data Security:

- RDS:SQL Explorer
- DMS for SQL audit
- Security Center (server Guard)

TABLE 4.4 Basic Classification of Application Systems

Application Name	Application Type	Function Description	Tech. Stack
Index creation service	Basic services	Synchronize RDS data to ES for data conversion. You can receive Kafka messages from DTS and convert them into ES operations to insert data.	Mainstream technology stack, such as Java, EDAS and microservices
Search service	Foreground application	Provides External services directly based on ES.	Mainstream technology stack, such as Java, EDAS and microservices
SOA	The third-party application	Old American system.	Mainstream technology stack, such as Java, EDAS and microservices
ADFS	The third-party application	Microsoft ADSF for web authentication of internal systems	Mainstream technology stack, such as Java, EDAS and microservices
Alipay, fast money, UnionPay	The third-party application	Used for the third-party payment-related function.	Mainstream technology stack, such as Java, EDAS and microservices
QQ, WeChat, and Weibo	The third-party application	Used for party III login.	Mainstream technology stack, such as Java, EDAS and microservices
CMS	The third-party application	Content Management System (web site).	Mainstream technology stack, such as Java, EDAS and microservices

V. Data storage

Several situations need to be considered in the construction data of the Business Middle Office system: System creation, integration with external systems and replacement of the original system. For the transformation of the Business Middle Office, synchronization or switching of the original system data is inevitable. Therefore, the data of the existing system also need to be analyzed in detail in the technical research stage.

First, based on the scope of business planning, we need to focus on the data situation of the integrated system, such as the

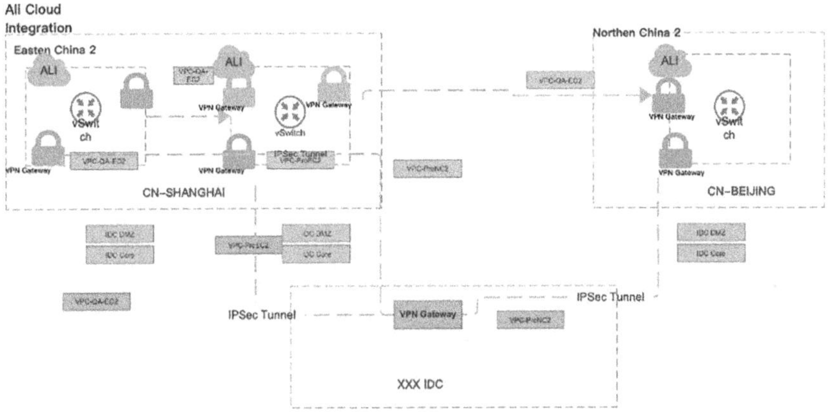

FIGURE 4.29 Network status diagram of an enterprise.

data volume of the current system, the matching rules between Chinese and foreign system models during initialization, detailed design is required in the integration solution.

In the case of switching between the old and new systems, it is suggested that an independent team will be able to participate in the cutting analysis of the system and focus on possible solutions in the early stage of the project, the data analysis is performed based on the evaluation scheme, and how to switch data will be demonstrated in the subsequent launch phase. The research work of data situation is the basis of subsequent design.

In the data survey process of some enterprises, data are synchronized from off-premises data to on-premises data, and some sensitive data need to be specially processed, for example, the encryption of personal data of distributors or customers in off-premises enterprises (Figure 4.30 and Table 4.5).

VI. System integration

Figure 4.31 shows the overall integration architecture.

VII. Status quo of other technologies

Performance status of the integrated system: The Middle Office system faces C-end requests and generally carries a large number of requests. During integration with other applications, whether performance bottlenecks will be formed can be sorted out in the early analysis process.

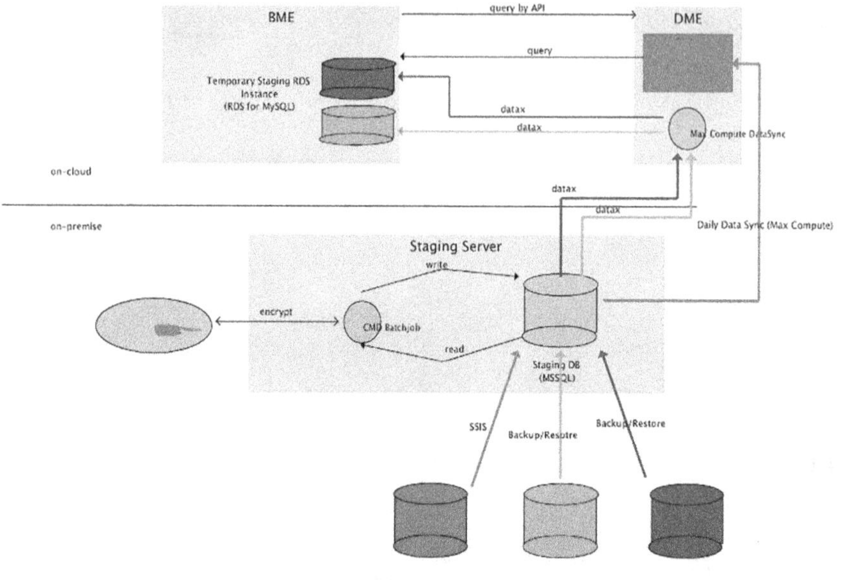

FIGURE 4.30 Cloud migration solution for Business Middle Office data.

TABLE 4.5 Cloud Migration List of Business Middle Office Data

SOURCE System	SOURCE Database Backup File (GB)	Size After Encryption/ Processing (GB)	Total Incremental Data (GB)
System A	XXX	XXX	XXX
System B	XXX	XXX	XXX
System C	XXX	XXX	XXX
Subtotal (GB)	XXX	XXX	XXX

4.5.3 Requirement Analysis

In this phase, we need to analyze the previous requirements survey, fully understand the requirements, and write the overall business flow and typical business scenario flow charts.

First, analyze the main process and design the business flow chart according to the actual business needs. The main process consists of several business scenarios. Next, you need to split the business scenarios in the process, draw a flow chart for each business scenario, and then draw a sequence diagram for each business flow chart to identify the entity objects and business operations in the business process through the sequence diagram, provides a data base for the following Middle Office capability center design (Figures 4.32 and 4.33).

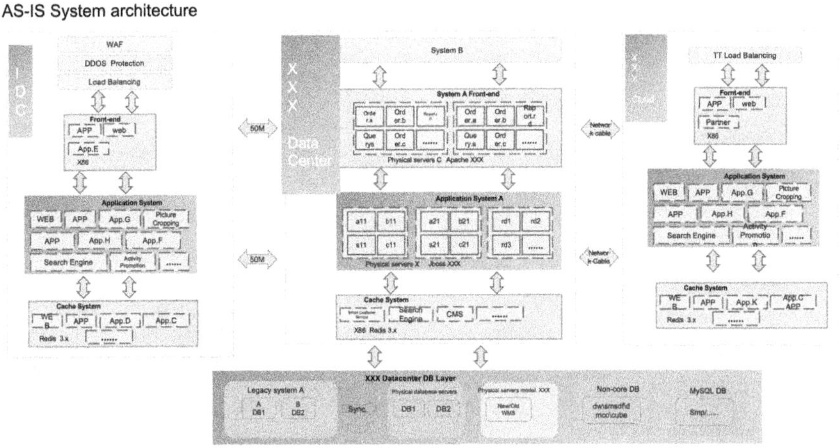
AS-IS System architecture
WAF
DDOS Protection
Load Balancing
Front-end
Application System
Cache System
50M
System B
System A Front-end
Application System A
Cache System
XXX Data Center
TT Load Balancing
Front-end
Application System
Cache System
XXX Datacenter DB Layer
Legacy system A
Non-core DB
MySQL DB

FIGURE 4.31 System integration architecture diagram.

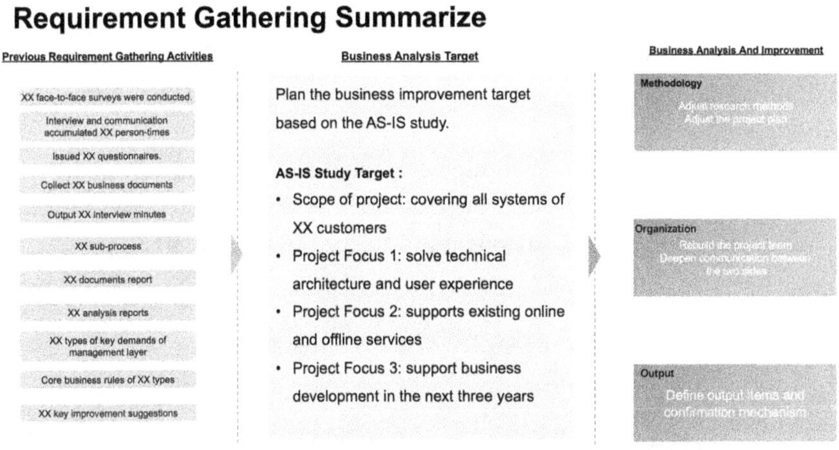
Requirement Gathering Summarize
Previous Requirement Gathering Activities
XX face-to-face surveys were conducted.
Interview and communication accumulated XX person-times
Issued XX questionnaires.
Collect XX business documents
Output XX interview minutes
XX sub-process
XX documents report
XX analysis reports
XX types of key demands of management layer
Core business rules of XX types
XX key improvement suggestions
Business Analysis Target
Plan the business improvement target based on the AS-IS study.
AS-IS Study Target :
• Scope of project: covering all systems of XX customers
• Project Focus 1: solve technical architecture and user experience
• Project Focus 2: supports existing online and offline services
• Project Focus 3: support business development in the next three years
Business Analysis And Improvement
Methodology
Adjust research methods
Adjust the project plan
Organization
Rebuild the project team
Deepen communication between the two sides
Output
Define output items and confirmation mechanism

FIGURE 4.32 Summary of research results.

4.5.4 Architecture Design

The construction of the Business Middle Office starts from business requirements and realizes the implementation of business functions through technical construction. After the Middle Office is implemented, it can support the initial business requirements and lay a solid foundation for the subsequent business development; at the same time, through a series of operation methods, the business is fed back from technology to achieve the development goal of a virtuous circle. Architecture design is

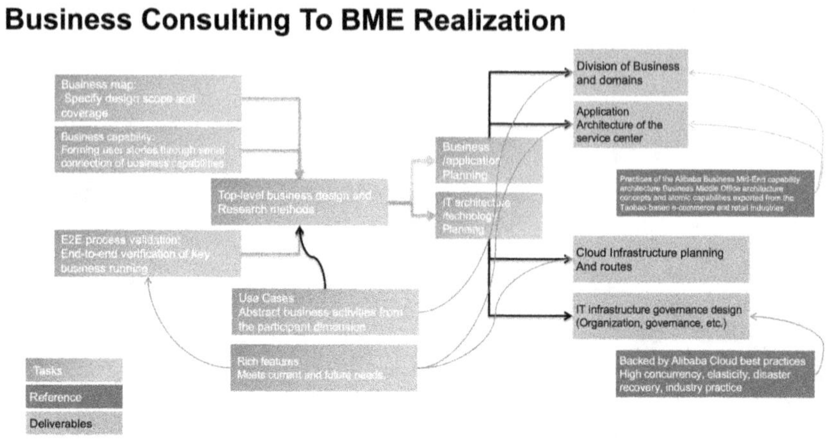

FIGURE 4.33 Connection relationship from research consultation to design.

the starting point of technology implementation. The process of architecture design is also the overall thinking process of Middle Office construction. A good architecture design will also provide a good prerequisite for Middle Office construction.

Figure 4.34 expresses the input factors that need to be considered in the architecture design phase and the characteristics of architecture iteration evolution. Among them, various input factors can be classified into three

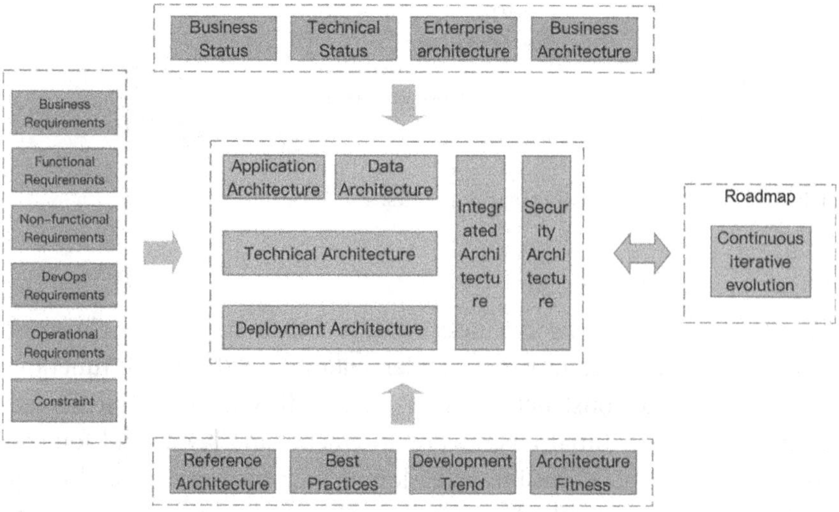

FIGURE 4.34 General Idea of architecture design.

categories: the first category is 'status quo' or objective facts, which can be the current status quo or the status quo that the enterprise is about to plan to achieve, in short, it is the current situation of the enterprise or business to which the system architecture of the Middle Office is applied; the second type is specific requirements, including specific business requirements, functional or non-functional requirements, O & M operation requirements or constraints, etc. These requirements can be implemented through Middle Office construction to meet the business development requirements within a specific period of time; the third category is the technical development planning section, the architecture design of the Middle Office should match the trend of technology development, ensure sufficient advancement and evolutionability, reserve space for the future Middle Office construction of the enterprise and ensure certain architectural adaptability, avoid being out of touch with reality due to excessive investment and excessive advance in the initial stage of Middle Office construction.

When the system construction of the Middle Office enters the stage of architecture design, the current situation of the enterprise, the current business situation and specific requirements have been clearly defined, therefore, the architecture design phase focuses more on how to implement the Middle Office system technology based on the above input. Generally speaking, a general-scale Middle Office system is a business system based on DDD and implemented through distributed technology after business analysis. However, this is not certain. The specific implementation form and architecture of the Middle Office depends on the business requirements. For example, for a small-scale enterprise, its technical capability is not strong enough and its R & D personnel are not enough. However, it is actually feasible to build its own system through the concept of Middle Office, as long as a framework with sufficient evolution capability or even a single service framework is planned, and the technical system will continue to improve as the business continues to develop; business capabilities can also be continuously innovated and broken through with this system, which is a good example of Middle Office evolution, or the Middle Office system is becoming larger and larger, and the overall linear expansion capability or the overall availability of the system can no longer be completely managed by a simple distributed system, so the Middle Office needs to continue to evolve its own unique capabilities, such as extended points, Remote Multi-Active or full-link stress testing. Therefore, how the specific Middle Office should be implemented should be determined in accordance with the actual situation. The architecture design is also the same. Do not blindly pursue after the most leading technology, while ignore the benefits and cost brought by new technologies, so as to achieve the benefits gradually to following up with business development trend.

Next, we will systematically explain the overall concept of architecture design, the process of implementation and the content of each architecture.

4.5.4.1 Architecture Design Concept

Generally speaking, architecture design is designed to solve the problems caused by business complexity. Therefore, business requirements and future evolution should be considered first in architecture design. If it is a simple system and there is no iterative development plan for a period of time, a relatively simple architecture, such as a single architecture, can be adopted. These benefits can save labor costs and avoid a series of maintenance and management problems caused by complex architectures. Even if it is a single architecture, as long as it achieves better layering and modularization, and avoids the problems of architecture wear caused by human beings, it can still support the business well, and after modularization, it also facilitates microservice splitting when business needs.

Single architecture is a good choice for simple systems. However, for such complex business requirements that require Middle Office construction, single architecture includes personnel organization composition and iteration speed, the disadvantages of scalability and other aspects are highlighted. With the development of business, we have to choose an architecture with higher overall complexity but more advantages in other aspects. The gradual maturity of Middle Office architecture has evolved with the needs of business and the growth of technology. The following describes several important technologies to promote Middle Office evolution from several aspects.

- Domain-driven design: Provides a theoretical basis for understanding complex business. Based on the concept of DDD, complex business scenarios can be split to generate business domains, aggregates, entities and key business definitions such as domain events. All of these are ready for the overall business dismantling to meet the needs of the field.

- Microservices: It breaks through the physical limits of traditional single architectures and solves a series of communication, link, configuration and other problems caused by microservice splitting through a series of microservice governance capabilities. Microservices also provide technical implementation basis for each business domain divided by DDD.
- Container: Provides an environment foundation for running microservices. After a business system is split into microservices, it may be divided into dozens or hundreds of microservices,[1] the environment dependencies of each microservice operation may be different, and the governance of the microservice operation environment has become a serious problem. Therefore, containers have the ability to isolate the running environment and resources, which solves the urgent problem of service operation for microservices.
- CI/CD (continuous integration/continuous delivery) system: The problems brought by microservices are not only in the governance and operation environment of the microservice system. When there are many microservices, the compilation and release of its own services have also become an extremely complicated and heavy task. Fortunately, these problems can be solved through tools; CI/CD system is invented for the automation to improve the overall release efficiency of the microservice system.
- Promotion of cloud computing: The important position of cloud computing has been gradually clarified in recent years and has now become the infrastructure of the information industry. Cloud computing can quickly provide computing, storage, network and other resources required for the operation of business systems. Rich cloud products can reduce the difficulty of system construction, at the same time, the O & M capabilities of cloud vendors can also make up for the shortcomings of general enterprise O & M capabilities, and promote the rapid incubation and launch of business systems. The construction of the Business Middle Office can maximize the advantages of cloud computing and promote the maturity of the Middle Office in various aspects such as construction speed, operation flexibility, system maintenance and security.

[1] The number of microservices on Taobao is far more than that.

Based on the above technical basis, the following basic concepts of Middle Office architecture design can be proposed:

- Distributed design: Use DDDs, microservices and containers;
- Configuration-oriented design: The configuration is isolated from the code. The service operation depends on the independent configuration of each environment. You can select Git Flow mode or compile and deploy multiple times at a time;
- Toughness-oriented design: Fault tolerance and self-healing;
- Elastic design: Elasticity based on time or system load;
- Performance-oriented design: Quick response, concurrency and efficient utilization of resources;
- Automation-oriented design: Automated DevOps system;
- Diagnostic-oriented design: Logs, measures and traces at the cluster level;
- Security-oriented design: Security endpoint, end-to-end encryption and overall security system.

4.5.4.2 Architecture Design Process

Generally speaking, the process of architecture design can follow the following figure 4.35. The design sequence shown in the figure is designed, because some subsequent designs need to be based on the previous architecture design as input. For example, the data architecture needs to plan the data distribution based on the application architecture, the technology architecture determines the technical components based on the functional definitions of the application architecture.

However, the process of architecture design may not be a waterfall. For example, in some cases, the technology architecture may affect the design of application architecture or data architecture in reverse. Therefore, the overall architecture design process is an overall iteration process with dependencies, and ultimately achieves an overall ideal goal.

Architecture design is the starting point of Middle Office design. The whole process actually simulates the construction process of Middle Office, allowing designers and readers to gradually understand the construction

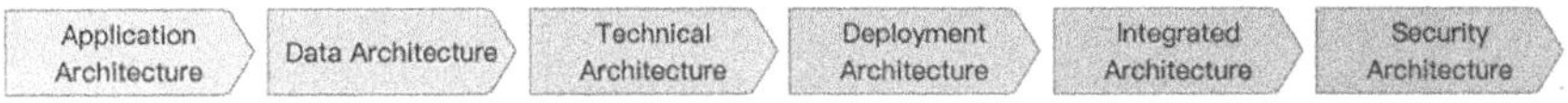

FIGURE 4.35 Process of architecture design.

objectives of the system from the beginning of business, data organization, technology implementation, overall deployment, internal and external interaction and security system. These can summarize all the main aspects that a system should focus on in a large scale, so that the Middle Office construction team can have an overall understanding on the macro level and have a basic basis in the process of landing (Figure 4.35).

4.5.4.3 Application Architecture

The application architecture analyzes and abstracts business processes and scenario use cases through the understanding of business and the analysis method of domain modeling, and completes business domains, aggregation, entities, field events and other field-related analysis results, and the specific business implementation according to the foreground application plus shared service center model to form an application architecture, accurately defines the application scope, functions and modules.

Application Architecture is the intersection of system requirements and technology implementation. Through the definition of application architecture functions and scope, it provides guidance for subsequent system construction. For example, in the subsequent design of data architecture or deployment architecture, the application architecture can be used as the design basis; or in terms of organizational structure, the organization can be divided according to the application architecture, in this way, the boundaries and responsibilities of the organization are clear at a glance.

As shown in the following figure, the service center defined in the application architecture is a division of domain capabilities from a business perspective, but the relationship between the service center and microservices may not be one-to-one. Microservices are divided into one or more domain-driven contexts, taking into account the complexity, Personnel organization, and hardware resource allocation, overall system management, and release complexity.

The business application defined in the application architecture, also known as the foreground, is the service closest to the business and directly undertakes the business. They can be quickly implemented based on the

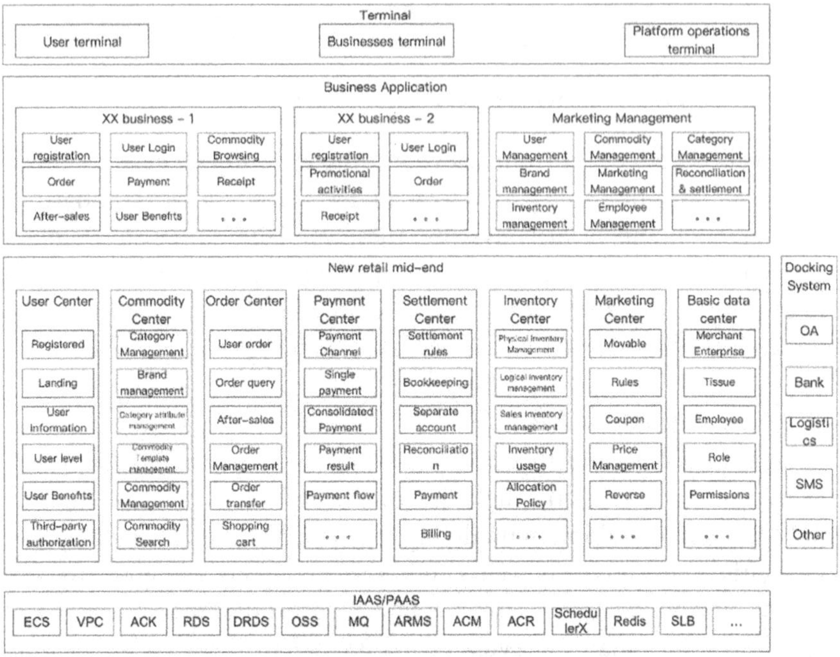

FIGURE 4.36 Application architecture design example.

capabilities provided by the Middle Office. The significance of the Middle Office is to help the business succeed quickly.

This application architecture is a highly abstract of the overall business. In actual design, it needs to be supplemented on the basis of this overall architecture to describe the specific responsibilities of each module as clearly as possible (Figure 4.36).

4.5.4.4 Data Architecture

The application architecture defines the domain model of each service center, so the data architecture is the entity (DO) to persist the model (PO) design the implementation in the form of storage. The definition of data architecture involves not only databases, but also data-related capabilities and models such as caching, object storage and search engines. It has completely planned the data storage form, function positioning and technology selection in each shared service center. Figure 4.37 shows an overview of the situation of each center. Each center can be refined based on the actual usage of data, such as the specific selection of data storage, the size of data storage, the link used by the data, the method of backup and recovery, etc.

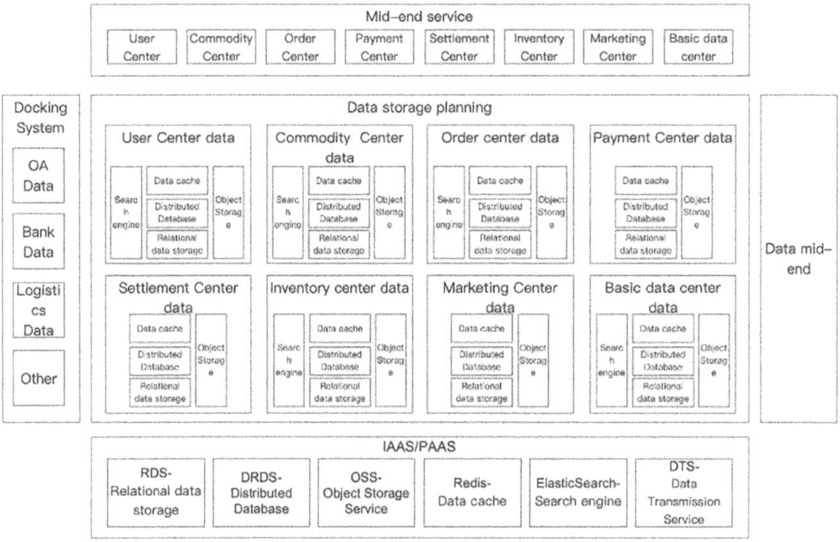

FIGURE 4.37 Data architecture design example.

4.5.4.5 Technology Architecture

The technology architecture is designed based on the application architecture and data architecture, taking into account the future development trend of technology, matching technology with business and data implementation, etc. As shown in the following figure, the technology architecture determines the overall technical selection. It describes in detail the application of the underlying basic resources of the Middle Office system, the selection of cloud product storage and middleware, and the technical points used by Middle Office services.

All Middle Office team members should pay attention to the technology architecture, which is the refinement of technical solutions in the implementation process of Middle Office personnel, and directly defines the technical-related work content of Middle Office personnel to avoid excessive use of Middle Office personnel, due to the excessive selection of technology, the overall Middle Office technology stack is too large; and the technical framework is also the technical requirement for Middle Office implementation personnel, the Middle Office implementation personnel are required to be able to fully understand and properly use the technical points within the selection scope related to the work they are responsible (Figure 4.38).

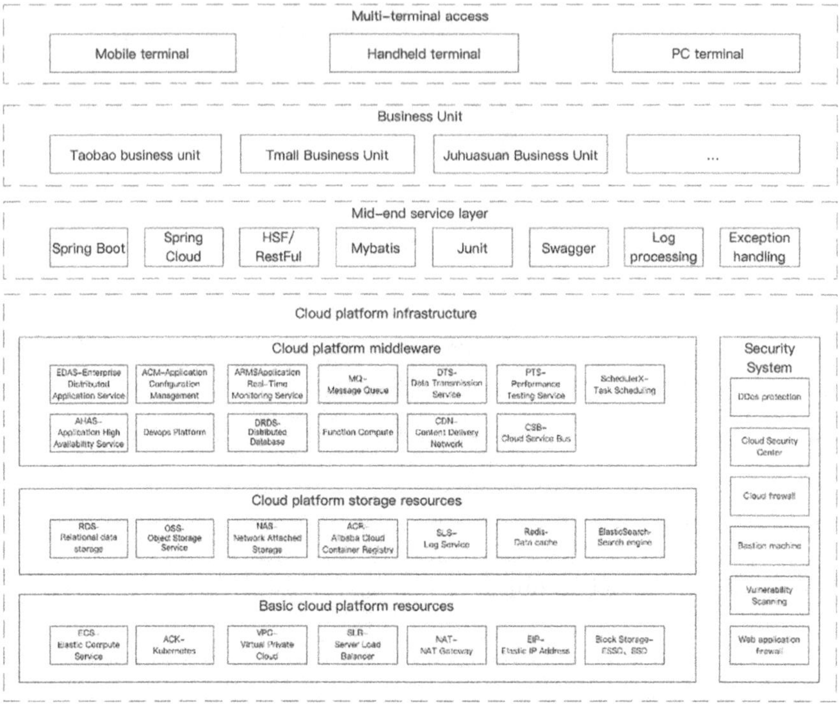

FIGURE 4.38 Technology architecture design example.

4.5.4.6 Deployment Architecture

The deployment architecture describes the relationships among all cloud products, services, networks and request links used by the Middle Office system. As shown in the following figure, the deployment architecture describes that all resources are in one zone, and the specific services run in Pod on, the situation of the cloud products used; all services Pod run on Alibaba Cloud Container Service for Kubernetes (ACK) in, ACK uses Elastic Compute Service (ECS) as a server resource; ACK and Enterprise Distributed Application Service (EDAS) in namespace is a one-to-one correspondence; the underlying cloud resources and EDAS the microservices in are all in the same VPC the access path from the client to the Middle Office service. It also describes some more detailed contents, such as services. Pod pass NAT access the Internet.

This figure clearly describes the deployment of all resources in a zone, but the specific deployment of the Middle Office system may not be in a zone. For example, the Middle Office system considers certain disaster

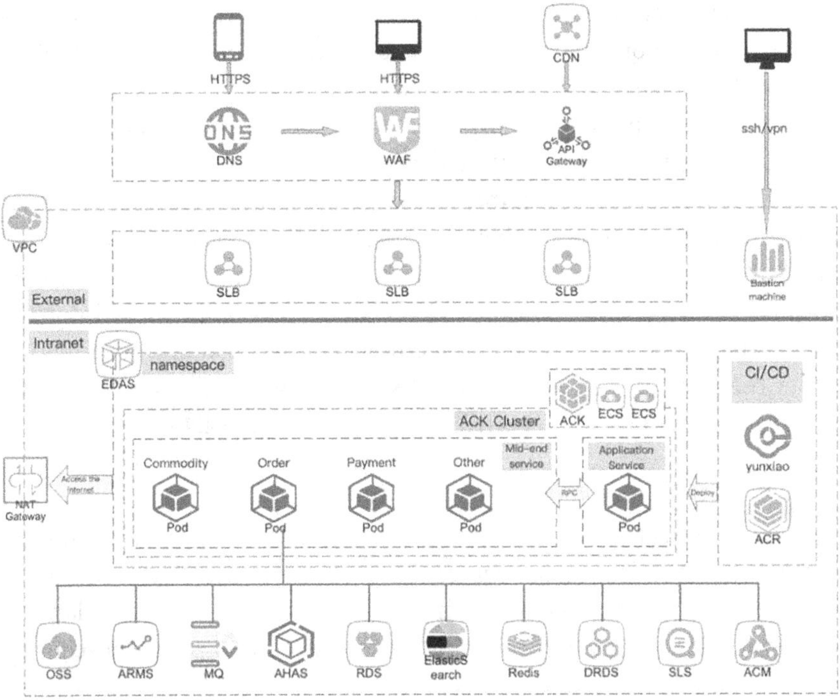

FIGURE 4.39 Deployment architecture design example.

recovery requirements, may be selected in a region[2] in the more available area[3] deploy; or deploy in multiple regions like Alibaba, and implement traffic switching between systems in multiple regions. This technology is called Remote Multi-Active in Alibaba, related implementations cannot be solved simply through deployment, but also take into account technical implementations such as service registration and data synchronization.

Therefore, the deployment architecture is not only concerned with the satisfaction of business functional requirements, but also with the technical implementation and non-functional requirements (Figure 4.39).

4.5.4.7 Integrated Architecture

Based on business requirements and context, the integrated architecture analyzes the business coupling relationship between the Middle Office

[2] The distance between regions is generally relatively long, with a span of nearly 1,000 km and obvious network delay.

[3] The distance between zones is generally not very long, about dozens of kilometers, and the delay is basically controllable on the premise that the request link does not jump frequently.

system and other business applications of the enterprise, as well as external applications of the enterprise, so as to realize the business coupling relationship between the Middle Office system and other business applications of the enterprise, planning of data integration and application integration between enterprise external applications.

As shown in following figure, it simply enumerates all docking systems and communication modes between systems. If there are many external docking systems in the Middle Office and different systems are connected with the same type of functions, or you want to describe more information through the integrated architecture, you can also add information about the functions and functions of specific communications and even major interfaces to the integrated architecture (Figure 4.40).

4.5.4.8 Security Architecture

Security architecture is based on business security requirements and relevant security compliance requirements, from business security, application security, data security, basic security, account security, plan the security of the Middle Office system in terms of cloud platform security, security monitoring and operation. As shown in the following figure, the security concerns of the Middle Office are listed in more detail, which

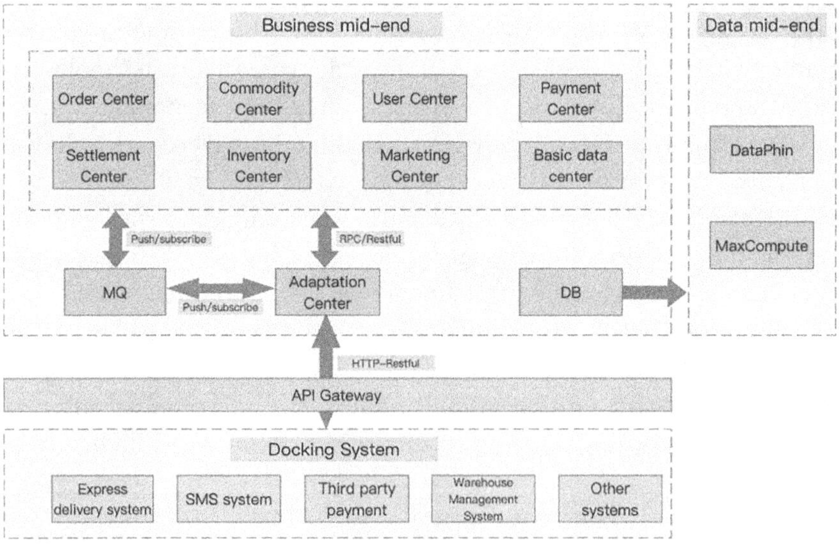

FIGURE 4.40 Integration architecture design example.

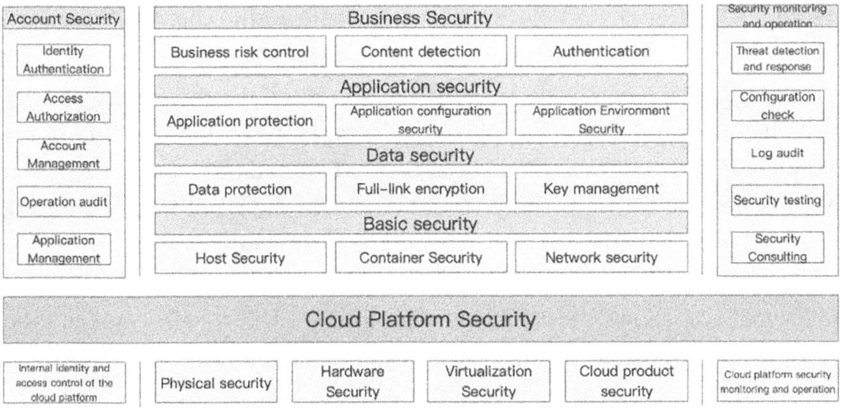

FIGURE 4.41 Security architecture design example.

need to be paid close attention to when building the Middle Office and implementing in combination with the actual business requirements and technical conditions of the Middle Office (Figure 4.41).

4.5.5 Business Middle Office System Construction

The system construction is based on requirement analysis, domain design and architecture design. On this basis, the detailed design of the system is further planned and refined. After the detailed design is completed, the specific capabilities of the service are constructed through code, complete the overall system construction through the construction of resources such as services and middleware. Generally speaking, there will be a lot of contents in the work of system construction, and a certain amount of manpower needs to be invested in each link to meet the needs of the work, it may not be suitable to use waterfall's research and development mode to build the whole system. Therefore, for Middle Office system construction, it is recommended to use agile methods to complete this process.

In an agile way, a minimum closed loop on the business is planned to be implemented and verified as soon as possible. The success of the implementation verification can basically demonstrate the feasibility of the overall architecture, and it can also expose and solve problems in advance; and this agile iteration method also conforms to the gradually mature concept of the Middle Office, and the subsequent enterprises can completely reuse this mode for the improvement of Middle Office capabilities. However, if the waterfall mode is adopted in the construction of the Middle Office, it may be in a hurry in the later stage of construction, and the ship is not easy to turn around, and the Evolution Mode of the Middle Office also lacks the opportunity to drill due to the use of the waterfall mode, the inertia of system construction may be buried. If the rapid iteration capability of the Middle Office is not well built, the ability to respond quickly to the system becomes empty talk, which is not conducive to the rapid maturity of the Middle Office, and also loses the advantage of the Middle Office to quickly support business construction.

Figure 4.42 shows the main iteration links in the system construction phase and the main work contents of each link. The following sections will explain the overall construction process of the Middle Office.

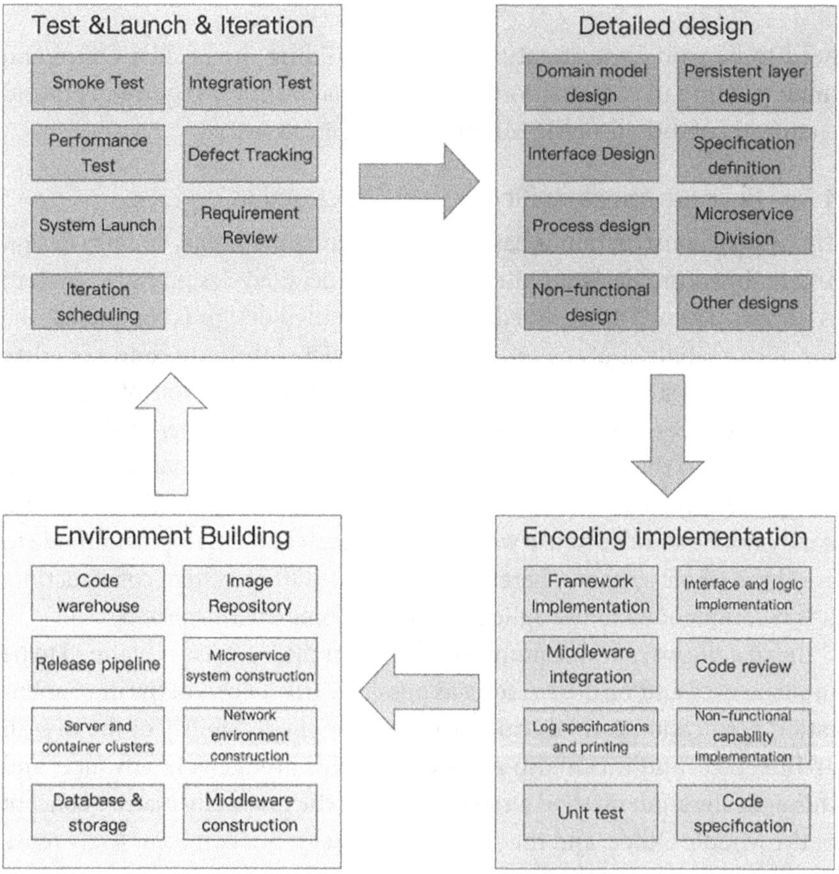

FIGURE 4.42 System construction closed loop.

4.5.5.1 Detailed Design

Based on the previous research and analysis of the business requirements, as well as the analysis of domain-driven model, the splitting of the center and the ability identification, the detailed design of the specific center starts from the relationship of the domain model, gradually refine the persistence model definition, interface definition, main process, event release and listening design content in the center.

1. Domain model

   Domain modeling is a way of system construction and analysis of business logic. Its greatest advantage is that it connects business language with technical language, it provides a translation approach for business-to-system implementation. Based on the domain model, domain capabilities are realized through technology research and development, that is, business logic is also realized.

   The detailed design of a specific center in the Business Middle Office starts with the definition of domain capabilities. As shown in Figure 4.43, in the business analysis and domain design phase X as shown in the model relationship, there is a one-to-many relationship between categories and commodities, and a one-to-many relationship between brands and commodities. This corresponding relationship provides the design basis for the persistent model of the system; the central capabilities extracted in the domain design phase can be implemented at the domain layer and application layer within the

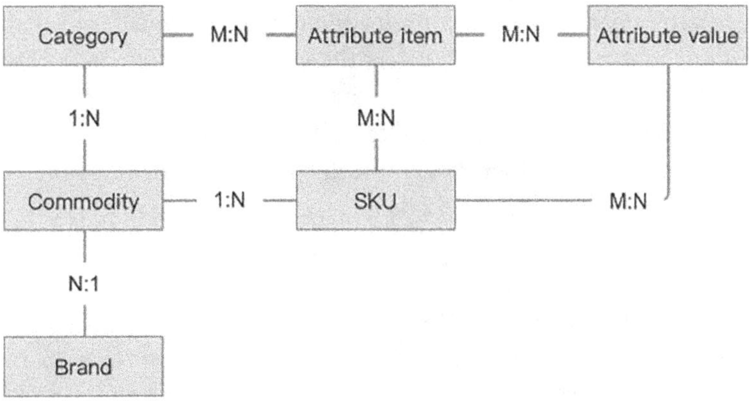

FIGURE 4.43 Category, commodity and brand model relationships.

service and exposed through interfaces, it is used for inter-service calls or front-end services. Domain events can be transmitted and monitored between services through messages to handle events. Therefore, the implementation of the domain by technical means can realize the implementation of business requirements.

2. Microservice Division

Domains are the understanding and division of business, while microservices actually need to be implemented with microservice development, so there is a problem of domain and microservice transformation. There are many factors that need to be considered in how to convert them. The following lists the effects of these factors on the division of microservices.

Inter-entity relationship: The division of domain entities is the result of domain modeling. Entities obtained through domain modeling have entity relationships such as inclusion and association. Generally speaking, the smallest aggregation unit between entities is the aggregation root and its related entities, so the entities in the aggregation root cannot be divided into different services. A larger level is the bounded context, which defines the smallest business boundary. Therefore, the bounded context can be used as the smallest unit for microservice splitting. However, in actual system construction, it is not necessary to split the system based on the bound context, and other factors should be considered.

Personnel and organizational structure: This is also a very important factor affecting the division of microservices. The number of microservices needs to consider the carrying capacity of organizations and personnel, as well as the communication costs between organizations. If there are too many microservices and the number of people and organizations cannot match them, additional costs will be incurred. Therefore, from the perspective of personnel, multiple bounded contexts will be merged into one microservice.

Non-functional considerations: For example, if there is a strongly consistent transaction between two bounded contexts, the implementation method of the transaction must be considered. For example, put two bounded contexts into a microservice; or multiple bound contexts are in a microservice. If a certain bound context requires frequent iterations or high-performance requirements, you can consider splitting the bound context into microservices independently.

Modularization and evolutionability: Even though many factors are considered in the design, the previous microservice division may be unreasonable after a period of actual operation. In fact, the emergence of this situation does not mean that there are problems in design, but that with the development of business, the adjustment of personnel and the change of technology implementation, microservices need to evolve accordingly. Therefore, this situation should be taken into account at the beginning of the design. At present, we should do a good job of modular encapsulation in microservices. When we need to split or merge microservices, we should use modules as units for splitting or merging, enable Microservices to grow together with the business.

For Figure 4.44 with business semantics analysis, categories, attribute items and attribute values can be divided into a bounded context. SKU can be divided into a bounded context; brands can be divided into a bounded context independently. There may be a question here, that is, SKU it is also associated with attribute items and attribute values. Why can't it be divided into a bounded context? This needs to be analyzed in combination with actual business scenarios.

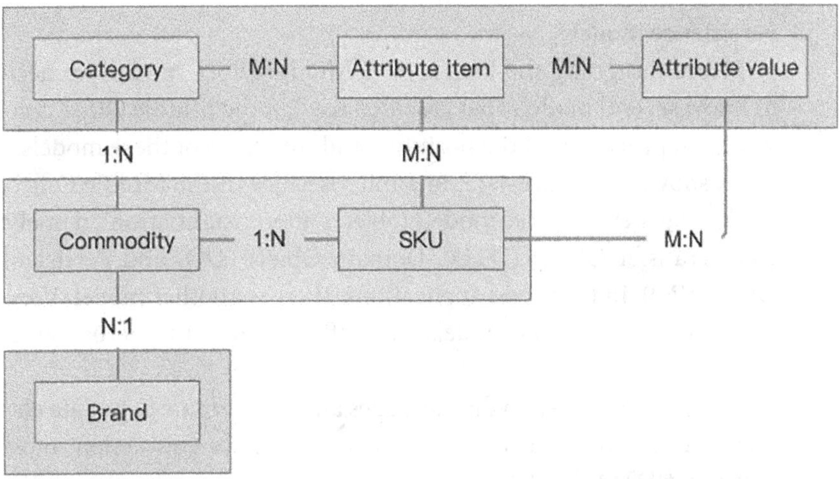

FIGURE 4.44 Category, commodity and brand domain model boundary demarcation.

In actual scenarios, you need to attach an attribute item to the leaf node of the category, and the attribute value needs to be attached to the attribute item. While SKU is different, after the commodity is attached to the category leaf node, you can select the attribute items to be associated with this category. Therefore, SKU and attributes are not divided into a bounded context.

After dividing the bounded context, consider how to divide microservices. If there is little business pressure or manpower shortage, you can consider dividing these three bounded contexts into one microservice, and the microservice is modularized to facilitate future microservice splitting; if you consider that commodity-related information will be frequently accessed or changed, then the category and brand can be classified as a microservice, commodity and SKU, this bounded context creates a separate microservice, SKU obtains property-related information through RPC calls. If you consider the loss of SKU remote calling properties, you can cache the properties in the product microservice (but pay attention to the update and synchronization of attributes). Of course, we can also set up microservices separately for these three bounded contexts. Therefore, the division of microservices needs to be analyzed according to the actual situation, rational trade-offs should be made and problems caused by different divisions should be solved.

3. Persistence model

Before designing the long-lasting Middle Office model, we need to know several models that are often used in the Middle Office construction process, and the positions and functions of these models.

As shown in Figure 4.45, in a microservice in the Middle Office, there are generally three models (objects after instantiation), namely, Data Transfer Object (DTO), Domain Object (DO) and Persistent Object (PO); in front-end applications, there is another model, View Object (VO). Different models play different roles in microservices or systems.

VO: Typically used to display pages and encapsulate page data elements. Data are usually composed of DTO after conversion, most fields and DTO the data content are the same, but considering the diversity of page display, page display is generally not directly reused DTO model, instead, use the newly defined View model (VO).

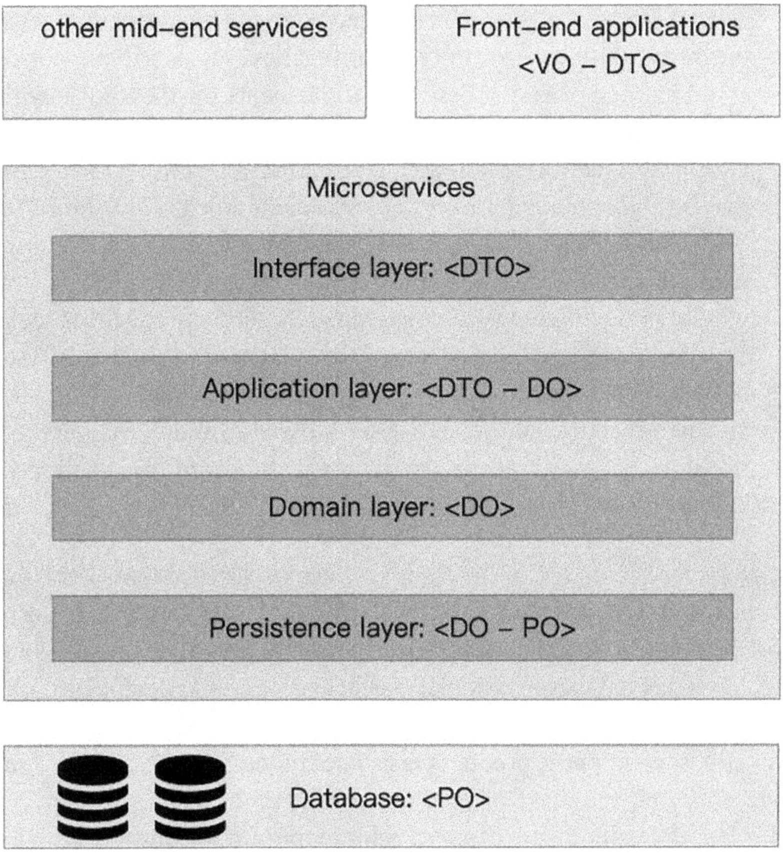

FIGURE 4.45 Commodity persistence model.

DTO: Mainly used for data transfer between services, DTO attention should be paid to the refinement of common features in the design. For example, when a front-end application uses the gender field, VO the display is 'male' 'female'; and another front-end application VO 'Little Brother' and 'Little Sister' need to be displayed. 'In this case you cannot adapt DTO to each front-end, so a DTO consider using 1' And '0' as a universal gender identifier, and personalized VO to implement. In DTO, there is another very important task, because DO the model may have many fields or many nests. Not all of them need to be transmitted during transmission. Therefore, DTO take the responsibility DO the function of transforming into a suitable transmission model.

DO: Domain object, which is characterized by business fields and business methods, can implement business logic. It does not care about storage or transmission. It implements business logic within entities and aggregates through domain-driven implementation.

PO: The real persistent database model, PO and DO sometimes may be highly matched, especially on business fields. PO and DO can be one-to-one correspondence, PO pay more attention to the implementation of the database and technical implementation, so there will be some differences. For example, in PO there are fields such as 'creation time' and 'update time' which is more related with system usage; or PO is to implement the optimistic lock, add 'version' field; or when the two DO models have a many-to-many mapping relationship, PO may create a separate table to record this relationship, but the model of this table does not exist in DO.

Therefore, for the design of the Middle Office persistence model, the main consideration is that the business model (DO) through a certain transformation, a persistent model that can be saved persistently, contains business attributes, contains database additional fields and retains model relationships through some associated fields or database tables. As shown in following figure, the content of the product persistence model is expressed. 'id', 'gmt_create' and 'gmt_modified' are additional fields in the database. 'name', 'url' and 'other_do_column' are required fields in the business model. 'brand_id' and 'category_id' are externally associated fields (Figure 4.46).

1. Interface Definition

   An interface is a way for microservice capabilities to be provided externally. The implementation of the interface capabilities is achieved through the application layer and domain layer within microservices, while the interface is exposed externally, making external services accessible starts with the interface definition. The general interface definition must include the interface name and function, Interface Service Name and path, RT and QPS, request parameters, interface responses and error codes. For some special interfaces, such as when the caller needs to pass in encryption information, you need to describe the encryption method in the interface definition so that the caller can implement encryption according to the description. Therefore, the definition of an interface is mainly to

	Column name	Type	Default value	Is empty	Description
Specification requirements	id	bigint		No	Unique Identifier
Model business fields	name	text		No	Commodity name
	url	text		No	Commodity details link
	other_do_column	...		...	...
External association	brand_id	bigint		No	Brand id
	category_id	bigint		No	Category id
Specification requirements	gmt_create	datetime		No	Creation Time
	gmt_modified	datetime		No	Modification time

FIGURE 4.46 Content of commodity persistence model.

enable the caller to know how to use the interface as much as possible and reduce the communication cost.

The following is a brief description of each part of the interface definition:

- name: commodity query
- note: through the commodity ID query commodity information
- service name: commodity
- path:/commodity/{id}
- key metrics:

RT	One Machine QPS
300 ms	200

- request parameters:

Field Name	Type	Required	Description
commodityid	bigint	Yes	Unique ID
option	xxxDTO	No	Additional query expansion of commodity information, such as reading a certain identification bit in an object to determine whether to query the current commodity inventory. (This field will not be demonstrated later.)

- response:

Field Name	Type	Required	Description
Result	CommodityDTO	Yes	Commodity details
Success	Boolean	Yes	Successful
Code	String	Yes	Error codes
Message	String	Yes	Additional information
ext	Map<String,Object>	No	Extended Information

- error code:

Error Code	Error Description	Solution
Commodity-x-001-01-001	Commodity does not exist	Check ID correct
Commodity-x-001-01-002	The commodity has expired	When querying a commodity, the value of the added commodity is still invalid

2. Main process

   For some important or special functions in the service, it is not enough to describe them only through interfaces. In this case, more detailed descriptions are needed, so timing diagram is a good choice. As shown in Figure 4.47, it describes the main process of obtaining commodity. Since reading commodity is a high-frequency action in the business, it is necessary to add a cache. In this sequence diagram, the lazy loading method is used to cache commodity. Of course, the cache loading method can also be replaced according to the actual system needs, for example, the preloading method is used. If a microservice contains domain events, you can also use the sequence diagram to describe the entire process.

3. Non-functional design

   Non-functional design mainly solves the business metric requirements clearly put forward and the requirements that are not explicitly put forward in the business requirements and need to be obtained through analysis of the business requirements.

   Business Metrics include a series of system non-functional requirements, such as RT (response time), QPS (load capacity), RTO (recovery time target), RPO (recovery point target), storage space and auto scaling.

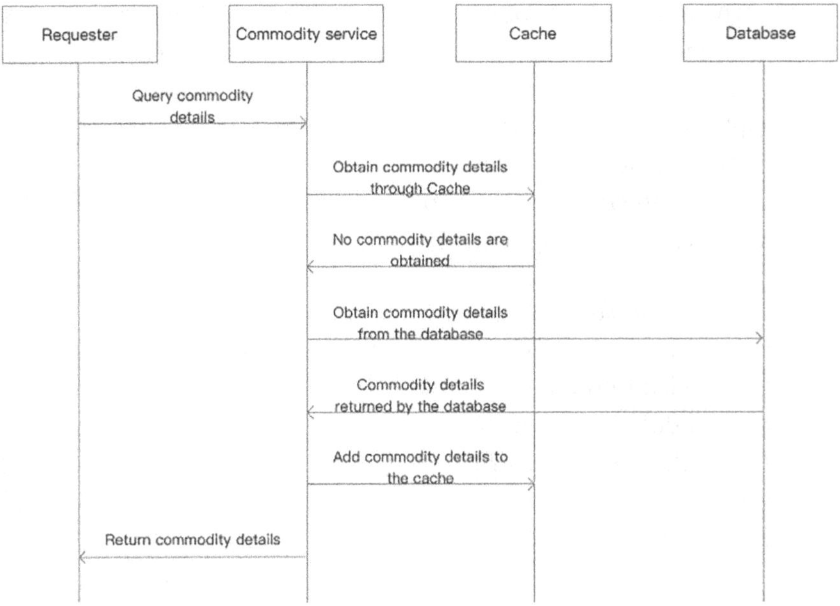

FIGURE 4.47 Main process of commodity detail acquisition.

The other one cannot be viewed intuitively in the business requirements. For example, the overall user size of the user Center is estimated to be 0.1 billion, it is obviously inappropriate to use MySQL single-table storage. Therefore, you need to consider database sharding and table sharding. If you need to search by user name or other information, the performance of searching directly in the database is not as good as expected, you must consider adding search engines to user information, or select strong consistency or final consistency for the operation of a transaction, or the idempotent implementation of an operation.

4. General protocol

The general protocol is not limited to the interior of a center. It is an agreement that each center in the overall Middle Office construction complies with in the detailed design stage, the content of the specific agreement mainly covers the common technology and design parts of each center. General specifications generally include database design specifications, coding specifications, log specifications, security specifications, unit testing specifications and design specifications. For more information, see Alibaba Java development manual.

There are many common protocol entries. If all of them are regulated and implemented manually, the cost is relatively high. Therefore, the protocol needs to be solidified through scaffolding or implemented through tool-assisted inspection. Scaffolding can refer to the instructions in the code implementation section of the following chapters; Auxiliary tools can be found in IDE to install the Alibaba protocol plug-in in 'Alibaba Java Code Guidelines'.

4.5.5.2 Coding Implementation

After the detailed design is completed, the design results of each center need to be coded into Java services that can be run. Generally speaking, during the Middle Office construction process, the coding phase takes up about half of the overall construction cycle, so how to improve the coding quality and efficiency is an important part of continuous exploration and optimization in the Middle Office construction process.

1. Standard scaffold

   Writing microservices usually starts with a basic scaffold, so if we can define general capabilities and framework standards in the scaffold, it can improve programming efficiency, improve programming quality, unify technology stack, convenient personnel scheduling and mutual undertaking, convenient code review and other effects. The following is a list of the recommended capabilities of scaffolding (Table 4.6).

2. Microservice framework

   Figure 4.48 shows the[4] summarized description of the microservices code project architecture: from module capability definition to module internal package function. In this way, the functions of technical implementation are separated from each other through modules. For example, api_module can be referenced by other services as a two-party package, and the start-up Module boot_module is separated from the business implementation Module domain_module; modularization within the business logic, namely domain_module, is also considered, which provides a technical basis for the splitting and merging of microservices.

---

[4] The module and package names in the figure are added module and package, this is to facilitate readers to read, but this content does not need to be added to the actual project.

TABLE 4.6 Standard Scaffolding Capacity List

Scaffolding Capability	Description
Module definition	Define the modules and layers of scaffolding; for example, through Module separate large functional areas; in business implementation Module internal pass package distinguish between business logic modules and layers
Standard use of storage and middleware	Inject the common middleware of each center by default, and provide standard writing or default code generation
Test-related capabilities	Unit testing; Mock code
Interface definition	Integrated Swagger convenient interface debugging The standard request response format If necessary, consider passing AOP the Interface section logs are printed
Log printing	The standard log output format Log output settings
Code check	Call relationship detection, code check, etc.
Unified dependency and packaging management	Pass maven the parent medium pom file to implement unified management of dependent versions Provides default dockfile file, unified packaging specifications
Other auxiliary capabilities	Unified error code management; unified exception capture

3. Interface and implementation

To facilitate local debugging and viewing request input and output parameters, you can introduce Swagger to help implement this capability. For more information about the usage of Swagger, see the official documentation.

In the microservice framework described earlier, use api_module as the interface definition module, this module uses a communication protocol to define the communication mode of the interface by default. If more communication protocols need to be met in the actual system construction process, consider splitting this module again and splitting it into a pure API definition module and multiple protocol interface definition modules. Each protocol module references API to define the module, and pure API module interface adds the protocol information.

As for how to implement the business logic of the interface, you can combine the previous scaffold architecture layers with the idea of domain encapsulation. Each module at each layer assumes corresponding responsibilities to implement the business logic of the interface:

```
▾ xxxMicroService
  ▾ api_module
      api_package
      dto_package
  ▾ boot_module
      config_package
      aop_package
      application
  ▾ infrastructure__module
      cache_package
      mq_package
      oss_package
  ▾ share_module
      exception_package
      util_package
      enums_package
  ▾ domain_module
      controller_package
      consumer_package
      service_package
    ▾ domain_package
      ▾ xx_bounded_context_package
          entity_package
          event_package
          repository_package
          service_package
        xx_bounded_context_package
```

FIGURE 4.48 Microservice module division.

- xxxController: The interface implementation class in the domian_module controller_package module inherits an assumece class in api_module. This file is mainly used to supplement communication protocol information, parameter verification and standardized return. This file contains the corresponding class instance xxxService in service_package, and calls the method in xxxService when implementing logic.
- xxxService: Called by xxxController is the implementation class of a specific interface. When a microservice has multiple bounded

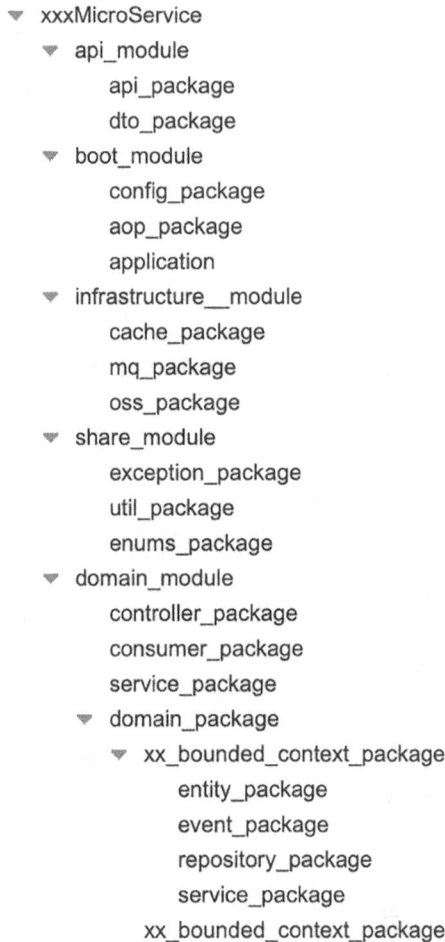

contexts and an interface needs to call multiple bounded contexts, you can use this file to assemble and call xxxDomainService down.

- xxxDomainService: The capability encapsulation in the bounded context. You can use it to load domain entities and orchestrate the capabilities of domain entities. You can call methods in xxxEntity.
- xxxEntity: Specific aggregate root and entity, with domain capabilities, can call such methods to implement the business logic of the entity.

After the API is executed, it is generally returned to the caller in the standard return format. For more information about the standard return, see the following code:

```
@ApiModel("standard response")
public class ResultResponse<T> {
    private static final String SUCCESS_MSG = "success";
    @ApiModelProperty(
        value = "business error codes, 如Commodity-x-
        001-01-001, Commodity-x-001-01-002",
        example = "Commodity-x-001-01-001"
    )
    private String code;

    @ApiModelProperty("additional messages")
    private String message;

    @ApiModelProperty("successful")
    private Boolean success;

    @ApiModelProperty("some meta information or extended
    information")
    private Map<String, Object> ext;

    @ApiModelProperty("actual data, generally DTO")
    private T Data;

    public ResultResponse(String code, String msg, T data) {

        this.code = code;
        this.message = msg;
```

```
            this.Data = data;
            this.success = true;
        }
        ...(omit other content)
}
```

1. Unit test

   Middle Office systems are generally highly complex, and require fast iteration speed in order to quickly respond to business requirements. Under this premise, how to ensure speed and quality is particularly important. Unit testing is a very important part of Middle Office construction. In short, unit testing is used to cover the main business logic and pass Mock simulate external responses. This service uses Assert to check whether the execution result of the unit test is correct. Run a unit test after each code is written and before the code is submitted to ensure that local modifications and submissions do not affect the existing logic and avoid the unavailability of the existing logic due to this modification, so as to achieve the goal of ensuring quality.

2. Scan

   Code scanning has different operation methods depending on the plug-ins and tools introduced. The following is a brief list of the usage of 'Alibaba Java Code Guidelines'.

   After installing the 'Alibaba Java Code Guidelines' plug-in in the IDE, restart the IDE. Then, right-click the project to see the newly added function menu as shown in Figure 4.49. Click code protocol scan to scan the selected project. After the scan is completed, you can see the scan result as shown in Figure 4.50, so that you can find the corresponding code to modify it according to the scan prompt.

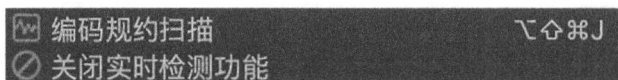

FIGURE 4.49 Add Alibaba coding protocol scan plug-in.

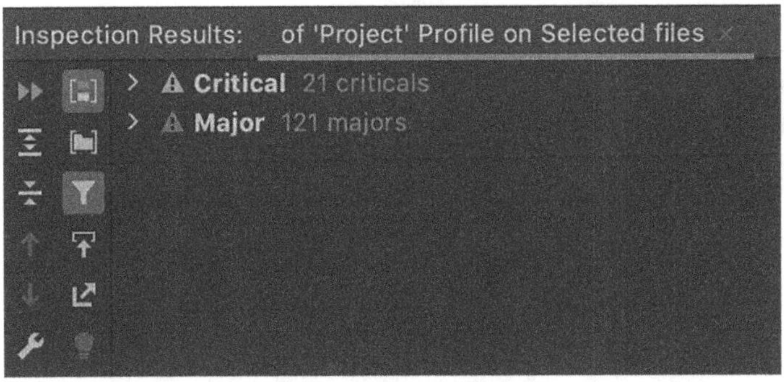

FIGURE 4.50 Scanning results of Alibaba coding protocol.

3. Environment building

During the development of the Middle Office to the launch of the system, four environments are usually used (of course, they can also be adjusted according to actual needs). The positioning of these four environments is different, as follows:

- DEV environment: The R & D environment used by R & D personnel for R & D and debugging;
- SIT environment: Integrated test environment, mainly used by testers for internal testing;
- UAT/SIM environment: User acceptance/simulation environment, mainly used for acceptance or experience of actual users;
- PRD environment: Production environment, the environment in which the system truly provides external services.

We recommend that these four environments have independent networks and operating resources.[5] As shown in Figure 4.51,[6] each environment is

[5] The service carrying capacity of the environment is different, and the construction is based on the actual needs. The capacity is not completely equivalent.

[6] Figure 4.51 shows the Deployment diagram of each environment. For more information about the deployment in the environment, see the deployment architecture. For more information about the construction of specific cloud products in the environment, see the official documentation.

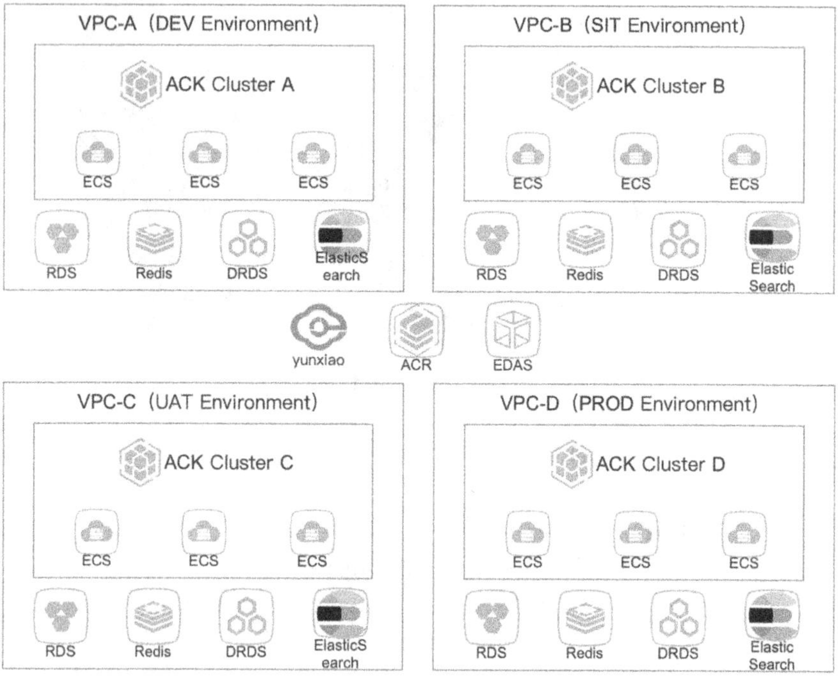

FIGURE 4.51 Business Middle Office standard 4 sets of environment.

isolated from each other,[7] and the code parts of these four environments are unified, release in different environments through CI/CD capabilities.

4. Code hosting and publishing

   The Middle Office construction is accompanied by a lot of code work. A good code warehouse can improve the overall code management capability of the team. You can choose to build your own code repository or choose some managed repositories. 'Codeup' is a code management product of Alibaba Cloud, it provides code hosting, code review, code scanning and quality inspection functions to help enterprises achieve secure, stable and efficient R & D management. You can select 'Codeup' as the Code Management Platform for Middle Office construction. For more information, see the official documentation.

[7] The specific isolation method depends on the characteristics of cloud products. For example, each environment corresponds to an ACK cluster; four environments correspond to an EDAS, and each environment corresponds to a namespace in EDAS.

After the code warehouse is selected, how to match the branches of the code warehouse with the actual work content is a problem to be solved. You can choose GitFlow workflow, which is used as a Middle Office code management method. GitFlow's several common branches are master, release, develop, feature and hotfix. The function and operation mode of each branch have its own characteristics. If this kind of information needs more space, readers can search and view it on the Internet. However, there are few discussions on the corresponding relationship between branches and environment on the Internet. The following are some suggestions on the corresponding relationship between branches and environment, hoping to be helpful to you (Table 4.7).

Generally speaking, there are many microservices in the Middle Office system. If you use manual publishing, it obviously does not meet the requirements of the Middle Office for fast and agile, therefore, you can choose Alibaba Cloud's 'Yunxiao' product. 'Yunxiao' has the function of pipeline and is the carrier of continuous construction. It completes the continuous construction from development to launch through construction automation, integration automation, verification automation and deployment automation. In addition to the pipeline capability, Yunxiao also has requirements management, defect management, knowledge base (KB) and other functions, which can fully manage the Middle Office R & D lifecycle from requirements to launch, for more information, see the official documentation.

TABLE 4.7 The Corresponding Relationship between Gitflow Branch and Environment

Branch	Environmental	Note
Master	PRD	After the production environment goes online, master merge branches into the latest code to ensure that PRD the environment is unified with this branch
Release	UAT	It is the branch to be released of PRD, which is consistent with UAT; after the PRD environment is released, the code is merged into the master branch
Develop	DEV/SIT	Branch of R & D environment and internal test environment
Feature	DEV	The branch that the R & D personnel are developing
Hotfix	PRD	Emergency bug the branch of the repair. After the repair, go online PRD environment, and merge into other branches

4.5.6 Business Middle Office Project Commonly Used Alibaba Cloud Products

Alibaba Cloud provides a wealth of base products for the construction of Business Middle Office projects. Here are several commonly used products for projects: EDAS, MSE, Relational Database Service (RDS), Distributed Relational Database Service (DRDS), Enterprise Redis, RocketMQ, CSB, SLS, Application Real-Time Monitoring Service (ARMS), etc. The following is a brief introduction to each product by quoting the contents of Alibaba Cloud official website, for more details, please refer to Alibaba Cloud official website.

4.5.6.1 EDAS

EDAS is a platform as a service (PaaS) service for application hosting and microservice management, providing full-stack solutions such as application development, deployment, monitoring and O & M. It supports Dubbo, Spring Cloud and other microservice runtime environments, helping you to easily migrate applications to Alibaba Cloud. Its main functions are as follows:

1. Microservice solution

   EDAS supports three mainstream microservice frameworks: High-speed Service Framework (HSF), Apache Dubbo and Spring Cloud. HSF is an efficient built-in microservice framework that has been tested by Alibaba's Double 11 Shopping Festival. It is incubated with best practices in many Alibaba business scenarios. EDAS supports Apache Dubbo and Spring Cloud. With zero code intrusion, you can migrate Apache Dubbo and Spring Cloud applications to Alibaba Cloud, effectively reducing O & M costs. EDAS supports multiple advanced features such as phased release and throttling, helping you easily build microservice-oriented applications on Alibaba Cloud.

   - Rapid application construction based on a mature microservice framework: You can build microservice-oriented applications on Alibaba Cloud by using Alibaba's time-tested HSF.
   - Migration of Apache Dubbo and Spring Cloud applications to Alibaba Cloud: You do not need to build microservice dependencies such as ZooKeeper, Eureka and Consul, greatly reducing O & M costs.

- Enterprise orientation: EDAS provides advanced enterprise-level features such as phased release, throttling and environment isolation.

2. Application hosting solution

   The application hosting solution saves you the trouble of logging on each server to deploy it. Simply by logging on to the EDAS console, you can quickly deploy applications through WAR packages, JAR packages or images, and then manage applications anytime, anywhere. For example, you can create, deploy, start, roll back, scale in, scale out and delete applications.

   - More cost-effective than building on-premises clusters independently: No server purchase is required. You only pay for your usage of CPU and memory resources.
   - Significant reduction in O & M costs: No IaaS O & M or cluster O & M is required, greatly reducing related human resources.
   - Application lifecycle management: You can visually manage the application lifecycle to know the application running status.

3. Containerized application hosting solution

   EDAS allows you to manage containerized applications and seamlessly interconnect with Container Service for Kubernetes, without having to know the underlying details of container service. You can manage the lifecycle of applications in containers in the EDAS console. For example, you can monitor and diagnose applications. You can easily use new container technologies and maximize resource utilization.

   - Seamless support for Kubernetes: After hosting Kubernetes clusters to EDAS, you only need to focus on application lifecycle management.
   - Optimal combination of containers and microservices: You can quickly build a containerized microservice architecture based on Kubernetes.
   - No image build required: You can deploy applications through WAR and JAR packages. EDAS automatically builds images and deploys applications in Kubernetes clusters. This simplifies the deployment process and improves ease of use.

4.5.6.2 RDS

ApsaraDB for RDS is a stable, reliable and scalable online database service. Based on Apsara Distributed File System and high-performance SSD storage of Alibaba Cloud, ApsaraDB for RDS supports the MySQL, SQL Server, PostgreSQL, PPAS (highly compatible with Oracle) and MariaDB database engines. It provides a portfolio of solutions for disaster recovery, backup, restoration, monitoring and migration to facilitate database operations and maintenance.

ApsaraDB for RDS allows you to quickly build a stable and reliable database system. It has the following advantages compared with user-created databases:

- Cost-effective and easy to use. You can choose flexible billing methods, change database configurations on demand and obtain an out-of-the-box database service.
- High performance, including suggestions on parameter and SQL query optimization.
- High-availability architecture and multiple disaster recovery solutions.
- High security. Various preventive measures are used to protect data.

ApsaraDB for RDS has significant advantages in cost-effectiveness, availability, reliability, ease of use and performance. Its cost is one-third of the expense to build databases on ECS instances and one-tenth of the expense to build databases on physical servers.

4.5.6.3 DRDS

DRDS is developed by Alibaba Cloud. This service is integrated with the distributed SQL engine DRDS and the self-developed distributed storage X-DB. Based on the integrated cloud-native architecture, this service supports up to tens of millions of concurrent connections and hundreds of petabytes of mass data storage. DRDS aims to provide solutions for mass data storage, ultra-high concurrent throughput, performance bottlenecks for large tables and efficiency for complex computing. DRDS has been tested in each Tmall Double 11 Shopping Festival and in the business of Alibaba Cloud customers in various industries. This service boosts the digital transformation of enterprises.

DRDS adopts standard relational database technologies to provide core features. The databases are deployed with the comprehensive management, O & M and product-based capabilities. This makes the databases more stable, reliable, scalable, maintainable and operable as in a traditional single-instance MySQL database.

DRDS has been used on Alibaba Cloud and Apsara Stack for many years, and has been tested in core transaction services of each Tmall Double 11 Shopping Festival and in the business of Alibaba Cloud customers in various industries. DRDS supports core online business for a large number of users across many industries, such as the Internet, finance and payment, education, communications and public utilities. DRDS is an industry standard for all the core online services of Alibaba Group and business of Alibaba Cloud customers to connect to distributed databases. Features are as follows:

- Stability

  For most applications, relational databases are the core foundation of the data management system. The database performance affects user experience on services and protects business data. Therefore, stability is the core factor for you to select a database.

  Proper use of the time-tested MySQL databases ensures the stability of DRDS. However, single-instance MySQL databases provide low performance in the high-concurrency, large-volume data storage, and complex computing scenarios.

  DRDS distributes data to multiple ApsaraDB RDS for MySQL instances. To ensure stable services, each instance undertakes a proper number of concurrent requests and computing loads, and stores a proper amount of data. DRDS implements distributed logic at the computing layer. This forms a stable, reliable and highly scalable distributed relational database system.

  Compared with self-developed distributed NewSQL databases, DRDS focuses on continuous stability and O & M availability. By using standard database technologies, DRDS databases can be operated in a similar way as single-instance databases. You can get started with DRDS in a simple way and increase your business value.

- High scalability

  Compared with traditional single-instance relational databases, DRDS uses a hierarchical architecture to ensure linear scaling in concurrency, computing and data storage. DRDS allows you to scale out computing and storage resources.

Compared with new cloud-native databases based on distributed storage, DRDS databases can be scaled out without limits. This eliminates the worries and O & M pressures on database scalability during rapid business development.

- Continuous O & M availability

  For most applications, relational databases must work around the clock in a stable way. Therefore, continuous O & M availability is the key capability for relational databases.

  DRDS has been used on Alibaba Cloud and Apsara Stack for many years, and provides a variety of product capabilities and a complete O & M system. Services can be automatically scheduled and integrated based on a complete set of API operations.

4.5.6.4 Redis

ApsaraDB for Redis is a database service that is compatible with native Redis protocols. It supports a hybrid of memory and hard disks for data persistence. ApsaraDB for Redis provides a highly available hot standby architecture and can scale to meet requirements for high-performance and low-latency read/write operations. Benefits are as follows:

- Hardware and data are deployed in the cloud. ApsaraDB for Redis is a fully-managed cloud database service provided by Alibaba Cloud. Alibaba Cloud manages infrastructure planning, network security and system maintenance. This allows you to focus on business development.
- ApsaraDB for Redis supports various data types, such as strings, lists, sets, sorted sets, hash tables and streams. The service also provides advanced features, such as transactions, message subscription and message publishing.
- ApsaraDB for Redis Enhanced Edition (Tair) is a key-value pair cloud cache service that is an upgraded version developed based on ApsaraDB for Redis Community Edition. ApsaraDB for Redis Enhanced Edition (Tair) provides the following series of instances: Performance-enhanced instances and Hybrid-storage instances.

4.5.6.5 RocketMQ

Message Queue (MQ) for Apache RocketMQ is a distributed message-oriented middleware service that is built by Alibaba Cloud based on Apache RocketMQ and features low latency, high concurrency, high availability and high reliability. MQ for Apache RocketMQ provides asynchronous decoupling and load shifting for distributed application systems, and supports features for Internet applications, including massive message accumulation, high throughput and reliable retry. Scenarios are as follows:

- Load shifting

  Large activities, such as flash sales, red envelope snatching and enterprise success, may cause high traffic pulses, and the system may become overloaded or even stops responding due to a lack of proper protection. The user experience may be affected because excessive requests fail upon many limits. MQ for Apache RocketMQ provides the load shifting feature to solve this problem.

- Asynchronous decoupling

  As the core system of Taobao and Tmall primary sites, the transaction system can attract the attention of hundreds of downstream business systems, including logistics, shopping carts, credits and stream computing analytics when each transaction order is created. The overall business system is large and complex. MQ for Apache RocketMQ supports asynchronous communication and application decoupling to ensure the continuity of services on the primary sites.

- Sending and subscription of ordered messages

  Several scenarios need to ensure the sequence in daily life, such as the time-first principle of securities trading, order creation, payment, and refund in the trading system, and handling of boarding messages of passengers on flights. Ordered messages in MQ for Apache RocketMQ are sent and received in the first-in-first-out order.

- Consistency of distributed transactions

  Final data consistency must be ensured in scenarios such as transaction systems and payment envelopes. MQ for Apache RocketMQ distributed transactions can implement decoupling between systems and ensure final data consistency.

- Big data analytics

  Data create value during movement. Traditional data analytics is mostly based on the batch computing model and cannot be performed in real time. The combination of MQ for Apache RocketMQ and Stream Compute of Alibaba Cloud allows you to conveniently analyze your business data in real time.

- Distributed cache synchronization

  During Double 11, changes to commodity prices in different activities need to be perceived in real time. A large number of concurrent access requests to the database result in slow page response. The centralized cache restricts the traffic for commodity data changes due to bandwidth bottlenecks. To solve this problem, MQ for Apache RocketMQ provides a distributed cache to notify commodity data change in real time.

4.5.6.6 SLS

Log Service (SLS) is a one-stop logging service developed by Alibaba Cloud that is widely used by Alibaba Group in big data scenarios. You can use Log Service to collect, query and consume log data without the need to invest in in-house data collection and processing systems. This enables you to focus on your business, improving business efficiency and helping your business to expand.

1. LogHub

   The LogHub component collects real-time log data from ECS, containers, mobile terminals, open-source software and JavaScript. The log data can be metrics, events, binary logs, text logs and clicks.

   LogHub provides a real-time consumption interface to connect with Real-time Compute (formerly StreamCompute).

   Scenarios: Data cleansing (ETL), stream computing, monitoring, alert, machine learning and iterative computing (Figure 4.52).

2. LogSearch/Analytics

   The LogSearch/Analytics feature allows you to index, query and analyze log data in real time.

   - Keyword, fuzzy, context and range queries are supported.
   - A variety of statistical methods are provided. For example, you can use SQL aggregate functions to obtain log data statistics.

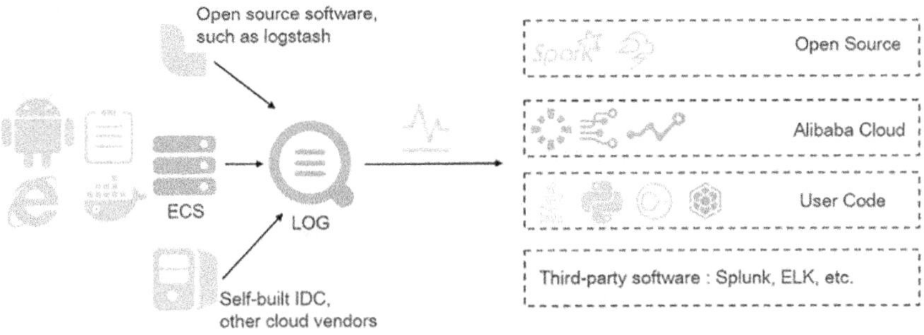

FIGURE 4.52 Loghub real-time log collection and consumption.

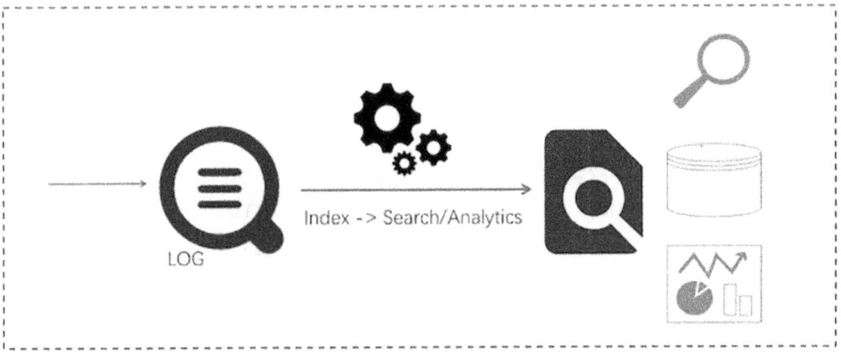

FIGURE 4.53 Loghub real-time log query and analysis.

- You can use dashboards and charts to visualize log data.
- Log Service supports seamless interconnection with Grafana based on the JDBC and SQL-92 protocols.

Scenarios: DevOps, online O & M, real-time log data analysis, security diagnosis and analysis, business operation systems and customer service systems (Figure 4.53).

3. LogShipper

LogShipper ensures stable and reliable log shipping. With LogShipper, you can ship logs from LogHub to storage services. LogShipper allows you to store log data in compressed files, user-defined partitions, rows and columns.

Scenarios: Data warehousing, data analysis, data auditing, product recommendation and user profiling (Figure 4.54).

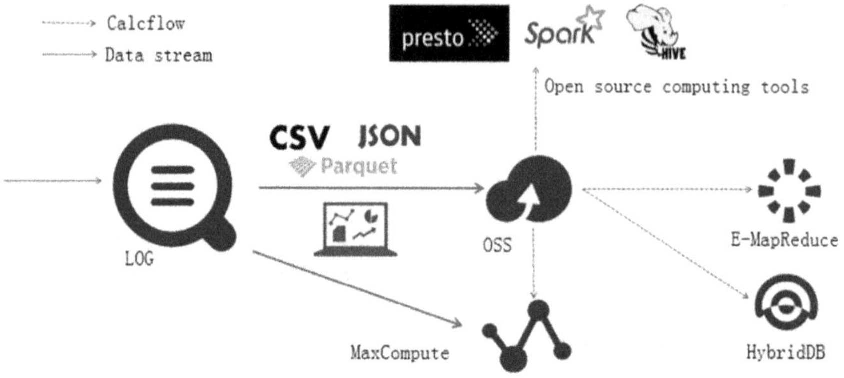

FIGURE 4.54 Loghub real-time log post database.

4.5.6.7 ARMS

ARMS is an application performance management (APM) product of Alibaba Cloud. With ARMS, you can quickly and conveniently build business monitoring capabilities with few-second response time for businesses and enterprises based on custom dimensions such as the browser, application and business.

1. Workflow

   Figure 4.55 shows the ARMS workflow.

   - Data collection: ARMS supports capturing logs from ECS instances, MQ and LogHub through configuration.
   - Job definition: ARMS allows you to define jobs such as real-time processing, data storage, presentation and analysis, data API, and alerts through job configuration, to define your own application scenarios.

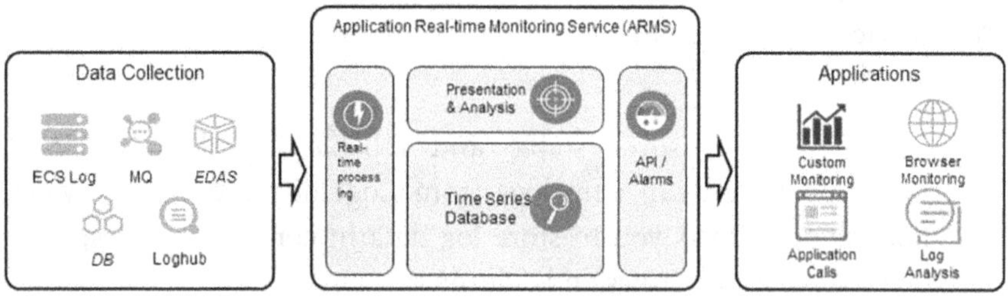

FIGURE 4.55 ARMS main functions.

ARMS directly performs business monitoring with preset scenarios, such as browser monitoring and application monitoring.

- Application scenario: In addition to custom monitoring, ARMS also provides ready-to-use preset monitoring scenarios, including browser monitoring and application monitoring.

2. Scenarios

- Highly tailored business monitoring: You can create real-time monitoring alerts and dashboards based on business characteristics and requirements. Business scenarios include the e-commerce scenario, logistics scenario, and airlines and tour scenarios.
- Browser experience monitoring: ARMS can show the performance and errors of pages you visited by region, channel, link or other dimensions.
- Application performance and exception monitoring: ARMS provides the APM capabilities to monitor performance exceptions and query traces for distributed applications.
- Central alert and report platform: Custom monitoring, browser monitoring and application monitoring are integrated on a central alert and report platform.

With ARMS, IT engineers can build and start a big data platform-based real-time application monitoring system within minutes, which maximizes the timeliness of data monitoring and improves the efficiency of IT engineers.

4.5.7 System Migration

There are two options for migrating traditional systems to the Middle Office. One is to build a new Middle Office technical framework system; the other is that business parts are migrated to the Middle Office technical system one by one in units of domains or microservices. This migration method has little impact at a time but lasts for a long time; the other is to develop and build a new Middle Office system. All the docking systems are migrated at one time. This method has a great impact on the overall system, but the overall transformation of the Middle Office system is relatively rapid and thorough. Both methods have their own advantages and disadvantages, but from the perspective of fast Middle Office transformation, the second method is relatively fast, and the second method is also highly complex, this topic describes how to implement full migration.

For one-time full migration, considering that the new Middle Office is generally greatly changed, it has an impact on technology architecture implementation, data model and the connection between peripheral systems and foreground systems, therefore, we recommend that you choose to stop the service for a short time. During downtime migration, the main tasks include historical data migration, peripheral system update and docking, and full regression verification.

4.5.7.1 Data Migration

Data migration is to retain historical data of users to achieve consistency in business use. For example, you can view the order records in the old system in the new system. To achieve this goal, you need to carefully analyze and design the data migration scope, migration impact, migration scheme, data correction and migration verification.

1. Migration scope

   To understand the migration scope, you need to understand it from both business and technical perspectives. First, at the business level, confirm the business scope of the migration, which need to be migrated, which do not need to be migrated, how to process the data that do not need to be migrated, whether to discard or otherwise undertake the system; second, at the technical level, you need to know what tools are used to store data in the old system, such as databases and files, as well as the distribution location of data and the corresponding link address.

2. Migration impact

   The impact of data migration is the impact of data migration on business and other factors on data migration.

   When data are migrated from the old system to the new Middle Office, the old system may contain some business data that cannot be used by the new Middle Office. These types of data are generated for various reasons. For example, a historical function once existed; however, the latest old system has removed this feature, so how to process the historical data caused by this feature, whether to migrate to the new Middle Office or not, and if you do not migrate to the new Middle Office, you need to decide how to do it, whether to save it separately or discard it.

The impact of other factors on data migration refers to the data generated by scheduled tasks or MQs. After the system shuts off external traffic, the number of new data records or status changes will still occur in the old system. If the data of these latency types are not processed and the migration solution does not consider the incremental situation, data loss or status error may occur.

3. Migration solution

As shown in Figure 4.56, the core content that needs to be paid attention to in the data migration solution is listed. In fact, the migration solution needs to be customized according to the actual migration requirements. It needs to take all aspects of the migration action into account.

Data source and destination: Check where the source data of the old system are stored and where they need to be migrated. For example, you can migrate data from the Oracle database of the old system to the DRDS of the new Middle Office. You can migrate data from FastDFS of the old system to the OSS of the new Middle Office. This type of migration is relatively simple. However, there are also special cases. For example, if you migrate data from the old system database to the OSS in the new Middle Office, you need to find the corresponding solution for migration.

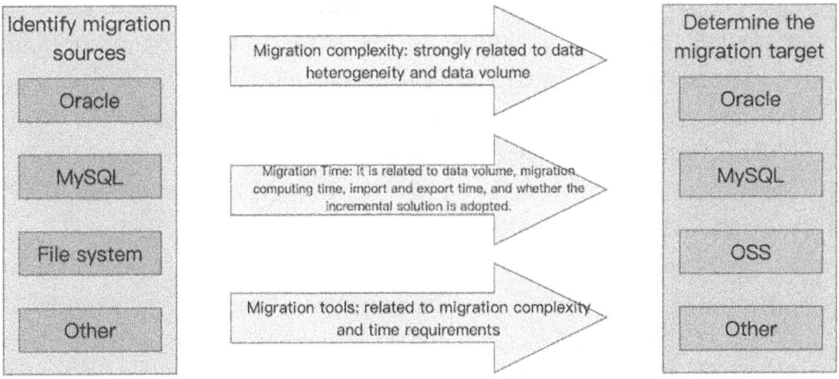

FIGURE 4.56 System migration solutions.

Complexity of data migration: If data are migrated horizontally, the complexity is relatively low. For example, table a of the old system is migrated horizontally to table A of the new Middle Office. In a slightly complex case, it is necessary to simply convert the data in the old system Table b and migrate it to the new middle Table B; in more complicated cases, the fields in the old system Table c, after complex conversion (such as hash), are migrated to Table C in the new Middle Office. In more complicated cases, the data in Table d of the old system and some columns are merged, migrate to Table D in the new middle table. A more complicated situation is to split the data in the old system Table e into multiple pieces, or merge multiple pieces of data into one piece, and migrate the data to the new Middle Office Table E. The complexity continues to be upgraded. Data from multiple tables in the old system are converted into data from a new table and stored in the new Middle Office through a field association; therefore, the implementation difficulty of the migration solution will increase as the above situations occur. It is better to have a special standard solution to solve the above problems to ensure that the migration implementers can work under control.

Migration time: The downtime migration time is limited and cannot be stopped for too long. Therefore, how to ensure that data migration can be completed within the specified time becomes particularly important. If there are too many historical data that cannot be completed within the specified time, you need to consider incremental migration instead of simple one-time full migration. When designing an incremental migration solution, you need to analyze the data characteristics. The basic data and operation data of the system can be migrated in advance because they will not be added or changed as users use them. Moreover, the early historical data of users cannot change their status, you can also migrate data in advance. Generally speaking, in order to reduce the migration time, we should try our best to finish the pre-existing work first.

Migration tool: Select a migration tool based on the actual data status. If the system is less heterogeneous and the data model is basically the same, you can consider using migration tools such as DTS. If the data difference is not large, you can consider using DTS and intermediate Library conversion to implement migration; if the degree of heterogeneity is serious, the amount of data is large, and the amount of computation is large, you can consider using big data processing tools such as DataWorks or Dataphin, which can import and export data, complex computing, linear scale-out computing power, custom functions, managed data computing dependencies and scheduling capabilities, so you can be competent for complex migration scenarios.

Migration drill: The overall data migration process must be rehearsed to ensure the time and quality of data migration.

4. Data correction

   If the old system runs for a long time, it is inevitable that some dirty data will occur. If the data are migrated to the new Middle Office, system failures may occur. Therefore, it needs to be fixed according to the requirements of the new Middle Office. For example, some data need to be deleted, some empty fields need to be assigned values and some fields need to be changed in status. These problems must be exposed to the corresponding products and technologies for joint decision-making, and then be dealt with accordingly after decision-making.

5. Data verification

   Data verification is a verification method to ensure the correct execution of data migration, which generally requires the participation of technicians and testers. Technicians verify at the data level, and testers verify at the business and performance levels.

   The technical personnel check includes the number of data records, data type, number of data records in a certain state, garbled code, digital calculation errors, database sharding errors and so on. The verification of testers is mainly based on data migration test cases, covering system usage, intermediate process connection and historical data viewing.

4.5.7.2 System Cutover

System cutover includes data migration, the launch of a new Middle Office and the docking and switching of peripheral systems. It is not a pure technical work, but also includes a lot of communication and coordination work. The system cutover work for shutdown mainly includes the following contents.

Related party identification: To ensure that all possible impacts can be taken into account during cutover, the identification of related parties mainly includes system, business and personnel. The system includes old systems, new Middle Office systems and peripheral systems (including connected systems and systems that may affect each other); the business includes the business scope that may be affected by the cutover action; personnel includes all system-related personnel, business-related personnel and other personnel who may affect cutover.

Pre-preparation: In order to ensure that the cutover work can proceed smoothly, the pre-preparation work includes cutover announcement and cutover maintenance page; the construction of various environments, for example, the new Middle Office PRD environment is a new environment, the construction can be completed in advance without having to be built on the day of cutover; network construction of related systems; application for new resources; environment accounts and operation permissions; and synchronization of plans of relevant personnel.

Refinement of cutover scheme: The actual operation on the day of cutover needs to be refined to the minute level. All operations have clear time points, clear executors, clear operation inspection methods and inspectors; try to practice before all operations have something to do, and it is better to do two full drills before the actual operation of cutover to ensure smooth cutover; all cutover operators have a clear understanding of their work content and ensure the correct operation.

Cutover rollback: Cutover is a meticulous and complicated task. Any problem in any link that cannot be quickly solved may lead to the failure of the overall cutover action. Therefore, the cutover solution must have the rollback capability to ensure that when cutover fails, it can be rolled back to the old system so that the business can continue to run. The rollback scheme of cutover also needs to be rehearsed to ensure that the rollback is in good order.

Cutover verification: Cutover verification needs to be considered in combination with actual system conditions and verification methods of each system. For example, some systems can be verified in advance; some systems must be verified synchronously; some systems can be verified based on the whitelist; some systems must start the public network process to verify; some systems have a long verification time, how to shorten; conflicts that may occur due to inconsistent verification methods of various systems; time nodes for data migration verification; and how to handle dirty data verified.

Organizational assurance: The entire lifecycle of organizational assurance is accompanied by cutover. There are many parties involved in the cutover action, and the organizational guarantee mechanism can promote the formulation and implementation of the schemes of each system when the cutover scheme is formulated; it plays a unified and coordinated role in drills and actual cutover; and in case of cutover problems, it plays a role in decision-making.

4.5.8 System Release and Guarantee

The importance of the release point, as the final stage of the results of the R & D phase, is self-evident. After a tense project cycle and everyone's hard work, the time has finally arrived. In order to show better results to the 'audience', we need to make all preparations before going on stage. Generally speaking, we can divide the release process into the following four important stages: release preparation, release execution, release guarantee and release maintenance.

4.5.8.1 Stage One: Release Preparation

As the saying goes, 10 minutes on stage and 10 years of work under stage, a worker must first sharpen his tools if he wants to do well. This fully shows that we need to spend more energy on preparations if we want to implement a smooth launch. Generally speaking, preparations can be made according to the following four dimensions:

1. Organization Preparation

   Involved roles: Chief Technical Officer, Chief Business Officer and Project/Product Manager

   These roles together form the 'release management team' to jointly ensure the successful execution of the release actions and also have the right to make decisions in unexpected situations during the release process. We need to elect a commander in chief of the entire release time.

   Organization preparation checklist:

   - The team responsibility is listed in the table of release process (refer to Table 4.8) We will use this form to clarify the release formation and achieve the goal of rapid response in joint operations. Pay attention to the position of the key person in charge, it is recommended to define AB role to prevent unexpected situations.

TABLE 4.8 Division of Responsibility for System Go-Live

Parent Team	Sub-Team	Principle A	Principle B	Responsibility	Team Info
Tech team/ Business team/...	User center team/account center team/...	Jack/ Rose/...	Tom/ Jerry/...	Cooperate user center team to release, test, code merge/...	Phone/ seat/...

- In the team responsibility sub-table of the release, each team refers to the parent table to clarify the person in charge of the internal module of each team, or the person in charge of the release steps.
- The release guarantee organizations chart needs to define release formation according to system guarantee requirements and SLA (Service Level Agreement), and specifies the key contact person and contact methods. We can use the same method to define it, as the 'team responsibility summary table of the release' (Figure 4.57).

2. Process Preparation

   Involved role: Release management team

   Process preparation checklist:

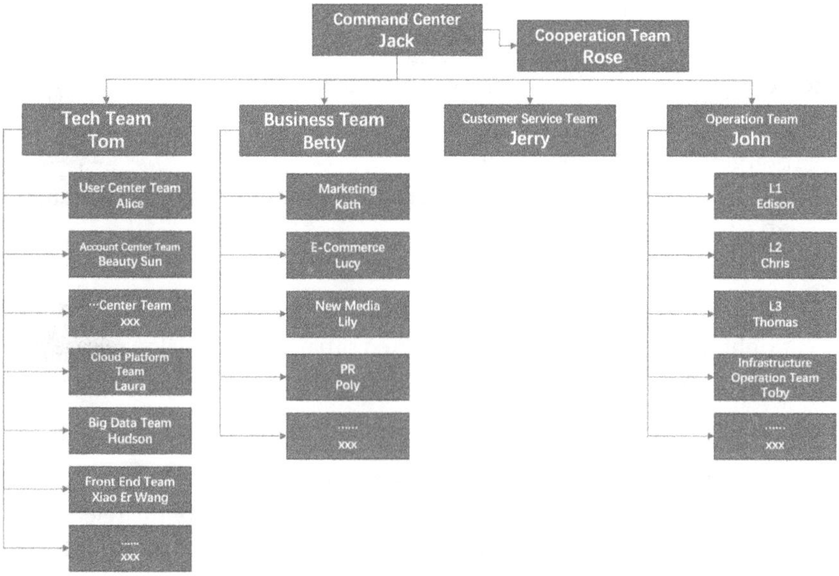

FIGURE 4.57 Organization design of system go-live.

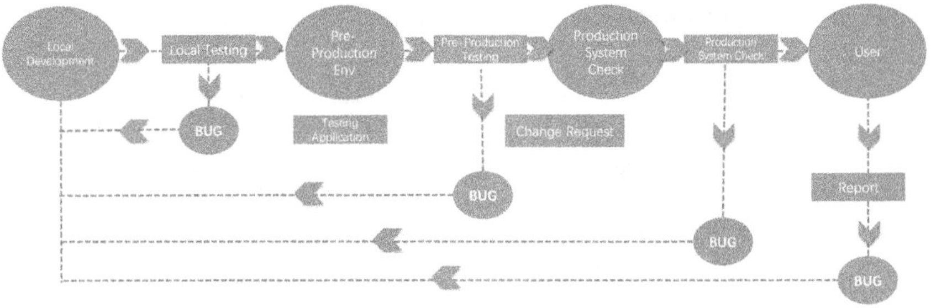

FIGURE 4.58 Preparation flow chart of system go-live.

I. The release preparation process is shown in Figure 4.58. There are several basic principles:

- General release time and frequency: Once a week (the network will be limited during critical release), after 21:00 on Friday, version control.
- General release process: Approval process must be followed. In principle, approval will be completed on Thursday.
- Exceptional release process: Live system critical bug/fault fix, emergency business needs. Only when (supplier+general responsible party+customer) tripartite consensus and the impact evaluation are controllable, the approval process can be initialized.
- Define release red lines, service quality red lines, and business failures: If crossing red line causes business failures, monthly quality reports shall be adopted and the senior management shall be involved (Figure 4.58).

II. In the release action time table (refer to Table 4.9), the first half of the release actions include all technical preparations and business preparations, and the second half include the execution actions during the release process. Some common release actions, such as confirmation of the team responsibility summary (sub) table of the release, confirmation of the release action schedule, confirmation of the completion of the pre-production environment release test, acceptance and signature of the test report, system change requests and confirmation of the release note, release

TABLE 4.9 General Schedule of System Go-Live

Release Action No	Action Description	Action Group	Execution Start Time	Execution Finish Time	Emergency Plan No	Completion
001	Pre-production testing confirmation	User center team/ Account center team/...	2020/8/7 21:00 GMT+8	2020/8/7 23:00 GMT+8	Plan#001	/

confirmation from all parties, confirmation of release plan, data backup and cutover plan and drill confirmation, emergency plan confirmation, release notice and statement, code branch merging, code packaging, maintenance page preparation or business data flow cutover preparation, code deployment and release, technical team testing confirmation, Business team testing confirmation, maintenance page removal or traffic switching completion, release finish notification, release report, release guarantee activation and release issue tracking.

The author has had a number of global project/product launch experiences, usually the projects/products jointly launched by the marketing department and the public relations department. The easiest way to describe it is that at a fixed point in time, Bang! Global cooperative media advertisements are launched at the same time, and official website promotional videos are pushed. If the release process is not completed with the established time or communication and coordination is not done well, it is conceivable that users in the local market of a certain country will be very embarrassed to see the 404 page or the notification that the product is not available. The damage to companies that have invested heavily in PR is self-evident. Among the e-commerce scenarios, the most representative one is the 'big sales' promotion, the same reason. Therefore, in the table, if it involves cross-time zone joint publishing, the start time and end time are very important, and the time zone identifier of GMT is added to make the time description more accurate.

III. The release action time sub-table, referring to the parent table specification, clarifies the sub-processes and steps that each team needs to perform, so that the tasks are arranged to the specific people and specific time. In addition, practical considerations need to be made as to whether execution can be completed within a predetermined time or whether there is enough buffer time to deal with emergencies, etc. If you find that the time for the release action planning is insufficient, you need to immediately report back to the release management team to make overall adjustments and notifications to avoid information that is out of sync.

3. Tech Preparation

Involved roles: R & D Leader, Testing team, Operation Team

Tech preparation checklist:

- Whether the release organization formation and release process are clear.
- Whether the official release confirmation has been obtained.
- Whether to formally submit and confirm the team responsibility summary table of the release, release action time schedule and release guarantee organizations chart.
- Whether there is a clear release operation manual.
- Whether the target system passes the final code quality audit.
- Whether the entire team has a clear code freeze time window.
- Whether the pre-production environment has passed the key quality check points such as full regression test, security test and performance test, and obtained the officially confirmed test pass report.
- Whether the code passing the quality check point is merged into the production branch and deployed to the pre-production environment again, and the code freezing is enabled.
- Whether to formally submit and confirm the system change and online function description list, it is recommended to use Word to describe the information in the form in more detail (such as Table 4.10).

TABLE 4.10 Go-Live Function Descriptions Sheet

Change Request No/ Requirement No/ User Story No	Affected Product/ Module/Center	Configuration Change	Code Branch	Rollback Plan	Principle
US#1001 Adding new User gender field	User center	Env Config adding xxx	Git/xxx/ xx	xx	Jack

- Whether officially confirmed various plans: Release and deployment plan, data backup and cutover plan, emergency and rollback plan, release technical testing plan, stability monitoring plan during the release period, and tech solution for 'big sales' when applicable, etc.

4. Business Preparation

Involved roles: All relevant business departments and business chiefs

Business preparation checklist:

- Whether the release organization formation and release process are clear.
- Whether the official release confirmation has been obtained.
- Whether to formally submit and confirm the team responsibility summary table of the release, release action time schedule and release guarantee organizations chart.
- Whether prepared and confirmed the release maintenance notice and PR announcement, and synchronized the release arrangements of upstream and downstream companies/organizations/departments.
- Whether each business department clearly and formally confirms the system change requests and release note.
- Whether the pre-productions environment has passed the acceptance test of all services.
- Whether the release business test plan and business emergency plan been formally confirmed.

4.5.8.2 Stage Two: Release Execution

With full preparations, the release execution process can be proceeded step by step according to the schedule. Generally, the system cutover will involve the following steps, refer to Figure 4.59.

Notice in Cutover:

- Even small cutover may cause failure.
- Half of the failures are caused by cutover.
- Even if the system is running well in the testing environment, there may be problems when deploying to production.
- Grayscale is very necessary.
- The scope of impact must be fully assessed before cutover.
- Adequate preparation and testing can reduce the possibility of failure.
- It is necessary to summarize the faults.

In the above notices, it is said that 'grayscale is very necessary', so the author here briefly talks about some common release methods. The four common release methods are: full-volume release, blue-green release, rolling release and grayscale/canary release.

The simplest and most straight-forwarded release method is full-volume release. It could be described as deployment in one shot, which

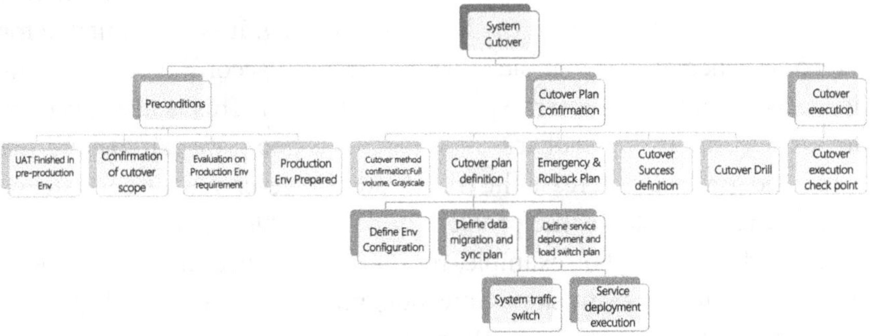

FIGURE 4.59 Action decomposition of system go-live.

usually requires some beautiful and friendly system maintenance pages to tell users that the system is being updated. This is generally used for systems that do not require high business continuity.

The blue-green release method is a bit more elegant than the full-volume release; however, the difficulty is still relatively low, that is, the two systems are switched. This means that generally the reverse proxy/router is used to directly switch between two identical systems. In theory, there is no or very little business unavailability time, and if there is a problem, it will immediately switch back to the original production system. This kind of system is more suitable for less transactional systems, such as official website type projects. If there are more transactions or the architecture of the new system and the old system are completely different, a large part of time may be spent on data consistency and dirty data issues.

Rolling release method is different from the simple switching approach of blue-green release. It uses batches of machine stopping and then starting after updating. The advantage of this is that it saves resources compared to blue and green, but it brings more problems. For example, when a problem occurs when the release reaches 80%, it needs to be rolled back batch by batch, plus the auto-scaling mechanism in the cloud ecosystem. If scaling occurs during the rolling process, it will be more troublesome to manage. So, we should consider carefully when choosing this method.

Grayscale release method is the recommended method in the update of existing business systems. This method is smoother and the solution is not difficult. It mainly relies on the adjustment of load balancing to remove the released machine from the load, perform release and deployment, then test, and then add it back to the load after passing. It is very suitable for the coexistence of new and old versions, but split according to user traffic, such as some angel user experience promotion. This is also easier to do A/B testing. If the feedback for version B is good, we can expand the scope and gradually increase the traffic rate. However, for the coexistence of new and old systems, the problem of data double writing needs to be particularly noticed. For example, both systems consume the same MQ (Messaging Queue) at the same time, but among the new and old systems, the data structure has changed. The architecture in the case of an independent environment is shown in Figure 4.60.

When using EDAS to manage the whole system chain, refer to Figure 4.61.

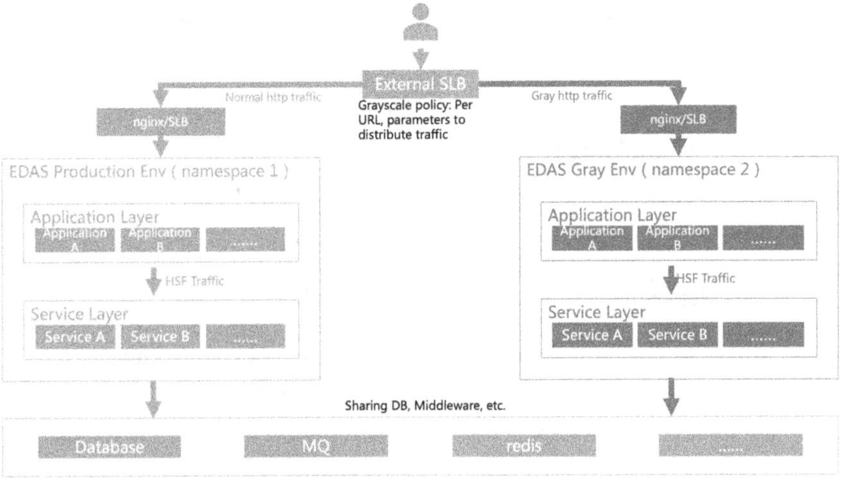

FIGURE 4.60 Grayscale release.

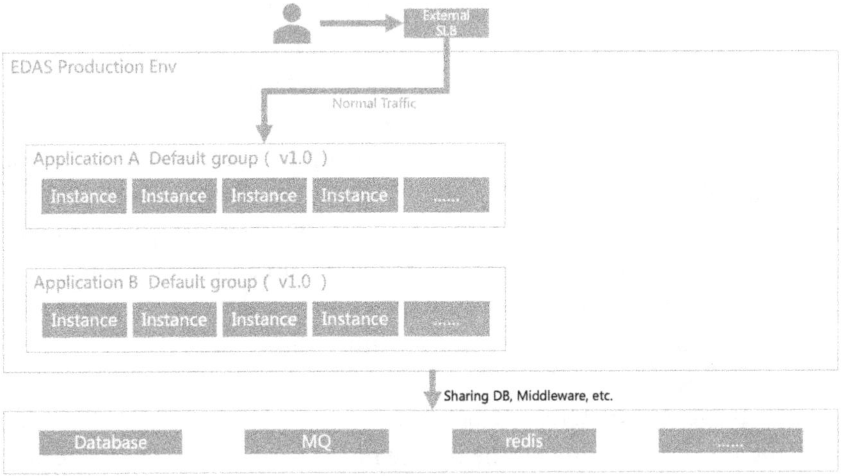

FIGURE 4.61 Grayscale release based on Alibaba Cloud EDAS.

When performing release actions, it is recommended that the relevant people in charge are in a physical space, such as a war room. If you are in different regions and time zones, you must ensure that the relevant people are in the same conference meeting. Everyone should complete the release actions seriously, efficiently and carefully in accordance with the instructions of the release commander. At the same time, the release process should meet the following general principles:

- The release instructions require the technical person in charge to review first, and all submissions should have been reviewed in advance.
- 'Free rider' is not allowed, and any publish of irrelevant code to this release should be prevented.
- The testing team provides the test results, and the technical person in charge conducts risk assessment and gives a conclusion.
- The release process must be traceable and returnable.
- In principle, the authority of the production environment is led by the business responsible party to avoid excessive authority of the technical team. If it is really needed, apply for a single release authority as needed.

If there is a failure or problem during releasing, for example, some problems that were not discovered during the test phase are found. The business/technical team need to conduct preliminary judgment and characterization of the problem, then record it (refer to the defect description requirements). For non-blocking issues, they will be summarized to testers after going online and repaired during the guarantee period. For blocking issues, it is necessary to immediately synchronize to the release management team to make a unified decision. If it can be repaired immediately before the end of the action schedule, it can be repaired; if not, it is necessary to immediately activate technical and business emergency plans. Refer to Figure 4.62 for a more intuitive process.

4.5.8.3 Stage Three: Release Guarantee

Starting from the sending of the release finish notification, the entire system will officially enter the guarantee period. A successful launch cannot

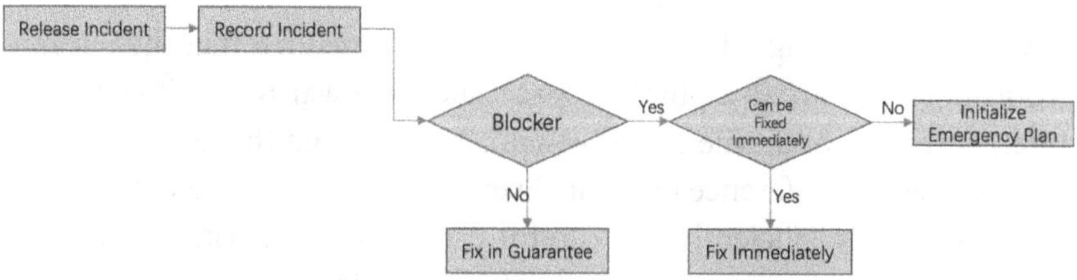

FIGURE 4.62 Emergency response process of system go-live.

be equally treated as the completion of the release action list. We emphasize that the success of the launch must be based on the stable support of the business as the standard. Generally speaking, the guarantee period can be divided into two stages: the hyper care period and the online stable guarantee period.

According to the author's experience, the general hyper care period adopts a 7×24-hour monitoring and support model, and the duration varies according to the scope of release influence and the importance of the business. Generally, 1 week is the average length. During the hyper care period, blocking and severe production problems are given priority to stop bleeding and ensure business continuity. If you need to modify the code to fix the problem, generally take the emergency change process (note the Hotfix branch merging problem). For general problems or non-blocking problems found during the release process, the R & D team will fix them in parallel to ensure that as many problems as possible are resolved in the next normal release window.

The online stability guarantee period generally adopts a 5×8-hour monitoring and support model, and generally takes half the time of the hyper care period as the duration. After entering the stability guarantee period, it indicates that the blocking problems of the system have been resolved, so the initiation of the emergency change process should be reduced as much as possible, and change back to the normal release process. In the middle of the stability guarantee period, it is recommended to a normal release to push the fixed non-blocking issues live once, in order to prepare the business acceptance of transfer to maintenance phase (Table 4.11).

In order to better organize the guarantee period with quicker response, we will designate a clear duty schedule with contact person information and problem escalation route. All guarantee-related parties, including the people in charge and the escalation contacts, need to ensure that their contact channel is unblocked and respond quickly. As shown in Table 4.11:

TABLE 4.11 Support Contact List of System Go-Live

Guarantee Module	Duty	Contact	Phone	Escalation Contact	Phone
XX Center/infrastructure/ XX application	9:00–17:00	Jack	150xxxx	Rose	155xxxx

4.5.8.4 Stage Four: Release Maintenance

After the successful launch, the system will enter the long-term operation and maintenance stage. General operation and maintenance will be based on a system support contract (SLA), in which response time and resolution time are specified according to different incident levels/severities, for example, Table 4.12. Generally speaking, corresponding penalties are specified in the contract also. For example, the timeout percentage of P1 incident tickets is not allowed to exceed 10%, otherwise a 5% penalty fee of the total support contract amount will be imposed. It will also specify that the annual service availability should reach 95% or more.

In order to solve problems more efficiently and meet SLA requirements, a pyramid-shaped support model is normally used, namely L1, L2 and L3. A good hierarchical support model needs to have a good KB system to avoid pressure passing through directly to the last layer from the front layer.

L1: Front-line support team. Large enterprises usually use call centers or desktop support organizations, which are responsible for incident clarification, issue distribution and basic usage problems and simple configuration problems resolve.

L2: Second-line operation and maintenance team, the main force for problem solving, can be grouped according to different support fields, such as Middle Office, cloud platform and front-end applications. For a stable system, most of the problems usually appear as the configuration issue or usage of advanced functions. To resolve those, support team

TABLE 4.12 Incident Ticket SLA Level Description

Incident Ticket Level	Incident Description	Response Time	Resolution Time
P1 – Blocker	Blocking problems, server hang-up, services completely unavailable, system down, serious security leaks, etc.	10 minutes	30 minutes
P2 – Critical	Serious problems, core link unavailable, business continuity severely affected, such as the inability to place orders and pay in the e-commerce system	30 minutes	2 hours
P3 – Major	Important issues, daily functional issues, but do not block business, such as failure of certain verification rules, etc.	1 hour	Differs by case
P4 – Minor	Non-important issues, not affect the use, more common in page display issues, such as wrong lines of text, unsightly pictures, etc.	4 hours	Differs by case

mainly rely on own experience and the improvement and precipitation of KB. Generally speaking, the team's KPI will also include the KB building and precipitation figures.

L3: Third-line R & D team, are mainly to solve functional defects or new improvements found by L1 and 2, or incident tickets related to product principles. The resolution of the problem usually requires the system release process.

The author has also worked in global support centers before. For 7×24-hour service centers, generally speaking, the P3 and P4 incidents are to follow the rules of the sunset rule (Follow the Sun), that is, the support center staying in this time zone will not circulate to keep those incidents tracking. However, for P1 and P2 incidents, they are required to continuously help the proposer to solve the problem, so the support team in the previous time zone will be required to record and transfer the problem. In addition, for all incident tickets, in order to ensure customer satisfaction, ticket follow-up rules need to be defined. For ordinary incident tickets of P3 and P4, the support team usually keep proposer updated at least every 2 days. If the incident ticket solution involves system release, or version changes, before closing the incident ticket, the release time and corresponding version number must be provided to the proposer.

In the meanwhile, if the proposer's incident ticket is very urgent, it can also initiate an escalation process, which will get the attention of the support team management. However, it should be noted that some SLAs will state that additional charges will be conducted for unreasonable upgrade incident tickets when the annual support fee is settled.

Therefore, for the operation and maintenance support team, the problem tracking interval and the number of escalations can also be used as an important evaluation metric. A sound operation and maintenance support system are important guarantees for system stability and customer satisfaction.

4.5.9 Business Middle Office Operation

The founder of GrowingIO once said a famous saying, 'If your enterprise does not grow, it is going to die', which is similar to a Chinese proverb, 'sailing against the current, no progress simply means regression' has the same effect, which shows that growth is the eternal truth of the enterprise. Coase, a famous economist, had a classic definition of an enterprise: The nature of an enterprise is a mechanism of resource allocation, and the enterprise and the market are two alternative ways of resource allocation.

Enterprises have economic organizations that aim at making profits, thus carrying out production and operation and providing products and services for the society. Whether it is a start-up or a listed enterprise, how to optimize the allocation of resources to achieve the goal of growth is the core metric to measure the enterprise.

4.5.9.1 What Is Operation?

It is believed that readers have more or less been exposed to the word 'operation', especially in the Internet industry, there are generally content operation, user operation, activity operation and product operation.

1. Content operation

   The core problem to be solved by content operation is the closed-loop system of production and consumption formed around content, and at the same time, it continuously improves content-related metrics, such as content quality, visits, transmission times and click-through rate. The core issues to be solved by content operation include but are not limited to the following points:

   - Content management, including basic and extended attributes of content.
   - How to display the content?
   - In what channels can the content be displayed to reach customers more effectively?

2. User Operations

   The core of user operation is a closed loop of user management built around the whole life cycle of users, including user recruitment, user management, user retention, sleeping user activation and user hierarchical management, at the same time, the metrics of users are continuously improved, including user profiling, user satisfaction, total number of users, unit price of customers, number of active users and user stay time. Core issues to be solved for operations include but are not limited to the following points:

   - Who is our user and where does he come from? What are the channels for online users, local marketing activities, and retaining user recommendations?

- What is the user's behavior after the user enters our system? How do we maintain the relationship with the customer, and recommend new products and discounts through regular email updates?
- What should we do to retain our customers? Incentive system, user credits?
- What is the cause of the loss of users? If an early warning mechanism for the loss of users is established? How do I recall lost users?

3. Activity Preparation

   The core problem to be solved in activity operation is a management closed loop built around activity planning, resource confirmation, promotion and effect evaluation. An activity must have a clear goal. From planning to advertising, it needs to continuously pay attention to the efficiency and effect evaluation of the activity.

4. Product Operation

   Product operation is mainly through a variety of means to enhance the popularity, market share, user access times, etc., of a certain kind of goods.

   From the perspective of the responsibilities assumed by the operation position, the more subdivided operation positions are divided into copywriting, Weibo blog, WeChat articles, activity copywriting, press release/soft writing, encyclopedia creation, Q & A Marketing, scheme planning, activity planning, microblog activities, offline activities, user operations, data operations, topic marketing, event marketing, community marketing, short video marketing, script writing, video shooting and editing as shown in Figure 4.63.

4.5.9.2 The Essence of Operation

Based on the analysis of the above operation types, the author believes that the core essence of Internet operation or the underlying logic is to 'meet the needs of the target users'; in other words, operation is a process of connecting products and users and achieving each other at the same time.

It can be divided into three parts: target users, what they think and meeting requirements. It can be said that all the operation work should be carried out around these three parts. Under these three parts, the

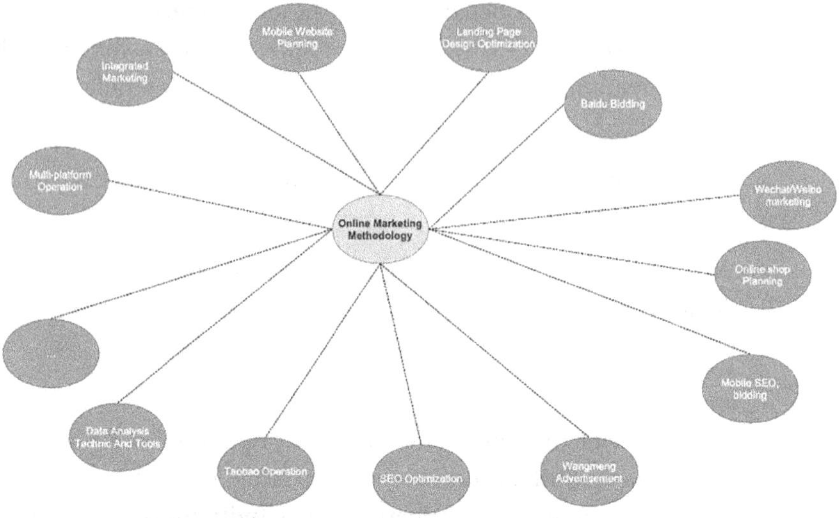

FIGURE 4.63 Functional division of product operation positions.

operation work will be divided into more detailed work items. Let's briefly introduce these three parts.

1. Target users

   That is, find the target user groups. It is basically a part of the work of user operation, such as user analysis, research and so on, to determine 'what is the target user'; after confirmation, promote, that is, 'find those target users', this is basically most of the work of channel operation and marketing promotion.

2. 'thoughted'

   What we think here can be explained as 'the user originally has it, but he is not clear about it. After operating activities, he understands his own needs, so he completes the purchase'. Note, there is but not knowing it, not without it, because if there is no, it is a demand that cannot be stimulated in any case.

   It can also be understood as exploring potential needs in a deeper level than 'product operation'. If you want to do it well, you need to deeply understand users' needs and psychology, and even explore the weaknesses of human nature.

   Meet requirements. The requirements here can be understood as regular requirements, rigid requirements and requirements that can be slightly explored. More in terms of functionality, take the medical industry as an example, a sick online registration user may have the need to buy medicine online later, which is basically in this category, if an online registered user needs to purchase insurance or physical examination.

To sum up, operation is a series of virtuous circle intervention to promote the connection between products and users for the sake of user growth, user activity and user realization on the premise of meeting user needs, this is the so-called operation essence in the Internet environment; the purpose of all operations is enterprise growth, or simply to increase revenue.

4.5.9.3 Operation of Business Middle Office

After introducing the operation and operation classification, we return to the main body of this section, the operation of the Business Middle Office. In terms of operation purposes, there is no essential difference between the operation of Business Middle Office and Internet operation, but the operation of Business Middle Office is different from that of traditional operation. The details are as follows:

1. Based on Business Middle Office

   The Business Middle Office is responsible for abstracting the core business capabilities of the enterprise and forming standard services. The construction of the Business Middle Office generally starts from the overall height of the enterprise and combines the development strategy of the company, integrates the core business scenarios of enterprises to form reusable business capabilities, and is exposed through APIs for internal and external users to call. Business capability reuse rate and new business construction TTM are core metrics to measure Business Middle Office effectiveness.

2. Service capability reuse times

   Business capability reuse Times refers to the number of times that reusable business capabilities provided by the Business Middle Office are reused within an enterprise. For example, an enterprise builds a login center, which provides login and authorization capabilities, if we want to measure the service capability reuse metrics of the logon center, we can count which new systems have reused the service capability of the service Middle Office logon center in the past cycle, the total number of times is the number of times that log on to the center.

3. Build new business TTM

   TTM refers to the product launch cycle. We can compare the TTM of a new business system before the Business Middle Office was launched with the TTM of a new business system based on the Business Middle Office, to measure whether the reusable business capabilities provided by Business Middle Office have significantly reduced the TTM of new business systems of enterprises and how much has been reduced.

   Therefore, the core of business operations based on Middle Office is to improve the reuse times and new business construction of business capabilities of Middle Office through various means.

4. Data-driven

   After enterprises build Business Middle Office, enterprises can easily collect a lot of business data, and the database will gradually accumulate a large amount of such data, including user access data, Page access data, order data, shopping cart data and commodity purchase data. These data contain great value and are gold mines for enterprises to be exploited.

5. Business operation dashboard

   With the development of Business Middle Office, enterprises have accumulated more and more data. How to gradually exert the power of data through big data technology and guide the production and operation activities of enterprises is an important issue that enterprises need to solve from CEO. Generally speaking, for non-technical personnel, big data only stay at the end of the year Alipay application's annual bill, WeChat New Year's personal social analysis and other scenarios. For enterprise operators, it is necessary to show the power of data and the core metrics of the enterprise through media. As a kind of big data display medium, it has been widely used in various exhibition halls, exhibitions, conferences and various carnival festivals, among which there are some common solutions: Alibaba Group DataV products. Its large screen has a variety of themes and provides a variety of templates. Especially in Alibaba, DataV is also

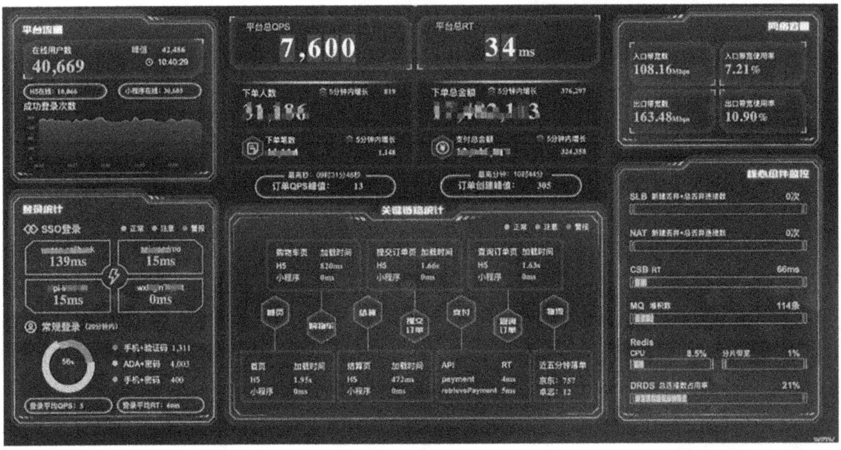

FIGURE 4.64 Example of digital large screen.

used to show the events of double 11 over the years. Figure 4.64 shows the Business Middle Office dashboard of a direct selling enterprise. As shown in the figure, the enterprise uses the data dashboard technology to show the status quo and health status of the business and technology platform during the promotion.

Therefore, the operation of Business Middle Office can be defined as: based on the business platform, it can improve the reuse times of business capabilities of Business Middle Office and shorten the TTM of new business construction to make products and users connect more accurately and achieve a process of each other.

4.6 BUSINESS MIDDLE OFFICE TECHNICAL PRACTICES

In addition to the support of a set of concepts and architecture methodology, the construction of Business Middle Office also requires a corresponding technical platform or framework system to ensure that Business Middle Office is not a castle in the air, but has a solid technical foundation.

The technical practices of the Business Middle Office refer to the decentralized SOA concept and the best practices of microservice architecture, from service governance, transaction processing, distributed database and cache technology, high concurrency and high availability, phased release, end-to-end full-chain stress testing and monitoring, big promotion assurance and new serverless architectures can all be applied in the practice of Business Middle Office, which reflects that the Business Middle Office technology system is inclusive, the feature of keeping pace with the times.

4.6.1 Distributed Service Framework and Governance

With the increasing business scale and complexity, applications have developed from single applications to distributed applications. With the development of distributed application architecture, the repeated problems in distributed applications have been continuously studied and solved by experts in the industry, gradually forming a framework and solution for common problems that can be reused in the field of distributed services, this framework is called distributed service framework. For example, the well-known Spring Cloud framework and Dubbo framework.

Distributed service governance is a must for developing and maintaining distributed services, and the distributed service framework is the main tool used for this work. Next, we will introduce the main problems faced by distributed services and the solutions adopted by the distributed service framework.

4.6.1.1 Service Registration

When it comes to distributed services, it must involve service providers and consumers. To call a service provider, service consumers need to know the specific address of the service provider. This problem is the most basic and common problem in the development of distributed services. Currently, the solutions in the industry are similar. Basically, a service registration service is provided to store the addresses of all service providers, and push the address to the subscriber. The specific implementation form varies according to different distributed service frameworks.

The Dubbo framework is a distributed service framework developed by Alibaba. Currently, the project has become the top-level project of Apache. It uses ZooKeeper as its service registry and uses dubbo rpc protocol to implement mutual calls between services.

SpringCloud started from packaging Netflix related components of the company, and has developed all the way. Now, many series have been developed, Netflix Eureka registry, and services use HTTP protocol to call each other.

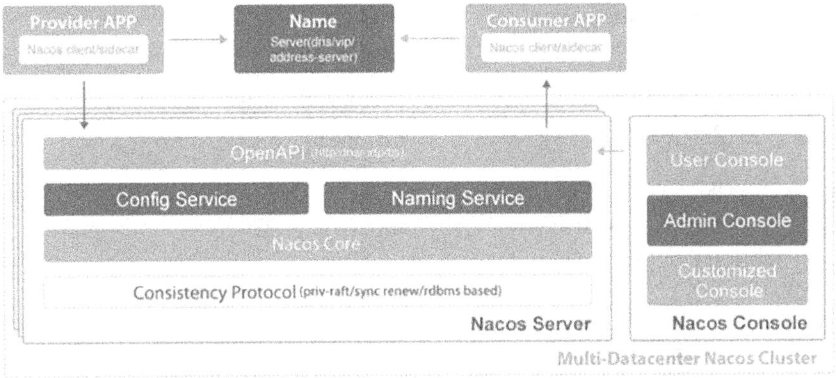

FIGURE 4.65 Nacos architecture diagram.

SpringCloud Alibaba is a series of open-source SpringCloud solutions provided by Alibaba. Its Service Registry is Nacos, and inter-service calls support dubbo rpc protocol. It performs well in both performance and functionality.

Alibaba Cloud distributed service platform EDAS provides a commercial version registry of Nacos. Applications developed using Nacos as the registry do not need to modify any code. After being deployed to EDAS, you can use the shared registry provided by EDAS.

The architecture and concept of Nacos are as follows (Figure 4.65):

1. Service

   A service refers to one or a set of software functions (such as retrieval of specific information or execution of a set of operations), the purpose is that different clients can be reused for different purposes (for example, through cross-Process Network calls). Nacos supports mainstream Service ecosystems, such as Kubernetes Service, gRPC|Dubbo RPC Service or SpringCloud RESTful Service.

2. Service Registry

   The service registry is a database of services, instances and metadata. The service instance is registered with the service registry at start-up and logged off when it is closed. The client of the service and router queries the service registry to find available instances of the service. The service registry may call the health check API of the service instance to verify whether it can process requests.

3. Service Metadata

   Service metadata include service endpoints, service tags, service version numbers, service instance weights, routing rules and security policies.

4. Service Provider

   Refers to the application provider that provides reusable and callable services.

5. Service Consumer

   The application that initiates a service call.

6. Configuration

   In the system development process, some parameters and variables that need to be changed are usually separated from the code and managed independently, and exist in the form of independent configuration files. The purpose is to better adapt static system artifacts or constructions (such as WAR and JAR packages) to the actual physical runtime environment. Configuration management is generally included in the system deployment process, which is completed by the system administrator or O & M personnel. Configuration changes are one of the effective methods to adjust system runtime behavior.

7. Configuration Management

   In the data center, all configuration-related activities such as configuration editing, storage, distribution, change management, historical version management and change audit are collectively referred to as configuration management.

8. Naming Service

   Provides mapping management services between the names of all objects and entities in a distributed system and associated metadata, such as ServiceName -> Endpoints Info, distributed Lock Name -> Lock Owner/Status Info, DNS Domain Name -> IP List, service discovery and DNS are the two major scenarios of Name service.

9. Configuration Service

   A service provider that provides dynamic configuration or metadata and configuration management during service or application running.

4.6.1.2 Service Configuration

When a single service uses local configuration to manage application configuration, the management complexity of related configurations increases with the increase of distributed services and the number of instances of the same service. The following problems occur:

- Static configuration needs to be modified or even republished, at least need to restart.
- Configuration files are too scattered, so you need to manage the configuration file locations of different systems.
- Configuration modification cannot be traced back. If an error occurs, rollback is troublesome.

Due to the above problems, all service configurations are unified into one configuration service center for unified management. The configuration center generally meets the following four capability requirements:

- It is applied to configuration separation. You do not need to modify the source code to modify the configuration.
- Centralized configuration management, convenient and efficient O & M.
- The configuration service must be stable and highly available.
- Configuration modification can be traced back to facilitate management and rollback.

The configuration center with the above capabilities can effectively improve the operation and maintenance efficiency and reduce the complexity and cost of configuration management.

SpringCloudNetflix uses spring-cloud-config-server as the configuration center. SpringCloudAlibaba uses Nacos as the configuration center.

Alibaba Cloud EDAS platform provides the commercial version configuration center ACM of Nacos.

The overall introduction of Nacos has been mentioned in the previous section. Here is the data model of the service configuration section (the picture is from the official website https://nacos.io) (Figure 4.66).

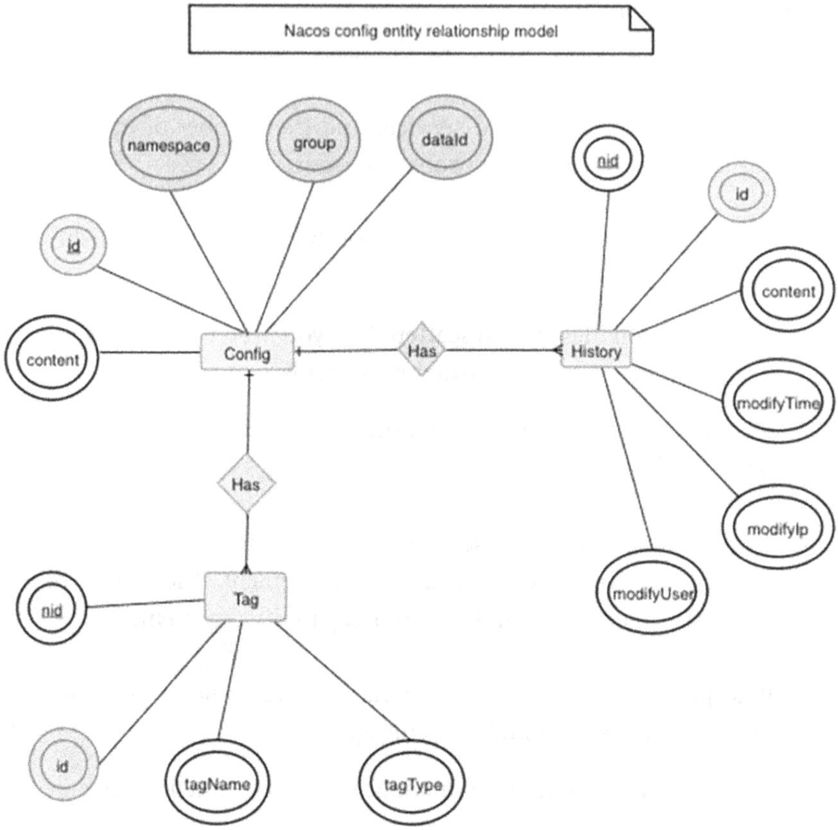

FIGURE 4.66 Nacos configuration partial data model.

The configuration consists of three main entities: Config, History and Tag.

4.6.1.3 Service Protocol

Calls between distributed services need to use specific RPC communication protocols. Generally speaking, they can be divided into HTTP protocol and TCP protocol. The specific comparison is as follows (Table 4.13):

Generally, we recommend that you use TCP communication protocols for internal communication, such as communication between internal services of Business Middle Office. We recommend that you use http communication protocols for external communication, such as communication between front-desk applications and Middle Office services.

TABLE 4.13 Comparison between HTTP and TCP Protocols

	TCP	HTTP
Case	Dubbo, motan, grpc, thrift, etc.	Springcloud, WebService, etc.
Serialization	Binary	Text
Performance	High	Low
Interface description	Thrift, etc.	Swagger, etc.
Client	Strong-type clients Generally, it can be automatically generated Supports multiple languages	Support for multiple languages
Developer friendliness	The client uses TCP directly The readability of binary content is not high	The browser is directly accessible High readability of text content
Openness	Convert to HTTP	Open directly to the outside world

SpringCloudNetflix uses the http protocol, SpringCloud Alibaba supports the dubbo protocol for internal calls, and supports the http protocol externally to better meet the requirements of performance and ease of use.

Dubbo uses a single long connection and NIO asynchronous communication by default, which is suitable for service calls with large amounts of small data and concurrent calls, and the number of service consumer machines is much larger than the number of service provider machines. Otherwise, the Dubbo default protocol is not suitable for services that transmit large amounts of data, such as files and videos, unless the request volume is very low. The architecture diagram is as follows (the picture is from the official website http://dubbo.apache.org/) (Figure 4.67).

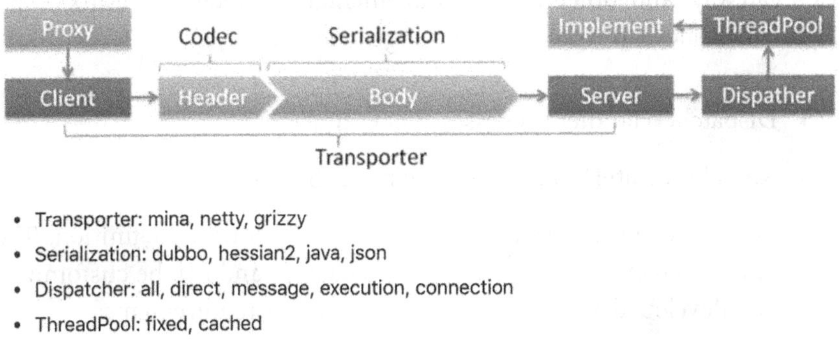

FIGURE 4.67 Dubbo architecture.

Dubbo uses tbremoting based on mina 1.1.7 and hessian 3.2.1. It has the following features:

- Connections: Single Connection
- Connection mode: Long connection
- Transmission protocol: TCP
- Transmission mode: NIO asynchronous transmission
- Serialization: Hessian binary serialization
- Scenario: The number of incoming and outgoing parameter packets is small (less than 100K is recommended), the number of consumers is larger than that of providers, and a single consumer cannot fill up the providers, try not to use dubbo to transfer large files or large strings.

4.6.1.4 Service Gateway

Service Gateway=Service Route+filter. Why do you need a service gateway? First, there are multiple services in the distributed service system. If each service is exposed to the external environment, it is not conducive to operation and maintenance management, and it is also not conducive to load balancing. In addition, cross-cutting functions such as service authentication, traffic limiting and monitoring do not need to be implemented once for each service. Therefore, service gateways are necessary in distributed service systems.

SpringCloud uses SpringCloudGateway as the recommended service gateway, and other related implementations include Netflix Zuul. Figure 4.68 shows the process of SpringCloudGateway:

- DispatcherHandler processes client requests.
- RoutePredicateHandlerMapping matching route.
- FilteringWebHandler runs the filter in the routing definition. The Filter is an extendable point of the Gateway and can be customized and developed according to specific business requirements.
- Runs specific business services.

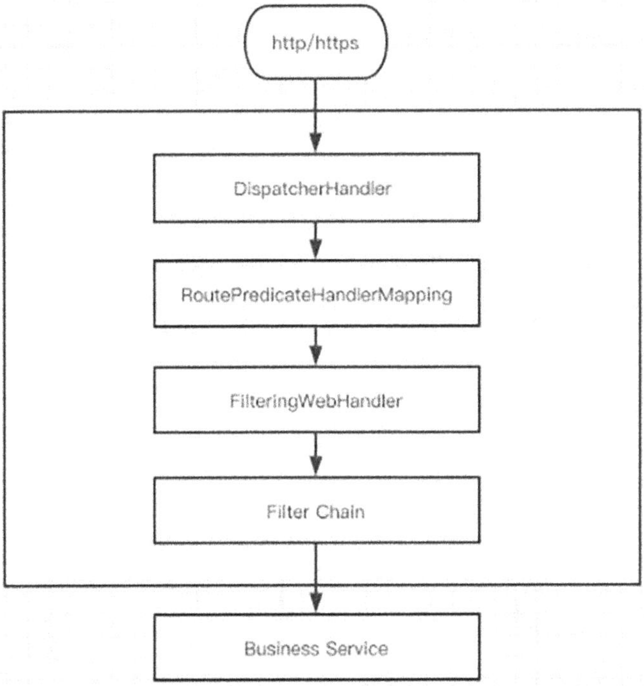

FIGURE 4.68 Springcloudgateway process.

4.6.1.5 Traffic Monitoring

Distributed service systems form call links due to mutual calls between services, which is different from single services. With the increasing scale and complexity of distributed service systems, it is important to call link monitoring. SpringCloudAlibaba uses Alibaba's open-source Sentinel as a link monitoring tool. Alibaba Cloud AHAS provides Sentinel commercial SaaS service version. Its main functions are as follows:

1. Architecture awareness

   - Automatically detects the application topology.
   - Visually displays the application dependencies on the infrastructure and the dependencies between components.
   - Continuously record the preceding dependencies.

2. Traffic protection
    - Specialized and diversified throttling methods.
    - Real-time monitoring in seconds.
    - Rule management that takes effect immediately.
3. Fault drill
    - Provides high-availability drill services based on real online faults.
    - Recommends fault drill scenarios based on your application architecture.
4. Function switch
    - Defines and manages business runtime function switching in a unified manner.
    - You can define any type of configuration items and automatically disassemble them based on their types. The code layer does not need to pay attention to type information, which simplifies the process of using configuration items at runtime.

4.6.1.6 Service Monitoring

Service monitoring is an important aspect of service governance. SpringBoot provides a lightweight monitoring solution SpringBootActuator and SpringBootAdmin. SpringBootActuator provides monitoring of a single Springboot application, including status, memory, thread, stack and so on. When the number of our distributed services is not convenient to use and the data format returned by the Actuator monitoring interface is Json, it is not convenient to read, SpringBootAdmin is born in this context, and it encapsulates and beautifies SpringBootActuator interfaces. It can query monitoring information in all SpringBootActuator and modify the level of logger.

The logs monitored by the SpringBootAdmin are still the logs of the service itself. When the service is down and cannot be accessed, the specific logs cannot be read. The Log Monitoring of distributed services generally adopts an independent monitoring mechanism, we usually use the ELK suite to monitor logs. The specific architecture is shown in Figure 4.69.

In addition to the monitoring solutions provided by the framework, the production environment generally has independent third-party monitoring

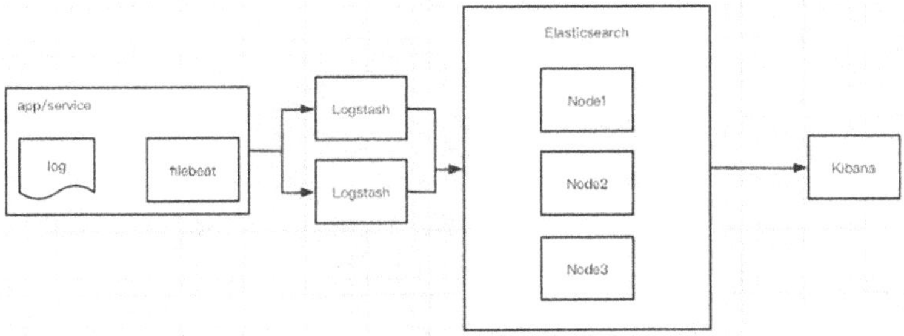

FIGURE 4.69 ELK overall architecture.

platforms, such as Prometheus. Alibaba Cloud provides a set of easy-to-use distributed service log collection services (SLS) and a distributed service monitoring platform Arms to provide powerful log and monitoring services for cloud services. For more information about SLS and Arms, see Alibaba product base.

The preceding sections describe the main problems and common solutions of the distributed service framework, Alibaba Cloud provides a comprehensive range of products and services for the entire life cycle of distributed services, for more information, see Alibaba product chapter.

4.6.2 Distributed Transaction Processing

Business Middle Office consists of different service centers at the business architecture level, which is reflected in the technology architecture level. Each service center has an autonomous service that supports private data in its own domain. No matter between service centers or aggregation applications built on top of a service center, there must be interaction between them to complete an actual business transaction or scenario. This means that Business Middle Office technical practices are inseparable from transactions in a distributed system.

4.6.2.1 Concept of Distributed Transactions and Business Middle Office Application Scenarios

Conceptually, a transaction is a group of non-grouping operation sets. These operations are either executed successfully or all canceled. The most typical transaction scenario is the transfer between bank accounts.

For example, if account A wants to transfer 100$ to account B, account A needs to deduct 100$ and account B needs to add 100$. Only when the data of both accounts are successfully changed can the transfer be regarded as a successful transfer. More strictly speaking, the transaction can be expressed in the atomicity, consistency, isolation, durability features:

- Atomicity: all operations in a transaction must be successfully executed, or all operations must be canceled.
- Consistency: Transaction Consistency is derived from a specific set of statements for the data, and the invariants must always be valid. The maintenance of data invariants is essentially an application, rather than an essential part of database transactions. Consensus here is essentially different from distributed consensus. For example, account A and account B each have 100$. No matter how many times the two accounts transfer money to each other, the total amount of funds in the two accounts is still 200$.
- Isolation: Indicates the degree of Isolation between concurrent transactions. The ANSI/ISO SQL standard (SQL92) defines four transaction isolation levels: Read Uncommitted, read committed, repeatable read and serialized.
- Durability: Indicates the persistence property. After a transaction is completed, data changes made to the database are retained. From a technical perspective, there is no perfect persistence solution.

In Business Middle Office technical practice, if a service in the service center can only access a single logical database or physical database instance of the service center, which does not involve remote calls across service centers or multiple databases or data sources within the service center, the transaction in this case is called a local transaction. Traditional databases allow you to process local transactions. If a service in the service center involves remote calls across service centers or across multiple databases or data sources in the service center, consider distributed transaction processing.

In a distributed transaction, the initiator, participant, resource server and transaction manager (TM) of the transaction are located on different nodes of the distributed system. In summary, distributed transactions

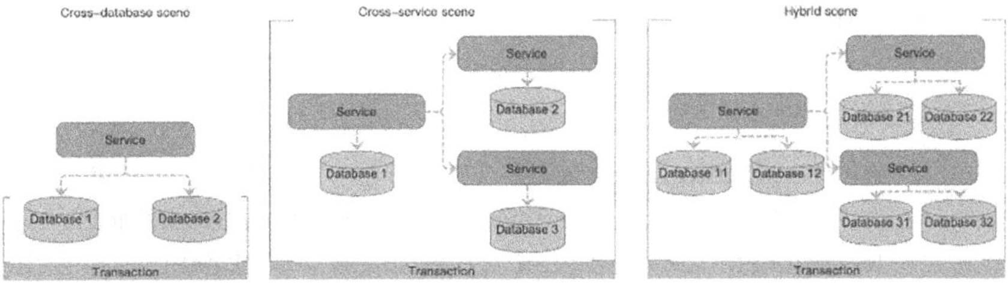

FIGURE 4.70 Dubbo architecture.

can be used in three scenarios: cross-database distributed transactions, cross-service distributed transactions and hybrid distributed transactions (Figure 4.70).

In distributed transactions, participants are distributed in asynchronous networks. These participants achieve distributed consistency through network communication. Network communication, failures and timeouts inevitably occur. Therefore, it is more difficult to implement distributed transactions than local transactions. The following are several common distributed transaction solutions.

4.6.2.2 Solution for Distributed Transaction

1. XA specification

   In the 1980s, a group of distributed transaction middleware represented by tuxedo (transactions for UNIX, extended for distributed operations) appeared, which focused on solving the distributed transaction processing of OLTP system in telecommunication field. Later, the standardization organization X/Open absorbed and adopted the design ideas and some interfaces of these distributed transaction middleware, and introduced the distributed transaction specification: XA specification.

   The XA specification defines a distributed transaction processing model that includes four core roles:

   - RM (Resource Manager):

     RM provides interfaces for data resource operation and management to ensure data consistency and integrity. A database management system is the most representative. Of course, some file systems and MQ systems can also be considered RM.

- TM (Transaction Manager):
  TM coordinates the actions of all RMs associated with cross-database transactions.
- AP (Application Program):
  The AP calls the RM interface according to the business rules to complete the change of the business model data. When the data change involves multiple RMs and the transaction is to be guaranteed, the AP will define the boundary of the transaction through the TM. The TM is responsible for coordinating each RM involved in the transaction to complete a global transaction.
- CRMs (Communication Resource Managers):
  Used to transmit transactions across services.
  In the XA specification, a distributed transaction is built on the basis of the RM's local transaction, which is then considered a branch transaction. The TM is responsible for coordinating these branch transactions, and either all of them are committed successfully, or all of them are rolled back. The XA specification divides the distributed transaction processing process into two phases, which are also called two-phase commit protocol (two-phase commit):

I. Preparatory phase:

   - TM records the transaction start log and asks each RM if commit preparation can be performed.
   - After receiving the instruction, the RM evaluates its own state and tries to perform preparatory operations for the local transaction, such as reserving resources, locking resources and performing operations. However, the RM does not commit the transaction and waits for a subsequent instruction from the TM. If the attempt fails, inform the TM that the current phase fails and rolls back its own operation, and then the TM is no longer involved in the transaction. Take MySQL as example, if resources are locked, the redo logs and undo logs are written in this phase.
   - The TM collects the RM's response and records the transaction preparation completion log.

II. Commit/rollback: In this phase, the transaction is committed or rolled back based on the coordination result of the previous phase.

III. If all RMs returned success in the previous step,

   - TM records the transaction commit log and initiates the transaction commit instruction to all RM.
   - After receiving the directive, the RM commits the transaction, releases the resources and returns the commitment to the TM.
   - If TM receives a response from all RMs, the transaction end log is logged.

IV. If RM returns an execution failure or does not respond to the timeout in the previous step, TM will handle it as an execution failure.

   - Record the transaction abort log and send the transaction rollback instruction to all RMs.
   - After receiving the instruction, RM rolls back the transaction, releases the resources and responds to TM that the rollback is completed.
   - If TM receives all RM responses, the end of transaction log is logged.

The XA specification defines the interaction interfaces between core components in detail. For more information, see https://pubs.opengroup.org/onlinepubs/009680699/toc.pdf. Figure 4.71 shows the interactions between TM and RM in a global transaction.

The XA two-phase commit protocol is designed to implement the four ACID characteristics of a transaction like a local transaction.

- Atomicity: Ensure that the transaction is atomic in the prepare and commit phases.
- Consistency: XA protocol implements strong consistency.

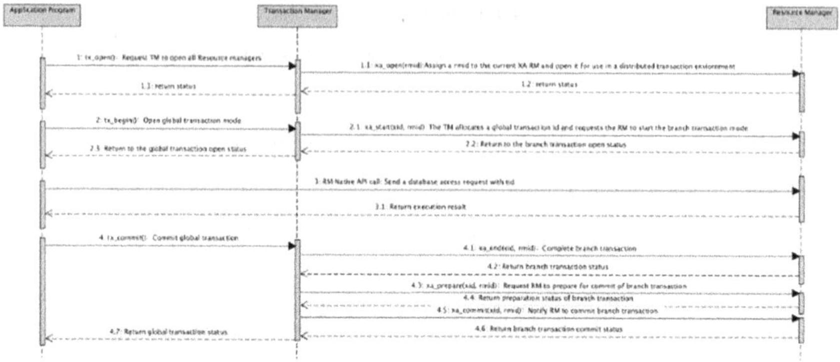

FIGURE 4.71 Interactions between transaction manager and resource manager.

- Isolation: XA transaction always holds the lock of resource before completion, so transaction isolation can be achieved.
- Persistence: Based on local transaction implementation, persistence is guaranteed.

XA is the earliest distributed transaction specification. Mainstream databases, such as Oracle, MySQL and SQL server, all support XA specification. JTA specification in J2EE is also written according to XA specification, which is compatible with XA specification. XA is a distributed transaction model implemented at the resource management level, which has low intrusion to business logic.

XA two-phase commit protocol can cover three scenarios of distributed transaction, but in the technical practice of Business Middle Office platform, the practical application scenarios are very limited due to the following reasons:

- In the process of global transaction execution, RM always holds the resource lock, and the call keeps the synchronous blocking state. If there are too many participating RMs, especially in the cross-service scenario, the number and time of network communication will increase sharply, so the blocking time will be longer, the throughput of the system will become very lower, and the probability of transaction deadlock will increase, so it is not suitable for the Business Middle Office to adopt distributed transaction pattern in the cross-service scenario, especially for high-concurrency case.

- For each TM, because it is a single point, there is a single point of failure risk. If TM hangs up after phase 1, the participating RM will not receive the request of phase 2 for a long time and holds the resource lock for a long time, which will affect the throughput of business transaction. At the same time, there are up to eight interactions between TM and RM in a complete global transaction, which greatly affects the processing performance of the system.
- XA two-phase protocol may cause blocking issue. If the TM goes offline while transactions are waiting for its final decision, RM will be stuck and holds their database locks until the TM comes online again and issues its decision. This extended holding of locks may be disruptive to other applications that are using the same databases.

2. TCC

TCC (Try, Commit, Cancel) is a compensation-based distributed transaction processing mode. In this model, each service of the Business Middle Office application or service center is required to provide three interfaces, that is, try, confirm and cancel. Its core idea is to release the resource lock as earlier as possible through the reservation of resources. If the transaction can be committed, the system completes the confirmation of the reserved resources. If the transaction is to be rolled back, the reserved resources are released.

TCC is also a two-phase commit protocol, and it can be considered as a variant of 2PC/XA. However, TCC does not hold resource locks for a long time. The TCC model divides the commitment of transactions into two phases:

I. Phase 1: Completes the business check (consistency) and reserves business resources (quasi-isolation), that is, try in TCC.

II. Phase 2: If the reservation of all business resources is successful in the try phase, the confirm operation is performed. Otherwise, the cancel operation is performed.

   - Confirm: Performs service operations only on the reserved resources, and keeps retrying if the operation fails.
   - Cancel: Cancels the execution of a service operation to release the reserved resources. If the operation fails, the system will continue to retry the operation (Figure 4.72).

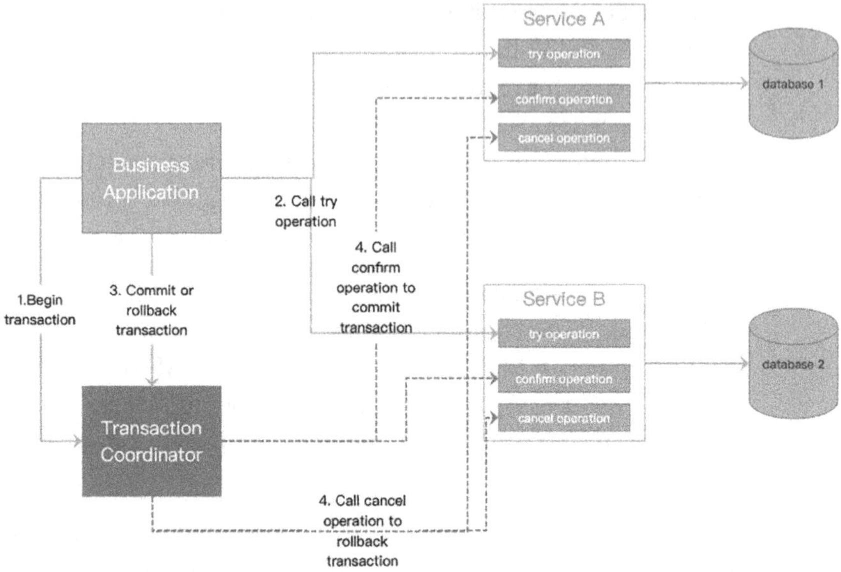

FIGURE 4.72 TCC transaction processing mechanism.

In the TCC model, the initiator and participant of a transaction need to record the transaction log. The initiator of a transaction needs to record the status and information of the global transaction and each branch transaction. The participant of a transaction needs to record the status of the branch transaction.

The TCC transaction may encounter errors, such as downtime, restart and network interruption, at any phase in the execution process. When the transaction is in a non-atomic state and in a non-eventual consistent state, the remaining branch transactions need to be committed or rolled back based on the logs of the master transaction record and branch transaction record. Ensure that all exhibitions in the whole distributed transaction reach a final consistent state and implement the atomicity of transactions.

TCC transactions have four features:

- Atomic: The initiator of the transaction coordinates all commit or rollback of each branch transaction.
- Consistency: TCC provides eventual consistency.

- Isolation: Implements data isolation through trying to pre-allocate resources.
- Persistence: Implemented by each branch transaction.

TCC implements distributed transactions from the resource layer to the business layer. This allows businesses to flexibly choose resource locking granularity, and locks are not held throughout the global transaction execution process. Therefore, the system throughput is much higher than that of the 2PC/XA Mode. Open-source frameworks that support TCC transactions include Seata, ByteTCC, Himly and TCC transaction.

The TCC transaction model is intrusive for the business side. The business side needs to split the function implementation from one interface to three, resulting in high development effort. At the same time, in order to avoid exceptions caused by communication failures or timeouts in asynchronous networks, TCC transactions require the service side to follow three policies in the design and implementation:

- Nullable rollback: The exception occurs in phase 1, and some participants do not receive a try request, thus triggering the cancel operation of the entire transaction. If the try operation fails or the participant that does not perform the try operation does not receive the cancel request, an empty rollback operation is required.
- Idempotence: The exception occurs in phase 2. For example, when the network connection times out, the confirm and cancel methods are repeatedly called. Therefore, the two methods must ensure idempotence in implementation.
- Prevent resource suspension: Network exceptions may cause two phases to fail to ensure strict sequential execution. In case that participants try to arrive later than the cancel request, cancel will perform a null rollback to ensure the correctness of the transaction, but the try method cannot be executed at this time.

3. Saga

Saga also serves as a mechanism to maintain data consistency between different services in Business Middle Office technical practices.

A Saga represents a system operation used to update data in multiple Business Middle Office services. It is a compensatory distributed transaction processing mode, like the TCC. However, it does not have a try phase, but considers distributed transactions as a transaction chain consisting of a group of local transactions. Each forward transaction in the transaction chain corresponds to a transaction operation that is reversible. The Saga transaction coordinator executes branch transactions in sequence in the transaction chain. After branch transactions are executed, resources are released. If a branch transaction fails, the transaction compensation operation is performed in the opposite direction.

If a Saga distributed transaction chain has n branch transactions to form $[T_1, T_2,\ldots, T_n]$, the distributed transaction can be executed in three ways:

- $T_1, T_2,\ldots, T_n$: n transactions are all executed successfully.
- $T_1, T_2,\ldots, T_i, C_i,\ldots,C_2, C_1$: The execution of the i $(i \leq n)$ transaction fails, and the compensation operations of all previous local transactions are invoked in the reverse execution order of the ith transaction. If the compensation fails, try again all the time. Compensation operations can be optimized for parallel execution.
- $T_1, T_2,\ldots,T_i$ (failure), T_i (retry), T_i (retry),…, T_n: Applicable to scenarios where the transaction must succeed. If the transaction fails, the transaction is retried, and no compensation operation is performed.

Based on the actual execution process of Saga transactions, a typical Saga structure model can contain three types of transactions, as shown below.

- Compensatory transactions: Transactions that can be compensated for the rollback of transactions.

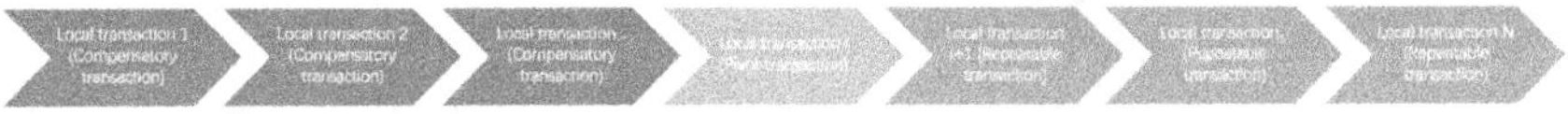

FIGURE 4.73 Saga processing flow.

- Pivot transactions: The key points of Saga's execution process. If the key transaction is executed successfully, Saga will run until completion. A critical transaction is not necessarily a compensatory or repeatable transaction, but it can be the last compensatory or first repeatable transaction.
- Repeatable transactions: Transactions after critical transactions to ensure success (Figure 4.73).

The implementation of Saga needs to include logic that coordinates Saga steps. When Saga starts, coordination logic must select and notify the first Saga party to execute a local transaction. Once the transaction is completed, Saga coordinates to select and call the next Saga party. This process continues until Saga has completed all the steps. If any local transaction fails, Saga must execute the compensation transaction in reverse order. There are two ways to build Saga's coordination logic.

Collaboration: Saga's roles and execution sequence are logically distributed among Saga's participants, and they communicate by exchanging events.

Orchestration: The Saga decision-making and execution sequence logic is centralized in a Saga orchestration device. The orchestration device sends imperative messages to each Saga participant to instruct them to complete local transactions.

Saga transactions guarantee three features of the transactions:

- Atomicity: The Saga coordinator can coordinate that local transactions in the transaction chain be both committed and rolled back.
- Consistency: Saga transactions can implement eventual consistency.
- Persistence: Persistence based on local transactions.

However, Saga does not guarantee the isolation of transactions. Therefore, the change will be visible to other transactions after the local transaction is committed. If other transactions change the data that have been committed successfully, the compensation operation may fail. This kind of scenario should be considered and the issue should be avoided in terms of business design.

Saga transactions are the same as TCC transactions. They have high requirements for service implementation and require the application design and implementation to follow three strategies:

- Allow nullable compensation: Due to network exception, the transaction participants have merely received the compensation operation. Null compensation is required because no normal operations have been performed.
- Idempotence: Forward operations and compensation operations of transactions can be repeatedly triggered. Therefore, the idempotence of operations must be ensured.
- Prevent resource suspension: If the forward operation of a transaction arrives later than the compensation operation due to network exception, this normal operation must be discarded. Otherwise, resource suspension may occur.

Although Saga and TCC are compensation transactions, they are also different due to different commit phases:

- Saga is an imperfect compensation: The compensation operation will leave traces of the original transaction operation, so the impact on the business must be considered.
- The TCC is perfect compensation: The compensation operation thoroughly cleans up the original transaction operation, and the user is not aware of the status information before the transaction is canceled.
- TCC transactions can support asynchronization better, but the Saga mode is generally more suitable for asynchronization in the compensation phase.

The Saga mode is suitable for business process with long transaction scenarios. At the same time, the Saga uses one-phase commit mode, which does not lock resources for a long time and has no cask effect. Therefore, this architecture has high system performance and high throughput, which is very suitable for microservice architecture.

The Saga mode is supported by many frameworks, including Alibaba's open-source Seata project and Chris Richardson's Eventuate Tram Saga.

4. Distributed Transaction based on Transaction Message

The core idea of the message-based distributed transaction model is to notify other transaction participants of the transaction execution status through the message system.

The introduction of message systems can more effectively decouple transaction participants and allow each participant to execute asynchronously. The difficulty of this solution is to ensure the consistency between the execution of local transactions and the sending of messages. The execution must be successful at the same time or be canceled at the same time. There are two main ways to implement this:

- Solution based on transactional messages of MQ
- Solution based on transactional message with local database

The current section would introduce the above first one.

I. Distributed transactions based on transactional message of MQ

Ordinary messages cannot solve the problem of consistency between local transaction execution and message sending. Because message sending is a network communication process, an error or timeout may occur during message sending. If the transaction times out, the message may be sent successfully or fail, and the message sender cannot determine the message. Therefore, whether the transaction is submitted or rolled back, the message sender may have inconsistency.

To solve this problem, you need to introduce transactional message. Transactional message and ordinary message are different because they are in the prepared state and cannot be consumed by the subscriber after the transactional message is sent successfully, and can only be listened to by the downstream subscriber after the transactional message is changed to the consumable state.

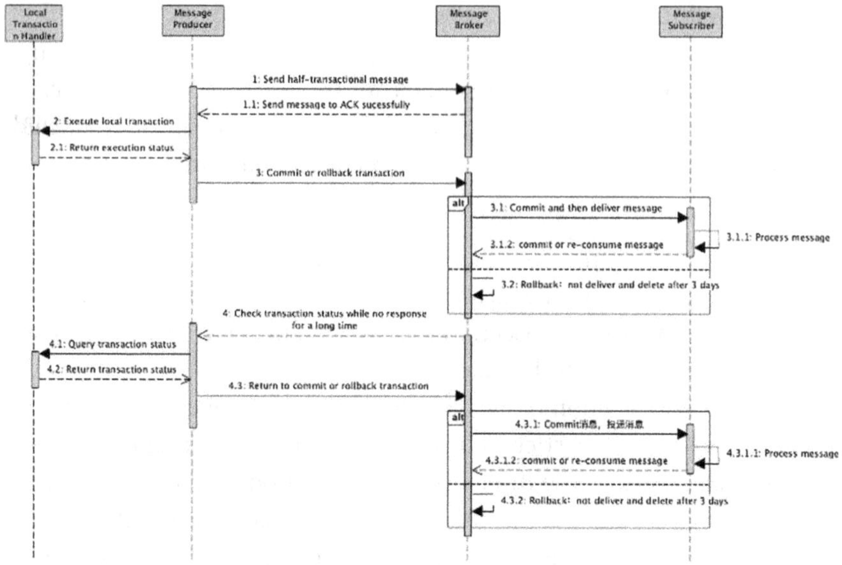

FIGURE 4.74 Process of transaction messages.

Figure 4.74 shows the processes for sending local transactions and transactional messages.

- The transaction initiator sends a transactional message in advance.
- After receiving a transactional message, MQ persists the message, sets the message status to pending for sending and sends an ACK message to the sender.
- If the transaction initiator does not receive the ACK message, the execution of the local transaction is canceled. If the ACK message is received, the local transaction is executed, and another message is sent to the MQ system to notify the local transaction execution status.
- After receiving the message notification, the MQ system changes the status of the transactional message based on the execution of the local transaction. If the execution is successful, the message is changed to 'consumable' and delivered to the subscriber. If the transaction fails, the transaction message is deleted.

- The notification message sent to MQ after the execution of the local transaction may be lost. Therefore, MQ, which supports transactional messages, provides a scheduled scan logic. This logic scans messages that are still in the 'pending sent' status and sends a query to the message sender to query the final status of the transactional message and update the status of the transactional message based on the query result. Therefore, the initiator of the transaction needs to provide the MQ system with an interface for querying the transactional message status.
- If the status of the transactional message is 'ready to send', The MQ system pushes the message to downstream participants. If the push fails, the system will keep retrying.
- After receiving a message, the downstream participant runs the local transaction. If the execution of the local transaction is successful, an ACK message is sent to the MQ system. If the execution fails, the message can be re-consumed.

5. Distributed Transaction based on Transaction Message in Local Database

The transactional message-based model has strong requirement on the MQ system. Not all MQ systems support transactional messages. RocketMQ is one of the few MQ systems that supports transactions. If the MQ system does not support transactional messages, the local message can be considered.

The core idea of this model is that the transaction initiator maintains a local message table, and both the SQL execution of service and the creation of local message record are in the same local database transaction. If the service is successfully executed, a message in pending status is also recorded in the local message table. The system starts a scheduled task to regularly scan records in the local message table that are in pending status and sends them to the MQ system. If the sending fails or times out, they will always be sent. Upon the successful sending, those related message will be deleted from the local message table. The subsequent consumption and subscription process is similar to that based on transactional messages.

The local message-based distributed transaction model supports the following features of atomicity, consistency, isolation and durability:

- Atomicity: Both branch transactions and non-branch transactions can eventually be executed.
- Consistency: Eventual consistency is provided.
- Isolation: Isolation is not guaranteed.
- Durability: Guaranteed by local transactions.

Message-based distributed transactions can effectively decouple distributed systems from other systems. The calls between transaction participants are not synchronous calls. It has strong requirements for MQ system, and it is also invasive to business implementation. Either it provides transaction message status query interface, or it needs to maintain local message table. Message-based distributed transaction is suitable for business scenarios that require final consistency, and it is also widely used in the technical practice of Business Middle Office.

6. Best Effort Notification-based Distributed Transaction

   The best effort notification-based distributed transaction solution is also based on MQ systems, but does not require MQ to be reliable. After completing business processing, the initiative of the business activity sends a notification message to the passive party. The initiative party can set a time ladder notification rule, repeat the notification rule after the notification fails, and no notification is sent after N times. The active party provides an interface for checking the query. The passive party checks the query on demand to recover lost service messages. The passive party to the business activity must ensure its idempotence.

   Suppose that the customer recharges the mobile phone bill through the client provided by the online business hall of the telecom operator, and the recharge method chooses Alipay payment. The workflow of the operation is as follows (Figure 4.75):

- The customer chooses to recharge the amount of 50$, and the payment method is Alipay.

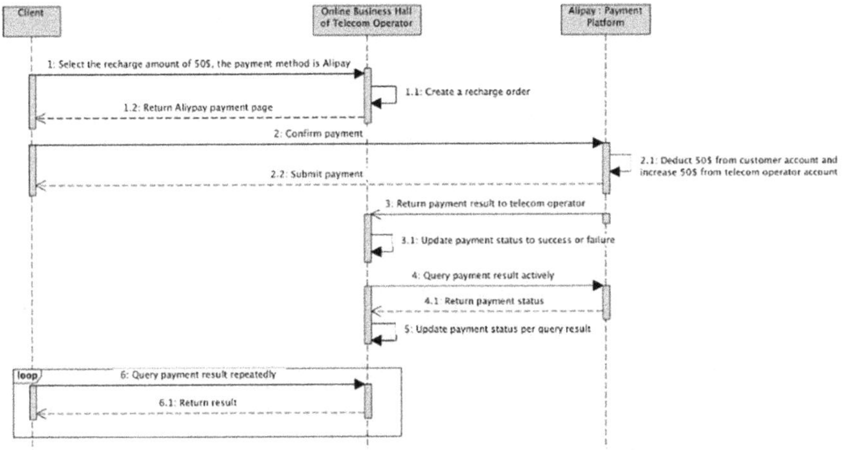

FIGURE 4.75 Alipay pays online fee recharge process.

- The telecom operator's online business hall creates a recharge order with a status of payment and jumps to Alipay's payment page (entering Alipay's system at this time).
- After Alipay confirms the customer's payment, it deducts 50$ from the customer's account and adds 50$ to the telecom operator's account. After the execution is completed, the recharge result will be notified to the telecom operator's online business hall. After the telecom operator's online business hall system is notified, the status of its recharge order is updated to success or failure. In the notification stage, the notification may fail due to some abnormalities, such as network failure or abnormal service of the telecom operator's online business hall system. Usually, Alipay will re-notify in stages, but there will be a maximum number of such notifications, so there is no guarantee that it will be successfully notified to the letter operator's online business hall. This is also the meaning of doing our best to inform.
- If Alipay's notice exceeds the upper limit, the telecom operator's online business hall can scan the order in payment and initiate a request to Alipay to verify the payment result of the order.

- The telecom operator's online business hall updates the status of its recharge order according to the results of the inquiry.
- The client can poll the status of the recharge order during this process.

Based on the workflow described above, from a technical point of view, the ACK mechanism of the MQ can be used to realize the best effort notification. Typical products that support message consumption retries, such as the Alibaba RocketMQ.

Taking the ACK mechanism of RocketMQ from Alibaba Cloud as an example, there are two solutions to realize the best effort notification for distributed transaction. The business initiatives and passive parties mentioned in the two schemes refer to the party who initiated the notice and the party receiving the notification, respectively. Taking the above-mentioned business scenario as an example, Alipay is the initiative of the business and the online business hall of the telecom operator is the passive party of the business.

The first scheme is that the active party and the passive party implement the best effort notification based on the message and ACK mechanism of RocketMQ. The key process is as follows (Figure 4.76):

- Notify the initiator to send the notification to the user-defined topic of RocketMQ through the ordinary message mechanism. If the message is not sent, the notification receiver can actively query the service execution result of the notification initiator.

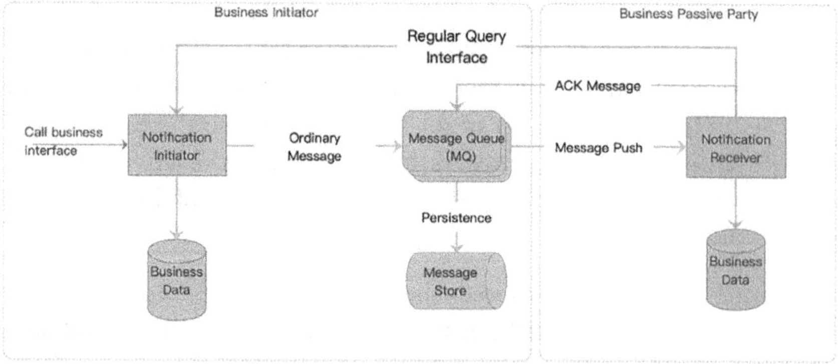

FIGURE 4.76 Best effort notification solution one.

- Inform the receiver to listen to RocketMQ's custom topic.
- Inform the receiver to receive the message and respond to ACK after the business processing is completed, otherwise the response needs to be re-consumed.
- If the receiver does not respond to the ACK, RocketMQ will repeat the notification. RocketMQ will gradually increase the notification interval in a way similar to the interval of 1, 5, 10, 30 minutes, 1, 2, 5 and 10 hours until it reaches the upper limit of the time window required by the notification.
- The receiver can check the consistency of the message through the message checking interface.

The second solution is to take the initiative service as the consumer of MQ for RocketMQ messages and provide the notification gateway to interact with the passive service through an interface to implement best effort notification. The key process is as follows (Figure 4.77):

- The notification initiator sends the notification to the custom topic of RocketMQ through the ordinary message mechanism. If the message is not sent out, the notification recipient can actively query the service execution result of the notification initiator.
- Send a notification to the gateway to listen on the custom topic of RocketMQ.

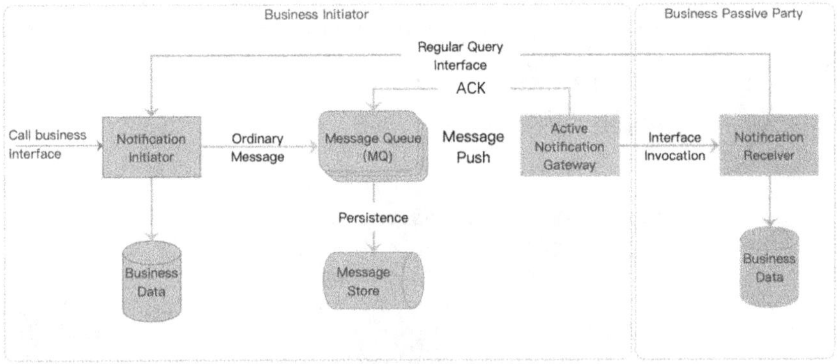

FIGURE 4.77 Best effort notification solution two.

- The initializer notifies the gateway to receive the message, and calls the corresponding method to notify the API provider to complete the business processing and then return the ACK. Otherwise, the consumer must re-consume the message. The Here interface can be a traditional Web Service, RESTful API or RPC Interface.
- If the Gateway fails to respond with an ACK message, RocketMQ notifies the Gateway again. MQ for RocketMQ gradually increases the notification interval by intervals such as 1, 5, 10, 30 minutes, 1, 2, 5, and 10 hours, until the required time window is reached.
- Notification recipients can proofread the consistency of messages through the message proofreading interface.

The key feature of the best effort notification solution is to introduce a periodic verification mechanism to guarantee the eventual consistency. This provides simple implementation that is less intrusive for the business and less demanding for the message system. It is suitable for scenarios with low sensitivity to real-time strong consistency, such as cross-platform business interaction and cross-enterprise systems.

4.6.2.3 Technical Best Practice for Distributed Transaction

1. Use Transactional Message of RocketMQ to Handle Distributed Transaction

   Perform the following steps to send transactional messages:

   I. Send a half message and execute a local transaction. An example of the code is as follows:

```
import com.aliyun.openservices.ons.api.Message;
import com.aliyun.openservices.ons.api.PropertyKeyConst;
import com.aliyun.openservices.ons.api.SendResult;
import com.aliyun.openservices.ons.api.transaction.
LocalTransactionExecuter;
import com.aliyun.openservices.ons.api.transaction.
TransactionProducer;
```

```
import com.aliyun.openservices.ons.api.transaction.
TransactionStatus;
import java.util.Properties;
import java.util.concurrent.TimeUnit;

public class TransactionProducerClient {

 private final static Logger log = ClientLogger.getLog();
 public static void main(String[] args) throws
 InterruptedException {
     // local business
     final BusinessService businessService = new
     BusinessService();
     Properties properties = new Properties();
        // Group ID created from console
     properties.put(PropertyKeyConst.GROUP_ID, "XXX");
        // Alibaba Cloud AK created for your account via
        console
     properties.put(PropertyKeyConst.AccessKey, "XXX");
        // Alibaba Cloud SK created for your account via
        console
     properties.put(PropertyKeyConst.SecretKey, "XXX");
        // RocketMQ naming server address from console
     properties.put(PropertyKeyConst.NAMESRV_ADDR, "XXX");

     TransactionProducer producer = ONSFactory.createTrans
     actionProducer(properties,
             new LocalTransactionCheckerImpl());
     producer.start();
     Message msg = new Message("MyTopic", "TagA", "Hello
     MQ transaction===".getBytes());
     try {
           SendResult sendResult = producer.send
           (msg, new LocalTransactionExecuter() {
               @Override
               public TransactionStatus execute(Message
               msg, Object arg) {

                   Object businessServiceArgs = new
                   Object();
                   TransactionStatus transactionStatus =
                   TransactionStatus.Unknow;
                   try {
                       boolean isCommit =
                           businessService.execbusinessSer
                           vice(businessServiceArgs);
                       if (isCommit) {
```

```
                        // Commit message if local
                        transaction is committed
                        transactionStatus = Transaction
                        Status.CommitTransaction;
                    } else {
                        // Rollback message if local
                        transaction is rolled back
                        transactionStatus =
                        TransactionStatus.
                        RollbackTransaction;
                    }
                } catch (Exception e) {
                    log.error("Message Id:{}", msgId, e);
                }
                System.out.println(msg.getMsgID());
                log.warn("Message Id:{}
                transactionStatus:{}", msgId,
                transactionStatus.name());
                return transactionStatus;
            }
        }, null);
    }
    catch (Exception e) {
            // If exception occur, it recommends to
            retry or store for compensation later
        System.out.println(new Date() + " Send mq
        message failed. Topic is:" + msg.getTopic());
        e.printStackTrace();
    }
  }
}
```

II. Submit transaction message status

After the execution of a local transaction, which can be successful or failed, the Broker must be notified of the transaction status of the current message. There are two notification methods:

- Commit after executing the local transaction.
- The execution of the local transaction has not been submitted, waiting for the server to check the status of the message transaction.

There are three types of transaction status:

- TransactionStatus.CommitTransaction to commit the transaction, allowing the consumer to consume the message.
- TransactionStatus.RollbackTransaction rolls back the transaction, and the message is discarded and cannot be consumed.
- TransactionStatus.Unknow: The status of the MQ for Apache RocketMQ is unknown, and the Broker is expected to query the status of the local transaction that corresponds to the message from the message producer.

```
public class LocalTransactionCheckerImpl implements
LocalTransactionChecker {
    private final static Logger log = ClientLogger.
    getLog();
    final  BusinessService businessService = new
    BusinessService();

    @Override
    public TransactionStatus check(Message msg) {

        Object businessServiceArgs = new Object();
        TransactionStatus transactionStatus =
        TransactionStatus.Unknow;
        try {
            boolean isCommit = businessService.checkbusines
            sService(businessServiceArgs);
            if (isCommit) {
                // Commit message if local transaction is
                committed
                transactionStatus = TransactionStatus.
                CommitTransaction;
            } else {
                // Rollback message if local transaction is
                rolled back
                transactionStatus = Transaction
                Status.RollbackTransaction;
            }
        } catch (Exception e) {
            log.error("Message Id:{}", msgId, e);
        }
        log.warn("Message Id:{}transactionStatus:{}",
        msgId, transactionStatus.name());
        return transactionStatus;
    }
 }
```

If the Half-message is sent, but transaction status is unknown whether returned or no local transaction of any status is committed because the application exits, the status of the Half message is unknown to the Broker. Therefore, the Broker periodically requests the sender to check and report the status of the Half message.

The Check method for transactional messages needs to contain the logic of checking transactional consistency. When sending a transactional message, MQ for RocketMQ requires LocalTransactionChecker to respond to the request of the Broker for local transaction status check. Therefore, the Check method for transactional messages needs to complete two things:

- Check the status of the local transaction corresponding to the half message (committed or rollback).
- Submit the status of the local transaction corresponding to the half message to the Broker.

After submitting and confirming the transactional message, MQ for Apache RocketMQ can deliver the message to subscribers or consumers. A message consumer can obtain the message according to the message subscription method. During message processing, ensure the idempotence of its service logic. MQ for RocketMQ supports the following two methods for obtaining messages:

- Push: Pushes a message from MQ for Apache RocketMQ to a Consumer.
- Pull: The Consumer actively pulls the message from MQ for Apache RocketMQ.

Pull Consumer provides more options for receiving messages. The following example code demonstrates only the most common push modes. For more information about pull modes, see the official documentation of Alibaba Cloud RocketMQ.

```
import com.aliyun.openservices.ons.api.Action;
import com.aliyun.openservices.ons.api.ConsumeContext;
import com.aliyun.openservices.ons.api.Consumer;
import com.aliyun.openservices.ons.api.Message;
import com.aliyun.openservices.ons.api.MessageListener;
import com.aliyun.openservices.ons.api.ONSFactory;
import com.aliyun.openservices.ons.api.PropertyKeyConst;
```

```
import java.util.Properties;

public class ConsumerTest {
   public static void main(String[] args) {
       Properties properties = new Properties();
        // Group ID created from console
     properties.put(PropertyKeyConst.GROUP_ID, "XXX");
        // Alibaba Cloud AK created for your account via
console
     properties.put(PropertyKeyConst.AccessKey, "XXX");
        // Alibaba Cloud SK created for your account via
console
     properties.put(PropertyKeyConst.SecretKey, "XXX");
        // RocketMQ naming server address from console
     properties.put(PropertyKeyConst.NAMESRV_ADDR, "XXX");
       properties.put(PropertyKeyConst.NAMESRV_ADDR,
"XXX");
         // Clustering subscription model (default)
         // properties.put(PropertyKeyConst.MessageModel,
PropertyValueConst.CLUSTERING);
         // Broadcast subscription model
         // properties.put(PropertyKeyConst.MessageModel,
PropertyValueConst.BROADCASTING);

       Consumer consumer = ONSFactory.
createConsumer(properties);
       // Subscribe multiple tags
       consumer.subscribe("MyTopic", "TagA||TagB", new
MessageListener() {
           public Action consume(Message message,
ConsumeContext context) {
               System.out.println("Receive: " + message);
               return Action.CommitMessage;
           }
       });

       consumer.start();
       System.out.println("Consumer Started");
   }
}
```

2. Distributed Transaction Middleware – Seata

After opening up Seata by Alibaba in early 2019, the project has received great attention. Seata aims at high performance and zero invasion to solve the distributed transaction problem in the field of microservice, which is in the process of rapid iteration.

The Seata supports four modes: AT, TCC, Saga and XA.

- The AT pattern is a non-intrusive distributed transaction solution. In AT mode, users only need to pay attention to their own 'service SQL' and use the 'service SQL' as the first phase. The Seata framework automatically generates the second-phase commit and rollback operations for the transaction. The AT mode is optimized for developers, but remains to be tested in terms of performance and scalability, and the overhead on performance cannot be underestimated. Therefore, it is not recommended in high-concurrency scenarios.
- In the TCC model, users need to perform the Try, Confirm and Cancel operations based on service scenarios. The transaction initiator executes the Try method in one-phase commit, the Confirm method in two-phase commit and the Cancel method in phase 2.
- In the Saga mode, a distributed transaction has multiple participants, each of which is a forward compensation service. You need to implement your forward operation and reverse rollback operation based on the business scenario. Forward operations are executed during a distributed transaction. If all forward operations are successfully executed, the distributed transaction is committed. If any forward operation fails, the distributed transaction rolls back the committed participants in the reverse order of the distributed transaction.
- In XA Mode, the Seata service only supports data sources that already support the XA protocol.

In the following sections, focus on the Seata support for TCC and Saga.

I. Using Seata to implement TCC only requires a few key steps described below.
   - Defines the configuration class. After the Seata framework injects the relevant elements, it automatically injects the DataSourceProperties to obtain JDBC information, and then injects a GlobalTransactionScanner to scan TCC transactions.

```
@Configuration
public class SeataAutoConfig {

    private DataSourceProperties dataSourceProperties;

    @Autowired
    public SeataAutoConfig(DataSourceProperties
    dataSourceProperties){
        this.dataSourceProperties = dataSourceProperties;
    }

    /**
     * init durid datasource
     * @Return: druidDataSource  datasource instance
     */
    @Bean
    @Primary
    public DruidDataSource druidDataSource(){
        DruidDataSource druidDataSource = new
        DruidDataSource();
        druidDataSource.setUrl(dataSourceProperties.
        getUrl());
        druidDataSource.setUsername(dataSourceProperties.
        getUsername());
        druidDataSource.setPassword(dataSourceProperties.
        getPassword());

          druidDataSource.setDriverClassName(dataSourcePro
          perties.getDriverClassName());
        druidDataSource.setInitialSize(0);
        druidDataSource.setMaxActive(180);
        druidDataSource.setMaxWait(60000);
        druidDataSource.setMinIdle(0);
        druidDataSource.setValidationQuery("Select 1 from
        DUAL");
        druidDataSource.setTestOnBorrow(false);
        druidDataSource.setTestOnReturn(false);
        druidDataSource.setTestWhileIdle(true);
        druidDataSource.
        setTimeBetweenEvictionRunsMillis(60000);
        druidDataSource.
        setMinEvictableIdleTimeMillis(25200000);
        druidDataSource.setRemoveAbandoned(true);
        druidDataSource.setRemoveAbandonedTimeout(1800);
        druidDataSource.setLogAbandoned(true);
        return druidDataSource;
    }
    /**
```

```
     * init datasource proxy
     * @Param: druidDataSource  datasource bean instance
     * @Return: DataSourceProxy  datasource proxy
     */
    @Bean
    public DataSourceProxy dataSourceProxy(DruidDataSource
    druidDataSource){
        return new DataSourceProxy(druidDataSource);
    }

    /**
     * init mybatis sqlSessionFactory
     * @Param: dataSourceProxy  datasource proxy
     * @Return: DataSourceProxy  datasource proxy
     */
    @Bean
    public SqlSessionFactory sqlSessionFactory(DataSourceP
    roxy dataSourceProxy) throws Exception {
        SqlSessionFactoryBean factoryBean = new
        SqlSessionFactoryBean();
           factoryBean.setDataSource(dataSourceProxy);
           factoryBean.setMapperLocations(new
           PathMatchingResourcePatternResolver()
                .getResources("classpath:/mapper/*Mapper.
                xml"));
        factoryBean.setTransactionFactory(new
        JdbcTransactionFactory());
        return factoryBean.getObject();
    }

    /**
     * init global transaction scanner
     * @Return: GlobalTransactionScanner
     */
    @Bean
    public GlobalTransactionScanner
globalTransactionScanner(){
        return new GlobalTransactionScanner("${spring.
        application.name}", "my_test_tx_group");
    }
}
```

- Defines the TCC interface for a specific business activity in a distributed transaction and marks it as @ LocalTCC, which identifies the TCC as the local mode, that is, the transaction is a local call and not a RPC call; in addition, add @

TwoPhaseBusinessAction annotation to identify TCC mode, which defines commitMethod and rollbackMethod. As shown in the following code, define one inventory deduction activity.

```
@LocalTCC
public interface InventoryDecrementBusinessAction {

    @TwoPhaseBusinessAction(name =
    "InventoryAction",commitMethod =
    "confirm",rollbackMethod = "cancel")
    boolean try(BusinessActionContext actionContext, @Busi
    nessActionContextParameter(paramName = "productDTO")
    ProductDTO productDTO);

    boolean confirm(BusinessActionContext actionContext);

    boolean cancel(BusinessActionContext actionContext);
}
```

- Implements the business activity interface. In the try phase, the pre-occupation logic or real business logic can be executed; in the confirm phase, the real business logic or an empty operation can be executed and the cancel phase executes the rollback logic.

```
@Component
public class InventoryDecrementBusinessActionImpl
implements InventoryDecrementBusinessAction  {

    ...

    @Override
    public boolean try(BusinessActionContext
    actionContext, ProductDTO productDTO) {

        ...
        int storage = baseMapper.
        decreaseInventory(productDTO.getProductCode(),
        productDTO.getCount());

        if (storage > 0){
            return true;
```

```
        }
        return false;
    }

    @Override
    public boolean confirm(BusinessActionContext
    actionContext) {
        return true;
    }

    @Override
    public boolean cancel(BusinessActionContext
    actionContext) {
        ...
        ProductDTO productDTO = JSONObject.
        toJavaObject((JSONObject)actionContext.getActionCo
        ntext("productDTO"),ProductDTO.class);
        int storage = baseMapper.
        increaseInventory(ProductDTO.getProductCode(),
        productDTO.getCount());
        if (storage > 0){
            return true;
        }
        return false;
    }
}
```

- Defines and implements business service interfaces. In the implementation of the business service interface, call the implementation of the business activity interface to complete the global distributed transaction.

```
@Service
public class InventoryServiceImpl implements
InventoryService {

    @Autowired
    private InventoryDecrementBusinessAction
    firstTccAction;

    @Autowired
    private InventoryOtherBusinessAction otherTccAction;

    @Override
```

```
    @GlobalTransactional
    public boolean decreaseInventory(final ProductDTO
    productDTO) {
        // inventory decrement, first phase
        boolean result = firstTccAction.try(null,
        productDTO);

        if(!result){
            // fail to decrease inventory, cancel local
            and global transaction
            throw new BusinessRuntimeException("The
            inventory decrement failed. ");
        }

        // other tcc action, first phase
        result = otherTccAction.try(null, productDTO);

        if(!result){
            throw new BusinessRuntimeException("The other
            tcc action failed. ");
        }
        return true;
    }
}
```

II. Implement Saga using Seata

Using Seata to implement TCC only requires a few key steps demonstrated below.

- Defines an interface for a specific business activity in a distributed transaction. As shown below, define two business activities: inventory deduction and business deduction. Each activity needs to include forward operations and its corresponding inverse or compensation operations.

```
public interface InventoryAction {

    boolean reduce(String businessKey, BigDecimal amount,
    Map<String, Object> params);

    boolean compensateReduce(String businessKey,
    Map<String, Object> params);
}
```

```
public interface BalanceAction {

    boolean reduce(String businessKey, BigDecimal amount,
    Map<String, Object> params);

    boolean compensateReduce(String businessKey,
    Map<String, Object> params);
}
```

- Defines the state machine corresponding to the business scenario. The state machine language is the JSON-based DSL customized by the Seata client. Figure 4.78 shows a state machine and the corresponding DSL language. The state

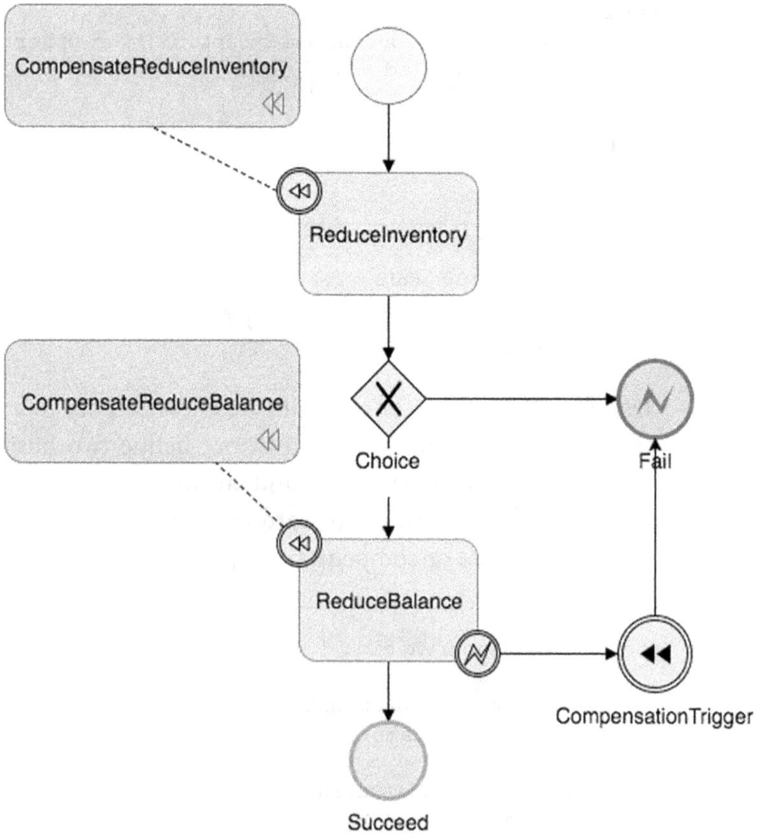

FIGURE 4.78 State machine and DSL language.

machine can be designed by using the visual designer provided by Seata. Details can be found in the Seata official documentation: https://seata.io/zh-cn/docs/user/saga.html.

```
{
    "Name": "reduceInventoryAndBalance",
    "Comment": "reduce inventory then reduce balance in a
    transaction",
    "StartState": "ReduceInventory",
    "Version": "0.0.1",
    "States": {
        "ReduceInventory": {
            "Type": "ServiceTask",
            "ServiceName": "inventoryAction",
            "ServiceMethod": "reduce",
            "CompensateState": "CompensateReduceInventory",
            "Next": "ChoiceState",
            "Input": [
                "$.[businessKey]",
                "$.[count]"
            ],
            "Output": {
                "reduceInventoryResult": "$.#root"
            },
            "Status": {
                "#root == true": "SU",
                "#root == false": "FA",
                "$Exception{java.lang.Throwable}": "UN"
            }
        },
        "ChoiceState":{
            "Type": "Choice",
            "Choices":[
                {
                    "Expression":"[reduceInventoryResult]
                    == true",
                    "Next":"ReduceBalance"
                }
            ],
            "Default":"Fail"
        },
        "ReduceBalance": {
            "Type": "ServiceTask",
            "ServiceName": "balanceAction",
            "ServiceMethod": "reduce",
            "CompensateState": "CompensateReduceBalance",
```

```
            "Input": [
                "$.[businessKey]",
                "$.[amount]",
                {
                    "throwException" :
                    "$.[mockReduceBalanceFail]"
                }
            ],
            "Output": {
                "compensateReduceBalanceResult": "$.#root"
            },
            "Status": {
                "#root == true": "SU",
                "#root == false": "FA",
                "$Exception{java.lang.Throwable}": "UN"
            },
            "Catch": [
                {
                    "Exceptions": [
                        "java.lang.Throwable"
                    ],
                    "Next": "CompensationTrigger"
                }
            ],
            "Next": "Succeed"
        },
        "CompensateReduceInventory": {
            "Type": "ServiceTask",
            "ServiceName": "inventoryAction",
            "ServiceMethod": "compensateReduce",
            "Input": [
                "$.[businessKey]"
            ]
        },
        "CompensateReduceBalance": {
            "Type": "ServiceTask",
            "ServiceName": "balanceAction",
            "ServiceMethod": "compensateReduce",
            "Input": [
                "$.[businessKey]"
            ]
        },
        "CompensationTrigger": {
            "Type": "CompensationTrigger",
            "Next": "Fail"
        },
        "Succeed": {
```

```
            "Type":"Succeed"
        },
        "Fail": {
            "Type":"Fail",
            "ErrorCode": "PURCHASE_FAILED",
            "Message": "purchase failed"
        }
    }
}
```

- Implements the defined business activity interface. This is related to business.
- Implements the defined business service interface, declares the global transaction and calls the implementation of the business activity interface in the interface method. This is similar to TCC.
- Configures a StateMachineEngine in the Spring Bean configuration file and starts the custom state machine.

```
<bean id="dataSource" class="...">
...
<bean>
<bean id="stateMachineEngine" class="io.seata.saga.engine.
impl.ProcessCtrlStateMachineEngine">
        <property name="stateMachineConfig"
        ref="dbStateMachineConfig"></property>
</bean>
<bean id="dbStateMachineConfig" class="io.seata.saga.
engine.config.DbStateMachineConfig">
    <property name="dataSource" ref="dataSource"></
property>
    <property name="resources" value="statelang/*.json"></
    property>
    <property name="enableAsync" value="true"></property>
    <property name="threadPoolExecutor"
    ref="threadExecutor"></property><!-- The thread pool
```

```
    used by event-driven execution. If all state machines
    are run synchronously, this is not required.
 -->
    <property name="applicationId" value="saga_sample"></
    property>
    <property name="txServiceGroup" value="my_test_tx_
    group"></property>
</bean>
<bean id="threadExecutor"
        class="org.springframework.scheduling.concurrent.
        ThreadPoolExecutorFactoryBean">
    <property name="threadNamePrefix" value="SAGA_ASYNC_
    EXE_" />
    <property name="corePoolSize" value="1" />
    <property name="maxPoolSize" value="20" />
</bean>

<!-- The Seata Server needs to use this Holder to obtain
the stateMachineEngine instance to recover the
transaction.
 -->
<bean class="io.seata.saga.rm.StateMachineEngineHolder">
    <property name="stateMachineEngine"
    ref="stateMachineEngine"/>
</bean>
```

- Gets StateMachineEngine bean and starts the custom state machine.

```
ApplicationContext applicationContext = new ClassPathXmlAp
plicationContext("classpath:saga/spring/statemachine_
engine_test.xml");
        stateMachineEngine = applicationContext.getBean
        ("stateMachineEngine", StateMachineEngine.class);

        Map<String, Object> paramMap = new HashMap<>();
        paramMap.put("a", 1);
        String stateMachineName =
        "simpleChoiceTestStateMachine";
stateMachineEngine.start(stateMachineName, null,
paramMap);
```

3. Distributed transaction based on transactional messages in local databases

   Its core design idea is to split distributed transactions into a series of local transactions for processing. The design is originated from Ebay's classic BASE solution. Ebay's complete solution is https://queue.acm.org/detail.cfm?id=1394128. The core of this solution is to asynchronously execute tasks that require distributed processing by using message logs. Message logs can be stored in local text files, databases or MQs. Service rules can be used to automatically or manually initiate a retry.

   The solution based on local database transactional messages is shown in Figure 4.79. The upstream of the Service Center 1 needs to complete a partial transaction processing before the downstream Service Center 2 triggers the transaction. In the figure, MQ is the general term for message queues and does not represent any specific product. The following key steps must be taken into account throughout the solution:

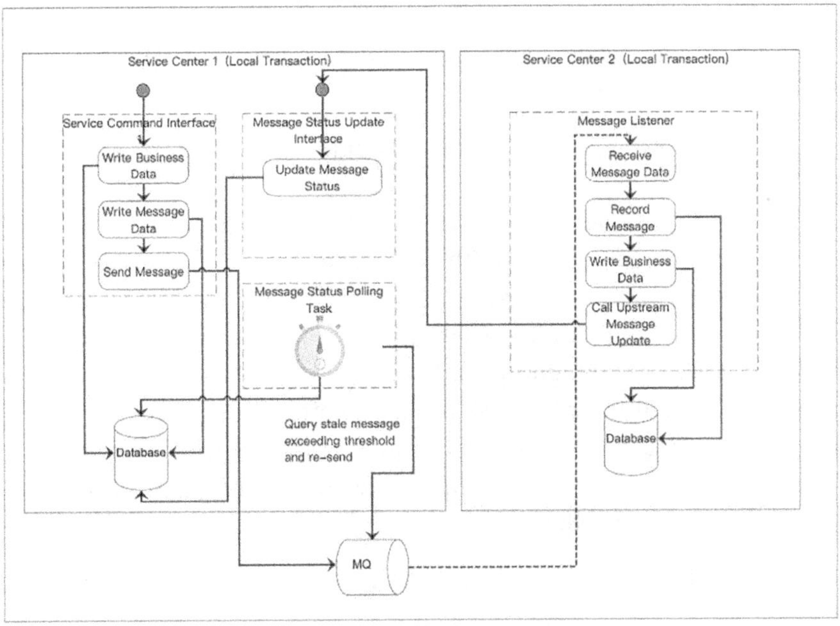

FIGURE 4.79 Distributed transaction scheme based on local database transaction message.

- As the message producer, the service center needs to create an additional message Table in addition to the business database tables and record the message sending status. The message Table and business data must be committed in the same transaction, that is, in the same database.
- The message will be delivered to the consumer of the message through MQ. If the message fails to be sent, the system retries to send the message according to the message status check polling task. The message status check polling service checks for messages in the message table that are not in the expected status. For example, the message may have a status that is longer than expected. Then, it re-sends the message to MQ.
- The second service center is the message consumer. After receiving the message, it records the message and completes its own business logic. If the local transaction is successful, the message status update operation of Service Center 1 is called and the status update operation is completed. If the transaction fails to be processed, the system retries the execution. If it is a service failure, you can call the API of Service Center 1 to notify the service of compensation.
- The consumer logic of Service Center 2 must be idempotent.

Imagine that there are now two service centers: a telemarketing center and a dispatching center. The telemarketing center handles specific marketing cases and depends on the scheduling center to assign cases to people for processing. The telemarketing center uses table marketing_task for maintaining marketing cases and the dispatching center uses table dispatching_task for dispatching different business cases, such as cases from telemarketing system and the approval system.

The business table structure of the marketing center and scheduling center is as follows:

marketing_task(id, status, description)

dispatching_task (id, case_no, case_type)

A message log table to the marketing center can be added.

marketing_task_log (xid, case_no, status, gmt_create)

Another message receiving table to the dispatch center can also be added.

trans_receive_log(xid,case_no, gmt_create)

When a marketing case is added to the marketing center, a message log entry is added at the same time. The two operations are performed in the same local transaction:

```
begin;
insert into marketing_task(id, $case_no, $description);
insert into marketing_task_log($xid, $case_no, $status,
now());
commit;
```

When the scheduling center receives a message, it adds a message receiving log when a scheduling task is added. The two operations are performed in the same local transaction:

```
begin;
insert into trans_receive_log($xid,$case_no, now());
insert into dispatching_task (id, $case_no, $case_type,
now());
commit;
```

A scheduled task can be launched, which polls marketing_task_log records whose gmt_create records in the scan table exceed the specified threshold to resend the message.

4.6.3 Sales Promotion Supporting

4.6.3.1 The Definition of Sales Promotion

Sales promotion is the main marketing form of large e-commerce platforms. In essence, marketing attracts more users' attention through the form of small profits but quick turnover. With the booming development of the Internet industry in China, the competition in the domestic e-commerce field is becoming more and more fierce, and Alibaba's advantages in the e-commerce field are becoming less and less obvious. Figure

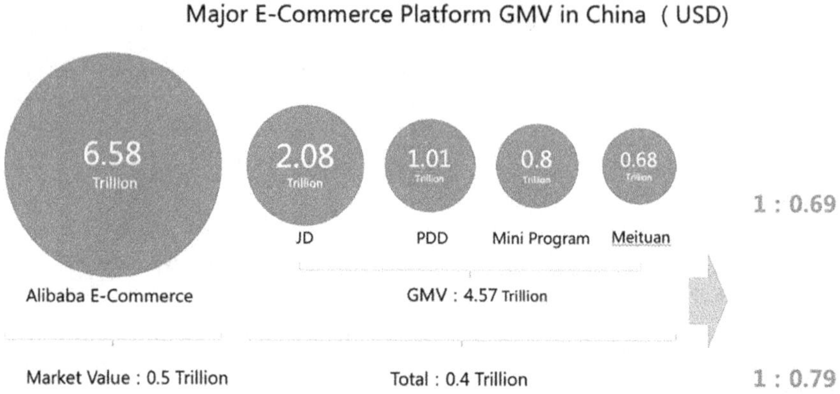

FIGURE 4.80 2019 China's e-commerce platforms' GMV.

4.80 shows the total turnover (GMV) of domestic mainstream e-commerce platforms in 2019. Comparison between those platforms:

As can be seen from the figures in the figure, compared with the turnover and total market value of other major domestic e-commerce competitors, Alibaba has no big advantage, that is to say, the competition in the field of domestic e-commerce has developed from the blue sea 10 years ago to the Red Sea where the sword hits the blood. Before introducing sales promotion in depth, I will start with Why to talk about Why e-commerce needs sales promotion or what benefits sales promotion can bring to e-commerce.

Sales promotion (or promotion, the two concepts in this book can be exchanged) originates from the Christmas promotion of American shopping malls. On this day, American shopping malls will launch a large number of discounts and promotions, in order to carry out the last large-scale promotion at the end of the year. Because American shopping malls generally record deficits with red pens and make profits with black pens, and on this Friday after Thanksgiving, people's crazy snap-up has greatly increased the profits of shopping malls, so it is called Black Friday by merchants.

There are many kinds of sayings about the origin of 'Black Friday', one of which refers to a group of people under the pressure of black going to the shopping mall in a long line on Friday after Thanksgiving. A more common view is that since this day is the first day of opening after Thanksgiving Day (the fourth Thursday in November), followed by the traditional and grand Christmas in the United States, people usually start to spend large amount of money at Christmas, and many shops will have a large amount of money because of customers. Traditionally, different colors of ink are used to keep accounts. Red indicates deficit, while black indicates profit. Therefore, this Friday is called Black Friday to indicate that the day is expected to be profitable. Because the discount activities in the store usually start at midnight (that is, midnight on Friday) after Thanksgiving Day, and start the next day of Thanksgiving Day, people who want to buy cheap goods must rush to the mall in the dark to queue up to buy cheap goods, this can also be seen as a origin of 'Black Friday'. This kind of behavior has a very vivid saying, called Early Bird.

In China, everyone should have heard of the Double 11, 618 sales promotion, especially the Double 11, abruptly, the year-end promotion of an Alibaba enterprise has evolved into the most important annual promotion for all e-commerce platforms to cope with the national carnival. Readers and friends who have bought things during the sales promotion period should notice that the price of goods during the sales promotion period is far lower than the usual price (of course, if the price is higher than usual, it is estimated that not many people will pay attention to such promotional activities as Double 11). I don't know if you have such questions, when sales promotion, can merchants really make money? To answer this question, you need to introduce the purpose of sales promotion, or why merchants do sales promotion. The author believes that sales promotion is a self-defense mechanism to prevent loss, especially after the demographic dividend disappears, sales promotion is a zero-sum game. Overall, sales promotion has the following purposes:

- Encourage customers to buy more and quickly increase the overall sales.
- Bring freshness to customers and deepen their impression on the platform and the products sold on the platform.
- Strive for potential customers, try to buy on their own platforms and familiarize customers with new platforms and products sold.
- Grab customers and crack down on competitors.

The Objective of Promotion sales:

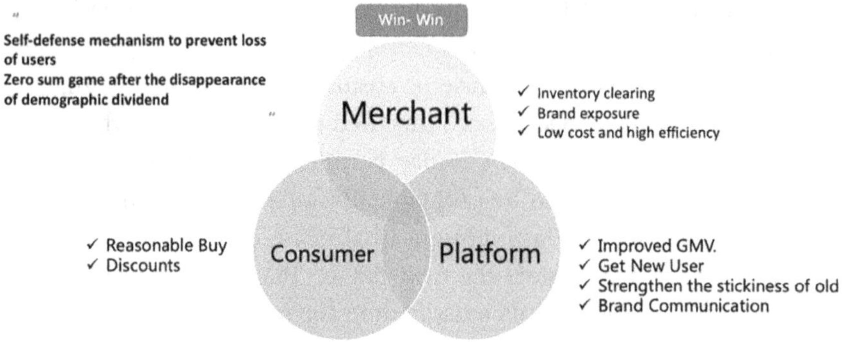

FIGURE 4.81 The purposes of promotion.

Figure 4.81 has a very important information to emphasize: Sales promotion is a win–win promotion for merchants, consumers and platforms, which is also the main reason why sales promotion can last for a long time. By giving profits, merchants attract more buyers to buy goods, more buyers to buy goods, clear inventory, improve GMV, increase the stickiness of old customers and obtain new customers at the same time; because more users buy on the platform, the brand value of the platform is improved and the brand is spread; buyers buy products of the same quality at a lower price, as you can see, through sales promotion, a win–win situation among merchants, consumers and platforms is perfectly achieved.

4.6.3.2 The Promotion Sales Process

E-commerce platform operation has three very important goals: attracting new products, promoting active activities and transforming. On the day of the large-scale promotion, many users will log in to the system to complete the purchase. From the perspective of the system, this also means that the traffic and concurrency on the day of the sales promotion will be blown out. Therefore, during the sales promotion, the requirements for system stability, security and capacity are very high. This requires us to do a good job in technical support for sales promotion. We must make a detailed analysis of every detail of the operation and understand the business gameplay in the promotion process. Based on the business gameplay, determine the critical path of the business. Based on the critical path, determine the pressure-bearing components and the capacity configuration of each component. For possible middleware and system service

FIGURE 4.82 Sale promotion guarantee closed loop.

interruptions, you need to plan specific contingency plans so that when services are interrupted, you will not be in a hurry. Therefore, readers can see that sales promotion guarantee is not a simple task, but a systematic project, covering project management, organization and coordination, technical support, risk identification, emergency plan, smelting planning, fault drills, system monitoring and business and technical review after sales promotion, as shown in Figure 4.82.

First of all, from the perspective of end-to-end process, sales promotion guarantee is divided into confirmation of business play, preparation before sales promotion, system guarantee during sales promotion, resource contraction after sales promotion, review and the final business review to analyze whether the sales promotion meets the five stages of the set business objectives. Next, we will introduce each stage and the core work of each stage in detail, the purpose is to let readers have an intuitive understanding of the work to be completed at each stage, and a very important point, why do these works be done.

1. Determine the marketing methods and business objectives

   Determine the business gameplay of the promotion, including sorting out the business target group, business volume and business gameplay of the promotion, and pay special attention to the traffic distribution of various channels. Different business methods lead to different pressures on core links. For example, common red packets and red packet fission have completely different pressures on components such as Redis. Common red packets can withstand the first wave of pressure, as the number of red envelopes in the prize pool decreases until zero is cleared, the system can be said to have withstood the postgraduate entrance examination and the whole process was successfully completed. However, the red envelope fission is different. After the first wave of red envelopes is grabbed, if users place orders and share them on social networks, they can have the opportunity to generate a list of red envelopes, for example, fission into three. In this way, the second wave and the third wave of red packet creation requests may come continuously. In extreme cases, the first wave of concurrent requests is 10w, and the second wave of concurrent requests may reach 30w. From the above example, you can see that the core links may be different for different business scenarios, and the pressure on the core components of the system may be different for different business scenarios, therefore, we must have a clear understanding of the business methods of promotion.

We need to evaluate the business volume of the promotion and the maximum number of concurrent users. Capacity estimation is usually from coarse to fine. The business owner needs to estimate the number of users who may be online at the same time for this activity, and provide an estimate of the number of customers who are online at the same time, such as 20 and 5 w, based on these simultaneous online user data, we can continue to evaluate fine-grained system capacity estimates. Of course, not all business departments can provide such data. When business departments cannot provide coarse-grained data, we can still analyze the data of past marketing activities, obtain the number of simultaneous online customers of the latest similar promotion and then estimate the number of simultaneous online customers of this promotion based on the consideration of business development. For example, in the project where the author is currently working, when we are preparing for 618 sales promotion, we analyzed the 618 sales promotion data in 2019, and obtained the number of simultaneous online users in 2019, the maximum number of concurrent orders, and the maximum number of concurrent coupons. Based on these data, as well as business development and changes, especially the impact of the novel coronavirus epidemic in 2020, more people will purchase goods through e-commerce channels, we accurately estimate that both traffic and concurrency will increase significantly. We estimate that the resource allocation will increase by 100% compared with last year. The final result has little deviation from our estimation, which helps customers to achieve a new high in revenue, at the same time, it also ensures the user experience and maintains the stickiness of old customers. At the same time, because of such innovative gameplay as fission, got 20% of new customers.

Through the above introduction, we can see that the first step of sales promotion is very critical. It determines whether our follow-up work is effective, whether we have accurate coarse-grained estimation data, and directly determines the capacity planning, effectiveness of technical support, risk assessment, etc. I have contacted too many enterprises and think that the guarantee of sales promotion is technical work, and technicians should be responsible for making the system stable and safe, and recover quickly in case of failure. I am not familiar with this practice, sales promotion and guarantee work are often incomplete and problems occur frequently. Therefore, it is important to have a full-year understanding of the sales promotion business scenario and give a reasonable coarse-grained capacity estimation for the promotion, laying a solid and stable foundation for the preparations before the second sales promotion.

2. Promotion sales preparation

After having a clear understanding of the sales promotion business scenarios and coarse-grained capacity, we enter the pre-sales promotion preparation phase. The main goal of the preparation phase of sales promotion guarantee is to provide fine-grained capacity planning for coarse-grained capacity estimation, including network bandwidth, DDOS bandwidth, gateway capacity, the maximum TPS, Redis specifications, RocketMQ maximum number of concurrent messages, distributed database DRDS specifications and RDS Database specifications, and if you use ES as a query engine, you also need to estimate the specifications of ES. After determining the specifications of resources, you also need to determine the scalability and scalability bottlenecks of critical path resources, for example, if we use an SLB to load traffic from the Internet, but the maximum number of connections to an SLB is 5w, if the number of connections to the Internet exceeds 5w at the same time, some users may encounter connection exceptions or wait, and 5w is the bottleneck of scalability. When considering the specifications, we must consider a certain margin. The second is to clarify the system deployment architecture diagram and the access link of the core business. For example, I had several painful experiences of promotion guarantee. At the beginning of the promotion, the traffic flooded into the system like a flood. At this time, the customer service reported that many users had 500 access errors and could not add shopping carts, because there is no detailed deployment architecture diagram. It took 20 minutes to find that the root cause of the problem was that the private SLB (load balancing) specification in front of the microservice gateway was too small, causing the number of connections to explode, this hidden SLB has never appeared in any deployment architecture diagram or integration architecture diagram. Everyone thinks that traffic directly enters the microservice gateway from the front end. In addition, before sales promotion, a complete full-link stress test is required to measure the maximum QPS that a single server or POD can support, the peak QPS capacity of a single POD is the key information for computing resource planning. Finally, the guarantee plan needs to be determined during the preparation phase of sales promotion guarantee, including personnel organization structure, division of responsibilities, system risks, pre-plans, pre-plan drills, system monitoring and alarms on the day of sales promotion, fusing and throttling schemes.

There are many tasks in the preparation phase before sales promotion, which are worth introducing in detail the core tasks. The author will focus on full-link stress testing, system deployment architecture diagram, capacity planning, the risk plan and personnel organization structure that are introduced in detail.

I. End-to-End performance testing

Before introducing the full-link stress testing and the relationship between the full-link stress testing and sales promotion, let's start with the technical pain points of sales promotion. I think the main pain point of sales promotion is how to solve the influence of uncertain factors on the promotion system. Figure 4.83 shows the technical pain points encountered by the author in the process of sales promotion and guarantee.

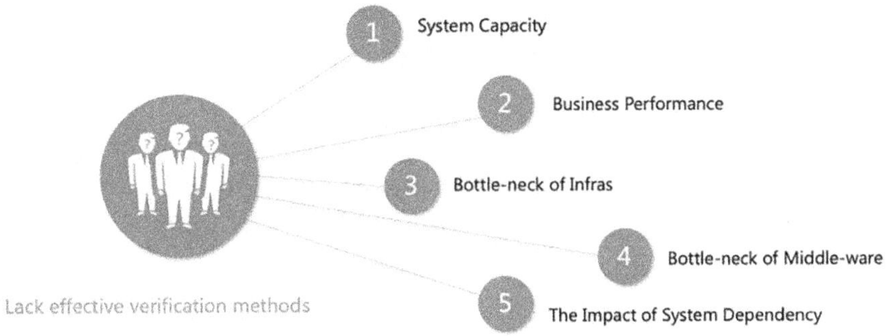

FIGURE 4.83 Technical pain points in the process of promotion guarantee.

As the owner of sales promotion, the author needs to answer such questions as how many pods should be planned for our application, what specifications should be selected for the database and how many shards should be required for Redis. Some students compare the mechanism, and may say that this is simple. We can follow the strategy of preferring to abuse. The heap has much more hardware resources than the actual situation for emergencies, this kind of thinking can solve the problem, but the cost is too high. The budget of any company should be limited. Then how can we answer the first few questions gracefully, and find a balance between resource planning and cost, which is the Favorable position for the use of skills of full-link stress testing.

Let's first look at the definition and purpose of the full-link stress testing? The author believes that the full-link stress testing is based on the full-link business model and integrates the entire system environment such as the front-end system, API interfaces provided by the back-end service centers, microservice gateways CSB, and DB, it is fully included in the stress testing scope, uses http requests as carriers, simulates real user behaviors, and constructs real large-scale concurrent user access traffic in the testing environment, tests the configured business scenario until the target peak is reached. During the stress test, system bottlenecks, including middleware bottlenecks and infrastructure bottlenecks, are found, and performance problems are found, the dependencies between systems and the overall throughput of the system are used to test the stability of the e-commerce system under the configured concurrency pressure. For more information about the full-link stress testing, see the next section.

II. System Deployment Architecture

The configuration and deployment of the hardware and software elements that make up the application.

Accurate system deployment architecture diagram can help us find the risk points of the system and formulate corresponding plans in the preparation stage of sales promotion guarantee. In sales promotion guarantee, the system deployment architecture diagram can be used when system problems occur, helps us quickly sort out access links, converge areas where problems occur, and quickly restore system services. Figure 4.84 shows the deployment architecture (simplified) of the project.

III. System capacity planning

Generally, after the business objectives are determined, the coarse-grained business objectives need to be converted into applications and middleware on important links, depending on

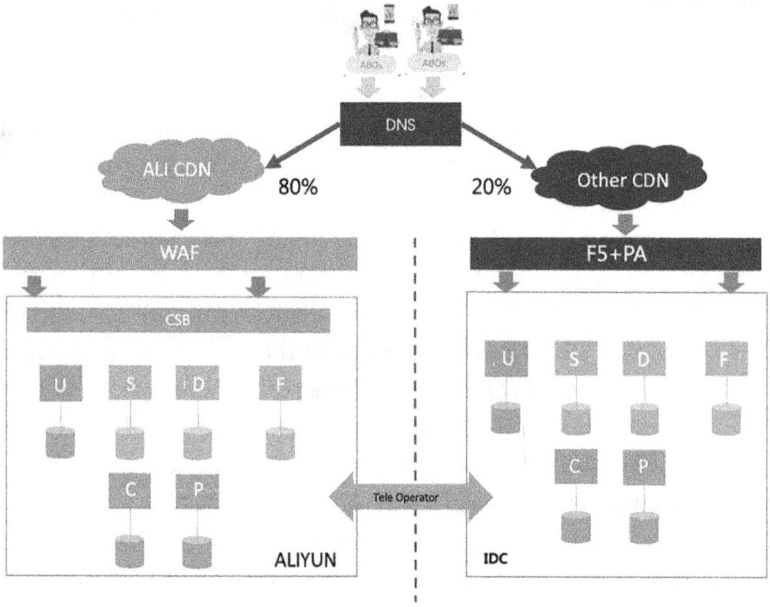

FIGURE 4.84 System deployment architecture diagram.

the capacity of the system. Capacity evaluation needs to evaluate the capacity plan of multiple nodes from the beginning of user access to the database, including applications, networks, storage, middleware, content delivery network (CDN), load balancing, microservice gateways and databases. Before the sales promotion, the clients that the author currently serves will analyze all the nodes on the access link of the system. Based on the business objectives, the planning office supports the capacity and specifications of specific promotion activities.

The capacity planning of an application is based on the maximum TPS supported by a single POD. For example, for a promotion, the marketing center needs to support 5 W TPS. If a single POD can support 1,000 TPS, we need 50 pods to support this promotion. The capacity estimates of databases, microservice gateways, MQ and server load balancer are based on the magnification caused by the access link and the business model of the system. For example, a user orders request, this will result in a three-time inventory center solution, so similar amplification needs to be considered.

In the process of promotion, the system has two important capabilities, one is the ability of horizontal expansion and the other is the ability of emergency plan. The emergency plan will be introduced in the next chapter, and we will focus on the horizontal expansion capability.

Horizontal scaling capability refers to the linear increase of machines to improve the throughput of the system, with large instantaneous traffic during large promotion and high requirements on the peak processing capability of the system. One of the most important tasks of sales promotion is to ensure the processing capability of peak values and ensure that the system can easily cope with instantaneous peak values. For a single data center, it needs to analyze the traffic flow link layer by layer, from layer-2, layer-3 routers, switches, cabinets, servers, layer-4 soft load balancing LVS (F5), to the application layer of Layer 7, horizontal scaling is required. At the data center level, the architecture and deployment must support both the same-city dual data centers and Remote Multi-Active Multi-data centers. These architectures are the key to horizontal expansion.

IV. Risk management and mitigation plan

Sales promotion assurance generally conducts risk assessment when you sort out the deployment architecture, and the earlier the risk assessment is, the better. Risk sorting and risk plan design are the core work of risk assurance in the sales promotion guarantee system.

Risk sorting is a risk sorting process based on the business objectives guaranteed by sales promotion. The direct dependencies of the e-commerce system are sorted out based on the key links of the application, and then based on the business conditions on the key links, assume that when a direct dependency is down or the service is unavailable, observe what impact it will have on the system and what impact it will have on users of the system. For the project where I am currently working, risk sorting generally starts from the user's perspective. The links include the homepage, login page, Commodity Search, commodity details, shopping cart, Settlement page, order submission, payment page, order details page and order query. For the dependency of the core pages on the link, assume that the service is unavailable, and then give priority based on the impact on the system, determine which risks need to be guaranteed and which have little impact on core links (Figure 4.85).

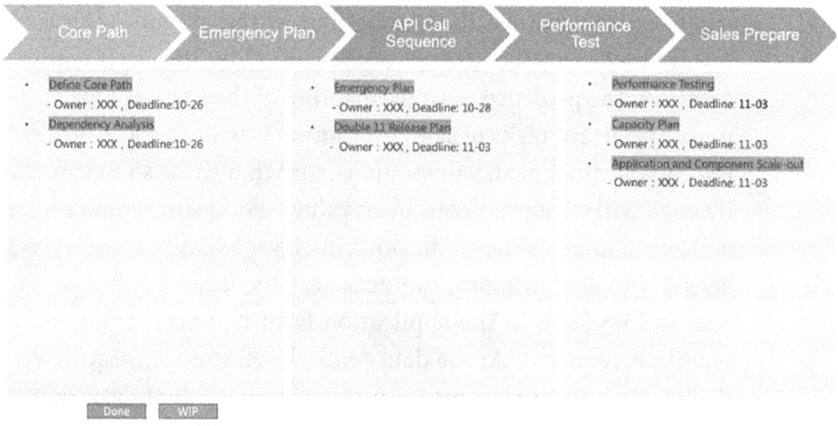

FIGURE 4.85 Risk analysis based on promotion guarantee process.

Based on the sorted risk priorities, a plan is formulated for the risks that need to be guaranteed. The plan requires clear risk definitions and implementation conditions. Usually in the process of sales promotion, many problems will be reported, and the problem decision-making mechanism needs to be determined before sales promotion. Only after the decision-making of the problem decision-making mechanism, the plan can be implemented. Generally, data recovery is involved, high-risk operations such as system updates. Plans are generally divided into regular plans and emergency plans. Regular plans can be seen as preparations before sales promotion, such as shutting down unnecessary data synchronization and complex recommendation engines, instead of using the list of recommended products. Emergency plan is an easy-to-understand plan type. In the process of sales promotion, for example, we found that no coupons were created after the red envelope rush started. At this time, we need to use emergency plan and manually use asynchronous methods, create coupons, or customers may give up shopping because they cannot see the red envelopes they have grabbed, which will directly damage the sales promotion objectives and indirectly reduce the stickiness of old customers.

Figure 4.86 shows the problem decision chain of my project during sales promotion.

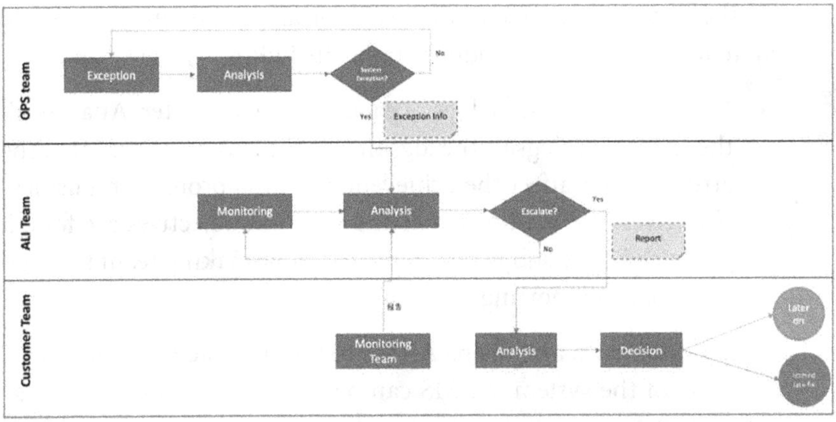

FIGURE 4.86 Decision chain of sales promotion period problem.

2. Sales promotion team structure

Organizational guarantee is an important part of sales promotion guarantee. To ensure the smooth progress of sales promotion, we need to reach a consensus from the management level of the enterprise to plan and tilt resources from the organizational level. The sales promotion guarantee organization includes the organization guarantee in the preparation stage of sales promotion and the organization guarantee on the day of sales promotion. The previous sections of the preparation phase of sales promotion have been described in detail. It can be seen that these tasks need corresponding resources to be implemented and guaranteed. Therefore, enterprises need to schedule appropriate resources to complete these tasks, prepare for sales promotion and guarantee. For example, full-link stress testing, emergency plan drills and network emergency plan drills all require special resources.

The organization guarantee on the day of sales promotion mainly establishes sales promotion headquarters, and reserves and arranges war rooms. All students who participate in sales promotion guarantee must sit together to facilitate communication and problem troubleshooting. Do a good job in link monitoring and analysis. When the system finds problems, someone can quickly find bottlenecks, report problems and determine whether to implement emergency plans according to the problem decision-making process.

3. Sales promotion system supporting

The system guarantee work in sales promotion mainly focuses on system monitoring and sudden traffic, including:

- Monitor the real-time logs of each log service center. Analyze all the Error-level logs in the logs in real time to make sure that the errors do not affect the achievement of the promotion business objectives, if exceptions affecting business objectives are found, they need to be submitted to the decision-making team through the problem reporting mechanism.
- Use tools such as Alibaba ARMS to monitor the real-time access links of the system. ARMS can be used to monitor the average impact time of each service, call link, database slow SQL and error rate. In particular, ARMS provides the integration of the

call stack for needles and error requests. Sales promotion can directly analyze the call stack through the graphical interface and analyze the code lines where errors occur, it is very effective to analyze and troubleshoot problems during sales promotion.

- In the process of sales promotion, the burst traffic may be larger than the designed traffic. To ensure that the system does not crash, you need to set the rate limiting and degradation, and if the burst traffic is monitored, through the scale-out capability of the system and the increase of processing resources, we can deal with such scenarios and ensure the quality of service. In particular, applications deployed based on Alibaba EDAS can use automatic scaling technology. When the traffic is higher than a certain metric, for example, the CPU usage of all processing capabilities of the service reaches 70%, the general pods are automatically expanded to cope with the traffic that may continue to grow. When the POD usage is idle for a period of time, EDAS scheduling capability will unpublish some pods without damage, ensure a balance between resource input and business products.

In the sales promotion guarantee, the core support personnel must be on-site, so that when problems occur, information flow is more efficient. In particular, each system should have a person in charge of the system's resource planning, configuration check before sales promotion, monitoring during sales promotion, problem collection and statistics, review after-sales promotion, only by assigning clear responsible persons to all the participating parts of the whole system can problems be discovered at the first time. When I was in charge of the promotion activities for the first time, it was because the monitoring tasks and systems were not assigned to people that everyone looked at the same tool and ignored the other one, as a result, an error occurred to the standby discovery. Ten minutes later, irreparable losses were caused.

4. Technical and business retrospective after-sales promotion

The end of sales promotion does not mean the end of sales promotion. Every sales promotion is an adventure. Especially in the process of sales promotion, if problems occur, these problems must be precipitated to form a problem analysis and review report. The project where the author works now requires all problems found online to form a problem analysis report, which includes:

- Problem description: Record the appearance, impact and preliminary analysis information of the problem through simple and straightforward language.
- The root cause analysis, based on the problem description and preliminary analysis information, further analyzes the root cause of the problem and provides sufficient information for the next step of designing a solution.
- Solution, based on the root cause of the problem, design a solution, solving many times may require a lot of manpower and material resources, the author suggests that we can consider from the perspective of short-term and long-term solutions. The short-term solution is to solve the symptoms and restore the business. The long-term solution is to solve the root cause of the problem and ensure that such problems do not occur in the future.
- Follow-up plan, many problems may not be a center, or an application problem, which requires inferences drawing, finding once, thoroughly solving once, therefore, the follow-up plan is not only to formulate short- and long-term repair plans, but also to solve similar problems based on the problems found and whether all centers need to float up.

After the sales promotion, technical and business reviews are required to sort out what has been done well and can be further improved in the process of sales promotion and guarantee, and suggest that the team work together in a brainstorm way, form continuous improvement. It should be noted that when conducting a review, remember not to open the review as a criticism meeting, because all the problems to be solved have been solved, and the review environment focuses more on the strength of the entire team, find out the improvement direction that everyone has reached a consensus. Only when the specified improvement items are discussed and agreed upon by everyone can they be implemented and the effect of continuous improvement can be brought into play.

After experiencing sales promotion for many times, how can we make a better guarantee plan for sales promotion and try to avoid most of the problems? To sum up, the main work of sales promotion is to ensure four capabilities: verification capability, horizontal scaling capability, rapid problem discovery capability and pre-planning capability. The sales promotion of large websites is a complex and systematic project, which requires strict project management and schedule management, as well as upstream and downstream communication. The key to sales promotion is to improve the horizontal scaling capability of the system, the premise of building a horizontal scaling capability is capacity planning, measurement and discovery and analysis of bottleneck problems. Finally, the horizontal scaling capability can be improved through a reasonable architecture mode.

4.6.4 End-to-End Full-Chain Stress Test and Monitor Solution

The full-chain stress testing can effectively help the pre-launch system to evaluate the system capacity and prepare for the actual sales promotion. The overall solution consists of three major directions: historical data analysis and traffic estimation, traffic and pre-release playback in the production environment, stress testing in the production environment, and shadow table construction.

- Solution 1 provides pressure suggestions and analysis for some typical scenarios through traffic analysis of historical sales promotion.
- Solution 2 mainly describes the preparation of the pre-release environment, traffic recording, playback and control script.
- Solution 3 mainly introduces a typical example of building scaffolding by Alibaba to evaluate the feasibility analysis and subsequent R & D Planning of the current XX project transformation. At the same time, solution 3 provides the description of the current production environment stress testing management platform output by Alibaba for overall evaluation by XX architects.

4.6.4.1 Main Technical Description

1. Solution 1: Historical traffic analysis and future traffic estimation (Figure 4.87).

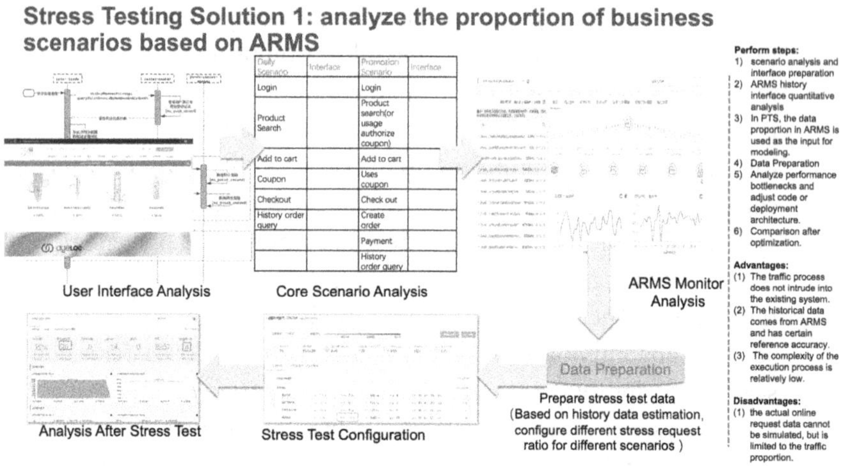
Stress Testing Solution 1: analyze the proportion of business scenarios based on ARMS
Daily Scenario
Interface
Promotion Scenario
Interface
Login
Login
Product Search
Product search(or usage authorize coupon)
Add to cart
Add to cart
Coupon
Uses coupon
Checkout
Check out
History order query
Create order
Payment
History order query
User Interface Analysis
Core Scenario Analysis
ARMS Monitor Analysis
Data Preparation
Prepare stress test data
(Based on history data estimation, configure different stress request ratio for different scenarios)
Stress Test Configuration
Analysis After Stress Test
Perform steps:
1) scenario analysis and interface preparation
2) ARMS history interface quantitative analysis
3) In PTS, the data proportion in ARMS is used as the input for modeling.
4) Data Preparation
5) Analyze performance bottlenecks and adjust code or deployment architecture.
6) Comparison after optimization.
Advantages:
(1) The traffic process does not intrude into the existing system.
(2) The historical data comes from ARMS and has certain reference accuracy.
(3) The complexity of the execution process is relatively low.
Disadvantages:
(1) the actual online request data cannot be simulated, but is limited to the traffic proportion.

FIGURE 4.87 Full-link stress testing solution 1.

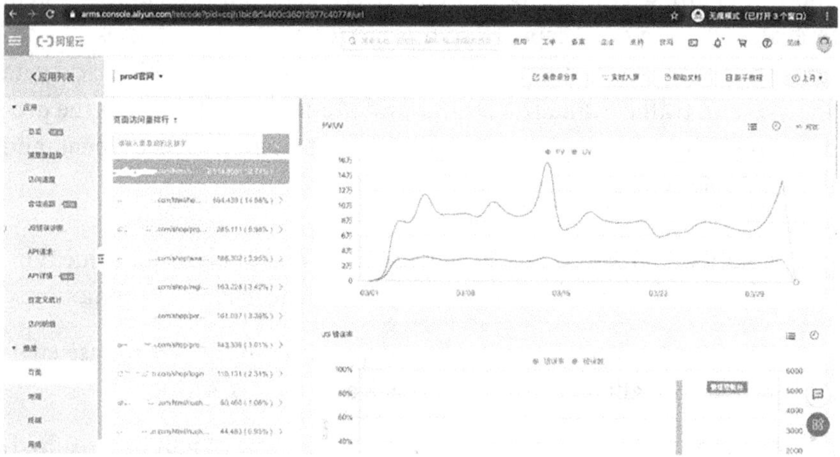

FIGURE 4.88 Daily API calls on the official website.

API calls on the official website (daily) (Figure 4.88):
Analysis of RPS and concurrent users (Figure 4.89):

I. Key technical concepts:

- Concurrent users: Refers to the users who operate business in real systems. In performance testing tools, it is generally called the number of virtual users (Virtual User). Note the number of concurrent users and the number of registered

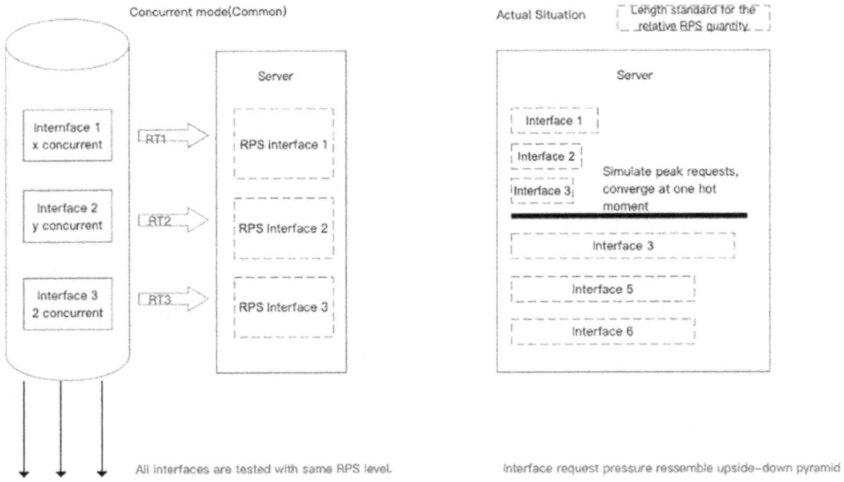

FIGURE 4.89 Comparison of actual and simulated RPS pressure.

users; there is a big difference in the number of online users. The number of concurrent users will definitely put pressure on the server, while the number of online users is only 'hung' on the system and does not put pressure on the server, the number of registered users generally refers to the number of users in the database.

- TPS: Transaction Per Second, the number of transactions Per Second, is a very important metric to measure system performance.
- RPS: Records Per Second, the number of Records Per Second.

II. VU and TPS conversion

- New System concurrent users (VU) Acquisition: No historical data are available for reference, and can only be evaluated by business departments.
- **Number of concurrent users in the old system (VU) Acquisition: For systems that have been online, you can select the number of users who use the system during a certain period of time during peak hours. These users are considered as the number of online users, 10% is enough for concurrent

users. For example, if the number of users using the system is 10,000 within half an hour, then 10% is enough for concurrent users.

- TPS acquisition of the new system: No historical data are available for reference and can only be evaluated by business departments.
- TPS acquisition of the old system: For systems that have been online, you can select peak hours to obtain the business volume and total business volume of each transaction in the system within 5 or 10 minutes, TPS is calculated based on the number of transactions completed per unit time, that is, the number of transactions/unit time (5 * 60 or 10 * 60).

III. Analysis of three modes

- Concurrent user mode:

  It is more suitable for qualitative analysis of the system, such as helping to locate performance bottlenecks and accumulating performance baselines for a single interface. (Compared with historical performance optimization or degradation), the difficulty of concurrency mode lies in the accuracy of RT.

- TPS mode:

  Quantitative analysis of the system has outstanding performance, such as service capacity planning and full-link performance baseline accumulation, which can also help locate performance bottlenecks. The difficulty of RPS model lies in the accuracy evaluation and prediction of the model. In terms of implementation difficulty, the former is more difficult and has lower control degree than the latter.

  The system performance is determined by TPS and has little to do with the number of concurrent users. The maximum TPS of the system is limited (within a range), but the number of concurrent users is not limited, which can be adjusted. We recommend that you do not set too long thinking time during performance testing to put pressure on the server in the worst case. In general, stress tests are performed on large systems (with a large volume of business and many machines), with 10,000–50,000 concurrent users. Stress tests are performed on small- and medium-sized systems, with 5,000 concurrent users.

- RPS mode:

  Applicable scenario: The RPS mode is the throughput mode. By setting the number of requests sent per second, you can directly measure the throughput of the system from the perspective of the server, eliminating the tedious conversion from concurrency to RPS, in one step.

  API operations (such as adding shopping carts to e-commerce and placing orders) mainly use TPS (Transaction Per Second, transactions Per Second) to measure the throughput of the system. If you select this mode, you can directly set the RPS according to the expected TPS. If you want to check whether the 'order' interface can reach the expectation of 500 TPS, set RPS to 500 and send 500 requests per second to check the throughput of the system.

  In this mode, high concurrency may occur when the request fails to respond in a timely manner. If an exception occurs, stop the request in a timely manner.

  This mode only supports non-automatic incremental stress testing (that is, manual speed adjustment is required during the stress testing).

IV. Stress testing scenario analysis

Currently, the stress test of XX is mainly based on the number of concurrent users. In some big promotion scenarios, abnormal traffic spikes cannot be restored. For example, in a sales promotion scenario of 4/10:

4/10 stress testing in a pre-release environment: Stress testing is performed mainly through the normal login and order placing process for about 400 concurrent users, and the stress results are relatively normal.

During the actual promotion process, the current product design needs to update the promotion page of the activity several minutes before the activity, which leads to the fact that CDN has not been preheated. After users cannot refresh the promotion page information, a large amount of traffic

Number	Promotion Scenario	Page	Peak PV/UV	Peak Time	Page Request Ratio VS. Total	Number	Promotion Scenario	Page	Peak PV/UV	Peak Time	Page Request Ratio VS. Total
1	Mobile homepage	china.****.com/html/home.app.html	1,0795/5176	11:00	60.38%	11	Innovation content center2	china.****.com/html/xxx channel/media-detail.app.html	120/60	19:00	0.76%
2	PC Homepage	china.****.com/html/home.html	3248/791	10:00	13.45%	12	Payment success page	china.****.com/shop/paysuccess	219/183	10:30	0.64%
3	Product List	china.****.com/shop/productlist**	864/312	10:00	4.59%	13	Discount list	china.****.com/shop/discountlist	137/102	10:30	0.54%
4	Personal info.	china.****.com/shop/personal	789/372	10:30	3.51%	14	discount details	china.****.com/shop/xxx	81/25	20:30	0.50%
5	Register	china.****.com/shop/register	519/187	11:00	3.10%	15	Innovation content center - detail page	china.****.com/html/nu channel/media-detail.html	120/14	22:30	0.42%
6	Wechat auto-login	china.****.com/shop/wxlogin	754/329	10:00	2.49%	16	Innovation content center3	china.****.com/html/nu channel.html	67/21	22:30	0.38%
7	Login	china.****.com/shop/login	285/133	10:00	1.85%	17	Order detail page	china.****.com/shop/orderdetail	111/108	10:30	0.33%
8	Product detail	china.****.com/shop/productdetail**	211/196	11:00	1.56%	18	Mobile landing page	china.****.com/products/XXXX	60/31	9:30	0.32%
9	Innovation content center1	china.****.com/html/nuchannel.app.html	301/184	19:00	1.22%	19	Mobile landing page	china.****.com/products/xxx	56/19	13:00	0.31%
10	Shopping cart	china.****.com/shop/cart	185/138	10:30	0.84%	20	Occupation	china.****.com/shop/career	54/36	19:00	0.30%

FIGURE 4.90 Stress testing scenario test results.

flow back to the homepage. This causes access bottlenecks at the application layer. Some users may be slow or inaccessible. After the application layer nodes are added urgently, the system access is normal.

Therefore, we need to analyze the business scenario based on the traffic in the big promotion scenario, and then plan the traffic pressure based on the previous data and expected business increment (Figure 4.90).

2. Solution 2: Production traffic recording and playback in pre-release environment (Figure 4.91).

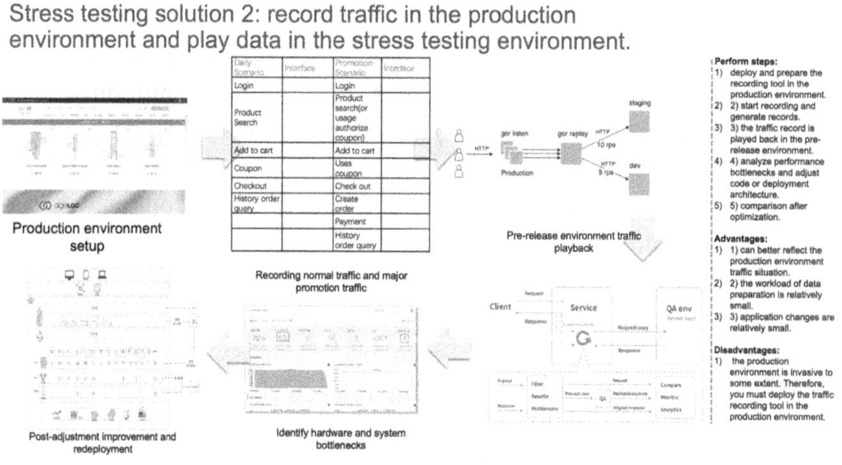

FIGURE 4.91 Production traffic recording and playback solution.

Data preparation is mainly divided into data synchronization content of RDS, DRDS, ElasticSearch and Redis. For traffic recording and playback, we mainly consider using GOREPLAY tools for stress testing.

The following content mainly uses GOREPLAY as a tool for traffic recording and playback.

Tools: As the application grows, the workload required to test it also increases exponentially. GoReplay allows you to reuse existing traffic for testing. GOREPLAY allows you to analyze and record application traffic without affecting application traffic. This eliminates the risk of placing third-party components in critical paths.

GoReplay increases developer confidence in code deployment, configuration and infrastructure changes. GoReplay provides a unique method for grayscale. The GoReplay in the background is not a proxy, but a listener for the traffic on the network interface. Instead, it does not need to change the production infrastructure, but runs GoReplay Daemon on the same computer as the service.

I. Deployment architecture

- Mount the NAS service to the production application server to store recorded files. This NAS service is still deployed in the production environment to avoid desensitization between environments.
- To avoid the pressure on the production environment caused by traffic playback, we recommend that you configure a separate traffic playback instance. At the same time, you can compile a script for batch control of machine recording and broadcasting in this instance to facilitate start and stop.
- The overall execution process is as follows: Install in the application layer and GOREPLAY the recording tool. Attach the NAS file system to each server and configure the script tool in the server where the traffic is played to enable and disable the traffic in batches.

II. Prepare environment

The demonstration environment uses an EDAS cluster with four ECS instances. A sample application is deployed in the cluster. The application layer provides restful interface and calls back-end services through RPC.

III. Application Deployment

Simulate a Middle Office center RX-BME-CENTER and an application layer: rx-trade-application.

IV. Mount NAS:

After activating the NAS service, you can use the Mount command provided to mount the storage space to the application instance.

V. Tool installation and configuration

Install the GOREPLAY tool on the application server:

```
Wget https://github.com/buger/goreplay/releases/download/
   v0.16.1/gor_0.16.1_x64.tar.gz tar xvf gor_0.16.1_x64.tar.gz
```

Listen to port 8090 and store the traffic file in the .gor file.

```
sudo./goreplay --input-raw :8090--output-file=rx_production_
   record-0001.gor
```

VI. Playback in the pre-release environment

Play the files stored in the NAS directory and forward the traffic to the SLB address of the staging environment.

```
sudo./goreplay --input-file rx_production_record-0001.gor
   --output-http="http://slb-address:8080"
```

VII. Request rewriting

During the stress testing of XX, some fields in the production environment are desensitized, and some information, such as user name and password, is reset to facilitate the testing.

```
gor --input-raw :80--output-http "http://staging.server" \

--http-header "user-name: xxxxx" \

--http-header "user-pass: xxxxx"
```

3. Solution 3: Stress testing in the production environment and shadow table construction

At present, in order to ensure the group's online business, especially the sales promotion business, many solutions are used. The typical example is online full-link stress testing. The full-link stress testing is divided into two parts: pts traffic stress testing and stress testing traffic entering the shadow table.

I. Technical principles

The project itself needs to be developed based on Mybatis and Springboot. It uses AOP to intercept traffic and Mark traffic on Springboot, and then writes the Mybatis plug-in to dynamically replace it to achieve basic functions. The automatic interception starts of MQ, Redis, and OSS will be added later. To achieve the full-chain mock pressure test.

II. The specific implementation

- Define the corresponding method and class annotation. If the specified restful interface has a head xxxx, it is marked.
- Each component will recognize this identifier and insert the data of the specified traffic into the shadow table. (In Redis and OSS scenarios, the data are prefixed.)

III. How to use

- Initialize the basic database. Add shadow to all tables.
- Introduce the maven two-party package.

```
<dependency>
        <groupId>com.aliyun.gts.bpaas</groupId>
         <artifactId>shadow-table-dependency-mybatis-
         starter</artifactId>
        <version>1.0-SNAPSHOT</version>
    </dependency>
```

- Add the annotation @ ShadowTable to the controller to be monitored.
- The request header x-shadow-table-code must be attached to the request.

4.6.5 Grayscale Release

4.6.5.1 Common Release Technologies

Table 4.14 shows commonly used publishing technologies.

TABLE 4.14 Comparison of Common Release Technologies

Release Technologies	What	Why	How
A/B testing	Methods for testing application performance, such as availability, popularity and visibility	Collect data to reduce or verify assumptions	Multiple technology stacks, including: Deployment slot Traffic Scheduling Feature switch Cyclic feedback
Canary releasing	Release new features to some end users (visible to users)	Verify that the new feature is valid Under the condition of limited capacity, we can only gradually promote the rolling upgrade of new features	Multiple technology stacks, including: Deployment slot Traffic Scheduling Feature switch Sampling
Dark launch	Very similar to Canary release However, new features are not visible to end users	We hope that the test of new features will not affect end users When it is expected to have a significant impact on online resources	Technology stack, including: Deployment slot Sampling technology
Blue/green deployment	Use a primary server cluster (blue cluster) and a standby server cluster (green cluster) to smoothly switch and migrate the old and new versions of the application code	Reduce the service stop time during publishing	Technology stack, including: Deployment slot Traffic Scheduling

4.6.5.2 Grayscale Release Concept

Grayscale Publishing refers to a publishing method that can smoothly transition between black and white. AB test is A grayscale release method, allowing some users to continue using A and some users to start using B. If users have no objection to B, then gradually expand the scope, migrate all users to B. Grayscale release can ensure the stability of the whole system, and problems can be found and adjusted at the initial grayscale to reduce its impact.

4.6.5.3 *Phased Release Scenario*

- Service Capability changes frequently and the release cycle is short.
- Compatibility risk, user churn risk and system down risk of frequent product upgrades.
- We need to make extensive changes and reconstructions to the original products.
- Customer first, allowing users to switch systems smoothly (Figure 4.92).

Before the application goes online, no matter how perfect the test is, all potential faults cannot be detected during offline testing. When version upgrade failures cannot be avoided, a controllable version release is required to control the impact of the failures within an acceptable range and support quick rollback.

4.6.5.4 *Phased Release Procedure*

- Select an appropriate grayscale policy and direct the access traffic that meets the grayscale policy to the new version of the application (the new version of the application is usually deployed on the grayscale machine).
- Analysis of grayscale release effect and operation data (Figure 4.93).

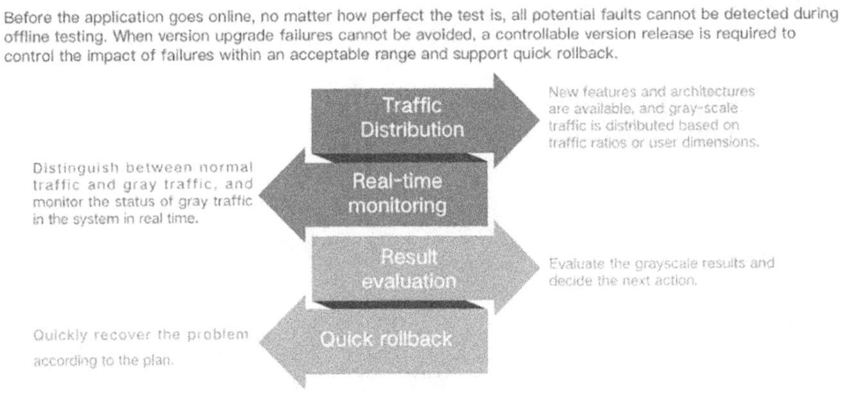
Before the application goes online, no matter how perfect the test is, all potential faults cannot be detected during offline testing. When version upgrade failures cannot be avoided, a controllable version release is required to control the impact of failures within an acceptable range and support quick rollback.
Traffic Distribution
New features and architectures are available, and gray-scale traffic is distributed based on traffic ratios or user dimensions.
Distinguish between normal traffic and gray traffic, and monitor the status of gray traffic in the system in real time.
Real-time monitoring
Result evaluation
Evaluate the grayscale results and decide the next action.
Quickly recover the problem according to the plan.
Quick rollback

FIGURE 4.92 Phased release scenario elements.

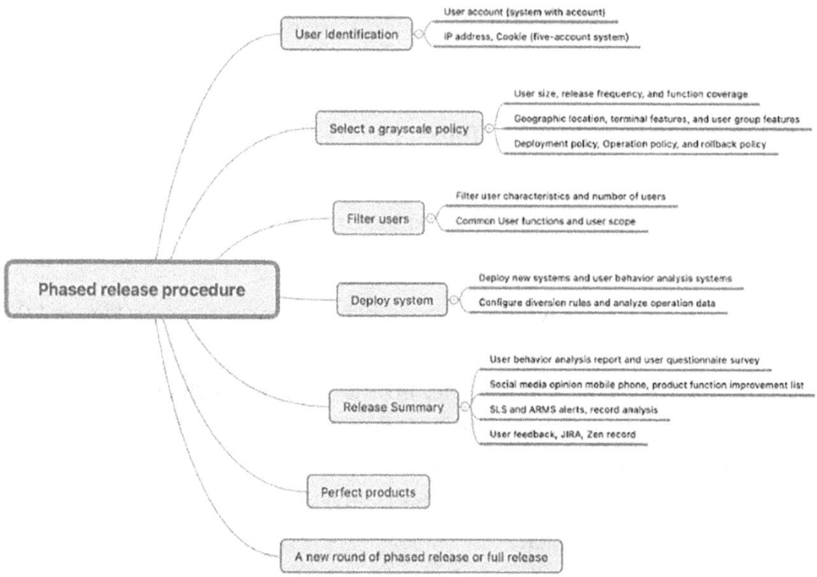

FIGURE 4.93 Phased release procedure.

4.6.5.5 Alibaba Gray Design Practices

Alibaba requires that production releases and changes must be grayscale, monitored and rollback. The following requirements are met for security changes (Figure 4.94):

4.6.5.6 Principles of Mainstream Microservice Grayscale Architecture

Currently, the main Java microservice frameworks of the Internet SpringCloud and Dubbo are implemented based on the Istio microservice

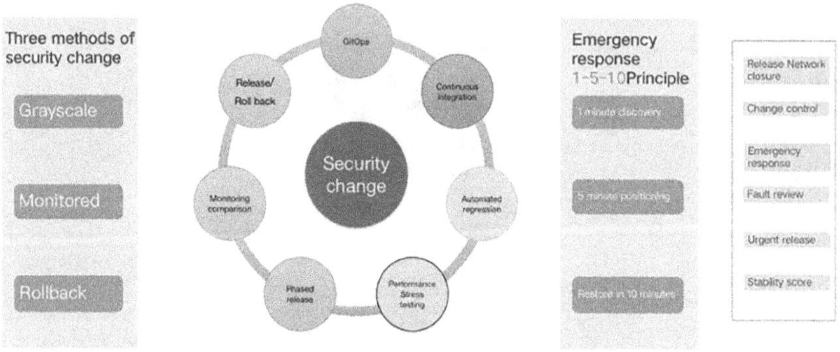

FIGURE 4.94 Grayscale release design practice.

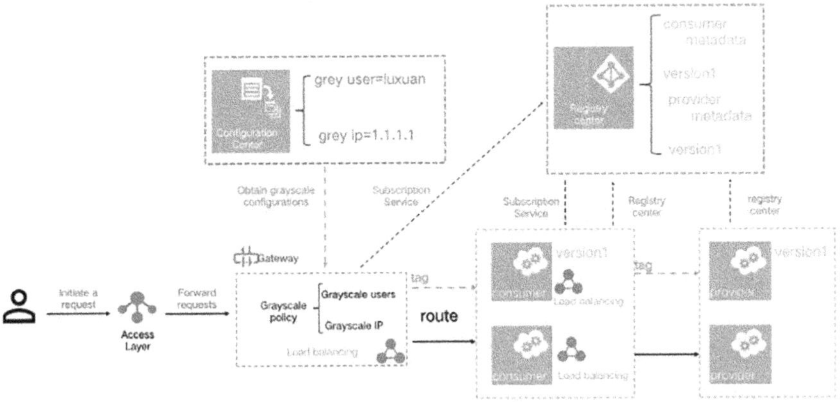

FIGURE 4.95 Principles of mainstream microservice grayscale architecture.

framework and Kubernetes. Generally, the implementation principle is that users initiate requests through grayscale policies, user accounts or grayscale IP addresses, through the gateway, the grayscale tags are routed to consumers and service providers to provide grayscale services, as shown in Figure 4.95:

It can be seen that the implementation of grayscale release requires the support of a complete set of systems and processes, but some of the capabilities required for grayscale release are common. Enterprises can also use other platforms to reduce their own R & D and O & M costs.

4.6.5.7 *Alibaba Cloud EDAS Grayscale Release Solution*

Alibaba Cloud EDAS has been developed and released version 3.0, which fully supports Canary release, that is, grayscale release. The capability is shown in Figure 4.96:

Currently, EDAS supports two grayscale release policies: Grayscale by traffic ratio and grayscale by request content. Spring Cloud: Includes Cookie, Header and Parameter. Set Dubbo according to actual requirements: Obtain expression settings based on actual application parameters and parameter values.

- Canary front- and back-end full-link gray release:

  Tags are first-class citizens in cloud-native environments. Based on EDAS publishing capabilities, you can deploy page rendering

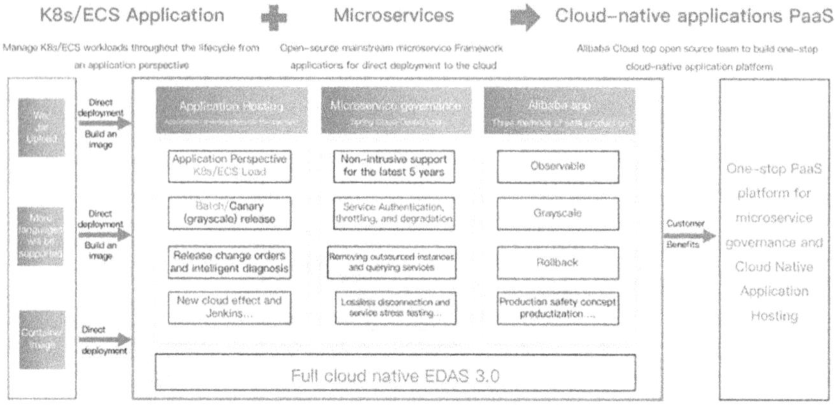

FIGURE 4.96 Alibaba Cloud EDAS 3.0 supports phased release.

servers in EDAS. The user requests the browser to pass the user id, add index.html based on the user's grayscale rule, and then load the corresponding grayscale version JS/CSS to the CDN back-to-source. Use the user id parameter to access the microservice gateway, configure the center grayscale rule and pull the pod version of the application to be accessed from the registry. For Link access, other access requests that are not configured with grayscale rules are dynamically routed to the corresponding pod based on RT (Figure 4.97).

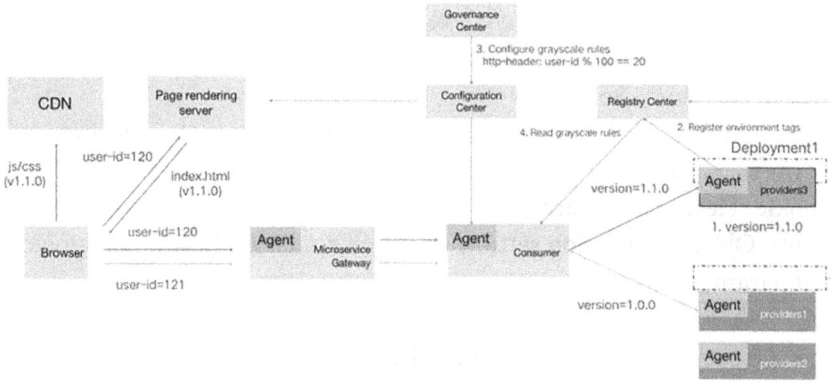

FIGURE 4.97 Canary release front and back-end full link.

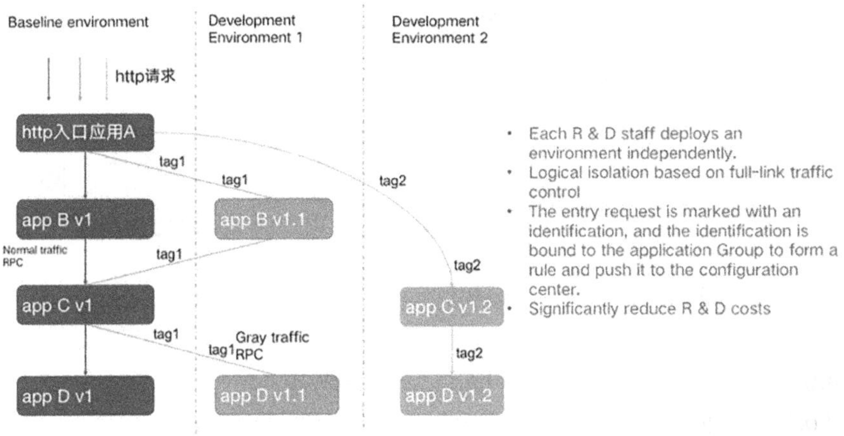

FIGURE 4.98 Full-link traffic control.

- Full-link traffic control

  Each R & D personnel independently deploys an environment for logical isolation based on full-link traffic control. The request is marked and bound to an application group to form rules and push them to the configuration center to significantly reduce R & D costs (Figure 4.98).

- Applications support lossless offline

  As shown in Figure 4.99, you do not need to modify the code when the application is offline. You only need to install the Agent (EDAS is automatically installed). When the client is offline, the client is notified to accelerate client awareness. In addition, Dubbo 2.5.8/Spring Cloud Edgware is supported and compatible with all registries (Figure 4.100).

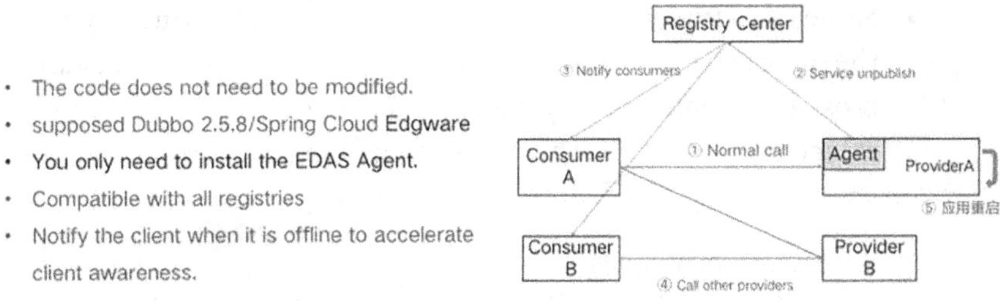

FIGURE 4.99 Application lossless offline.

4.6.5.8 Grayscale Release Value

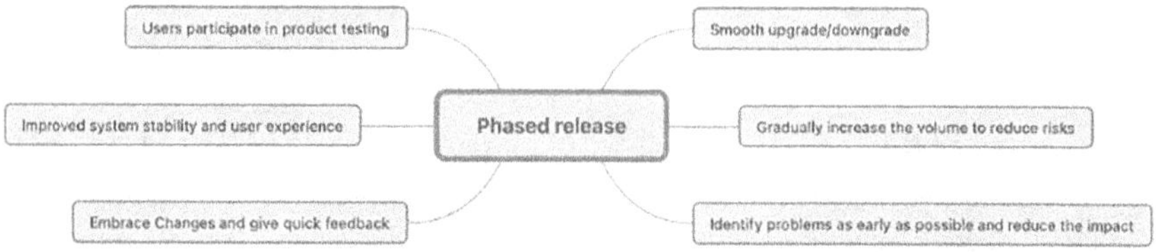

FIGURE 4.100 Gray release value summary.

4.6.6 High Availability and High Concurrency

4.6.6.1 High-Availability Architecture Panorama

Read a sentence and live more: Live More, English Multi-Site High Availability, as its name implies, is distributed across multiple sites to provide services at the same time. The main difference from traditional disaster recovery is that all sites in the active–active mode provide services to the outside world at the same time, which not only solves the problem of disaster recovery itself, but also improves business continuity, and the capacity is expanded (Figures 4.101 and 4.102).

4.6.6.2 Multi-Active High-Availability Solution

1. Dual active in the same city

   Split the meaning of 'dual active in the same city':

   - Active: Multiple peer-to-peer logical data centers provide services at the same time on a daily basis. However, in the event of a disaster or traffic adjustment in one data center, other data centers can take over the traffic in this data center, no sense of external business.
   - Same city: Typically, multiple data centers in the same city reuse the same set of infrastructure. Due to the small physical distance, cross-data center latency is acceptable. In the same storage infrastructure scenario, the business can accept the network overhead caused by physical distance, that is, 'city'.

   I. Design principles

   After understanding the concept of 'live more in the same city', we have determined some design principles:

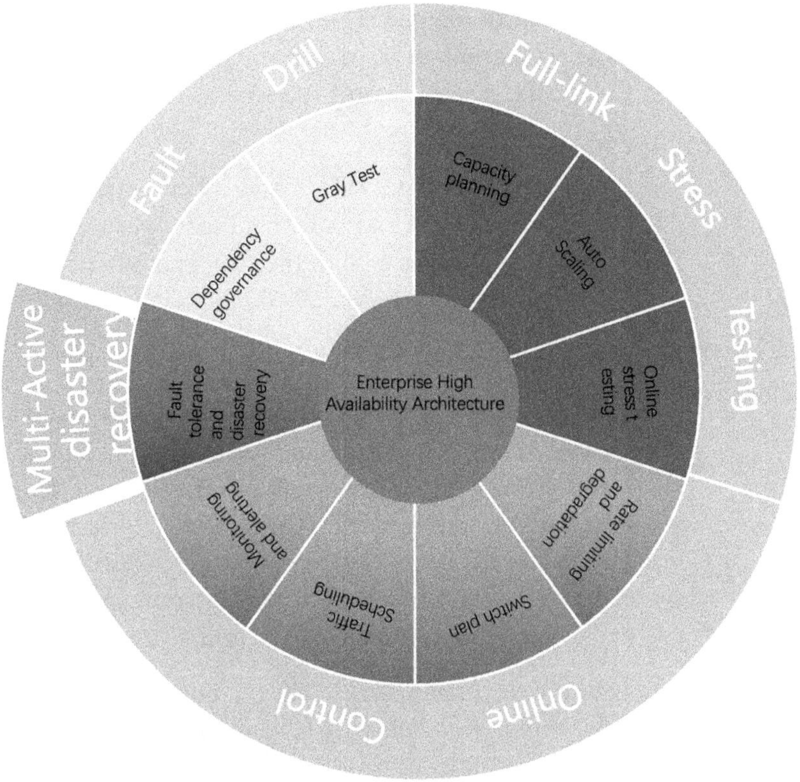

FIGURE 4.101 Enterprise high-availability architecture overview.

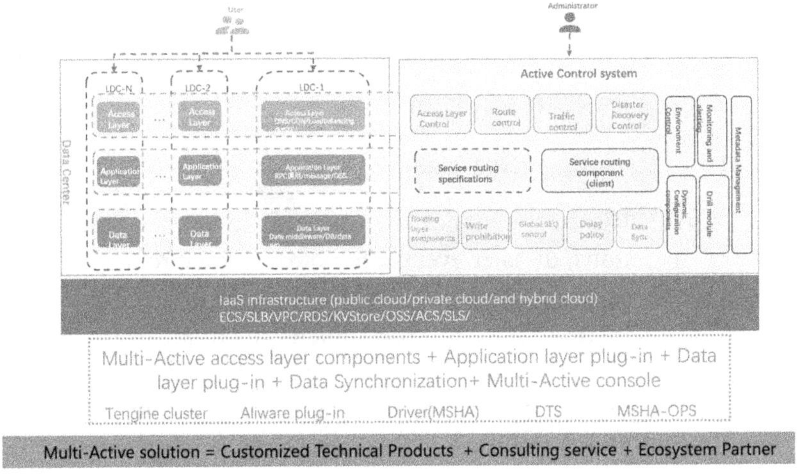

FIGURE 4.102 High-availability solution.

- Select partition range

  Select a group of resources and put them in the same Cell on the basis of reusing the same storage facility (DB).

- Cell first

  For RPC, message, HTTP and other traffic, the Cell priority policy is set to ensure that the traffic priority self-closing loop is completed in the Cell, and other cells can be used to cover the bottom.

II. Two models of Multi-Active scenarios in the same city

- Unit: Business traffic is in a logical data center that completes a strong self-closing loop based on the route id or identification. The data storage infrastructure is recommended to be independent of each other. However, if the business is at a cost perspective, it can be shared with other units. However, the disaster recovery capability and explosion radius are enlarged accordingly.

- Cell: a Cell. In the same Cell, the business traffic is as independent as possible. It is combined with other cells to form a Cell to provide strong routing capability and weak routing within the Cell, traffic flows based on priority policies. A logical data center distributes traffic at the front end based on weights and then switches traffic as weakly as possible based on priority policies. Data storage infrastructure must share one set, otherwise data quality problems may occur.

III. City-wide active value:

- Disaster tolerance is improved. In real business operations, disaster events are not only low-probability events such as earthquakes and optical fiber digging, but also high-probability events such as human causes; moreover, these can be solved through multi-activity in the same city. The following are some common scenarios (Table 4.15):

  To improve business continuity. An Internet company has achieved very good results through its own practical

TABLE 4.15 Common Disaster Scenarios

Scenario	Scenario Description
Human error	Common configuration errors and application publishing failures
Hardware failure	In common cases, network devices fail, causing multiple servers in the data center or cluster to be affected
Network attack	DDoS and other network attacks
Network disconnection/power failure	If Alipay optical cable is cut off
Natural disaster	For example, the power failure of the data center caused by Qingyun lightning strike

verification, which has decoupled the 'business recovery time' and 'failure recovery time'.

- Cost-saving. Under the same cost scenario, multiple logical data centers can bear external traffic, avoiding the waste of idle resources and making full use of resources.
- Grayscale verification: Based on the Multi-Active routing in the same city and strong routing capability by local standard, the business can implement custom grayscale verification (A/B Test or blue-green release), after the business party defines multiple cells:

  Single application: Based on the upstream distribution ratio or rule, it can complete the self-closing loop capability of the specified traffic within the specified application. Distributed application: Based on Alibaba Cloud middleware technology stack, it can complete the self-closing loop function of the specified traffic in the specified Cell link. When one application of the verification link is insufficient, other cells will be used for bottom-up, avoid link interruption (Figure 4.103).

2. Remote Multi-Active

   Remote Multi-Active means that multiple sites are distributed in different locations to provide services simultaneously. The main difference from traditional disaster recovery is that all sites in the active–active mode provide services to the outside world at the same time, which not only solves the problem of disaster recovery itself,

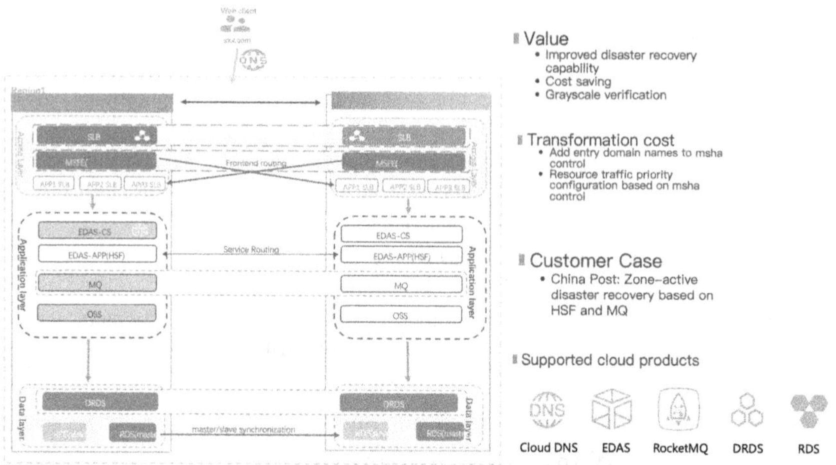

FIGURE 4.103 Distributed application architecture based on Alibaba Cloud technology stack.

but also improves business continuity and enables remote capacity expansion.

I. Problems that can be solved

- Traditional disaster recovery centers do not provide services on a daily basis, and the success rate of switching at critical moments is low.

- Traditional disaster recovery centers do not provide services on a daily basis, and resource utilization is not high.

- Traditional disaster recovery solutions cannot solve the problem of horizontal expansion of multi-region resources.

II. Architecture advantages

- Business continuity: All sites provide external services.

- Scalability: Supports site-level scalability.

- Accuracy: All platforms provide services at the same time, with high switching success rate (Figures 4.104 and 4.105).

III. Multi-Active (City+remote) solution

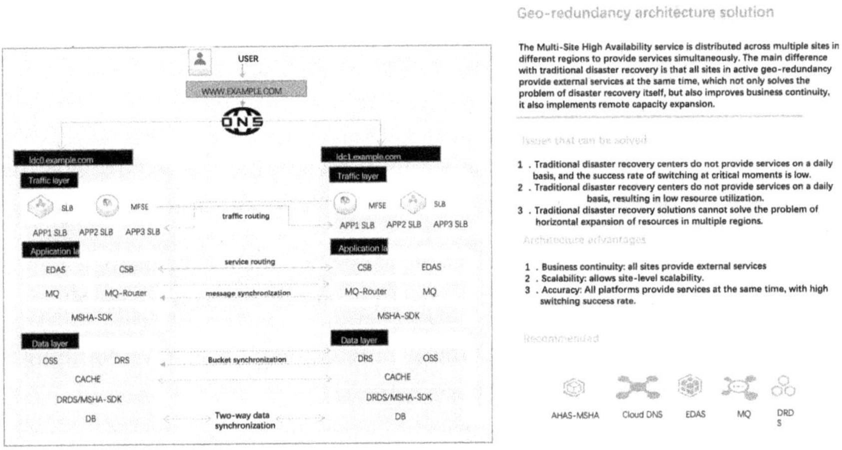

FIGURE 4.104 Remote Multi-Active architecture.

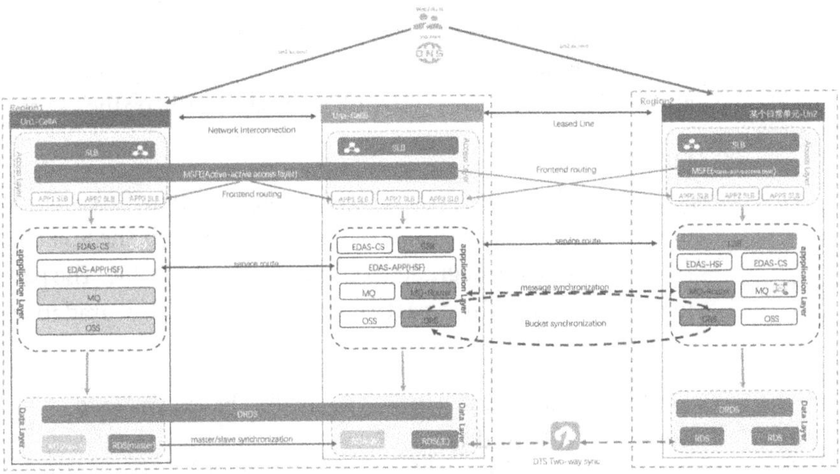

FIGURE 4.105 High availability between two locations.

3. AHAS

AHAS is a tool platform focusing on improving the high availability of applications and businesses. Currently, AHAS mainly provides application foot detection and detection, fault injection high-availability evaluation, the core capabilities of flow control, degradation, high availability and function switch can quickly improve business stability and toughness in marketing activities and core business scenarios at low cost through their own tool modules.

I. Applicable scenarios:

- Traffic protection: Provides comprehensive availability protection for business systems based on traffic.
- Architecture awareness: Automatically detects and displays the dependencies between the cluster topology and applications.
- Fault drill: Provides a wide range of fault scenarios to help distributed systems improve fault tolerance and recoverability.
- Function switch: Dynamically manage business configurations from a global perspective.

II. Common means of high-availability protection:

- Traffic control

  Response to peak traffic: Second kill, sales promotion, order placing and order return processing.

  Message-based scenarios: Peak shifting and valley filling, hot and cold start.Payment system: Pay based on the traffic usage.

- Fuse degradation

  Applicable to any service with complex structure. When unstable factors occur inside or outside the system, the unstable factors are quickly degraded to keep the service stable. The payment system calls third-party services to slow SQL degradation;

- System protection

  Dynamically adjust inbound traffic based on load, CPU usage and inbound QPS/RT.

4. Enterprise High-Availability Architecture

I. Application High Availability

EDAS is a PaaS platform for application hosting and microservice management. It is positioned as the preferred online business application hosting platform for distributed architecture and digital transformation to the cloud. Provides full-stack solutions such as application development, deployment, monitoring and O & M. It also supports microservice runtime environments such as Spring Cloud, Apache Dubbo (Dubbo) and Service Mesh (Istio), helps you to easily migrate your applications to the cloud (Figure 4.106).

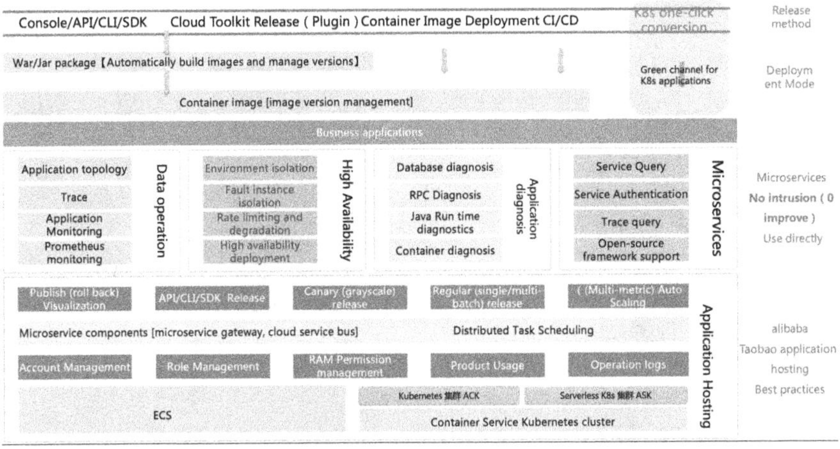

FIGURE 4.106 Enterprise high-availability architecture based on Alibaba Cloud EDAS.

II. Database High Availability

ApsaraDB for RDS is a stable, reliable and scalable online database service. It adopts a dual-machine hot standby architecture of one master and one slave, which is suitable for most user scenarios. When the primary node fails, the primary node and the secondary node are switched in seconds. The entire switching process is transparent to the application. When the secondary node fails, RDS automatically creates a new secondary node to ensure high availability. Single-zone instance: The primary and secondary nodes are located in the same zone. The primary and secondary nodes are located on two different physical servers. The cabinets, air conditioners, circuits and networks in the zone are redundant to ensure high availability. Multi-zone instances (also known as two data centers in the same region or disaster recovery instances in the same region): The primary and secondary nodes are located in different zones in the same region, providing cross-zone-disaster recovery capabilities, there is no extra charge (Figure 4.107).

PolarDB-X (formerly DRDS) is a cloud-native distributed database independently developed by Alibaba. It integrates the distributed SQL engine DRDS with the distributed self-developed storage X-DB and is designed based on the cloud-native integrated

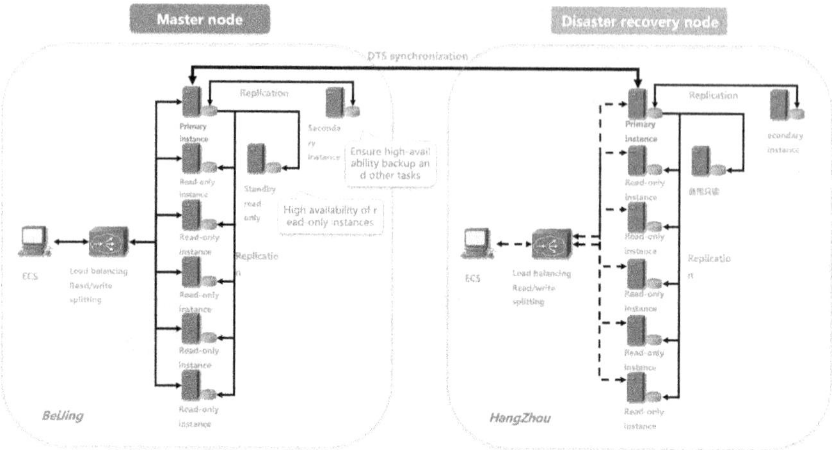

FIGURE 4.107 RDS disaster recovery architecture.

architecture, supports tens of millions of concurrent storage and hundreds of PB of massive storage. It focuses on solving database bottlenecks such as massive data storage, ultra-high-concurrency throughput, large table bottlenecks and complex computing efficiency. It has been tested by customers in various industries during the Double 11 Tmall and Alibaba Cloud, help enterprises accelerate their business digital transformation. PolarDB-X splits data into multiple MySQL stores, allowing each MySQL to bear appropriate concurrency, data storage and computing loads, and each MySQL is in a stable state. PolarDB-X layer to process distributed logic, a stable, reliable and highly scalable distributed relational database system is finally obtained (Figure 4.108).

The Redis database uses a master-replica architecture. The master-replica nodes are located on different physical machines. The master node provides external access. You can add, delete, modify and query data through the Redis command line and the general client. When the master node fails, the self-developed HA system automatically switches between the master node and the Slave node to ensure the smooth operation of the business. By default, data persistence is enabled and all data are on the disk. Supports data backup. You can roll back or clone backup sets to effectively solve data misoperations. At the same time, instances created in zones that support disaster recovery, such as

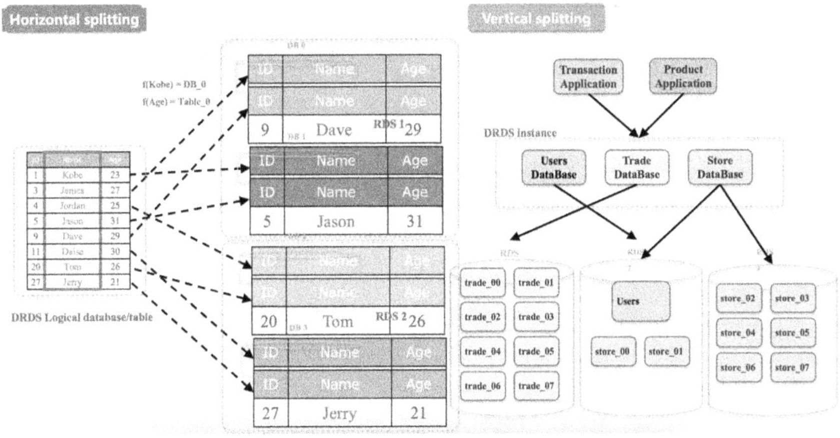

FIGURE 4.108 PolarDB-XSplitting principle.

Hangzhou zone H+I, also have the same-city disaster recovery capability (Figure 4.109).

There are many disaster recovery solutions for Alibaba Cloud ElasticSearch, such as dual-cluster standby, Multi-Active cluster and zone-disaster recovery. If the new cluster to traffic in the peak of the reality, for example, Double 11, 12, 618, etc.), it is recommended to use the Double Cluster mutual preparation of programs clustering disaster recovery to configure upgrade classification (Figure 4.110).

4.6.7 Distributed Database and Cache Technology

In the process of completing the planning of service center and entering the construction of service center, the overall performance requirements of the enterprise's business for the service center should be considered. The increase of business volume will inevitably lead to the increase of requests in the service center. Because the business logic layer of the service center is generally stateless and can achieve horizontal scaling, the challenges usually focus on the data layer. This section focuses on how to use distributed database and cache technology to meet the challenges at data layer.

4.6.7.1 Importance of Distributed Databases and Caches for Business Middle Office

Each service center in Business Middle Office can have its own independent database, which is equivalent to the vertical partitioning of a

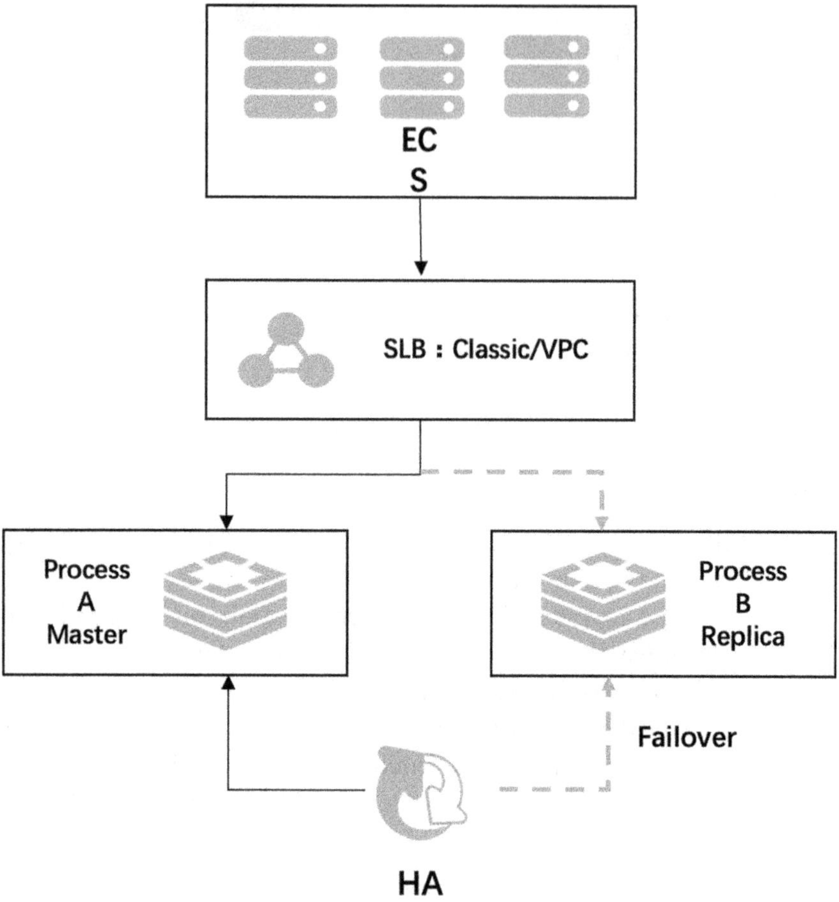

FIGURE 4.109 Alibaba Cloud Redis high-availability architecture.

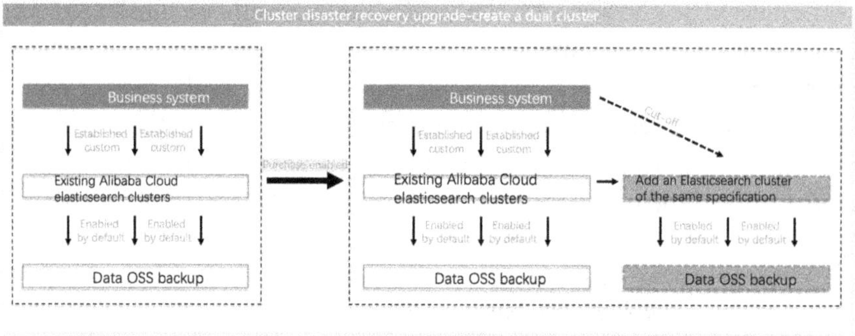

FIGURE 4.110 Alibaba Cloud ElasticSearch high-availability architecture.

single big database for a traditional monolithic application. Technically, it can alleviate the scalability bottleneck of a single database, such as the resource expansion limitation of database instance, the connection number limitation and the resource contention brought by the increase of tables. However, as the business access of a single service center increases to a certain scale, the service center's access to its own single database will also encounter a maximum capacity. This requires the construction of service center to solve the performance bottleneck of single database.

In addition, cache also plays an important role in performance optimization during the construction of Business Middle Office. Users' requests will be called from the front end to the application, from the application to the middle service center, and from the middle service center. The longer the access path, the bigger the access latency. In order to reduce the access latency, cache can be used to shorten the access path; at the same time, due to the hierarchical storage of data, such as register, hardware cache, local memory, hard disk, distributed cache and remote database, the access rate drops step by step. Therefore, the former component can be understood as the cache of the latter component, which improves the access efficiency of repeated requests (Figure 4.111).

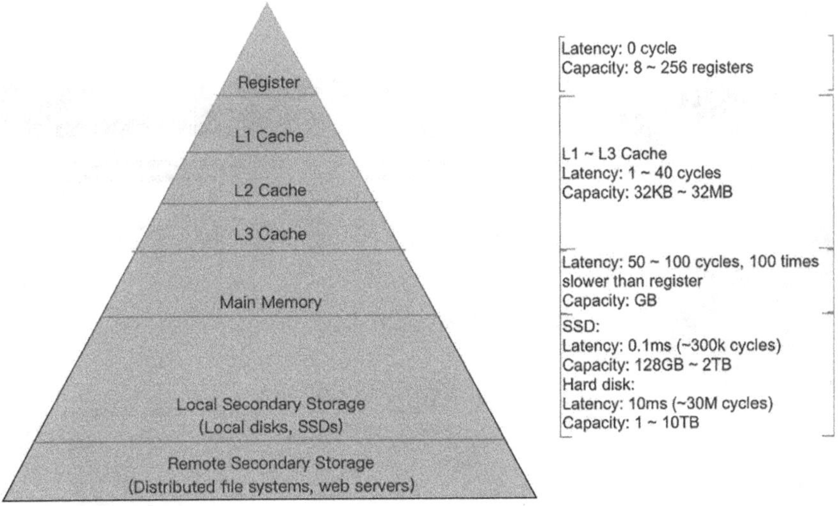

FIGURE 4.111 Latency and capacity of hierarchical storage.

4.6.7.2 Introduction of Fundamental Concept of Distributed Database

A distributed database is a collection of interconnected databases that are physically distributed across sites interconnected through a computer network. Distributed databases have the following features:

- The databases in the collection are logically related to each other and appear as a single logical database externally.
- Physical storages of data are located across multiple sites. The data in each site can be managed by DBMS independent of other sites.
- The processors in the site are connected through the network.
- A distributed database is not a loosely connected file system because it contains transactions, but the degree of support for transactions varies from system to system.

In terms of technology architecture, distributed database can be the following types (Figure 4.112).

Alibaba's e-commerce practice is constantly enriched and developed with the development of database technology architecture. The following lists several development stages of Alibaba database technology. From commercial database to open-source database, then to cloud-hosted database service, and finally to cloud-native distributed database (Figure 4.113).

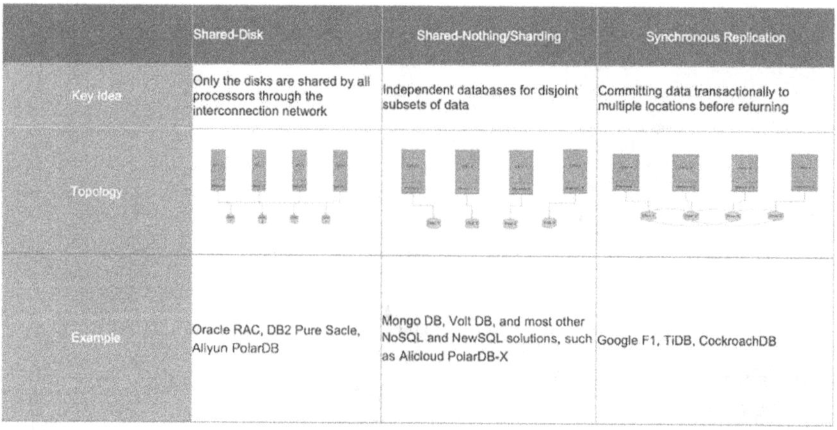

	Shared-Disk	Shared-Nothing/Sharding	Synchronous Replication
Key Idea	Only the disks are shared by all processors through the interconnection network	Independent databases for disjoint subsets of data	Committing data transactionally to multiple locations before returning
Topology			
Example	Oracle RAC, DB2 Pure Sacle, Aliyun PolarDB	Mongo DB, Volt DB, and most other NoSQL and NewSQL solutions, such as Alicloud PolarDB-X	Google F1, TiDB, CockroachDB

FIGURE 4.112 Three kinds of distributed database architecture.

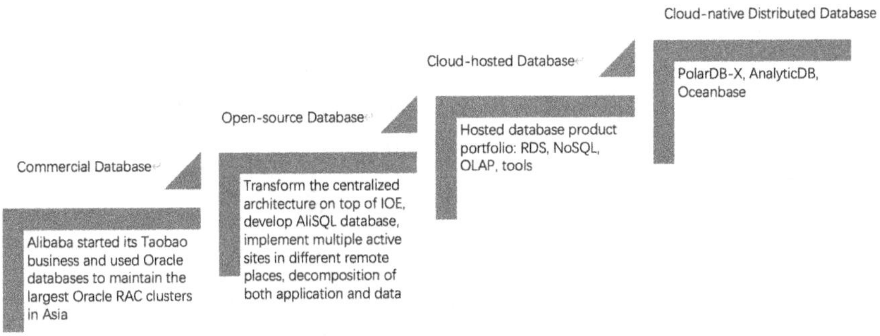

FIGURE 4.113 Alibaba Cloud database development stage.

PolarDB-X is a cloud-native distributed database independently developed by Alibaba, which integrates distributed SQL engine DRDS and distributed self-developed storage X-DB. Based on cloud-native integrated architecture design, PolarDB-X can support tens of millions of concurrent scale and hundreds of petabytes of mass storage. Focusing on solving database bottleneck issues such as massive database storage, ultra-high concurrent throughput, large table bottleneck and complex computing efficiency, it has gone through the test of business of customers in various industries of Alibaba Cloud and validation of Alibaba Tmall 11.11 promotion, and help enterprise to accelerate the completion of business digital transformation.

The core capabilities of PolarDB-X are implemented based on the standard relational database technologies. With its comprehensive management, operation and maintenance, and productization capabilities, PolarDB-X provides users with a stable, reliable, highly scalable, continuously operational and maintainable, and ease of use like traditional standalone MySQL database experience characteristic.

PolarDB-X has been in the public cloud and dedicated cloud environment for many years, and has been tested by the core transaction business of Alibaba Tmall 11.11 promotion and Alibaba Cloud customers' business of various industries. Carrying a large number of users' core online business, spanning Internet, financial payment, education, communication, public utilities and other industries, is the de facto standard for all Alibaba Group's online core business and many Alibaba Cloud customer business to access the distributed database.

- Stability

  For most applications, the responsibilities of a relational database are the core and basic in the entire data management system. Relational databases not only directly affect the service experience of end user, but are also the last guarantee of business data. Therefore, stability is the core factor for selecting a database.

  The stability of PolarDB-X is based on the reasonable use of proven MySQL. Single-instance MySQL is relatively weak in high-concurrency, large-volume data storage and complex computing scenarios.

  PolarDB-X to split data into multiple MySQL stores so that each MySQL store assumes a proper load for concurrency, data storage and computing while each MySQL is in a stable state. PolarDB-X processing of distributed logic at the layer is to ultimately obtain a stable, reliable and highly scalable distributed relational database system.

  Compared with the self-developed distributed NewSQL database, PolarDB-X products always focus on the continuous stability and O & M as the top priority. At the same time, it uses standard database technologies to make up for the experience difference from the single-server database, so that users can easily and quickly get started with it and give full play to the business value of the products.

- Highly scalable

  Compared with traditional single-server relational databases, PolarDB-X adopts a hierarchical architecture to ensure linear scaling in concurrency, computing and data storage. PolarDB-X can increase computing and storage resources to achieve horizontal scaling.

  Compared with the new cloud-native database based on distributed storage, ApsaraDB for MongoDB has no upper limits on the PolarDB-X scalability, which eliminates the worries and O & M pressure caused by database scalability after the rapid development of the business.

- Continuous operation and maintenance

  For most applications, relational databases must work stably for 7×24 hours. The key capability of a database is the ability to maintain data continuously.

  PolarDB-X has been developing on Alibaba Cloud and Apsara Stack for many years and provides a wide range of product capabilities

TABLE 4.16 Features of PolarDB-X

Features	Description
Horizontal sharding	It provides multiple data sharding methods that support OLTP services, allowing operations to focus on a small amount of data and improve overall concurrency and throughput performance.
Vertical sharding	It solves the problem that clients maintain a large number of data source configurations through connection proxies. At the same time, it provides strong consistent transactions across business database and online data analysis across business database.
Hotspot separation	It separates hotspot or skew data from others to balance storage and access of data.
Smooth scale-out	Scale-out data storage silently by adding new RDS instance without impacting online business access.
Read-write separation	Improve database query performance by adding RDS read-only instance.
Distributed transaction	Support distributed transaction with RDS5.7.
Global unique sequence	It provides efficient globally unique number sequence and supports automatic filling of auto_increment primary key.
Concurrent read-only instance	Handle concurrent queries of ultra-high traffic through physical resource and link isolation to ensure the stability of online business main links.
Analytic read-only instance	Accelerate the execution efficiency of complex analysis SQL in massive scenarios and greatly improve the response speed of complex analysis SQL. For multi-table join, aggregation, and sorting operations with 100 million data volumes, results can be returned in seconds.
SQL flashback	Provides row-level data recovery capability for SQL misoperations.

and a complete O & M System. It allows businesses to periodically integrate the services through a complete set of OpenAPI.

Table 4.16 shows some core product capability of PolarDB-X. PolarDB-X provides 1.0 middleware form and 2.0 integration form.

- PolarDB-X 1.0 middleware is composed of DRDS instances in the computing layer and RDS instances in the storage layer. It can be split horizontally by mounting multiple MySQL databases and tables.
- PolarDB-X 2.0 integrated form is composed of multiple nodes. Multiple nodes are deployed in the instance for horizontal expansion. Each node integrates computing resources and storage resources in a closed loop, which makes operation and maintenance management more convenient.

Like most traditional stand-alone relational databases, PolarDB-X is divided into computing layer and storage layer. The computing layer is divided into access layer, SQL Engine and processing engine. The access layer includes network and protocol layer, which provides support for customer access and is compatible with MySQL protocol. SQL Engine includes SQL standard support, SQL parsing and optimization layer. The optimization layer includes logical optimization and physical optimization. The processing engine includes single-machine two-stage execution, single-machine parallel execution and multi-machine parallel execution, which applies a variety of traditional single-machine database optimization and execution technologies; the processing engine interacts with the underlying storage engine through storage driver. The storage tier or storage engine can be a private-customized RDS instance or distributed storage. The consistency of data copies in distributed storage is guaranteed by PAXOS algorithm (Figure 4.114).

4.6.7.3 Best Practice for Distributed Database Technology

Each type of database is created to solve different types of problems. As a distributed database, PolarDB-X also has its own applicable scenarios.

Applicable application types

PolarDB-X products have good user accumulation and technical development in high concurrency, distributed transactions, complex SQL optimization, parallel computing, etc. They are applicable to the following scenarios:

- Internet online transaction business scenarios that require high concurrency and large-scale data storage.
- Traditional enterprise-level applications are experiencing explosive growth in Computation and Data volume due to business development. Online transactional database scenarios with stronger computing capabilities are urgently needed.

1. Select by capacity

   In the field of OLTP business, the capacity of database usually focuses on concurrency, data storage and complex SQL response time. If there is a bottleneck in any dimension of the current database, or for the consideration of planning the database selection in advance for the continuous high-speed development of

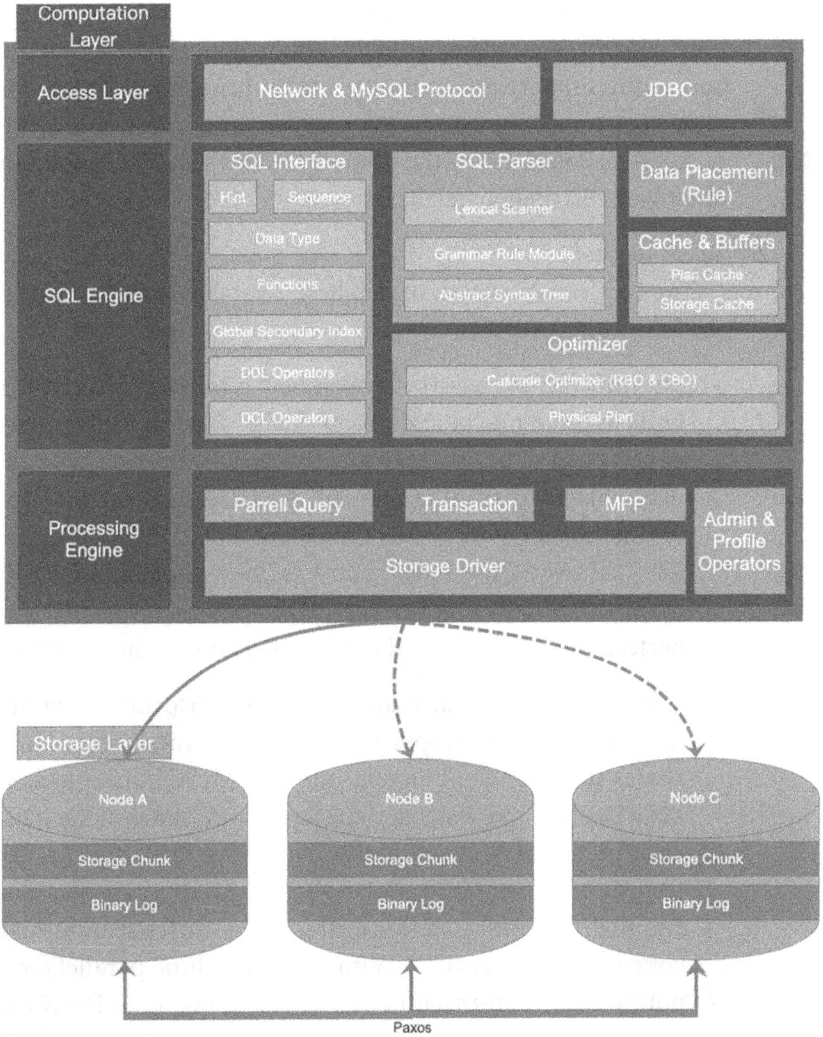

FIGURE 4.114 Overall architecture of PolarDB-X.

business, PolarDB-X is selected to build a distributed database, which can effectively reduce the pressure of database expansion and operation and maintenance in the later stage.

In the early stage of business development, many factors need to be considered when choosing a single-machine database or a distributed database. However, from the perspective of the database itself, the SQL statements, data types, transactions, indexes and other functions used by the business are all determined. For most businesses, as long as SQL syntax, data type, transaction and index support are complete, and it has effective means to scale horizontally in various extreme scenarios, for fast-growing businesses, PolarDB-X is the most viable and sustainable solution among all distributed databases.

2. Select by cost

   The cost consideration of database selection mainly includes the following two parts:

   - The difficulty of business development is too high, which often leads to project delay and unsatisfactory business effect. For a new type of database, how to effectively compatible with the usage habits of existing popular databases and the integrity of function support is very important. PolarDB-X is compatible with MySQL ecology, and has good compatibility with mainstream clients and drivers. SQL syntax is compatible and perfect, and business can be connected and adapted quickly.

   - The long-term stability and excellent performance of the database are very important for the business. PolarDB-X shares data and load among multiple MySQL instances, so PolarDB-X is more stable than large-scale stand-alone database in the face of gradually increasing load pressure. In terms of performance, PolarDB-X naturally supports distributed computing, and its strength is to resist ultra-high concurrency of services. With single-machine parallel computing and multi-machine DAG computing, PolarDB-X can cover the complex computing requirements of most online services.

3. Data architecture decision according to application life cycle

   PolarDB-X's various sharding modes can be seamlessly and smoothly opened up to meet the demands of database scalability in all business life cycles (Figure 4.115).

   I. Database sharding or table sharding

      With the continuous growth of business data scale in the business, the traditional stand-alone database is facing the

FIGURE 4.115 Evolution of PolarDB sharding mode.

bottleneck of scalability. This kind of problem can be solved by PolarDB-X's database and table sharding technology. By dividing databases and tables, business data and access pressure are allocated to multiple single database instances to solve the problem of super high concurrency of online business. At the same time, the data storage space can be linearly expanded by horizontal sharding, providing Pega-bytes-level storage capacity. Effectively solve the storage bottleneck of single database. Parallel query and MPP parallel acceleration capabilities are provided for online business, which can greatly improve the execution efficiency of complex analysis query under massive data of online business. If the number of single tables is too large, the throughput of the database will decrease and the overall performance will be slow. By dividing database and table, the single-table data are split into MySQL, which can effectively solve the problem of single-table data expansion. The service can perform joint query and transaction operation among multiple databases of different RDS instances by using database sharding or table sharding. It can effectively avoid the business side complicated hard code processing, and greatly improve the efficiency of business development.

II. Vertical sharding at database level

Database vertical sharding refers to grouping the tables in the database according to the business, and putting the tables in the same group into a new database (logically, not an instance). We need to start from the actual business and divide the big business into small business. For example, the user-related tables, order-related tables and logistics-related tables in the whole business of the mall are classified independently to form the user system database, order system database and logistics system database, as shown in Figure 4.116.

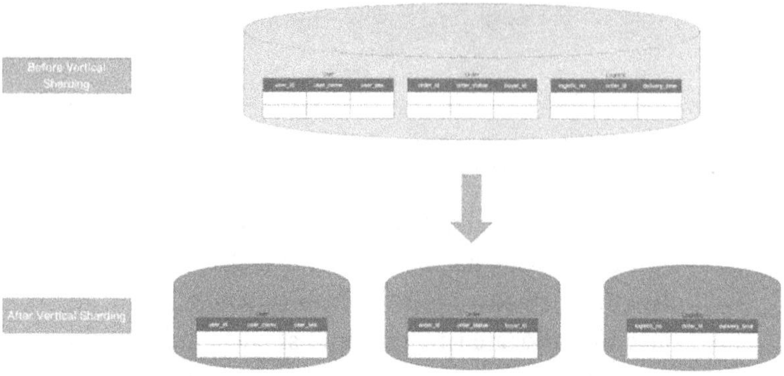

FIGURE 4.116 Database vertical sharding.

III. Horizontal sharding at database level

After the database is sharded vertically, if the amount of data continues to grow and a single shard also encounters a database performance bottleneck, the horizontal sharding at database level can be considered. The reason why the vertical sharding is first performed before the horizontal sharding is that the data are well-organized and distinct after the vertical sharding, which make it easier to specify horizontal standards. As user center in the e-commerce business an example, user-related tables are splitted from other non-user-related tables first, and then user-related data itself can be sharded according to the user registration time interval, or the user's area or the user ID range, etc.

Figure 4.117 shows one example to horizontally shard user-related tables with user_id range.

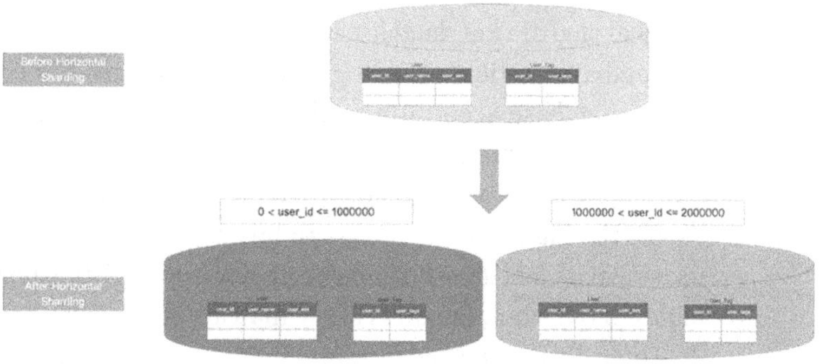

FIGURE 4.117 Database horizontal sharding.

IV. Vertical sharding at table level

Table-level vertical sharding can be described simply, that is, a large and wide table is split into multiple small tables. The vertical sharding of the data table is to vertically divide the columns in the table into multiple tables, changing the table from wide to narrow. Generally, follow the following points to split:

- Separate cold and hot, put frequently used columns in one table, and infrequently used columns in one table.
- Large field columns are stored separately.
- The columns of association relationships are closely grouped together.

Figure 4.118 shows one example to separate the frequently used from rarely used large fields in the user table into two tables.

VI. Horizontal sharding at table level

The table-level sharding principle is similar to the database-level sharding principle. Generally, for a database, hash sharding can be performed based on the primary key ID or a specified field. As shown in Figure 4.119, the user_log table is sharded based on the user_action_date field as the table partition key.

VII. How to determine sharding Key

The sharding key is a database sharding or table sharding field. It is a data table field used to generate a shard rule

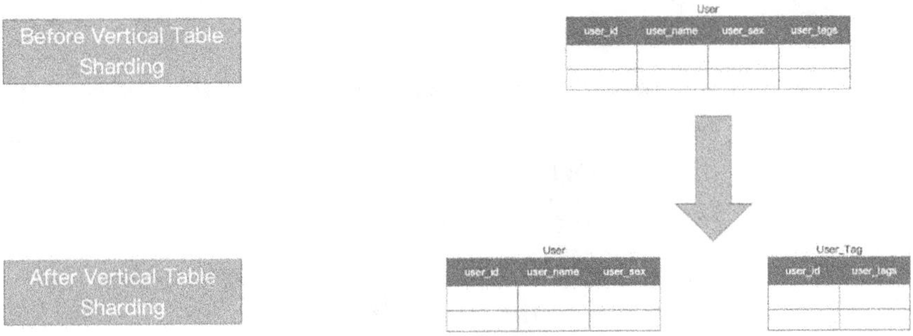

FIGURE 4.118 Vertical sharding of user table fields.

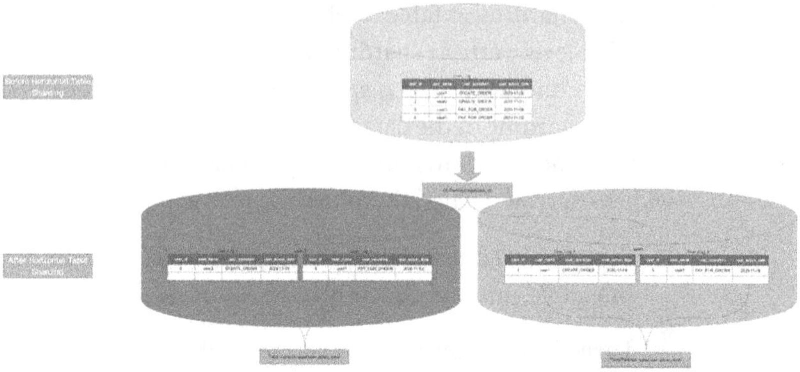

FIGURE 4.119 User table horizontally sharding.

during horizontal shard. PolarDB-X calculates the sharding key value by using the sharding function to obtain a calculation result, and then shards the data to a private custom relational database service instance based on the result.The primary principle of data sharding is to find the business logic entity to which the data belong as much as possible, and make sure that most of core SQL statements are performed around this entity, and then use suitable field belonging to the entity as the sharding key.Business logic entities usually vary with application scenarios. The following typical application scenarios have specific business logic entities, the identifier field can be used as a sharding key.

- User-oriented Internet applications perform various operations around the user dimension. The business logic entity is the user, and the user ID can be used as the sharding key.

- Focus on the seller's e-commerce application and perform various operations around the seller dimension. Then the business logic entity is the seller, and the seller ID can be used as the sharding key.

- Game-based online applications perform various operations around the player dimension, so the business logic entity is the player, and the player ID can be used as the sharding key.

- Online application of the Internet of Vehicle performs various operations around the vehicle dimension, then the business logic entity is the vehicle, and the vehicle ID can be used as the split key.
- Online Tax applications conduct front-desk business operations around taxpayers, then the business logic entity is taxpayers, and the taxpayer ID can be used as the sharding key.

For example, a seller-oriented e-commerce application needs to shard the order table horizontally:

```
CREATE TABLE sample_order (
  id INT(11) NOT NULL,
  sellerId INT(11) NOT NULL,
  trade_id INT(11) NOT NULL,
  buyer_id INT(11) NOT NULL,
  buyer_nick VARCHAR(64) DEFAULT NULL,
  PRIMARY KEY (id)
)
```

If the business logic entity is determined to be the seller, select the sellerid field as the sharding key, and then use the following distributed DDL statements to create the table:

```
CREATE TABLE sample_order (
  id INT(11) NOT NULL,
  sellerId INT(11) NOT NULL,
  trade_id INT(11) NOT NULL,
  buyer_id INT(11) NOT NULL,
  buyer_nick VARCHAR(64) DEFAULT NULL,
  PRIMARY KEY (id)
) DBPARTITION BY HASH(sellerId)
```

If there is no suitable business logic entity as the sharding key, especially for traditional enterprise applications, you can consider the following methods to select the sharding key.If there is no suitable field as a sharding key for the selected business logic entity, especially for traditional enterprise applications, the following methods can be considered to select a sharding key.

- Determine sharding keys to ensure the balance of data distribution and access, and try to distribute the data relatively evenly in different table partitions.
- According to the combination of number or string type and time type field as the split key, the database and table are divided, which is suitable for log retrieval application scenarios.

  For example, a log system records all user operations, and now it is necessary to split the following log table horizontally:

```
CREATE TABLE user_log (
  userId INT(11) NOT NULL,
  name VARCHAR(64) NOT NULL,
  operation VARCHAR(128) DEFAULT NULL,
  actionDate DATE DEFAULT NULL
)
```

At this time, the combination of user ID and time field can be selected as the sharding key, and split the table according to 7 days a week, then use the following distributed DDL statement to build the table:

```
CREATE TABLE user_log (
  userId INT(11) NOT NULL,
  name VARCHAR(64) NOT NULL,
  operation VARCHAR(128) DEFAULT NULL,
  actionDate DATE DEFAULT NULL
) DBPARTITION BY HASH(userId) TBPARTITION BY
WEEK(actionDate) TBPARTITIONS 7
```

For more choices of sharding keys and table sharding forms, please refer to the Alibaba Cloud help document about table creation-related content (https://www.alibabacloud.com/help/doc-detail/71300.htm?spm = a2c63.p38356.b99.124.79d91cabYJ2AYK).

VIII. How to determine number of shardings

There are two levels of horizontal sharding in DRDS: Database sharding and table sharding. By default, eight physical database shards are created on each RDS instance. One or more physical table shards can be created on each physical database shard. The number of table partitions is also called the number of shardings.

In general, it is recommended that the capacity of a physical table cannot exceed 5 million rows. Generally, you can estimate the data growth in 1–2 years. Divide the estimated total data volume by the total number of physical database shards, and then divide the recommended maximum data volume by 5 million, you can obtain the number of physical table shards to be created on each physical database shard:

Number of physical table shards on the physical database Shard = rounded up (estimated total data volume/(number of RDS instances × 8)/5,000,000).

Therefore, when the calculated number of physical table Shards is equal to 1, database shards are required without further Table shards, that is, a physical table shard on each physical database shard. If the calculated result is greater than 1, we recommend that you split databases and tables, that is, multiple physical tables on each physical database shard.

For example, if a user estimates that the total data volume of a table in 2 years is about 100 million rows and purchases four RDS instances, the calculation is based on the above formula:

Number of physical table shards on physical database shards = CEILING(100,000,000/(4*8)/5,000,000) = CEILING (0.625) = 1

If the result is 1, only database shards are required, that is, one physical table shard on each physical database shard.

If only one RDS instance is purchased in the preceding example, calculate according to the preceding formula:

Number of physical table shards on physical database shards = CEILING(100,000,000/(1*8)/5,000,000) = CEILING (2.5) = 3

If the result is 3, we recommend that you split the database into three physical tables.

4. Read-write Separation

It is a common database architecture that the database is divided into master and slave databases, one master database is used to write data, and multiple slave databases are used to read data by polling. The master and slave databases synchronize data through some communication mechanism. Figure 4.120 shows the structure of one master and two slaves.

When the main instance of PolarDB-X storage resource MySQL has many read requests and high read pressure, you can split the read traffic through the read-write separation function to reduce the read pressure of the storage layer.

PolarDB-X's read-write separation function is transparent to the application. Without modifying any code of the application, it is necessary to adjust the read weight in the console, so that the read traffic will be routed between the MySQL master instance and the multiple read-only instances according to the custom weight, and the write traffic will be routed only to the master instance.

After the read-write separation is enabled, reading from the master instance of MySQL is real-time strong consistent read; while the data on the read-only instance is asynchronously copied from the master instance with millisecond delay, so reading

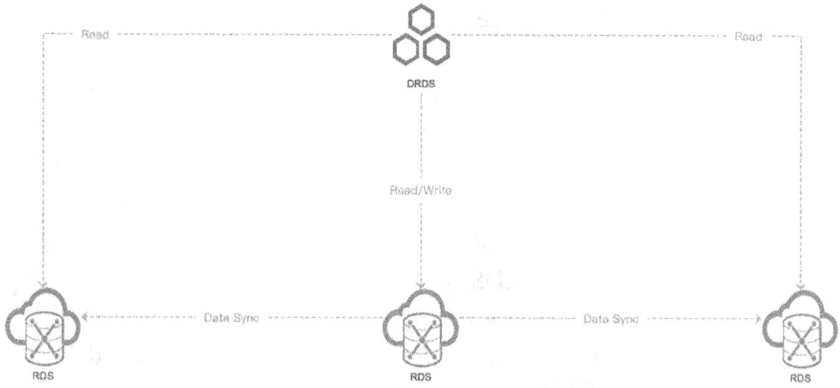

FIGURE 4.120 Database read-write separation architecture.

from the read-only instance belongs to weak consistent read. Specifying the read SQL that needs to ensure real-time strong consistency can be executed on the master instance through hint. For details, see separate read and write hint.

5. Practice of Heterogeneous Index GSI

After adopting the technology of database sharding and table sharding, the sharding key should be determined as discussed before. However, if the dimension of the query is inconsistent with the database or table sharding key, a cross sharding query will be inevitable. The increase of cross sharding query will lead to performance problems such as slow query and exhaustion of connection pool. In this case, the global secondary index (GSI) emerges. GSI supports adding splitting dimensions on demand and provides global unique constraints. Each GSI corresponds to an index table, and XA multi-write is used to ensure strong data consistency between the origin table and the index table.

The differences between GSI and local index are as follows.

- Global secondary index: Different from the local index, if the data row and the corresponding index row are stored on different sharding, this kind of index is called GSI, which is mainly used to quickly determine the data sharding involved in the query.
- Local index: In a distributed database, if the data row and the corresponding index row are stored on the same sharding, this kind of index is called local index. PolarDB-X specifically refers to the MySQL secondary index on the physical table.

The relationship between them. The two index types need to be used together. After PolarDB-X distributes the query to a single partition through GSI, the local index on the partition can improve the query performance within the partition.

Give one example case. First, create an order table, in which database sharding key is OrderID. If the current business needs to query the buyer's order list, a GSI can be created for the buyer ID (seller ID). After creating the GSI, PolarDB-X will create a heterogeneous index table in the database, which contains the specified index field and the database and table

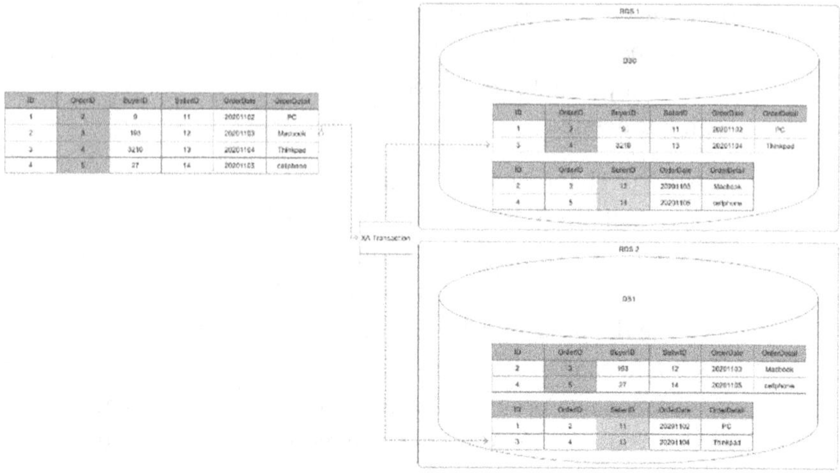

FIGURE 4.121 Global secondary index.

sharding key of the original table, and the index field is used as the sharding key of the index table. If covering columns are specified when creating GSI, these covering columns will also appear in the heterogeneous index table created for GSI. By using frequently used query fields as covering columns, the query overhead can be reduced in some scenarios, because only GSI heterogeneous index table will be queried. Creating GSI also brings overhead. When the original table has data added, deleted or updated, PolarDB-X will start distributed XA transactions and write to the original table and GSI heterogeneous index table (Figure 4.121).

```
CREATE TABLE orders (
    id bigint not null auto_increment,
    orderid int default 0,
    buyerid int default 0,
    sellerid int default 0,
    orderdate timestamp default now(),
    orderdetail varchar(200),
    primary key (id),
    global index gidx_seller (sellerid)
covering(id, orderid, orderdate, orderdetail) dbpartition
by hash (sellerid)
) ENGINE=InnoDB dbpartition by hash (orderid);
```

Restrictions

- The GSI increases the DML operation execution time of the table.
- The number of GSIs per table is not recommended to exceed 3.
- Frequent core queries require a GSI of coverage.
- It is not recommended to create a GSI for frequently updated fields because updates are involving GSI.

4.6.7.4 Introduction of Cache Technology

1. Classification of Cache

   According to the role of cache acting during request processing flow, cache can be classified as follows.

   I. Client cache

   The client cache is a kind of storage nearest to users in the network, which is often used together with server-side cache. The common client caches are as follows:

   - Web page cache: Web page cache refers to caching some elements of the static page to the local, so that the next request does not need to duplicate resource files. HTML5 supports the function of offline caching. The specific implementation can specify the manifest file through the page. When the browser accesses a file with manifest attribute, it will first obtain the information of loading page from the application cache, and through the inspection mechanism to deal with the cache update problem.
   - Browser cache: Browser cache usually opens up memory space to store resource copies. When users go back or return to the previous operation, they can quickly obtain data through browser cache, which can be stored in HTTP 1.1, browser caching can be well supported by introducing e-tag and combining the two features of expiration and cache control.
   - App cache: App can cache content to memory or local database. For example, some open-source image libraries have the technical characteristics of caching. When the image and

other resource files are obtained from the remote server, they will be cached, so that there will be no repeated requests next time, and the traffic cost of users can be reduced.

Client caching is an important direction of front-end performance optimization. After all, the client is the closest place to users and can fully exploit the optimization potential.

II. Network Cache

Network cache is located in the middle of client and server. It solves the response of data request by proxy and reduces the return rate of data request. There are several types of network caching.

- Web proxy cache: Common proxy forms are divided into forward proxy, reverse proxy and transparent proxy. Web proxy cache usually refers to forward proxy, which puts resource files and hotspot data on the proxy server. When a new request arrives, if the data can be obtained later on the proxy server, there is no need to repeat the request to the application server.

- Edge cache CDN: Like forward proxy, reverse proxy can also be used for caching. For example, Nginx provides caching function. Furthermore, if these reverse proxy servers can come from the same network as user requests, the speed of obtaining resources will be further improved. This kind of reverse proxy servers can be called edge caching. A common edge cache is CDN, which can put static resource files such as pictures on CDN.

III. Server cache

Server-side cache is the driving force of performance optimization in back-end development. Common back-end performance optimization is also solved by introducing cache, including database query cache, cache framework and application-level cache.

IV. Database query cache

The caching mechanism of MySQL is to cache the select statement and the corresponding result set. When a subsequent select request is received, if query is enabled in MySQL, the cache function hashes the select statement in the form of a string, and then queries from the cache. If the data are queried, it will be returned directly, eliminating the subsequent optimizer and storage engine IO operations, which can greatly improve the response time. How to optimize query cache needs to consider the following metrics:

- query_cache_size
  Set the size of memory area that can cache result set.
- query_cache_type
  Represents a scene where caching is used. 0 means that query cache will not be used in any scenario, 1 means that queries that explicitly specify that query cache will not be used can be used and 2 (demand) means that query cache will only take effect if it is explicitly indicated.
- Qcache hits
  Indicates how many queries hit the query cache.
- Qcache inserts
  Indicates how many times the query cache was missed and the data were inserted.
- Qcahce lowmem prunes
  Indicates how many queries are cleared due to insufficient space.
- Qcache free memory
  Indicates the size of the remaining memory.
- Qcache free blocks
  A large value indicates that there are many memory fragments that need to be cleaned up in time.

  In the process of qcache optimization, the above metrics can be comprehensively analyzed. For example, we can know that the cache hit rate of qcache=qcache hits/qcache hits+qcache inserts to judge the efficiency of the current qcache. It can also be combined with qcache low memory prunes, qcache free memory and qcache free blocks to judge the current memory utilization efficiency of qcache.

In addition, if you use InnoDB storage engine, you also need to focus on InnoDB_buffer_pool_Size parameter, which determines the index of InnoDB and whether the data have enough space to put into the cache. table_Cache determines the maximum number of tables that can be cached, and it is also a parameter that needs to be concerned.

4.6.7.5 Mainstream Cache Products

1. Redis

Redis is an open-source, BSD-compliant, memory-based, high-performance key-value database. Compared with other key-value caching products, Redis has the following features:

- Redis supports data persistence, and can save the data in memory onto disk, which can be loaded again when it is restarted for recovery.
- The key-value model of Redis not only supports the common string type, but also provides the types like list, set, ordered set and hash data structures. Additionally, there are some advanced data structures, such as hyperlog, geohash and stream.
- Redis supports high-availability deployment, and slave instance can take over upon failure of master instance. At the same time, Redis also supports cluster deployment, which can expand the read-write ability of Redis.
- The reading and writing speed of Redis is very high. In general scenarios, the reading speed of a single instance of Redis is about 100,000 times per second, and the writing speed is about 80,000 times per second.
- All operations of Redis are atomic, which means that they are either executed successfully or not executed at all. Individual operations are atomic. Multiple operations also support transactions, that is, atomicity, wrapped by multi and exec instructions. However, the transaction of Redis is different from that of relational database. If some intermediate operation fails, it does not support rollback.

- Redis runs in memory, but it can persist to disk. Therefore, when reading and writing different data sets at high speed, the memory capacity should be guaranteed, because the amount of data cannot be greater than the hardware memory. Another advantage of memory database is that compared with the same complex data structure on disk, it is very simple to operate in memory. In this way, Redis can do many things with strong internal complexity. At the same time, in terms of disk format, they are compact and generated by appending, because they do not need random access.
- Redis supports Lua scripts, and some key-value processing can be performed on the Redis server through Lua scripts. This feature is similar to the stored procedure of relational database, which can reduce the interaction with the client, and can play a great role in some high-performance scenarios. It should be noted that the Lua script of Redis has some restrictions on running in the Redis cluster version. Please refer to the relevant official documents for details.
- In addition to the community open-source version of Redis, Alibaba Cloud also provides the enterprise version of cache service, which has more powerful performance behavior.

2. Memcached

Memcached is a distributed cache system developed by Brad Fitzpatrick of livejournal, but it is now used by many websites. This is a set of open-source software, authorized by BSD license. Memcached lacks of authentication and security control, which means that the server should be placed behind the firewall. Memcached API uses 32-bit cyclic redundancy check (CRC-32) to calculate the key value, and then distributes the data on different machines. When the table is full, the new data will be replaced by LRU mechanism. Because Memcached is usually only used as a caching system, applications using Memcached need extra code to update the data in Memcached when they write back to slower systems (such as back-end databases).

3. Ehcache

Ehcache is a pure Java in-process caching framework, which is fast and lean and is the default CacheProvider in Hibernate. Ehcache is a widely used open-source Java distributed cache. It's mainly used for general cache, Java EE and lightweight containers. It has the characteristics of memory and disk storage, cache loader, cache extension, cache exception handler, a gzip cache Servlet filter, and it also provides support for REST, SOAP API, etc.

4. CDN

CDN is mainly used to cache data to the nearest location from the user, and generally cache static resource files (pages, scripts, pictures, videos, files, etc.). The domestic network is extremely complex, and the network access across operators will be very slow. In order to solve the problem of access latency cross operator, CDN applications can be deployed in important cities. Through it, the network congestion can be reduced and users can get the required content nearby, and response speed and hit rate can be greatly improved.

The basic principle of CDN is to widely use a variety of cache servers, and distribute these cache servers to the areas or networks where users visit relatively intensively. When users visit the website, the global load technology is used to distribute the user's access to the nearest normal cache server, and the cache server directly responds to the user's request.

Figure 4.122 shows the processing flow of HTTP request sent by users before deploying CDN application.

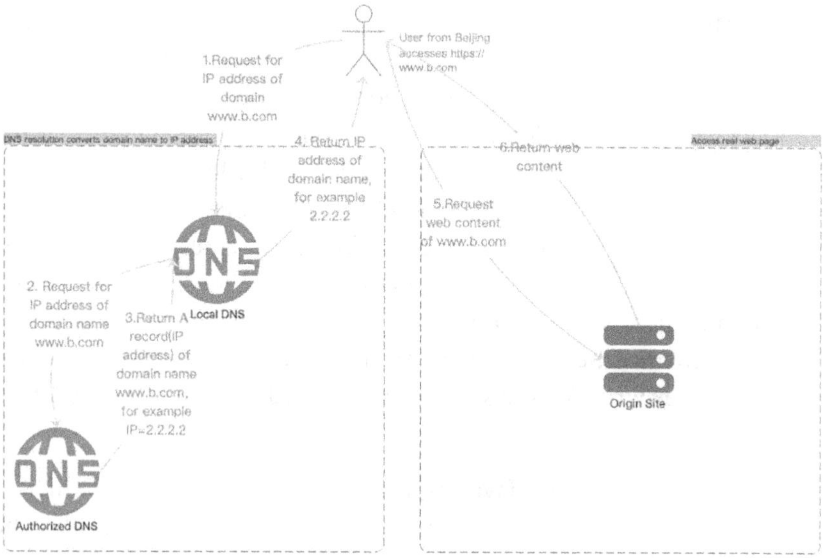

FIGURE 4.122 Application access link before deploying CDN.

- When end users from one location, for example, Beijing, initiate a request to resource under domain www.b.com, the browser would ask LDNS (local DNS) to resolve the domain name www.b.com.
- LDNS checks whether there is a local cache for IP address record of www.b.com. If yes, it will be returned to the browser directly; if not, LDNS will query from root DNS server to authorized DNS server recursively.
- When authorized DNS server returns the A record of the domain name www.b.com, that is, the IP address corresponding to the domain name, LDNS will return result to browser and then cache it locally. Assuming that the address returned to LDNS is 2.2.2.2.
- The browser obtains and resolves the IP address.
- The browser initiates the access request to the acquired IP address.
- The node returns the requested content of the resource.

Figure 4.123 shows the processing flow of HTTP requests sent by users after CDN deployment.

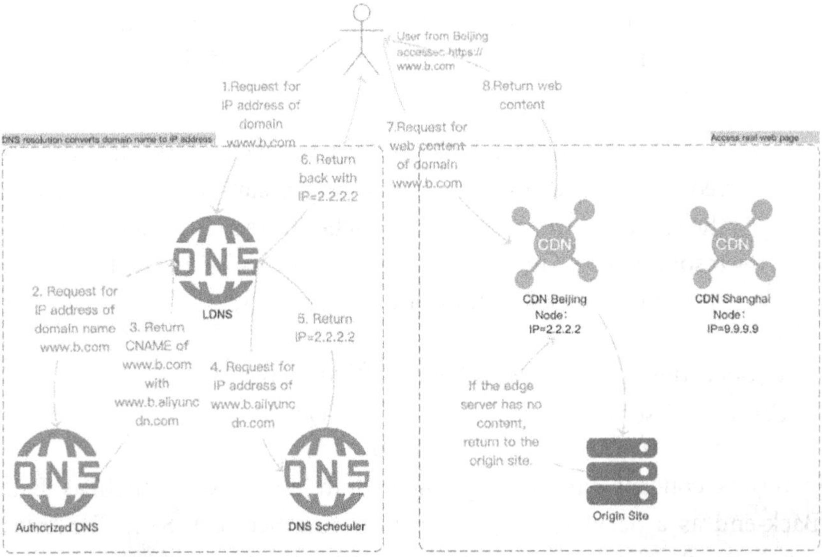

FIGURE 4.123 Application access link after deploying CDN.

- When end users from one location, for example, Beijing, initiate a request to resource under domain www.b.com, the browser would ask LDNS to resolve the domain name www.b.com.
- LDNS checks whether there exits local cache for IP address of domain name www.b.com. If yes, it will be returned to the browser directly; if not, it will be queried from the authorized DNS.
- When authorized DNS resolves the domain name www.b.com, it will return the CNAME record, www.b.aliyuncdn.com, of domain name www.b.com.
- The domain name, www.b.aliyuncdn.com, resolution request is sent to the Alibaba Cloud DNS scheduling system, and the best node IP address is returned to the request.
- LDNS gets the resolved IP address returned by DNS.
- The browser obtains the resolved IP address.
- The browser initiates the access request to the acquired IP address.
- If the CDN node corresponding to the IP address above has cached the resource, the data will be directly returned to the user, for example, steps 7 and 8 in the figure, and then complete the request.
- If the CDN node does not cache the resource, it will initiate a request for the resource to the origin station. After obtaining the resource, the user-defined cache strategy is used to cache the resource to the node, such as the Beijing node in the figure, and return it to the user. Now, the request ends.

4.6.8 Serverless in Business Middle Office Platform

4.6.8.1 Use Cases of Serverless Architecture in Business Middle Office Platform

Serverless concept can be implemented in many ways, such as BaaS (Back-end as a Service), FaaS (Function as a Service), SAE (Serverless App Engine), ECI (Elastic Container Instance), Serverless Kubernetes, Serverless Database and so on. When we discuss serverless architecture in Business Middle Office application architecture, we are usually talking about FaaS.

In the overall architecture of Business Middle Office system, serverless technology can be used in the front office, Middle Office and back-end systems. For example, use FaaS to implement a simple front-end application to provide service for App or H5 clients; use FaaS to build Middle Office center services; use FaaS to build a proxy service to communicate back-end products, third-party services or other cloud-native products.

Several typical use cases of serverless in Business Middle Office platform are listed as follows:

- To build light applications quickly

  For example, based on the capability of the Middle Office, you can use serverless technology to build web applications quickly, without complex server configurations and maintenance. The using of serverless can bring the strategy of 'large Middle Office and small front office' into full play.

- To build event-driven applications

  For example, after users upload photos and/or other qualification files to OSS, the serverless application will be automatically triggered to audit the pictures with artificial intelligence.

- To handle the high-visits scenarios

  For example, in the flash sale scenario, there are a large number of users swarming into the system, and the sudden increase of user traffic can be taken over through the serverless application.

- To run cron jobs

  For example, the message sending cron jobs can be implemented with serverless, and you do not need to occupy any other servers.

4.6.8.2 The Challenges of Serverless Architecture in Business Middle Office Platform

Serverless technology itself has brought great convenience for the development and operation, but it will face some difficulties and challenges when used in large-scale systems, especially in the Business Middle Office platform systems. Here are some difficulties and countermeasures of serverless architecture:

- Simplify serverless applications to avoid running timeout

  In order to avoid a single job taking up system resources for a long time, serverless products usually limit the running time for one job. The business logic needs to be split and lightweight to ensure that the serverless function is completed within the timeout period.

- Control concurrency to avoid current limiting

  Although serverless products have strong scalability, in order to avoid single user occupying too much resources at the same time, most manufacturers usually set the upper limit of concurrency for serverless functions. In order to avoid the failure of serverless applications due to exceeding the concurrency limit, users need to limit the concurrency or handle overlimit exceptions.

- Reserve resources or preheat to shorten cold start time

  Although the start-up and scaling of serverless is relatively fast, there is still a certain start-up time. For services with very sensitive delay requirements, it is necessary to preheat or reserve resources in advance to reduce the impact of cold start of serverless on services.

- Implement fault tolerance

  Using serverless technology in the production system, we need to consider timeout exceptions, current limit exceptions and other abnormal situations. Upstream services need to ensure the high availability of services, and implement fault-tolerant mechanisms such as failure retry, failure fast and circuit breaker.

- Overcome the difficulties of integration testing and debugging

  Serverless shields users from most of the basic operation and maintenance operations, while it's also usually difficult to diagnose and debug the serverless problems with the help of some external tools, it is still a better choice to print logs for debugging and diagnostic.

4.6.8.3 Alibaba Cloud FaaS Technology Practice

Alibaba Cloud FaaS is a typical serverless product. This chapter shows how to use Alibaba Cloud FaaS to build a serverless application, and how to implement auto build and auto deployment.

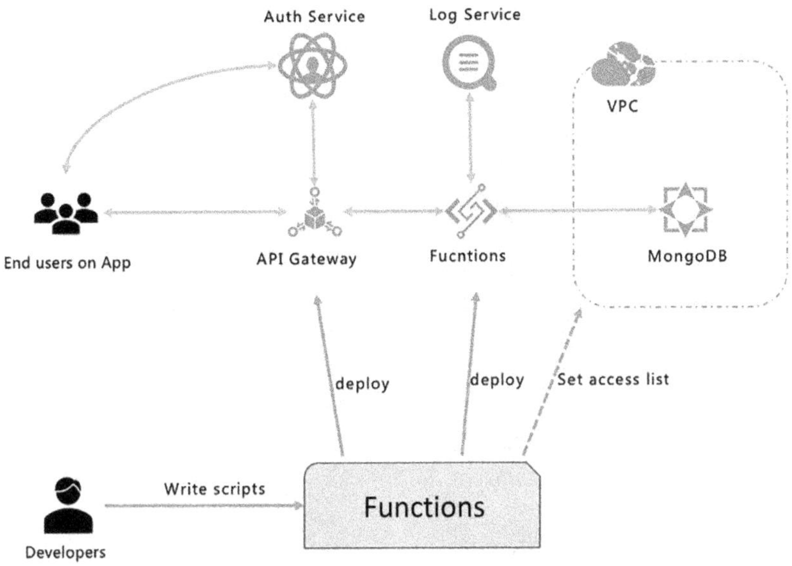

FIGURE 4.124 Funcraft architecture.

1. Objectives

   This case is to build a web application with nodejs. The application exposes restful interfaces through API gateway, and uses cloud-native MongoDB for data storage, then use Funcraft scripts to deploy and config service, function, API gateway, log service and authorizations.

   The system architecture is as follows (Figure 4.124):

2. Coding functions

   When creating a function, it's needed to specify the function entry from which function will start. The format of function entry is [file name]. [function name].

   With Node.js, for example, the handler specified when creating a function is index.handler, Then FaaS will load the handler function defined in index.js.

   The entry function is similar to the main() function in local development. The entry function needs to satisfy the FaaS programming model.

   The following example code shows how to handle an API gateway request and response:

```
module.exports.handler = function(event, context, callback) {
    var event = JSON.parse(event);
    var content = {
      path: event.path,
      method: event.method,
      headers: event.headers,
      queryParameters: event.queryParameters,
      pathParameters: event.pathParameters,
      body: event.body
    }

    // write your logic here

    var response = {
         isBase64Encoded: false,
         statusCode: '200',
         headers: {
           'x-custom-header': 'header value'
         },
         body: content
       };
    callback(null, response)
};
```

3. Coding Funcraft scripts

   Funcraft is a tool to support the deployment of serverless applications, which helps you to conveniently manage resources such as functions, API gateway and log service. Funcraft uses a resource configuration file template.yml to assist you in development, construction, deployment and other operations.

   The following is a sample code for configuring resources and deploying applications with Funcraft scripts:

```
ROSTemplateFormatVersion: '2015-09-01'
Transform: 'Aliyun::Serverless-2018-04-03'
Resources:
  demoService: #资源名称，根据需要命名
    Type: 'Aliyun::Serverless::Service'
    Properties: #属性设置
      Description: 'This is a faas demo ' #资源描述
      Policies:  #安全策略，系统自动根据策略为函数创建角色
```

```
        - AliyunOSSFullAccess
        - AliyunRAMFullAccess
        - AliyunLogFullAccess
        - AliyunApiGatewayFullAccess
        - AliyunFCFullAccess
        - AliyunMongoDBFullAccess
        - AliyunVPCFullAccess
        - AliyunECSNetworkInterfaceManagementAccess #特别说
          明，配置此策略才能创建弹性网卡进而打通函数以及VPC内连接
      VpcConfig: #允许函数访问的VPC配置
        VpcId: 'vpc-xxxxxx'
        VSwitchIds: ['vsw-xxxxxx']
        SecurityGroupId: 'sg-xxxxxx'
      LogConfig: #日志服务配置
        Project: sls-demo
        Logstore: logstore-demo
    demoFunction: #函数名称
      Type: 'Aliyun::Serverless::Function' #资源类型为服务，
      服务内可以挂载多个函数
      Properties:
        Handler: index.handler #事件处理入口
        Runtime: nodejs10   #程序运行环境
        CodeUri: './src'  #程序代码相对于当前脚本的路径
        EnvironmentVariables:   #环境变量设置，
          MONGO_URL: mongodb://userx:pwdx@dds-xxx.mongodb.
          rds.aliyuncs.com:3717/demoDb #Mongo内网连接地址
          RESULT_TABLE_NAME: demo_table
  demoGroup: # Api Group
    Type: 'Aliyun::Serverless::Api' #资源类型为API，每个API分
    组下可以挂载多个API接口
    Properties:
      StageName: RELEASE  #发布环境
      DefinitionBody:
        '/v1/recommendervera/[resultId]': # request path
          get: # http method
            x-aliyun-apigateway-api-name: demo_api # api
            name
            x-aliyun-apigateway-fc: # 当请求该 api 时，要触发
            的函数，
              arn: acs:fc:cn-shanghai:xxx:services/
              demoService.LATEST/functions/demoFunction
              timeout: 3000
            x-aliyun-apigateway-request-parameters: #设置参
            数类型
                - apiParameterName: 'resultId'
                  location: 'Path'  #传参方式，此处为在URI请求
                  路径中传参数
                  parameterType: 'String'
```

```
                 required: 'REQUIRED'  #设置为必选参数
        '/v1/recommendervera/': # request path
          post: # http method
            x-aliyun-apigateway-api-name: demo_api_post #
            api name
            x-aliyun-apigateway-fc: # 当请求该 api 时，要触发
            的函数，
              arn: acs:fc:cn-shanghai:xxxx:services/
              demoService.LATEST/functions/demoFunction
              timeout: 3000
            x-aliyun-apigateway-auth-type: APP  #设置鉴权类
            型，此处设置为简单的APP code类型鉴权
            x-aliyun-apigateway-app-code-auth-type: HEADER
            #鉴权加密方式，此处设置为通过Header传递授权后的app
ROSTemplateFormatVersion: '2015-09-01'
Transform: 'Aliyun::Serverless-2018-04-03'
Resources:
  demoService: # the resource name
    Type: 'Aliyun::Serverless::Service'
    Properties:
      Description: 'This is a faas demo '
      Policies:  #security policies
        - AliyunOSSFullAccess
        - AliyunRAMFullAccess
        - AliyunLogFullAccess
        - AliyunApiGatewayFullAccess
        - AliyunFCFullAccess
        - AliyunMongoDBFullAccess
        - AliyunVPCFullAccess
        - AliyunECSNetworkInterfaceManagementAccess
      VpcConfig: #allow function to access VPC network
        VpcId: 'vpc-xxxxxx'
        VSwitchIds: ['vsw-xxxxxx']
        SecurityGroupId: 'sg-xxxxxx'
      LogConfig: # log service configration
        Project: sls-demo
        Logstore: logstore-demo
    demoFunction: # the name of the function
      Type: 'Aliyun::Serverless::Function'
      Properties:
        Handler: index.handler # the function entry
        Runtime: nodejs10
        CodeUri: './src'
        EnvironmentVariables:
          MONGO_URL: mongodb://userx:pwdx@dds-xxx.mongodb.
          rds.aliyuncs.com:3717/demoDb #Mongo url
          configration
          RESULT_TABLE_NAME: demo_table
```

```
demoGroup: # Api Group
  Type: 'Aliyun::Serverless::Api' # api gateway
  Properties:
    StageName: RELEASE
    DefinitionBody:
      '/v1/recommendervera/[resultId]': # request path
        get: # http method
          x-aliyun-apigateway-api-name: demo_api # api
          name
          x-aliyun-apigateway-fc:
            arn: acs:fc:cn-shanghai:xxx:services/
            demoService.LATEST/functions/demoFunction
            timeout: 3000
          x-aliyun-apigateway-request-parameters:
              - apiParameterName: 'resultId'
                location: 'Path'
                parameterType: 'String'
                required: 'REQUIRED'
      '/v1/recommendervera/': # request path
        post: # http method
          x-aliyun-apigateway-api-name: demo_api_post #
          api name
          x-aliyun-apigateway-fc:
            arn: acs:fc:cn-shanghai:xxxx:services/
            demoService.LATEST/functions/demoFunction
            timeout: 3000
          x-aliyun-apigateway-auth-type: APP
          #authorization type
          x-aliyun-apigateway-app-code-auth-type: HEADER
```

4. Preparing cloud resources

   Services, functions and API gateways can be created in one stop with the Funcraft scripts above. External resources such as log service, MongoDB and authentication configuration need to be created or configured in advance.

   In particular, to access MongoDB, you need to specify the security group for FaaS into MongoDB whitelist. The diagram below shows how to config the security groups in MongoDB (Figure 4.125).

5. Building and deploying

   When all functions, scripts and related resource are ready, you can execute 'fun build' command to build and deploy FaaS applications.

   The command and the execution results:

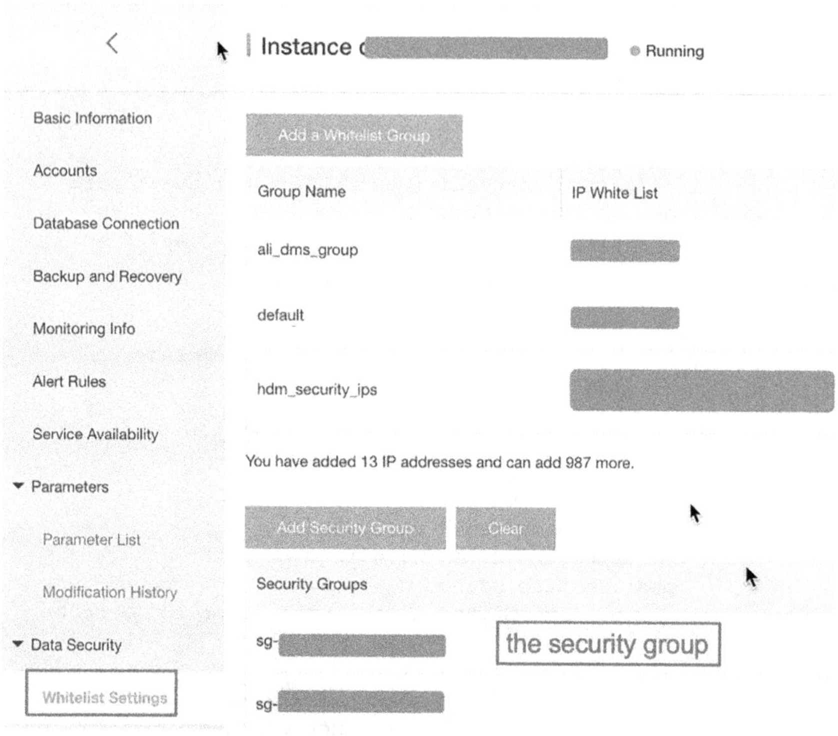

FIGURE 4.125 Alibaba Cloud MongoDB whitelist settings.

```
$ fun build
using template: template.yml
start building function dependencies without docker
building demoService/demoFunction
running task: flow NpmTaskFlow
running task: CopySource
running task: NpmInstall
Build Success
Built artifacts: .fun/build/artifacts
Built template: .fun/build/artifacts/template.yml
Tips for next step
======================
* Invoke Event Function: fun local invoke
* Invoke Http Function: fun local start
* Deploy Resources: fun deploy
```

After building completed, run "fun deploy" command to deploy the function and APIs onto cloud side:

```
$ fun deploy
using template: .fun/build/artifacts/template.yml
using region: cn-shanghai
using accountId: ***********3452
using accessKeyId: ***********1fap
using timeout: 60
Collecting your services information, in order to caculate
devlopment changes...
Resources Changes(Beta version! Only FC resources changes
will be displayed):
┌──────────────┬───────────────────────────┬────────┬────────
──┐
│ Resource     │ ResourceType         │ Action │ Property
│
├──────────────┼───────────────────────────┼────────┼────────
──┤
│              │                         │Description        │
│              │            │            ├───────────────────┤
│              │            │            │ Policies      │
│ demoService │ Aliyun::Serverless::Service │ Add ├──────┤
│              │                 │       │ VpcConfig     │
│              │                 │       ├───────────────┤
│              │            │            │ LogConfig          │
├──────────────┼───────────────────────────┼────────┼────────
┤
│              │                 │       │ Handler       │
│              │                 │       ├───────────────┤
│              │                 │       │ Runtime       │
│ demoFunction │ Aliyun::Serverless::Function │ Add ├──────┤
│              │                 │       │ CodeUri       │
│              │          │      │       ├──────────────────────┤
│              │          │      │  EnvironmentVariables  │
└──────────────┴───────────────────────────┴────────┴────────
──┘
? Please confirm to continue. Yes
Waiting for service demoService to be deployed...
    make sure role 'aliyunfcgeneratedrole-cn-shanghai-
    demoService' is exist
    role 'aliyunfcgeneratedrole-cn-shanghai-demoService'
    is already exist
    attaching policies ["AliyunOSSFullAccess","AliyunRAMFu
    llAccess","AliyunLogFullAccess","AliyunApiGatewayFullA
    ccess","AliyunFCFullAccess","AliyunMongoDBFullAccess",
    "AliyunVPCFullAccess","AliyunECSNetworkInterfaceManage
    mentAccess"] to role:
    aliyunfcgeneratedrole-cn-shanghai-demoService
```

```
    attached policies ["AliyunOSSFullAccess","AliyunRAMFul
    lAccess","AliyunLogFullAccess","AliyunApiGatewayFullAc
    cess","AliyunFCFullAccess","AliyunMongoDBFullAccess","
    AliyunVPCFullAccess","AliyunECSNetworkInterfaceManagem
    entAccess"] to role:
    aliyunfcgeneratedrole-cn-shanghai-demoService
    attaching police
'AliyunECSNetworkInterfaceManagementAccess' to role:
aliyunfcgeneratedrole-cn-shanghai-demoService
    attached police
'AliyunECSNetworkInterfaceManagementAccess' to role:
aliyunfcgeneratedrole-cn-shanghai-demoService
    Waiting for function demoFunction to be deployed...
        Waiting for packaging function demoFunction code...
        The function demoFunction has been packaged. A total of
        1675 files were compressed and the final size was 2.1 MB
    function demoFunction deploy success
service demoService deploy success
Waiting for api gateway demoGroup to be deployed...
    URL: GET http://xxx-cn-shanghai.alicloudapi.com/v1/
    recommender/[resultId]
      stage: RELEASE, deployed, version: 20200715144450426
      stage: PRE, undeployed
      stage: TEST, undeployed
    URL: POST http://xxx-cn-shanghai.alicloudapi.com/v1/
    recommender/
      stage: RELEASE, deployed, version: 20200715144453967
      stage: PRE, undeployed
      stage: TEST, undeployed
api gateway demoGroup deploy success
```

After the script executed, you can see that the relevant functions and APIs have been deployed in the cloud. Next, you can add the above Funcraft commands and scripts into Jenkins pipelines or other CICD tools (Figures 4.126 and 4.127).

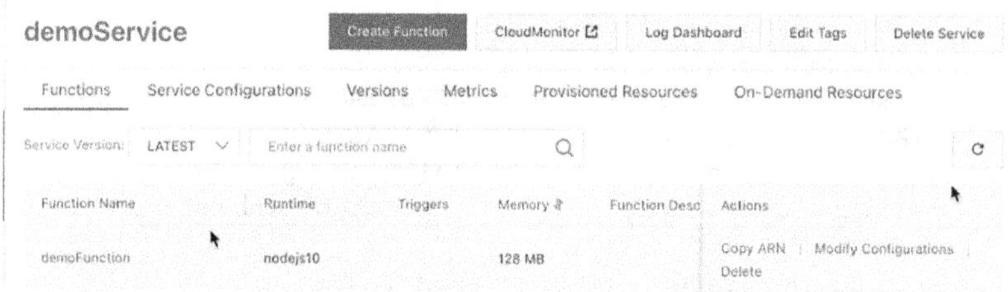

FIGURE 4.126 The service and function created by Funcraft.

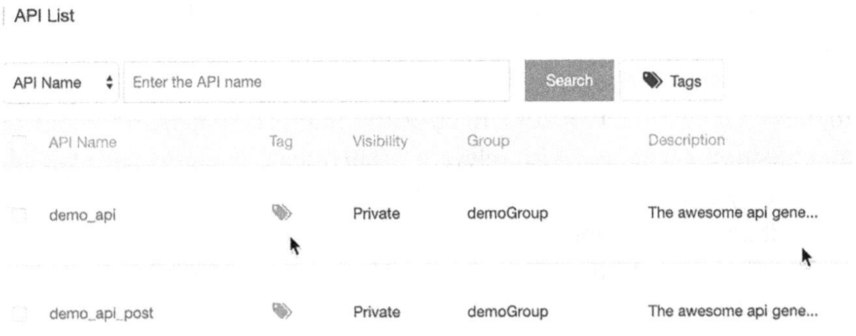

FIGURE 4.127 The APIs created by Funcraft.

4.7 ALIBABA BUSINESS MIDDLE OFFICE PROJECT CASES

4.7.1 Case of Direct Selling Industry

4.7.1.1 Project Background

One of the top companies in direct selling industry, with business in more than 80 countries and regions. There are more than 450 kinds of products produced, including nutrition and health food, beauty cosmetics, personal care products, home care products and household durable goods.

According to the IT construction target proposed in the latest medium and long-term plan of customers, we must change the traditional infrastructure construction idea with IDC as the core to adopt better flexibility, more advanced design concept and flexible resource scheduling, meet the flexible requirements of core e-commerce applications for computing, storage and network resources at different times and locations. The customer also mentioned the current IT O & M and monitoring requirements, as well as a reasonable IT input–output ratio. From the beginning of 2019, experts from Alibaba Cloud migration department were invited to attend the senior leadership meeting of the enterprise, Alibaba Cloud best practices based on digital transformation, and shared wonderful cases of cloud computing and big data with senior customers, based on the experience of optimizing the allocation of enterprise resources and accelerating innovation brought to Alibaba by the construction of Business Middle Office, it is recommended that customers immediately start the transformation strategy of IT systems based on Internet thinking, cloud computing platform and the concept of Middle Office. The management of the enterprise immediately carried out the strategic layout of IT transformation within the enterprise, and proposed a three-step IT transformation approach based on cloud computing and Middle Office platform.

The first step is to achieve the goal of fully migrating existing IT systems to the cloud by the end of 2019. In 2019, the enterprise had many technical and business exchanges with relevant departments of the Alibaba Cloud, and finally decided to carry the core IT system of the enterprise based on the Alibaba Cloud public cloud platform to solve the current poor scalability of the e-commerce system, low flexibility, low stability and slow response to new services. After 1 year's construction, the customer's core IT system has been launched throughout the year of 2019. Figure 4.128 shows the deployment architecture of the customer's legacy e-commerce platform (based on the traditional Hyacinth platform) after it is deployed on the cloud, by deploying multiple e-commerce cluster environments on the cloud, the pain points of low flexibility and scalability in IDC deployment are alleviated. In addition, when customers are on the cloud, AHAS is used to limit traffic and downgrade traffic, in previous sales promotion scenarios, the system avalanche was caused by sudden traffic. ApsaraDB for Redis was used to solve the problems of high O & M costs, poor scalability and stability of self-built database clusters. Because of these immediate results, the customer immediately made a high-level decision to plan the second phase after initially trying the cloud platform, not only to rebuild the core e-commerce platform of the enterprise based on Alibaba's shared service concept, it is called the new e-commerce platform within the enterprise and puts forward the target requirement of going online before the Double 11 at the end of 2020.

The second step is to rebuild the core e-commerce platform of the enterprise based on the concept of Alibaba shared service center, that is, the concept of Middle Office. The new e-commerce platform has three core goals to solve: First, the new e-commerce platform must be built based on a shared service center to meet the requirements of enterprises for IT flexibility and business innovation; second, introduce modern O & M and monitoring capabilities; finally, based on the concept of all business data, combined with DDD domain analysis methodology, it lays a solid foundation for Data Middle Office and subsequent double Middle Office strategies, ensure the success of the strategy of 'all business data and all data business' from the source. The customer's new e-commerce platform strictly follows the construction idea of Business Middle Office. The Alibaba Business Middle Office construction team starts to intervene from the demand analysis stage and guides the customer and supplier's business teams based on DDD methodology, the customer's business domains are subdivided step by step. On the one hand, the scope of problems that need to be solved by each final service is gradually reduced. On the other hand, the complexity of business understanding and system implementation is reduced. Event storm is the core methodology for building domain models. Through the analysis of customer business processes, user journeys, different purchase scenarios and different customer-level use cases, it establishes domain models and divides domain boundaries, builds the boundary context, finally determines the concerns of each object in the field, and finally designs seven centers and 200 different capabilities. The new e-commerce platform includes inventory center, commodity center, marketing center, User Center, transaction center, login center and payment center. Each center implements the capabilities of the center through one or more microservices (aggregation and aggregation root based on domain analysis). The atomic capabilities provided by the Service center are exposed to the front end through the aggregation layer. Before decoupling, the atomic capabilities are directly dependent on the front end, and the service capability level is reused externally. After a whole year of construction, the new e-commerce platform has supported the verification of 618 sales promotion scenarios, such as sales promotion in the golden autumn, to prepare for the upcoming double 11 sales promotion.

Data-driven Middle Office platform operations and maintenance. After completing the platform construction, customers need to further solve the operational efficiency, which has two connotations. First of all, many operations are still done through scripts. An efficient operation platform is needed to solve business-side problems such as promotion configuration, product shelving, business monitoring, business alerts, commodity prices, shelves and other operational requirements; second, how to form a closed loop, form business insights and support business decisions for the data accumulated in the operation process of the Business Middle Office, it is necessary to combine the data of Business Middle Office with the planning of Data Middle Office to form the effect that dual Middle Office their respective duties and complement each other. In the middle of the construction of the new e-commerce platform, the Alibaba Business Middle Office construction team began to contact with customers, planning the operation platform and the Middle

Office strategy, experts from Data Middle Office and customers were invited to hold several co-creation meetings to understand customers' real demands and complete the planning of the operation platform as well as the evolution strategy and planning of Data Middle Office and Business Middle Office.

In the process of communicating with customers, employees from the Alibaba Business Middle Office team introduced in detail the prototype design and technology architecture of building a new customer e-commerce platform based on the shared service center and the implementation process. Customers are very interested in the design scheme based on multiple service centers and centered on the construction of Middle Office. Customers are looking forward to better ECG providers based on the construction of Business Middle Office, quickly respond to the actual needs of enterprise business construction and changes. The new e-commerce platform built based on Business Middle Office not only improves the reusability of enterprise core business capabilities, but also the diversification of front-end business, which reversely nourishes the accumulation of business capabilities on the new e-commerce platform and forms a closed loop, better supports the development and innovation of customers' business in China, and provides a very important model for the next IT construction of customers' global business. The leaders of the customer information department have highly recognized this construction mode and structure, and have begun to plan and talk about IT best practices in greater China after showing immediate results in greater China, introduce global IT strategic planning to better support the global business of enterprises. Through the construction of the new e-commerce platform, the customer information department has gradually transformed from the business-supporting intelligence such as project management and system operation and maintenance to the accumulation and reuse of the core business capabilities of the enterprise, it has promoted the intelligent transformation of the information department and provided more opportunities for students in the information department to guide the digital transformation of enterprises.

4.7.1.2 Project Implementation

The project was officially launched in early January 2020, just before the novel coronavirus broke out, Alibaba's Business Middle Office construction team began in early January, with the current situation that customer business is almost understood as 0, together with customers, after more than 20 days of business research and analysis, we talked about the business carried by the legacy system that customers have been running for more than 10 years, conducted a process analysis, and cooperated with business analysts, business details are precipitated in the form of requirement specification. In the process of business research, the employees of Alibaba Cloud of the Middle Office construction team of GTS business conducted a technical research on the current IT situation of customers in combination with the best practices of Middle Office construction, the technology architecture, integration mode and infrastructure environment of the customer's existing e-commerce system are clarified. The entire technical research is arranged through the following tasks:

- Understand the current business situation of customers and the technology architecture solutions to support the existing business.
- Complete the technology architecture Panorama and output the technical specifications.
- Launch process and system monitoring and O & M System.
- Comprehend the existing technical challenges and problems of customers and understand the long-term technical strategic planning.
- External systems that depend on, including system functions, deployment architecture, integration methods and functional metrics.

Having a clear understanding of the overall IT environment of customers, forming a technical research report, avoiding technical risks and solving existing technical pain points for the design of the new e-commerce platform, and understanding the role of customer technical constraints. After completing the technical research, the employees of Alibaba Cloud of the Middle Office construction team of GTS business started the outline design of the customer's new e-commerce platform based on the requirements specification and the results of the technical research, summary design needs to complete the model design, capability list design and key technical scheme design of each center. The customer team, supplier team and Ali team have been at a pace of 997 since the first day.

The team started working from 9:00 a.m. to 12:00 p.m. every day, and went to work as usual on Saturdays and Sundays. The whole project adopted an agile construction mode with two iterations on 2 weeks. While carrying out the outline design, the construction work was almost started simultaneously. The whole project team was in a high-intensity working mode from the very beginning. Although it was very hard, we could hardly see the negative attitude caused by the endless overtime work to catch up with the progress. This is probably because both business personnel and technical personnel know the importance of this project well and cherish the opportunity to participate in such a large-scale digital transformation project based on the construction idea of Business Middle Office, closely observe and participate in the design and construction of Middle Office. Because I was responsible for the whole project, I almost participated in the whole process from requirement analysis to design, to construction to launch, and O & M. If you look at the customer-side investment from the perspective of an Alibaba employee, you will feel very surprised. The previous cognition said that non-Internet companies cannot accept the working style and intensity of Internet companies; however, during the whole process, all the clients involved in the project were very responsible and almost consistent with Alibaba's work intensity and market, most of the time, they may work harder than Alibaba employees. They need to report the progress to their bosses internally, control the progress and risks of the project as a whole and see the risks and dependencies that suppliers cannot see. Judging from the actual experience of this project, only under the guarantee of good mechanism and process can everyone have the same execution and combat effectiveness as Internet companies. Having a clear understanding of the overall IT environment of customers, forming a technical requirement report, avoiding technical risks and solving existing technical pain points for the design of the new e-commerce platform, and understand the role of customer technical constraints. After completing the technical research, the employees Alibaba Cloud the Middle Office construction team of GTS started the outline design of the customer's new e-commerce platform based on the requirements specification and the results of the technical requirements, high-level design needs to complete the model design, capability list design and key technical scheme design of each center. The customer team, supplier team and Ali team have been at a pace of 997(9 am-9 pm with 7 days' work) since the first day.

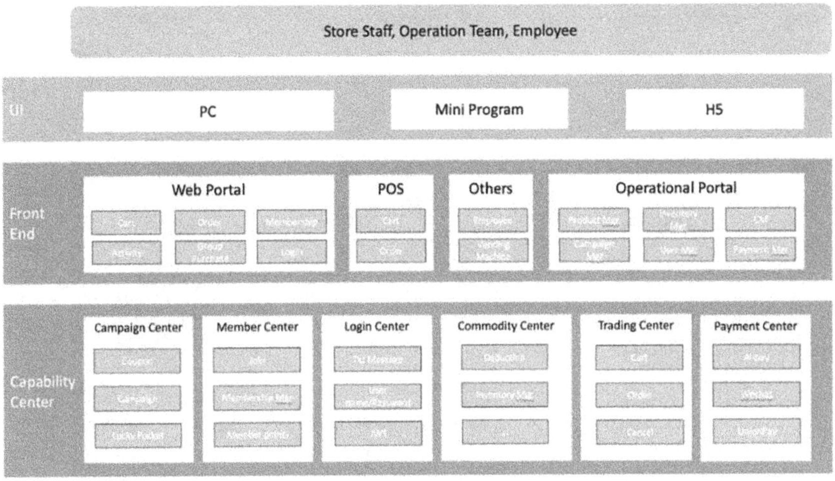

FIGURE 4.128 Business architecture of new e-commerce platform.

Below is Business Middle Office architect diagram.

After 4 months of system design and construction, based on the actual business requirements of the case and the concept of shared services, the Alibaba Business Middle Office team and the customer's technical team have jointly built seven shared service centers to support the customer's new e-commerce platform. Taking cloud-native support into account, combined with the Alibaba middleware team, complete the output of a series of middleware products and capabilities on Alibaba's public cloud platform, including but not limited to microservice gateway CSB, EDAS, managed container service, database service, MQ, monitoring, throttling and degradation, log service and alerts.

Figure 4.129 shows the overall deployment architecture of the customer's new e-commerce platform.

In the process of construction, technical solutions such as distributed transactions, final data consistency, how to prevent avalanche and hotspot keys, Database sharding and table sharding are implemented for the new e-commerce platform. In the end, the project meets the requirements of colleagues on the online timeline and helps enterprises build a new e-commerce platform based on the concept of shared service center, laying a solid foundation for the business development and innovation of enterprises in the next 10 years.

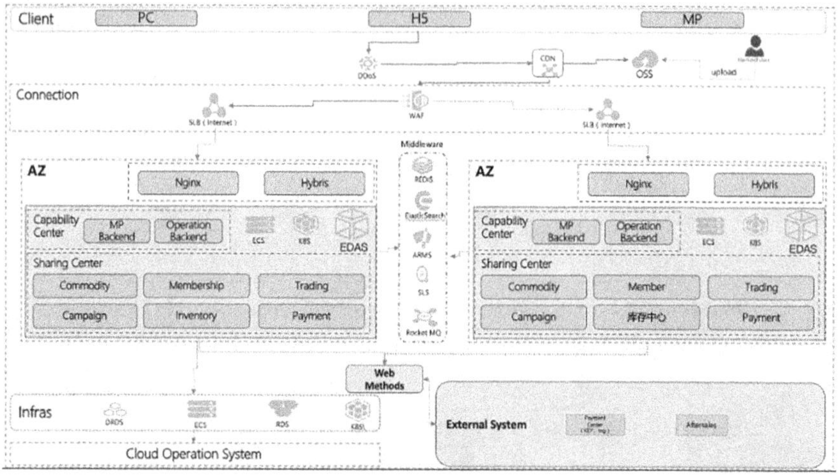

FIGURE 4.129 Entire deployment architecture of new e-commerce platform.

4.7.1.3 Benefits

The customer's new e-commerce platform has been gradually launched since May 22, 2020. The first capabilities to be launched include inventory center, marketing center and commodity center. Customers integrate the capabilities provided by the new e-commerce platform into the entire digital system by gradually switching business. The inventory center's first order locking and deduction started at the end of May. Although this small step is insignificant, it is a big step for customers, it is of special significance to the whole ECG platform based on Middle Office and digital transformation. Then the system gradually switched the core competence of the marketing center, the ability to create and use coupons to the new e-commerce platform. Until the author wrote this chapter, the inventory center and marketing center have successfully supported more than 20 promotional activities, including sales promotion at the end of 618 and sales promotion in the golden autumn and the capability center built based on Business Middle Office and microservices, again and again, they have verified the value they have contributed to customers: stability, flexibility and scalability, operation and maintenance convenience, performance, security, etc. The author is also thinking about a question, why do customers give such a high priority to the project and give higher satisfaction after the project reaches a milestone? Judging from the author's many conversations with the customer-side leaders, this project is not only the first attempt of enterprise digital transformation, but also the platform on which the enterprise will survive for many years in the future, it is also hoped that based on the idea of building Alibaba Cloud and shared service centers, enterprises can be reused in other regions of the world. Therefore, the harvest of this project will bring huge benefits to enterprises for a long time:

- The ECG platform based on Business Middle Office has helped enterprises build a new architecture of thick Middle Office, laying a solid foundation for enterprises to embrace digitalization in the next few years.
- The new e-commerce platform is based on Alibaba Cloud public cloud platform and Alibaba's middleware system that has been built for decades. It provides customers with a flexible, stable, secure and scalable system, technical support system for sustainable development.
- Through the project, we have trained a group of technical teams who understand business, technology, management, operation and maintenance and cloud computing, it provides a good talent reserve and training for the digital transformation of enterprises in the next few years.
- The exploration and practice of the agile transformation of customers as a whole. Through the new e-commerce platform, customers completely adopt the agile construction mode, which breaks the boundaries between organizations and improves the efficiency of organizations, it provides more reference cases for deepening organizational reform in the future.

4.7.2 Case of Fast Moving Consumer Goods Industry

4.7.2.1 Project Background

A company in FMCG industry has established a long-term cooperative partnership with a large international brand company by virtue of its strong strength. In the first 10 years of 2000, offline branches were established successively, and the market coverage area and sales volume increased continuously. With the vigorous development of e-commerce business in China, the company has opened a franchise store in Taobao

Mall, which has won market share in a few years of development and laid a dominant position in the industry, it realizes the effective combination of traditional industries and Internet e-commerce, and promotes the innovation of enterprise development model. With the continuous introduction of more new brands and the opening of flagship stores, business categories are more abundant and store operation management is more professional.

In recent years, relying on independently developed OMS and other systems, the company has realized the visualization and automatic processing of orders, supporting the rapid development of business in the past decade. With the changes in the market environment, the development and transformation of business require higher agility for the system. The traditional isolated system architecture has been difficult to support the sustainable development of its business. From the perspective of digital transformation, the company has made deep thinking about the overall business architecture in the future, and has made overall planning based on the current system status, drawing on the most effective Business Middle Office concept in the field of new retail and enterprise digitalization, the company started the design of Business Middle Office architecture.

4.7.2.2 Project Implementation

The whole project started at the end of 2019. In the early stage of the implementation of the project, it faced several challenges. The first is that the construction period of the project is very tough, and it is necessary to ensure that more than a dozen online stores of customers can smoothly migrate to the new system. Since the e-commerce industry has several big promotions every year, such as 618, 99, Double 11 and Double 12, the release of a major version must reasonably allocate the interval between these major promotion points to prevent the impact of changes on the major promotion. It is a great challenge to clarify the needs, organize limited personnel and arrange the development plan reasonably within a limited time without affecting customers' online business. The second is that the performance metric requirements of the project are very high, and the customer requires that the system can support the business volume of the Double 11 promotion at more than 10 million orders per day, in particular, the first hour of business accounts for about 50% of the total business volume, which is the peak period of business. The stability of the system must be able to withstand the test. At the same time, the business requires

5 million orders to be sent to the external WMS every hour through end to end. The third challenge is the large number of promotional activities and complex rules during the big promotion.

In order to meet the above challenges, the project team has taken the following measures.

- The project team has defined the core team around project managers, business architects, technical managers and quality assurance experts. The project manager formulates the overall construction milestone of the project, the progress of the project and the connection between the customer side. The Business Architect is responsible for the output and control of the business blueprint and business requirements specifications, the technical manager is responsible for the design and quality of the overall system technology architecture, the management of the R & D process and the overall technical responsibility. Quality Assurance experts are mainly responsible for system testing and quality control. The technical manager and the project manager build the organizational formation of the technical side, divide multiple research and development teams in the dimension of the service center, and each research and development team have a technical leader, responsible for the assignment, detailed design, core code writing and code review of research and development tasks within the team. The application architect cooperates with technical manager to complete the detailed design and implementation of application architecture. The delivery life cycle of the overall project covers business requirement and technical requirement research, requirement analysis, Business Middle Office architecture design, milestone construction, project acceptance and delivery transfer. Throughout the milestone construction phase, the overall research and development process is accelerated through less documentation, close communication and quick feedback in an agile iterative manner.
- The architecture design fully considers the business volume that needs to be supported to ensure that the division of service centers follows the principle of high cohesion and low coupling. The core service centers can be independently and horizontally scalable, this means that the services of the service center and the underlying data storage must be horizontally scalable.

- To ensure that the performance of the system that meets the required standards, continuous performance stress testing and optimization will begin after the first phase of deployment is launched, so as to cope with the following big promotion. With the help of the stress testing tool provided by Alibaba's Taobao JuShiTa platform, the order volume and order model of the Double 11 promotion are simulated in the stress testing environment, and then through Alibaba Cloud ARMS, AHAS, cloud monitoring and self-developed business monitoring tools observe the performance metrics of the Business Middle Office system under high load. By finding performance bottlenecks and continuously optimizing the system performance, the system performance metrics are finally reached. In addition, considering the inconsistency between the stress testing environment and the production environment, in order to ensure the stability and performance of the online system, the stress testing of the production environment is also implemented and the expected effect is achieved.
- In order to ensure the complex promotion rules that the system can support, on the one hand, through analyzing the promotion activities of previous years and configuring them during the stress test, the stress test similar to the real scene is completed, ensure that the system still maintains high performance in the promotion scenario. On the other hand, the business monitoring tool of self-developed promotion is developed to monitor the accuracy of promotion calculation. If any inaccuracy is found, it will be reported in time, and corresponding emergency plans will be provided to ensure timely correction (Figure 4.130).

1. Application Architecture

   Taking business blueprint planning and requirements specification as input, combined with Alibaba's best practices in the e-commerce field and domain-driven analysis technology, the entire Business Middle Office built by the project is divided into nine centers, the nine service centers are order receiving center, order center, commodity center, inventory center, Promotion Center, user Center, basic data center and interface adaptation center, together with the upstream Alibaba Tmall platform and downstream ERP system, CaiNiao WMS system, Alibaba intelligent after-sales service system AG/FBT, they support the core e-commerce business of customers (Figure 4.131).

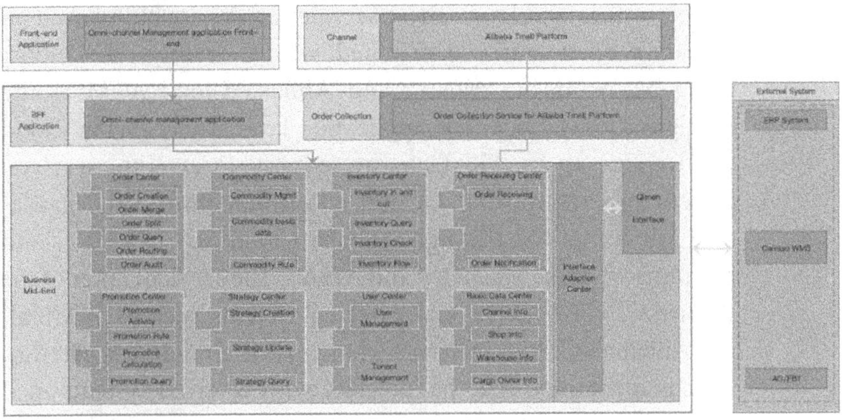

FIGURE 4.130 Application architecture.

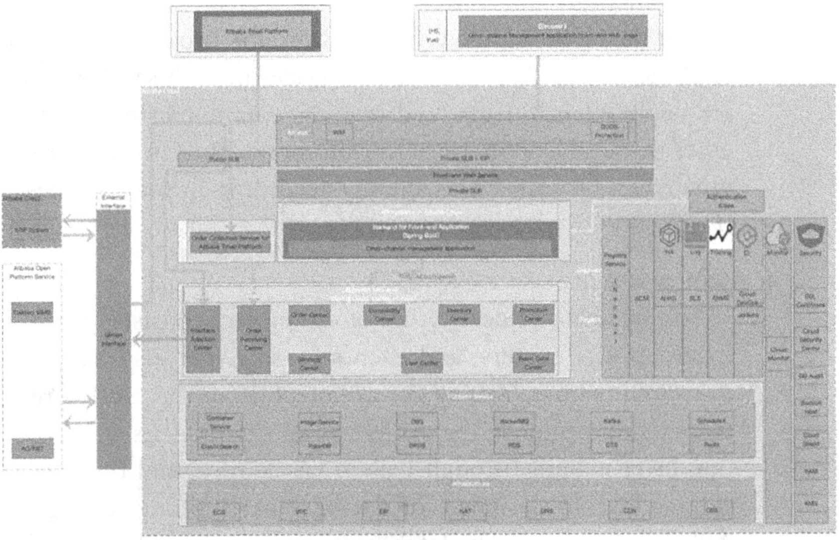

FIGURE 4.131 Technology architecture.

2. Technology Architecture

The Business Middle Office built by the project depends on Alibaba Cloud's IaaS and PaaS from the perspective of layered architecture style, covering a wide range of products and services. IaaS includes ECS, VPC, NAT and OSS. PaaS includes container services, DRDS, and RDS, RocketMQ, Redis, Kafka and ElasticSearch. Service centers of Business Middle Office are on top of IaaS and PaaS. These service centers are built based on Alibaba's open-source distributed service framework Dubbo, using Nacos as the registry and Alibaba Cloud ACM as the configuration service. The application layer uses Alibaba Cloud Service Bus as an API Gateway to provide external services. System logs, monitoring, CI, and system rate limit and degradation are completed through Alibaba Cloud SLS, ARMS, cloud monitoring, cloud efficiency and AHAS. The service center of the Business Middle Office and external systems such as WMS all use Alibaba's QiMen interface whose role is similar to a gateway, which can standardize the interface with external systems such as WMS, reduce the overhead when connecting multiple WMS systems.

3. Deployment Architecture

   The Business Middle Office built by the project is deployed on ACK cluster. Applications or services in each service center are deployed in images, and services communicate with each other through Dubbo RPC. The number of pods in each service center can be scaled horizontally according to business needs (Figure 4.132).

4. Architecture of Service Center

   The Business Middle Office of the project consists of different service centers, and each service center also has its own architecture. Taking the architecture of the most core transaction center in the project as an example, its application layer is built through Dubbo framework, and the data storage layer is designed for database partitioning through Alibaba Cloud's DRDS database, ensuring the scalability of the system. At the same time, in order to support complex queries in the order center, such as fuzzy search, read/write separation technology is adopted. First, DTS data subscription function is used to subscribe to data changes of orders, then write the order data to ElasticSearch; finally, complex query or fuzzy query will first find the database key of the order in ElasticSearch, and then use the database key of the order to query in DRDS, this improves the query efficiency (Figure 4.133).

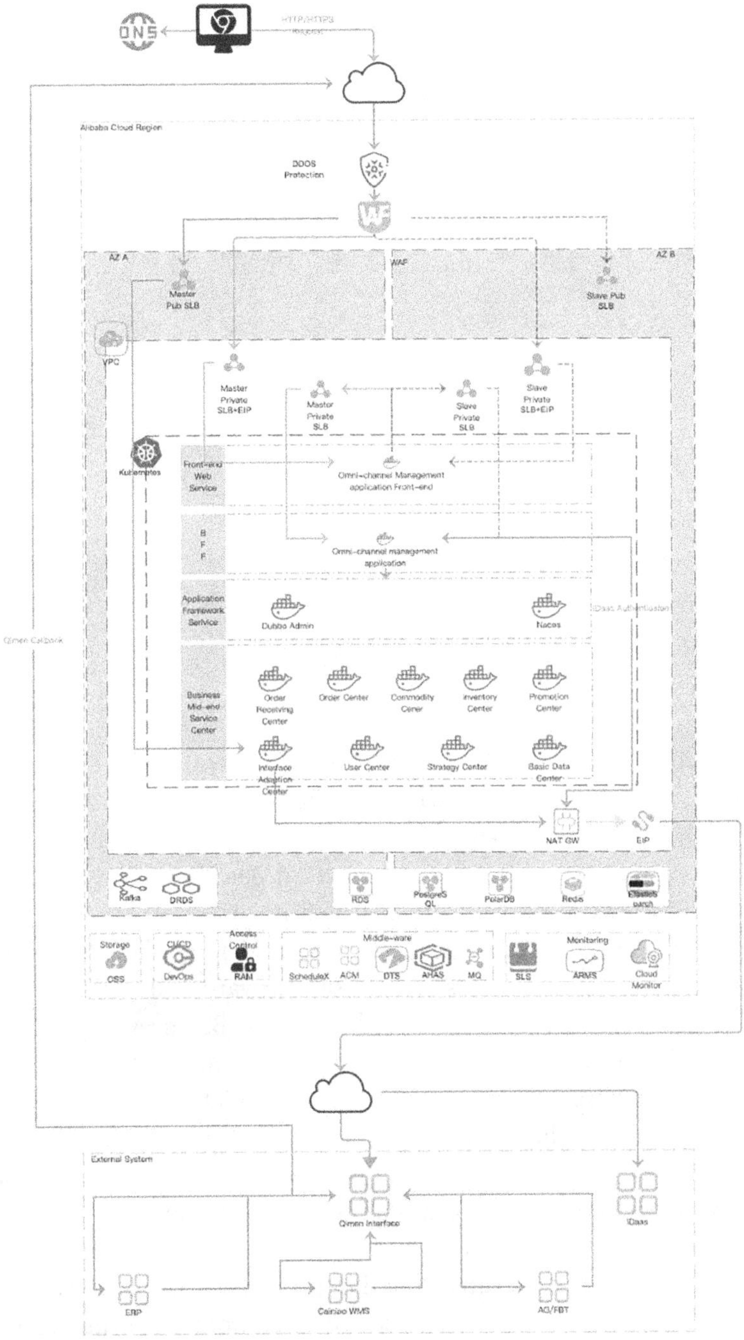

FIGURE 4.132 Deployment architecture.

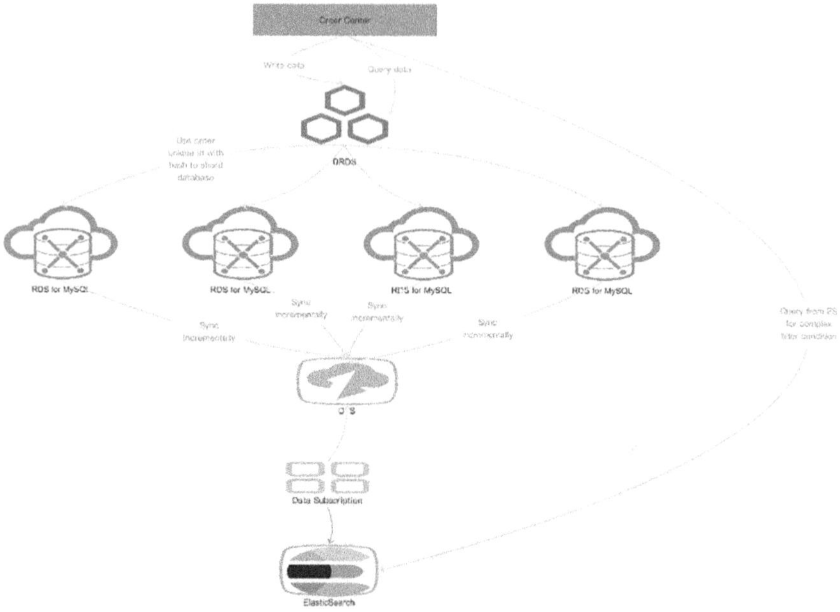

FIGURE 4.133 Architecture of service centers.

4.7.2.3 Customer Benefits

Compared with the original system, the project team has reconstructed the customized business platform from the business side and technology side.

From the business side, the project team has completed the deployment of Business Middle Office, migrated its online stores to smoothly access the online Business Middle Office, and integrated the product, marketing and inventory to support both forward and reverse transaction process, and unified the core data to the Business Middle Office platform for visual management. The Business Middle Office can fully support customers' real-time information circulation in multi-channel such as Alibaba Taobao or other platforms, supermarket/flagship store/live broadcast, procurement/finance/warehousing/logistics, and improve the refined operation control efficiency of business side. Among them, for the scenarios of long delivery time of orders, missing or wrong delivery of gifts and customer return and refund in the large-scale promotion activity, the new online Business Middle Office platform also fully supports business innovation, such as online pre-sales hot products sinking, self-service address modification, automatic cancel and refund, automatic interception, automatic review of order, multi-type promotion activity configuration and simulation verification. In this year's Double 11, without additional manpower, the promotion accuracy rate has increased from 95% to 99.95%, and the return rate has decreased by 4%, greatly reducing the customer's order fulfillment cost, and bringing a smooth shopping experience to the customers.

From a technical perspective, based on microservices and Alibaba Cloud middleware products, the project team also helps customers build an Internet-oriented architecture that meets high concurrency and high-performance requirements under the requirements of massive orders and flexible configuration. The shared service center is designed and implemented based on driving by business domain. Products, promotions, orders and inventory are independent of each other to implement quality attributes of architecture such as high availability and high scalability. At the same time, it considers the demands of refined operation and maintenance, supports flexible expansion and flexibly configures operation and maintenance costs.

This system reconstruction is also a technology upgrade for customers. Through the shared capability center, customers can quickly realize IT capability reuse and high business response brought by reuse.

In this year's Double 11, under the sophisticated planning and scheduling of the Alibaba Cloud escort team, the system was launched for the first time to withstand the traffic peak. The daily load of orders was tens of millions, and the peak business peak reached 24,000 TPS. The performance is comparable to medium-sized e-commerce platforms.

4.7.3 Case of Catering Industry

4.7.3.1 Project Background

A famous fast food brand enjoys a worldwide reputation for specialty food such as fried chicken. It has many popular Internet celebrity restaurants in many countries and regions around the world. The company has made plans to open many stores in many regions of the country in the next 10 years. In order to achieve this grand goal, the company's senior management has made sufficient research and preparation in the early stage. The CEO of the company in China pointed out that the catering environment has changed dramatically in the past 20 years, while the informatization level of the fast food industry is still relatively low, with business fragmentation and system isolation, resulting in larger scale and higher cost, the lower the collaboration efficiency. In order to solve these problems, the company chose to cooperate with Alibaba Cloud, introduce Middle Office technology and make use of resources, systems, service capabilities, technologies, experience, etc. from Alibaba Cloud's best practices, conduct comprehensive cooperation in the online and digital capabilities of business elements such as sales, marketing, channels, services, logistics supply chain, organization and information technology, so as to provide the company with innovative development, brand upgrading, process reengineering and marketing upgrading, as well as provides strong support to realize digital operation of the whole business.

4.7.3.2 Project Implementation

1. New catering business blueprint planning (Figure 4.134)

   The project implements the following main functional modules:

   - Shared service center: Based on the Alibaba Cloud Internet middleware, a shared service center is built for Business Middle Office, which mainly includes the product center, order center, marketing center, payment center, channel center and membership Center. In this phase and in the future, all applications to be provided on the Middle Office service are based on the Middle Office service.

   - Function center: Based on the Business Middle Office service center, Bastion host provides basic service capabilities, personalized service capabilities and third-party adaptation capabilities.

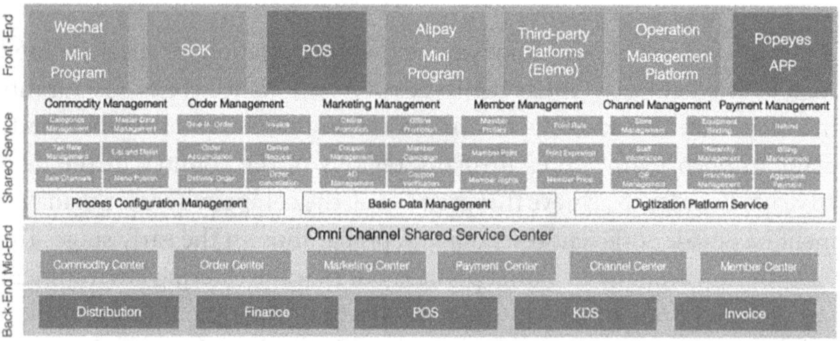

FIGURE 4.134 Blueprint for new catering business.

- Ordering mini-program and APP: Based on the Business Middle Office service center, the store ordering business is realized, including functions such as user registration, order placement, evaluation and points, so as to enrich application scenarios and improve user experience.

2. New catering Business Middle Office goals

- Build a new catering platform

  By integrating online and offline businesses, multi-terminal service integration and omnichannel scenario innovation are realized. A commodity inventory sharing and settlement system are built to enable multi-scenario and omnichannel operations.

- Build a membership marketing system

  To accumulate online and offline member data assets, realize member registration, member binding, point accumulation, member interaction, benefit exchange and other workflow, and provide data analysis, precision marketing and hierarchical operation capabilities for brand members.

- Accelerate the innovation of IT architecture and the establishment of a shared service system

  By integrating business data from various channels Business Middle Office, as well as Hall food and take-out services, data can be shared across all channels. Alibaba Cloud will build a unified and omnichannel shared business center to achieve centralized management, accumulation and sharing of core business capabilities. By building a Data Middle Office, DataWorks integrates and manages such data to make data readable, easy-to-use and unified. In addition, DataWorks converts data to assets and enables front-end applications to monetize data assets.

3. Technology architecture of new catering Business Middle Office (Figure 4.135)

   The technology architecture consists of the application front-end, application service capability center, Middle Office service layer, cloud infrastructure, security protection layer, data persistence layer and O & M service layer. In terms of architecture design, adhering to the concept of 'large Middle Office and small front office', public and general-purpose businesses in the front-end Business Middle

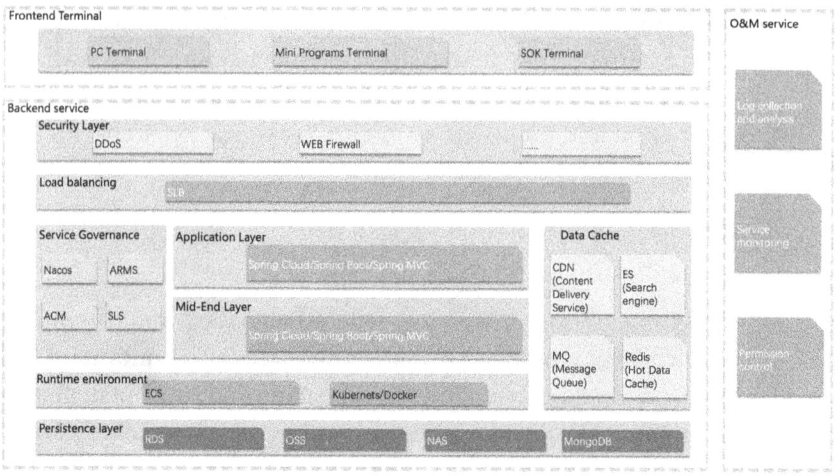

FIGURE 4.135 Technology architecture of new catering Business Middle Office.

Office be precipitated, such as order centers, product centers, marketing centers and other shared units, to form a 'thick platform'. The application service layer provides basic, composite, personalized and third-party capabilities to form a service capability pool based on domain models.

4.7.3.3 Customer Benefits

On the first day of opening the store, the system set off a wave of queuing. A few minutes after opening the store, the number of places on the day was all killed, including in the following scenes, such as card and coupon promotion, and peak meal ordering. The system has withstood many tests. The Business Middle Office architecture design breaks the traditional 'isolated island' model, reconstructs data and business resources and realizes the integration of members, products, transactions and marketing. Judging from the current launch effect, IT is of great help to user behavior analysis, user profiling, digital marketing and store selection.

After the launch, the ordering method is mainly mobile phone applet and ordering screen. Based on the ability of rapid iteration and update of the Middle Office, the company has quickly expanded the free take-out, third-party take-out (hungry, Meituan, etc.) and other platforms, which also reflects the positioning of the enterprise's younger brand. It also reflects the overall situation of embracing digitalization after entering the Chinese market.

The core of the new catering is to focus on consumers and marketing. In terms of consumer data accumulation, traditional IT systems are separated from each other, and the membership data cannot be integrated. Business Middle Office can integrate the membership data of all contacts (small programs, ordering screens, take-out platforms and so on), thus realizing comprehensive and detailed analysis of user profiles and user behaviors. In terms of marketing, the traditional marketing methods are mainly concentrated on members. Now the Internet is developing rapidly, consumers receive too many marketing information channels, and the reach of traditional marketing has been greatly restricted. The current Business Middle Office design can reach all consumers and realize fine marketing, which is a marketing method more suitable for the new catering era.

By building a Business Middle Office, a shared capability center is built based on the core competitiveness of enterprises. The Center accumulates business shared capabilities, improves business innovation efficiency and lays a solid foundation for the comprehensive digital operation of enterprises in the next few years. The launch of the new catering Business Middle Office greatly shortens the development cycle and helps enterprises implement the construction mode of agile development. Omnichannel order processing, full-process timeliness and flexible system expansion support provide strong technical service support for new stores, thereby promoting the overall transformation of enterprise business to digital and intelligent.

The new catering platform digitizes the contacts, interactions and transactions with consumers. Business Middle Office can accumulate the data, break the internal functional boundary of enterprises, accelerate innovation and provide sufficient ammunition for the subsequent Data Middle Office. Digital platforms allow enterprises to process and analyze the data, and then, in turn, support Business Middle Office. This implements a closed loop and comprehensively improves the marketing and operations capabilities of enterprises.

CHAPTER 5

Data Middle Office on Cloud

He Long, Li Linyang, HeQiang, Dai Shaoqing, Ma Xin, Wang Yaqing, Xu Fu, Pu Cong, Song Chunying, and Chen Li

5.1 OVERVIEW OF DATA MIDDLE OFFICE ON CLOUD

With the rapid development of mobile Internet and Internet of things, data explosive growth, a variety of data service needs continue to emerge. However, in the traditional IT construction mode, most of the enterprise's information systems are purchased or built independently, and it is difficult to get through the data precipitated in each system. There are many data islands and chimneys in the enterprise, which lead to data fragmentation, unable to form data services that can be shared and reused. It cannot meet the enterprise's demand of fine management through data, so as to achieve cost reduction and efficiency increase. It becomes a big pain point in the development of enterprises.

5.1.1 The Development History of Data Middle Office

The amount of data generated in 2015 is equal to the total amount of data generated by all human beings in history, which is the signal of data from multiplier growth to exponential growth. Massive data processing has become a challenge for all human beings. Enterprises also put forward higher requirements for data storage capacity and computing capacity.

DOI: 10.1201/9781003272953-5

Based on single machine, traditional database has been completed, which cannot meet the requirements of enterprises for data storage and computing. Domestic cloud computing companies, represented by Alibaba Cloud, provide enterprises with a technical base for efficient and low-cost storage and computing of massive data through cloud storage and computing capabilities.

In 2015, Alibaba formally put forward the Middle Office strategy based on the demand of its own business development, in which the construction of enterprise data asset system, enabling business and releasing data value were officially pushed to the stage of history. In 2018, based on the Data Middle Office methodology and products precipitated by Alibaba Group's massive business scenarios, Alibaba Cloud officially launched the Cloud Data Middle Office solution to help Alibaba Cloud customers build Data Middle Office based on Alibaba Cloud products and technologies.

5.1.2 Evolution Route of Data Middle Office

The construction steps of enterprise data platform should be divided into three stages, which are to look insight data from the whole business, to use the data for business, to fully digitize the business, to commercialize the data and to find new business models and opportunities for enterprises.

- Stage 1, the construction of global Data Middle Office and initialization of data assets

  Based on Alibaba Cloud Data Middle Office products and solutions, the global architecture of enterprise data platform is constructed, which lays a technical foundation for the further application of global Data Middle Office.

  Based on the global architecture of Data Middle Office, we should think from data up and business down synchronously, initiate data acquisition, data processing and data public layer construction, and synchronously start the most critical data application layer construction.

  Based on business requirements and business thinking, priority should be given to solving the most critical business scenario application of business looking insight data and using data.

- Stage 2, iteration and deepening application of global Data Middle Office

  Continue to iteratively optimize the global architecture and technical system of the Data Middle Office, enrich the data types and data scope and continuously optimize the construction of the iterative data public layer and data application layer. At the same time, it goes deep into business requirements, enriches data application scenarios, enables business and realizes data value.

- Stage 3, comprehensively promote business datalization and try data commercialization

  We will continue to promote the construction of the global Data Middle Office based on business needs, and comprehensively promote business datalization and business decision automation. While constantly expanding the applications of business scenarios, try to data commercialization, explore new business models of enterprises and make data continuously realize and generate economic benefits.

  In the three stages of the evolution of enterprise Data Middle Office, the organizational structure of the enterprise is also as shown in Figure 5.1, and has experienced three key stages from transformation, development to relatively mature and stable.

The value impact of Data Middle Office construction on customer organization structure can be divided into three stages:

- In the initial stage of Data Middle Office construction, we call it the transformation period architecture, which mainly focuses on the projects at the customer group level and exists in the situation of

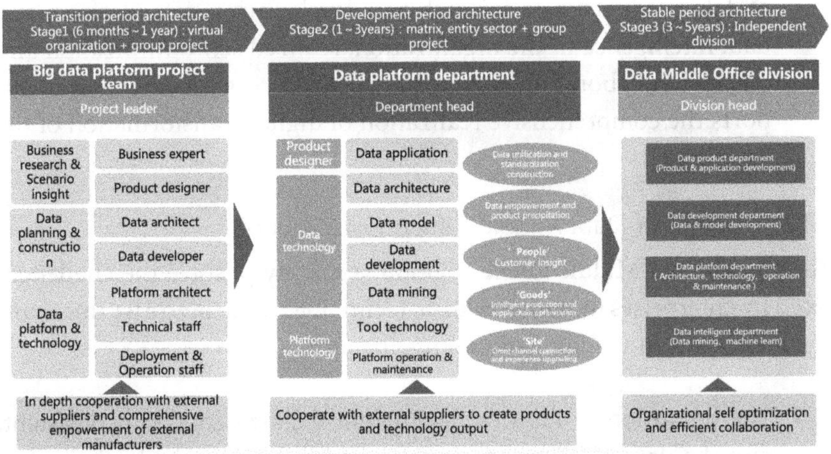

FIGURE 5.1 Construction of enterprise Data Middle Office and evolution of enterprise organization structure.

virtual project team. At this stage, Alibaba Cloud and customers cooperate in depth with the Data Middle Office project system. The project leader of the customer leads the whole project construction process and undertakes the personnel management of the virtual organization. Alibaba Cloud provides the big data platform for customer organizations and personnel, Data Middle Office construction methodology, business value exploration, etc.

- In the middle stage of Data Middle Office construction, we call it the development stage architecture. The customer organization architecture operates in the form of entity departments and group projects. At this stage, the data platform department is established, and the department head leads the whole data platform project construction process. At this stage, Alibaba Cloud fully cooperates and co-creates with customers, outputs Alibaba Cloud products and technical solutions for customers and helps customers realize data platform construction in an all-round way according to data unification and data capitalization, data-enabling business and data productization.
- During the continuous operation period of Data Middle Office, the customer organization has recognized the goal and value of Data Middle Office construction. The organization is upgraded from the data platform department to the Data Middle Office business unit. The head of this BU is in charge of the whole Data Middle Office construction from the group level, and realized the full-link division of data products, data research and development, data platform and data intelligence in the organization, realized self-optimization and efficient collaboration within the customer organization, and supports the comprehensive realization of digital transformation of the business.

5.1.3 Content of Data Middle Office Construction

The Cloud Data Middle Office solution is mainly to solve the challenges faced by enterprises in business, data, technology and enterprise fine management in the era of big data and intelligence:

- The definition of metrics in different departments is inconsistent, and the data consistency problem always puzzles the business and decision-makers.

- Chimney development mode has long cycle and low efficiency, which leads to slow response to business data requirements.
- Repeated construction leads to long data processing link, complex processing logic and poor maintainability, which eventually leads to the problem of data timeliness.
- The data of each department and each business system are fragmented, which cannot effectively integrate and get through the data of each department and each business system, and cannot provide unified data operation support for the business based on the integrated data asset system.
- Enterprise development from incremental market to stock market, in order to reflect the competitiveness of the stock market, is necessary to improve the enterprise fine operation and management ability.

In order to solve these problems encountered by enterprises in business and data processing, Alibaba Cloud proposed the Data Middle Office solution, which mainly includes the following aspects:

Intelligent cloud platform, based on the powerful computing platform provided by Alibaba Cloud, provides a multi-level and powerful data computing power for enterprises in many aspects, such as offline, real-time and OLAP (Online Analysis Processing) analysis. Time is the inherent enemy of data. Data value will decline rapidly in the advancement of time. Strong computing power can ensure that data serve business with the fastest time.

OneModel modeling products defines data specifications from business sources, standardizes data and capitalizes data. Data technical solutions such as data specification definition, data modeling and R&D, data processing task scheduling, operation and maintenance are tooled and produced to enable and improve the full life cycle of data R&D. It realizes the intelligent planning of the underlying calculation and storage by using the intelligent Data Middle Office solution that is driven by metadata.

OneID technology scheme technology drives data connection based on ID mapping and graph computing technology to achieve data homology identification and data connection. It drives technical value through business display, and it connects data and business islands through data, to achieve high-quality, high-value data sharing and service.

OneService products make data easier to use by shielding underlying complex physical data tables for users through thematic logical table services. To provide simple data query, real-time OLAP analysis, online data processing and analysis, and other forms of data services, convenient data use. Through unified data service interfaces and specifications, multiple complex and heterogeneous data sources are shielded, and lower the threshold for data using.

DataAssest that is known as global data asset management platform, centered on core data assets, establishes a manageable, verifiable, controllable, accessible and estimable data asset service center, supports the flow of data assets within the enterprise, promotes the awareness of data usage by all staff and assists the transformation of production mode driven by enterprise digitization.

Data Middle Office construction implementation methodology, standards and service system help enterprise customers build and use Data Middle Office well.

5.1.4 Data Middle Office Construction Method

Based on the practice of Alibaba Group and customers in many industries of Alibaba Cloud, we divide the construction of enterprise Data Middle Office into five stages as shown in Figure 5.2: requirement investigation,

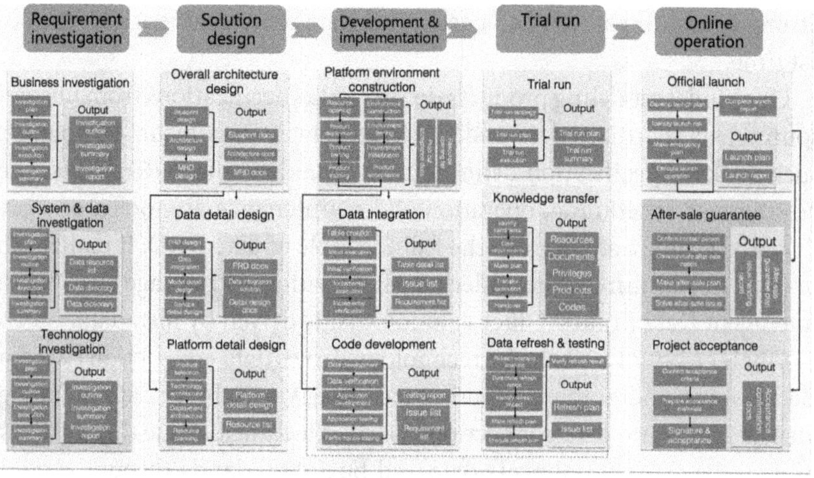

FIGURE 5.2 Five stages of Data Middle Office construction.

solution design, development and implementation, trial run and online continuous operation iteration.

5.1.4.1 Requirement Investigation Stage

Requirement investigation includes three work contents: business requirement survey, technical survey, and system and data survey.

Business requirements survey object is the core personnel of business departments of enterprises, generally including management and front-line core business personnel.

Business requirements survey mainly starts from the business objectives and scope of this project, understanding business model, business pain points and business needs for data support, breaking business objectives into specific business needs reports, providing input for project delivery implementation (MRD design, PRD design and business blueprint design).

The scope of technical survey is to inventory the existing IT systems of enterprises, among which big data platforms and data warehouses need more attention. The goal is to have a comprehensive understanding of the IT system, the level of IT construction, to prepare the physical environment in advance for project delivery, to plan resources for the deployment of Data Middle Office products and to provide sufficient input for detailed platform design.

The goal of system & data survey is to understand the core business process of the IT system, the basic architecture of the IT system, data content, data distribution, data flow, data type, data scale and data characteristics, so as to provide sufficient support from the system and data level for the detailed design of data such as data cloud and data modeling. Confirm that there is sufficient data support for the data application scenarios involved in the business requirement survey.

5.1.4.2 Solution Design Stage

Data Middle Office solution design includes blueprint design, architecture design and data application scenario PRD design.

Data Middle Office blueprint design, based on the input of requirement investigation stage, completes Data Middle Office business blueprint, management blueprint (optional) planning design.

Data Middle Office architecture design, based on business requirements, data and IT technology in the stage of requirement investigation, completes the overview design work. Architecture design mainly includes the Data Middle Office overall architecture, integration architecture, technology architecture, network architecture, deployment architecture, data architecture (including data hierarchy and data flow design), model architecture and data desktop standard specification design.

Data application scenario PRD design, based on business requirements and system data situation in the stage of requirement investigation, completes the PRD design of data application scenario. Common application types include data BI analysis scenario, data large screen visualization, PRD design of intelligent data application, PRD design to specific function points and implementation logic.

Detailed data schema design includes detailed data cloud schema design, detailed data model design and detailed data platform design, the core of which is detailed data cloud design and public data model design.

Detailed design of data integration on cloud, we make detailed design of data model on operational data store (ODS) layer based on input of system & data research, and operational implementation plan of data integration for each data source's tables.

Detailed design of public layer model, it is based on Alibaba OneModel modeling methodology, business research MRD and input of system & data research, complete detailed design of business section, data domain, business process, bus matrix, dimension tables (DIM) of CDM (common data model layer) public layer, fact tables (DWD) and mild summary tables (DWS). Based on Alibaba OneID methodology, complete the detailed implementation scheme design of data connection and connection of enterprise core entities.

Detailed design of data platform includes product selection and technology architecture design, deployment scheme design, resource planning, account and privileges planning and security scheme.

5.1.4.3 Development and Implementation Stage

After the corresponding detailed schema design of the Data Middle Office is completed, project delivery enters the stage of development and implementation. Development and implementation stage includes project-related product resources opening and environment deployment, data integration implementation, code development, history data refresh and testing.

5.1.4.4 Trial Run Stage

The main goal of Data Middle Office trial operation is to complete the preparation work before the system officially goes live, including Data Middle Office functions, data accuracy business validation and so on. Through trial operation, it finds the defects in project delivery, so as to handle them early, avoid the occurrence of failures, master the actual technical performance metrics of the system and formulate necessary response measures and management rules for go-live operation.

5.1.4.5 Online Operation Stage

In the online operation stage, we make maintenance plans in advance, ensure that Data Middle Office continue to provide services for business and ensure the stable operation of Data Middle Office.

5.1.5 Overview of Data Middle Office Values

5.1.5.1 Business Values

The value of Data Middle Office enabling business is mainly reflected in four aspects:

- Business global data monitoring, based on Alibaba Cloud products and technologies, provides convenient, fast and easy-to-use Data Middle Office solutions for business decision-making and data-based operation, minimizes the difficulty of data analysis and business decision-making and maximizes the effect of data analysis. At the same time, through data monitoring of the whole business process, business risks and opportunities can be discovered in time, and business decisions and data operations can be assisted.
- Data-based business operations, based on the establishment of full-link all-channel data, 'people' as the core of the data connection extraction solution, enterprise makes fine-grained management and operation of the entire life cycle of 'people' served by enterprises.
- Data implantation business, based on the standard, consistent and reusable data assets precipitated by the enterprise Data Middle Office construction, combines with business intelligence marketing promotion solutions, from pre-term population analysis, crowd selection, to mid-term customer reach, to late-term business data detection and business iteration optimization, to achieve the close

combination of data and business, to achieve full-link data marketing and marketing decisions.

- Data commercialization is the natural development and sublimation of Data Middle Office construction after capitalizing data precipitated by business. Using capitalized data for business or product itself mainly consists of two levels: data intelligence and data innovation. Data intelligence mainly uses big data technology to improve business efficiency and product experience, such as recommendation system and credit rating. Data innovation mainly uses the capitalized data to hatch and develop new business.

5.1.5.2 Technology Values

The value of Data Middle Office technology is also reflected from four aspects:

- Based on the powerful computing platform provided by Alibaba Cloud, it provides multi-level powerful data computing power for enterprises in many aspects such as offline, real-time and OLAP analysis. Time is the natural enemy of data, and the value of data will decline rapidly in the advance of time. Strong computing power can ensure that data can serve business with the fastest time.
- OneModel makes data specification definition from the business source, achieves data standardization and data capitalization. Data technical solutions such as data specification definition, data modeling, investigation and development, and data processing task scheduling, operation and maintenance are instrumented and productized to realize the empowerment and efficiency improvement of the whole life cycle of data research and development. It realizes the intelligent planning of the underlying calculation and storage of the data in the Data Middle Office by using the technology of metadata driving the scheme intelligent.
- OneID technology drives data connection, based on ID mapping and graph computing technology, it achieves data homology identification and data connection through. Through business display to drive technical value, through data connecting of business island, to achieve high-quality, high-value data sharing and service.

- OneService makes data easier to use by shielding the underlying complex physical data tables for users through the theme logical table service. To provide simple data query, real-time OLAP analysis, online data processing and analysis and other forms of data services, make data easy to use. The unified data service interface and specification can shield a variety of complex and heterogeneous data sources and lower the threshold of data usage.

5.2 ARCHITECTURE DESIGN OF DATA MIDDLE OFFICE

The idea of Data Middle Office is to enable enterprises to share the same set of data technology and assets, and provide unified and powerful algorithm, data, analysis and technical support for various business systems. This chapter mainly designs and explains the theoretical basis, product system, platform architecture, data architecture, development modeling specification and other aspects of data platform.

5.2.1 The Theory of Data Middle Office

5.2.1.1 OneModel

Data energy in DT era has become the water, electricity and coal in people's daily life. With the rapid development of business, the corresponding data are growing rapidly, data tables and business metrics are increasing, management data are becoming more and more complex, and the cost of operation is becoming higher and higher. It is urgent to guide a comprehensive, standardized and systematic big data construction system to help large data teams to build their own data system with high efficiency and high quality. Alibaba OneModel is a comprehensive, standardized and systematic big data construction system, which guides and helps big data teams to build their own data warehouse or data system with high efficiency and high quality, so as to achieve the efficiency of eliminating repeated construction of data chimney and reducing cost and energy saving.

OneModel is the methodology system of Alibaba's big data construction. In addition to a series of construction standards such as data standards, model specifications, data R & D specifications and data service specifications, it also includes a set of product toolsets related to the implementation of specifications to ensure the standardization and systematic construction of the whole data platform. OneModel system is manageable and traceable, and can avoid repeated construction from the normative definition of dimensions and metrics, data model design, data research and development to data services.

OneModel big data construction methodology system has significant value in three aspects: data standardization, technology kernel instrumentalization and metadata-driven intelligence.

- Data standardization, OneModel system data standardization, the implementation of the standardized definition of data from the source, rather than after data R & D, based on the carding of data metrics to achieve the standardized definition and standardization of data model and data metrics. Make every data unique, and the ambiguity of data is eliminated 100%.
- Technology kernel instrumentalization, in OneModel system, through Dataphin intelligent R & D management platform, the whole process of data access, data specification definition, data model design, data R & D, data scheduling, operation and maintenance and data quality management is instrumented and platformed, realizing the real full-link connection of data management, and realizing the management of every step and node Monitoring and analysis.
- Metadata-driven intelligence, implementation design as development based on Dataphin intelligent R & D management platform, which can realize automatic code generation at minute level after data modeling, greatly simplifies data R & D workload and improves R & D efficiency and quality. The important reason for the automatic code generation is that we have defined each metadata at the source, realized the atomization and structure of the data as much as possible and stored it all in the metadata center. It is these metadata that are of great significance to data computing, scheduling, storage, etc., in order to achieve the realization of intelligence from artificial to automation, and even evolution.

5.2.1.2 OneID

Alibaba OneID is a cross screen, cross domain natural person identity tracking and identification system. Each connected to the Internet natural person is assigned a virtual ID through the identification algorithm. OneID will aggregate the ID of natural person from various systems and fields, eliminate the ID that does not belong to the natural person, highlight the active ID and eliminate the dead ID, so as to provide help for business marketing, crowd selection, label aggregation, etc.

After years of practice, Alibaba OneID technology has been evolving and updating iteratively in its complex and huge ID group computing, and gradually precipitated a set of mature and efficient transplantation OneID technology. Alibaba based on self-developed large-scale graph computing and anonymous user identity matching algorithm to get the final natural person identity and its related ID.

Nowadays, every individual has the opportunity to participate in various social activities. In the process of participating in these activities, the behavior information of individuals is very valuable. It is the main basic information source for studying the characteristics, hobbies and labels of behavioral people. However, in our research, we find that in the process of these behaviors, the identity information recorded by a single system is often very rare, such as going to 4S store to see the car, maybe only the mobile phone number recorded, and maybe only credit card recorded in shopping in the supermarket.

A single system is not clear about other information of consumers, and even more ignorant about the behavior information of behavioral people in other places. For example, the store's Firmus guide doesn't know which milk powder you like, and the guide cannot recommend the latest promotion or appropriate product to consumers. Many traditional enterprises have a large number of business systems, recording a lot of customer information, but they have never been able to integrate it to play its due role. On the one hand, the system of these enterprises is complex and diverse, and their daily management is fragmented. On the other hand, because the enterprises are too large, the ID information is complex and diverse, and they cannot rely on the traditional information aggregation method to sort out the ID information. For example, Alibaba Group has dozens of ID data related to consumers, and the natural person of ID is in the hundred million level. Alibaba OneID is the technical solution to solve the practical business problems.

Alibaba OneID technology relies on the powerful computing power of the underlying MaxCompute, which naturally aggregates the relevant ID, reclassifies them according to the relationship between ID and behavioral people and logically sorts them according to the intimacy between ID and behavioral people, realizing the aggregation and splitting of the original complex ID information. OneID can more accurately represent a natural person. Through a unique OneID, it can establish a relationship with multiple real ID in real life or enterprise business system, and help enterprises to map all the real ID and consumer data of the consumer according to OneID, so as to help eliminate data islands, accurately depict the characteristics of consumers, and help precision marketing, crowd selection and other business scenarios.

5.2.1.3 OneService

OneService is for all BU and other business parties within Alibaba Group to connect with OneModel global data, build a unified data service platform, and provide stable, efficient and secure global data services. Its daily call amount is billions, and the QPS reaches more than 100,000, which transfers a continuous stream of data energy to various product lines. OneService enables data to serve the business online. There are four main features:

- API (Application Programming Interface), allowing data to connect to business system through API.
- Themed, allowing data to be organized as business theme objects, providing services, and making physical tables transparent.
- Configuration, simple configuration can complete the data API online.
- Serverless, no server trusteeship mode, high performance, availability guarantee.

The data value revealed by OneService covers the following aspects:

Metric output, including commodity domain, member domain, transaction domain, marketing domain, channel domain, log domain and supply chain domain; corresponding business processes such as order, merchant, transaction, delivery and evaluation.

User portrait tags, including basic attributes (such as basic characteristics, life stage, education information, occupation information and asset information), social relations (such as relatives relationship and pet information), geographical attributes (such as location based service, LBS for short, birthplace and receiving address), consumption behavior (such as consumption ability, registration and certification and credit attributes), interactive behavior (such as login, browsing, search, collection, evaluation, rights protection and compensation), preferences (such as online shopping preferences, online shopping habits and interest preferences) are six areas.

5.2.2 Data Middle Office Product System

At present, data platform product system includes intelligent asset platform Dataphin, agile Bi tool QuickBI, intelligent user growth QuickAudience and related cloud products. Dataphin mainly provides specification modeling, data connection and extraction, and data service functions in Data Middle Office, QuickBI mainly provides agile Bi capability, and QuickAudience mainly does intelligent user tag analysis, crowd selection and marketing.

5.2.2.1 Product Architecture of Dataphin

Dataphin intelligent data construction and management is the construction engine of intelligent big data platform, which aims to meet the demands of big data construction, management and application in all walks of life. By exporting the big data construction system OneModel, OneID and OneService of Alibaba Group, it provides one-stop full-link intelligent data construction and management services including data introduction, specification definition, data modeling, data R & D, data extraction, data asset management and data services, and helps enterprises build their own intelligent data system with unified standards, capitalization, service and closed-loop self-optimization.

Dataphin shields the differences between different computing and storage environments, helps users quickly introduce data, builds data in a standardized way, automatically develops data in a modeling way, extracts entity object-oriented data label system, precipitates business data and data assets, manages data quality problems and supports data table query for various types of data services.

Dataphin product architecture consists of four parts, which are technology kernel, tool side, data layer and management service layer, the details are as follows (Figure 5.3):

- Technology kernel, a set of technical framework to shield the differences between the underlying computing, storage and software systems, to ensure that data research and development can be

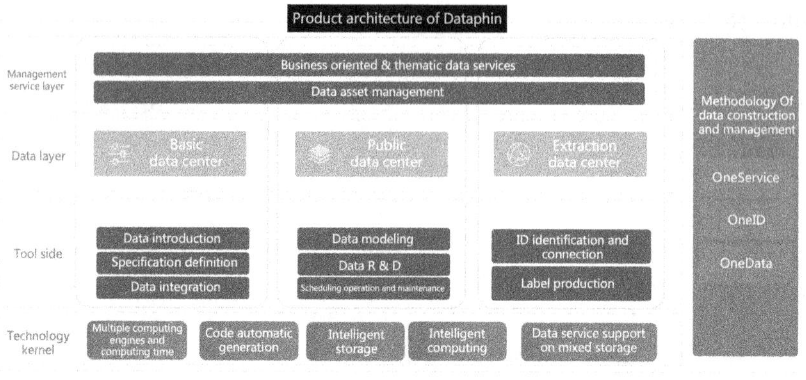

FIGURE 5.3 Product architecture of Dataphin.

compatible with multiple computing engines and computing time, code automatic generation and intelligent storage and computing, data service support mixed storage, etc.

- Tool layer, the tools for data construction and management for developers, including the standard specification and integration introduction of basic data, the definition of standard specifications of public data, intelligent modeling research and development, scheduling operation and maintenance and machine learning, ID identification and connection of extracted data and label production.
- Data layer: On the basis of the technical kernel, through tool processing and production, output three levels of structured data, build a high-fidelity, business-oriented basic data center, modeling, theme-oriented public data center, deep processing, entity-centered extraction data center.
- The management and service layer, managing the data and data services from the perspective of assets, so that the data R & D personnel and business personnel can obtain high-quality and unified data assets; from the perspective of business, the existing data are packaged and processed into thematic data services, so as to ensure that the business can query and call data uniformly.

5.2.2.2 Product Architecture of QuickBI

QuickBI is a product of Alibaba Cloud, it is a lightweight self-service BI tool service platform based on cloud computing and dedicated to efficient

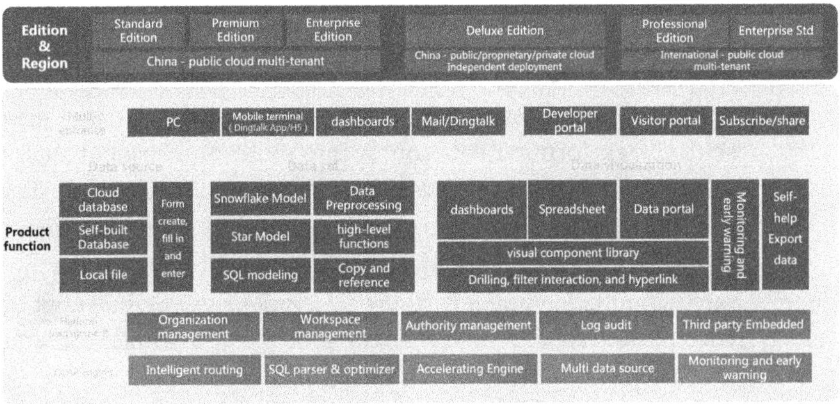

FIGURE 5.4 Product architecture of QuickBI.

analysis and presentation of big data. Through the connection of data sources and the creation of data sets, the data can be analyzed and queried in real time; through the spreadsheet or dashboard function, the data can be visualized by dragging and dropping.

QuickBI is based on the mission that everyone is a data analyst, it provides massive data online analysis, drag-and-drop operation and visualization to help you easily complete data analysis and business self-help exploration. It is not only a data analysis tool for business personnel, but also a booster for data operation, as well as an artifact to solve the 'last mile' of big data application.

The functional architecture of QuickBI product mainly includes four modules: data connection, data processing, data visualization and data permission processing, the details are as follows (Figure 5.4):

1. Data connection module

   Be responsible for adapting various cloud data sources, including but not limited to MaxCompute, RDS (MySQL, PostgreSQL, SQL Server), Analytic DB, HybridDB (MySQL, PostgreSQL), encapsulating metadata and standard query interface of data.

2. Data processing module

   - Query engine: Responsible for the query process of data sources.
   - Data preprocessing in advance: Responsible for lightweight ETL processing of data sources. At present, it mainly supports

MaxCompute's custom SQL function, and will be extended to other data sources in the future.

- Data modeling: Responsible for OLAP modeling process of data source, transforming data source into multi-dimensional analysis model, supporting standard semantics such as dimension (including date dimension and geographical location dimension), measurement and star topology model, and supporting calculation field function, allowing users to use SQL syntax of current data source for secondary processing of dimension and measurement.

3. Data display module

- Spreadsheet: Responsible for the related operation functions of online spreadsheet, including row and column filtering, general/advanced filtering, classification and summary, automatic summation, conditional format and other data analysis functions, and supporting data export, text processing, table processing and other rich functions.
- Dashboard: Responsible for drag-and-drop assembly of visual chart control into dashboard, supporting more than 40 charts such as line chart, pie chart, histogram, funnel chart, tree chart, bubble map, color map and metric; supporting five basic controls such as query condition, TAB, IFRAME, PIC and text box, supporting data linkage effect between charts.
- Data portal: Responsible for drag-and-drop assembly of dashboards into data portal, supporting embedded links (dashboards), and supporting basic settings of templates and menu bars.
- Sharing/disclosure: Support sharing electronic forms, dashboards and data portals to other login users, and support publishing dashboards to the Internet for non-login users.

4. Authority management module

- Organization authority management: Responsible for the two-level authority architecture system control of organization and workspace, as well as the user role system control under the workspace, to realize the basic authority management and realize different people to see different report contents.

- Row-level authority management: Responsible for row-level granular authority management and control of data, so that different people can see different data in the same report.

5.2.2.3 Product Architecture of QuickAudience

QuickAudience focuses on consumer operations, it completes multi-dimensional consumer insight analysis and multi-channel access through rich user insight model and convenient strategy configuration, and helps enterprises realize user growth. It has the following characteristics:

- Efficient model creation: Complete the user model for insight analysis through rapid model configuration.
- Multi-dimensional user insight: Multi-dimensional user insight analysis is completed through user model and 360 degree tag.
- Convenient strategy making: Multi-dimensional circle strategy and convenient plan making to quickly complete the crowd strategy.
- Multi-channel access: The integration of multiple delivery channels, especially the one click push of the audience to Alibaba economy, completes the global marketing closed loop.

QuickAudience product functional architecture includes six modules: data source access, data set creation, user insight, audience selection, audience analysis and content management, the details are as follows.

- Data source access: Provide access capability of multiple data sources and data sets, complete data source import and management and support access to AnalyticDB for MySQL 2.0, AnalyticDB for PostgreSQL and AnalyticDB for MySQL 3.0 databases.
- Data set creation: Provide model configuration capabilities of tag data set, behavior data set, AIPL model and RFM model, and independently configure scoring rules and thresholds for AIPL and RFM models.

- User insight: Provide the ability of crowd perspective analysis, RFM analysis, AIPL analysis and flow analysis. Aiming at the audience, through the function of label perspective and significance analysis, the audience is insight.
- Audience circle selection: Support users to quickly circle the specified number of target crowd with specified filter conditions in the process of crowd analysis.
- Audience management: Complete the management of selected audience, including audience analysis, editing, download, update, push and other functions.
- Audience analysis: Further insight analysis based on the audience selected by the circle, including perspective analysis, comparative analysis among audiences and significance analysis.
- Global marketing: Push consumer data of one party of the enterprise to Alibaba brand data bank with one button, establish brand global consumer data assets and comprehensively improve the effectiveness of brand-wide consumer operation.

5.2.3 Platform Architecture Design

This section takes the construction of a retail customer data platform as an example to elaborate the architecture design of data platform. For other projects, the content and method of data platform architecture design are basically the same, and the specific design scheme is different.

5.2.3.1 Overall Architecture Design

Combined with business requirements and data status, the overall architecture of a retail customer Data Middle Office consists of business system source data, intelligent big data platform (including Dataphin intelligent data research and development, data asset management and unified data center and data service) and data application, as shown in Figure 5.5:

- Data sources: They are divided into four types, including consumer business data, consumer behavior data, supply chain business system data and third-party data. These data are extracted in batches and synchronized to the vertical data center of the data center through

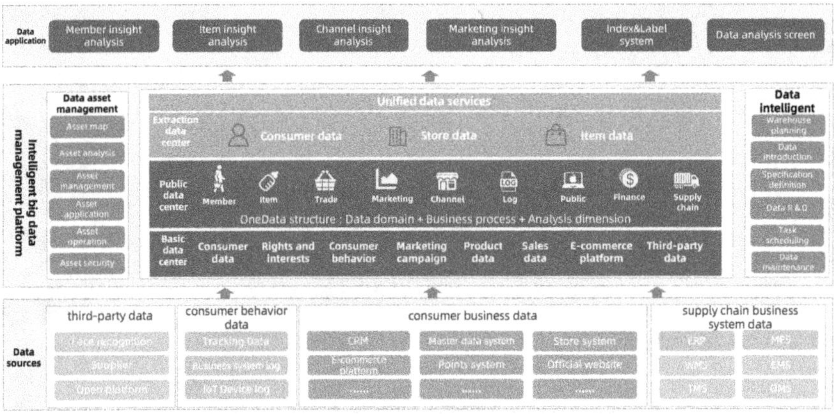

FIGURE 5.5 Overall architecture of Data Middle Office.

the data introduction function of Dataphin intelligent big data R & D platform.

- Intelligent big data management platform: It consists of Dataphin Intelligent Data R & D suite, Dataphin data asset management suite, vertical data center based on Dataphin products, public data center and extraction data center, as well as unified data services.

- Data application: Data analysis application and data platform enabling application for end users, including consumer insight of members, points and behaviors, marketing insight of marketing activities and different industry alliances, and terminal channel insight of stores, shopping guides and one-side LBS.

5.2.3.2 Technology Architecture Design

Based on the overall logical architecture of Data Middle Office, the selection and design of Data Middle Office technology are carried out, as shown in Figure 5.6, the selection of the whole technology architecture includes five levels: business data source storage technology, data source access technology, data storage and computing technology, data service and data application technology.

- Business system data source, the main storage technology is divided into: one is based on MySQL, SQL Server, Oracle relational database storage and calculation; one is based on unstructured and

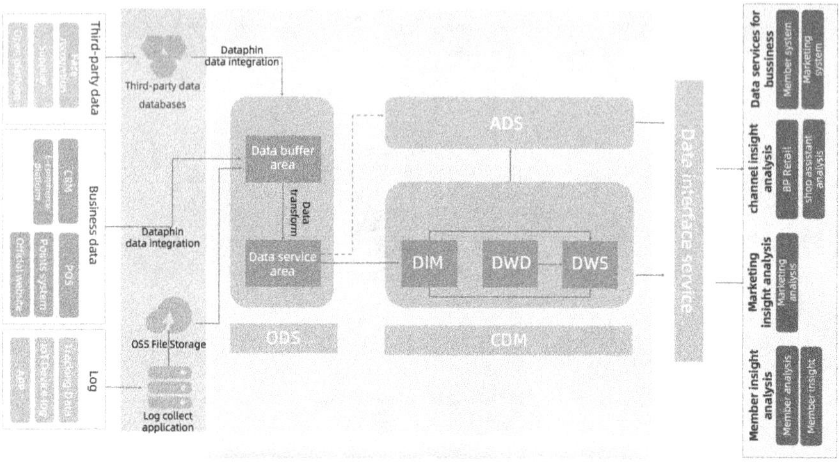

FIGURE 5.6 Offline data flow of Data Middle Office.

semi-structured storage of pictures, videos and user behavior log data; and the third party through the data API interface.

- In the data access layer, offline data processing can use Dataphin's data access function, or Alibaba's open-source batch data extraction tool DataX to realize batch synchronization of business system data to the Data Middle Office; real-time data processing can be realized by Alibaba Cloud DTS (Data Transmission Service) data real-time synchronization service or DataHub, and the business system data or real-time log can be synchronized to the Data Middle Office.
- Data storage and computing: MaxCompute offline computing engine is used for data storage and offline computing in Data Middle Office; Alibaba Cloud StreamCompute streaming computing technology is used for real-time computing in Data Middle Office; Alibaba Cloud Dataphin intelligent big data research and development platform is used for data development and management.
- Data service layer: Simple data query services are implemented by RDS relational database, such as daily, weekly and monthly reports; complex ad hoc multi-dimensional analysis query or result data with more than 10 million data are implemented and stored by Alibaba Cloud AnalyticDB analysis database; Alibaba Cloud elasticsearch service is adopted for data service scenarios of search.

- Data application layer, various customized data report analysis requirements are realized by using the intelligent report tool QuickBI of Alibaba Cloud; and personalized data application requirements are realized based on the technical system of Alibaba Cloud products.

5.2.3.3 Data Flow Design

On the construction of Data Middle Office, from business data generation to data middle station, it designs the data application and business enabling end-to-end data flow and technical implementation scheme, mainly including offline data flow scheme and real-time data flow scheme.

1. Offline data flow design

   For the source data access in the offline scenario, different technical solutions are adopted for different types of data sources. The source data are synchronized with the buffer of ODS in the Data Middle Office, and the data in the buffer are completely consistent with the source data. After the data enter the data buffer of ODS layer in the Data Middle Office, for the data of daily full synchronization or daily incremental synchronization, when entering the data service area of ODS layer, the data for full synchronization will be converted into a unified data type, and then stored into the daily full data partition of the data service area, and the data for incremental synchronization will be converted into a unified data type at the same time, the data of the whole data of the previous day and the incremental data of the current day will be merged and stored in the data service area.

   Data entering ODS layer of Data Middle Office can be provided to CDM layer or application data store (ADS) layer for data processing. The CDM layer will obtain dimension and fact data from the data service area of ODS layer, and then store them into DIM dimension table and DWD detail fact table according to the new data model after transformation. Based on the demand of public summary metrics, the DIM dimension data and DWD detailed fact data of CDM layer are processed to generate summary statistical data and store DWS summary fact table.

   The data of ADS layer generally come from the data of CDM general data model layer, but for the personalized data requirements or complex data requirements of some special applications, the data of

ADS layer can be directly obtained from the data service layer of ODS and processed to generate the business-oriented data of ADS layer.

Finally, the data of ADS application layer and CDM public layer are synchronized to the online storage product of data interface service layer, which provides unified data service for the upper data application and business system through the unified data interface service layer.

2. Real-time data flow design

For different types of data sources, Alibaba Cloud DTS real-time data transmission service and customized log collection application are used to obtain data from relational databases and other data source types (such as data API) in real time, and write the data to Alibaba Cloud Datahub in real time (Figure 5.7).

After the data are written into the datahub, for the business scenarios that need real-time calculation, real-time calculation is carried out through the real-time calculation engine, and the result data after calculation is pushed to the data interface service layer for data application and business system use.

Many scenarios need to save the data generated by the real-time data stream for offline computing, and archive the data of the datahub regularly to the data buffer of the ODS layer in the Data Middle

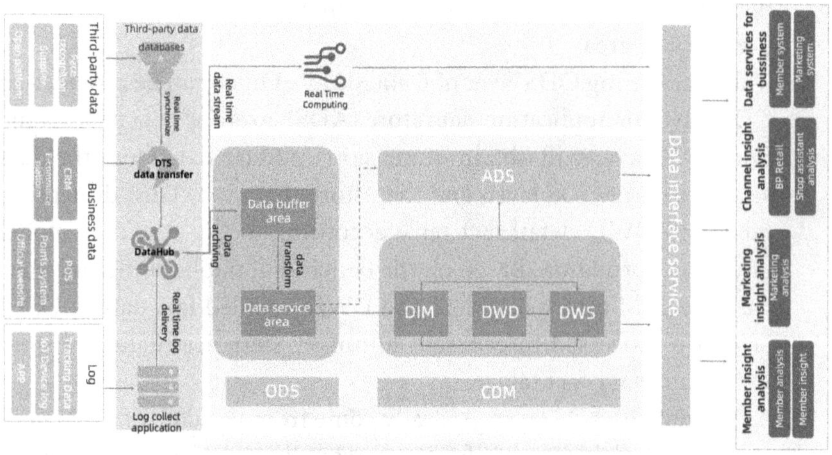

FIGURE 5.7 Real-time data flow of Data Middle Office.

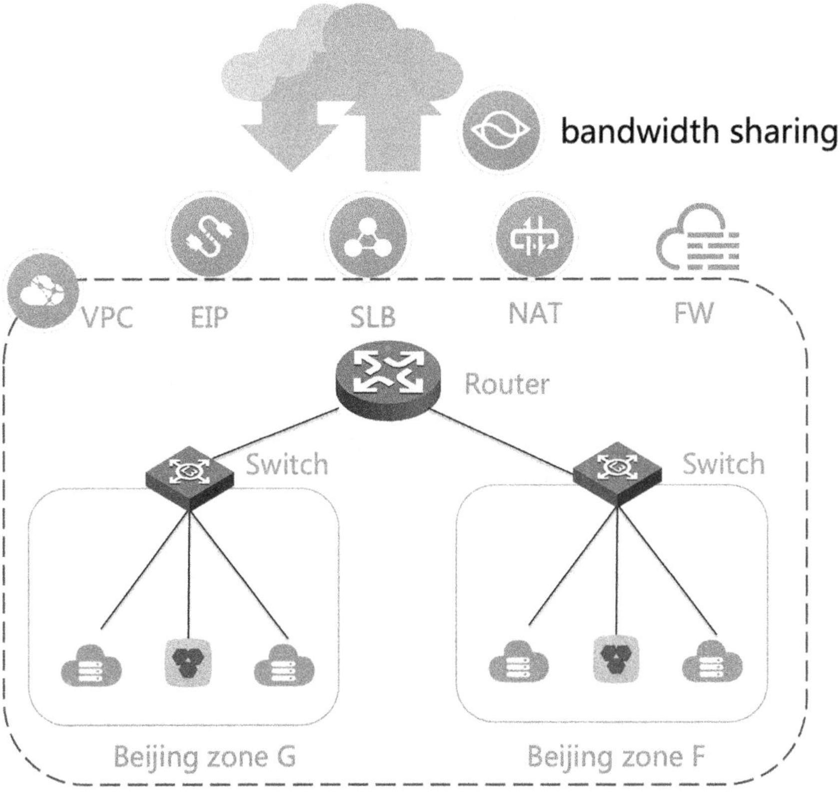

FIGURE 5.8 Inner architecture diagram of VPC.

Office. After the data are stored in the data buffer of ODS layer, the subsequent data processing and circulation process is the same as the offline data flow.

5.2.3.4 VPC Architecture Design

Taking a retail customer in Alibaba Cloud public cloud construction Data Middle Office as an example, the design of VPC internal network architecture is shown in Figure 5.8.

1. Overall architecture design

   - VPC is deployed in the same region, and must be in Beijing region and Qingdao region.

- VPC allocates network segments according to the services carried in VPC, and at the same time pays attention to the services of offline IDC to ensure no conflict.
- A logical router can be created in VPC, and multiple logical switches can be associated under a logical router.
- In order to prevent the conflict with the offline IDC website, 10.2.0.0/24 network segment is used in Data Middle Office VPC IP planning.

2. Reliability design

   For the deployment location of logical switch, we can balance the delay and reliability requirements of different services, and choose to deploy in the same zone or different zones according to the resource quota of the zone.

3. Security in VPC

   - VPC deploys a security group to control access to the same network segment.
   - Cloud firewall is deployed at VPC exit for L4-L7 security protection and access filtering.

4. Network connection in VPC

   The sub-systems in the VPC communicate with each other, and the owner of the VPC performs the association strategy of self-controller routing.

5. External service of VPC

   VPC exports deploy EIP, SLB, NAT gateway and other products, and associate with shared bandwidth packets to maximize bandwidth utilization.

5.2.3.5 Deployment Architecture Design

The product deployment architectures of Dataphin, QuickBI and QuickAudience in the data platform product system are designed as follows.

1. Dataphin deployment architecture design

   Dataphin privatization deployment technology architecture consists of five parts from bottom to top, which are storage cluster,

Mesos scheduling cluster, master cluster, worker cluster and the top access layer (Figure 5.9).

- Storage cluster, business data collected offline from various business system data sources are stored in Alibaba Cloud MaxCompute; all unstructured data involved in Dataphin platform are stored in OSS; Dataphin user behavior audit log and system error log are stored in SLS log service; Dataphin platform-related metadata such as accounts, permissions, various task configuration and operation, data specification definitions are stored in RDS relational database PostgreSQL; Redis is used to cache hot data such as account permissions that are often used by Dataphin platform to improve efficiency.
- The scheduling cluster is composed of three high-standard 24C/96G or higher cloud server nodes, which is composed of Apache mesos distributed resource management service and zookeeper responsible for the mixed distribution of application coordination service, and mesos is responsible for resource management and scheduling of the whole cluster, zookeeper is responsible for application configuration management and consistency synchronization.

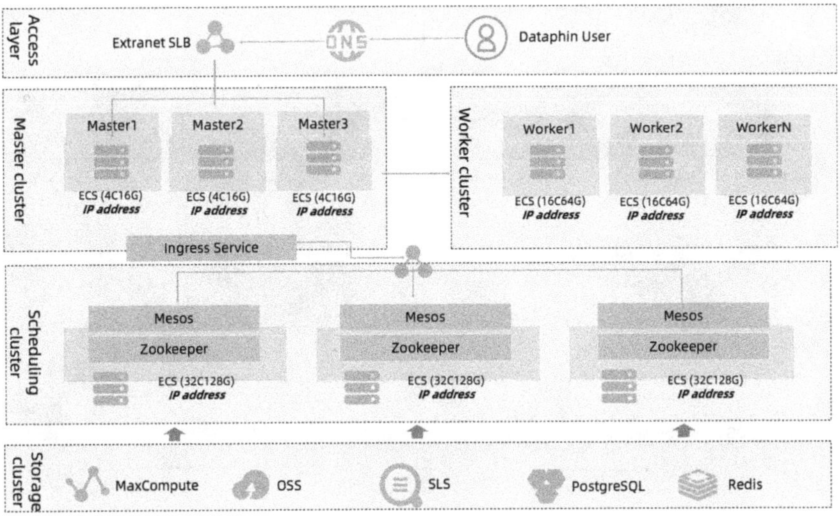

FIGURE 5.9 Deployment architecture of Dataphin.

- Master cluster, the three-node master control cluster is composed of 4C/16G cloud servers.

- Worker cluster, 16C/64G cloud server is used to form a multi-node worker work cluster, which is responsible for the execution of Dataphin tasks.

- Access layer, using SLB load balancing and DNS services, as the unified access portal of Dataphin platform.

2. QuickBI deployment architecture design

   The deployment architecture of QuickBI products is shown in Figure 5.10. The IP address is determined according to the actual situation, the details are as follows:

   - SLB: Load balancing service to improve the availability of QuickBI. One intranet IP and one public IP can be configured. Users can use QuickBI by using public or intranet IP.

   - QuickBI cluster: Consists of four ECS. Two are deployed as drivers and two as executors. The driver is responsible for the management of source data, session and other functions. The master and slave nodes of Redis are deployed on the two driver machines

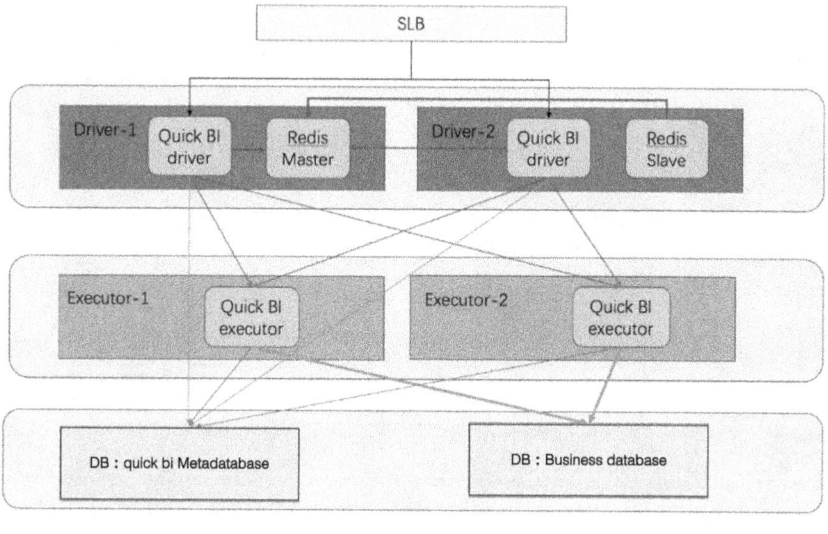

FIGURE 5.10 Deployment architecture of QuickBI.

at the same time. Two executors are responsible for specific task execution.

- Meta database instance: The stored information includes workspace information, user role information and permission information, so it is a necessary node. In the pre-deployment preparation phase, two libraries such as QuickBI (metadata) and quickbi_demo (for users to learn data) need to be created on the instance.
- Business database: Store business metric data, which is used as data source of spreadsheet and dashboard. You can use QuickBI to add data sources.

3. QuickAudience Deployment architecture design

The deployment architecture of QuickAudience product is shown in Figure 5.11. The IP address depends on the actual situation, the details are as follows:

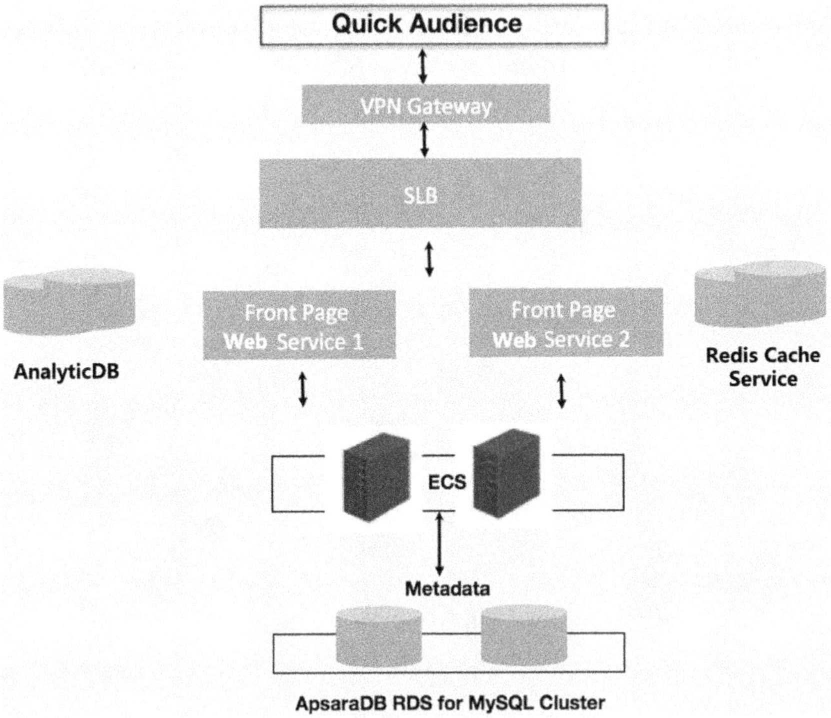

FIGURE 5.11 Deployment architecture of QuickAudience.

ECS is used to deploy web services for front-end pages. SLB is required for this, the front-end protocol of SLB is HTTP or HTTPS, which is forwarded to port 80 of the back-end HTTP protocol and deployed through docker.

RDS (MySQL) is used as the metadata storage of system runtime. Only one RDS needs to be purchased to deploy QuickAudience products. The two RDS in the above deployment architecture diagram mainly refer to the internal disaster recovery mechanism of RDS products.

ADB analytical database is used to store business data.

5.2.3.6 *Security Management Design*

Using the function of Dataphin business section, the data platform environment is designed as two independent business sections of development and production. In each independent business segment, the same user needs to use different independent accounts to access different logical environments. In each independent logical environment, different data table permissions are designed for different users to isolate and manage users' permissions. The overall architecture is shown in Figure 5.12:

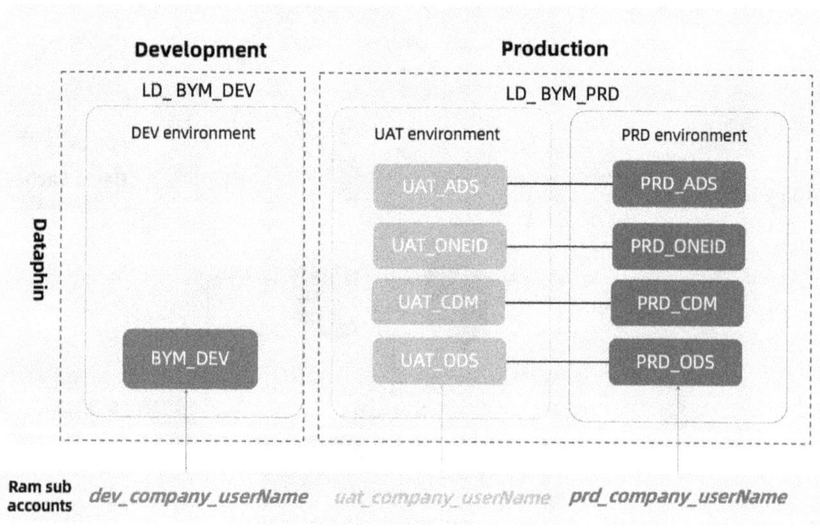

FIGURE 5.12 Business section overall design of Data Middle Office.

- The development environment (DEV) corresponds to the Dataphin business segment LD_ BYM_DEV: No real data or part of test data. It is used to be familiar with Dataphin products and debug code. Developers, operation and maintenance personnel and management personnel need to be given the corresponding data table permission.
- The user acceptance test (UAT) environment corresponds to Dataphin business section LD_BYM_PRD development environment: Development and testing use. Limit the scope of users, core developers use. Developers, operation and maintenance personnel, managers need to be given the corresponding data table permission.
- The production (PRD) environment corresponds to the Dataphin business segment LD_BYM_PRD production environment: Real production data environment. Fewer users, core developers use it in the online stage, management and operation and maintenance personnel as administrators after the official online. Developers, operation and maintenance personnel, managers need to be given the corresponding data table permission.
- At the same time, DEV environment, UAT environment and PRD environment use different cloud RAM sub-accounts to access. A RAM sub-account has and can only access one environment. Developers, operation and maintenance personnel, managers need to be given the corresponding data table permission.

5.2.4 Data Architecture Design

5.2.4.1 Theoretical Guidance

The design concept of data model in Data Middle Office follows the theory of dimensional modeling. Alibaba OneData public layer data unified modeling theory is the improvement and enrichment of dimension modeling theory in the context of the wide application of big data technology. The data model design under OneData theory system mainly includes dimension table, fact table and metric definition.

The dimension design of data model is mainly based on the theory of dimension modeling, and constructs consistent dimensions and facts based on the bus architecture of dimension data model.

The fact table design of the data model is based on the fact of the dimension model, combined with the specific practice of the data usage scenario, it is expanded to a certain extent, and the wide table design method is adopted.

The metric definition of data model is in the form of component. The metric is defined in the form of standardization, first, the standard definition is followed by production, the whole life cycle control ensures the unification of data caliber, reduces redundant construction, and emphasizes data reuse and sharing.

5.2.4.2 The Implementation Processes

The data architecture design is generally divided into three steps. The input of bus matrix design and detailed data design is data architecture design, as shown in Figure 5.13:

- Data asset inventory, based on technical research and system data research, sorts out the enterprise's core IT architecture and system architecture, and outputs data asset catalog and business process.
- Business process carding, based on data asset inventory and its output, carding the enterprise's core business nodes and actions, and combining with data asset catalog, producing business process and its corresponding data table.
- Consistency dimension, identifying consistency dimension information based on business process carding and its output.
- Data domain design, abstracting the core data domain of the enterprise based on business process carding and consistency dimension information.

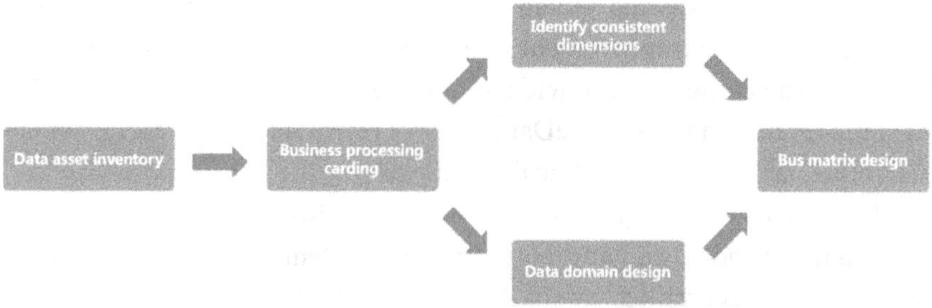

FIGURE 5.13 Data architecture design process.

- Bus matrix design, designing enterprise bus matrix based on data domain design, consistency dimension and business process.

5.2.4.3 Design Principles

Generally speaking, the data model construction of Data Middle Office follows the following basic principles:

- High cohesion and low coupling. What records and fields are the logical and physical models composed of should follow the most basic principles of high cohesion and low coupling in software design methodology. Mainly from the data business characteristics and access characteristics of two angles to consider: design the data with similar or related business and the same granularity as a logical or physical model; put together the data accessed at the same time with high probability, and the data with low probability of simultaneous access are stored separately.
- The core model and expansion model is separated, and the core model and the extended model system are established. The fields included in the core model can support core business. The fields included in the expansion model can support the needs of personalization or small reusability, and the expansion fields cannot be excessively intruded into the core model, which undermines the simplicity and maintainability of the core model.
- Public processing logic is sinking and single. The more common processing logic at the bottom, the more it should be encapsulated and implemented at the bottom of data scheduling dependence. It is not allowed to expose the public processing logic to the application layer to realize it, and avoid the common logic being realized simultaneously in multiple places.
- Balancing cost and performance, and appropriate data redundancy is used to exchange query and refresh performance, and redundancy or data replication technology should not be used too.
- Data can be rolled back, processing logic is unchanged, and the data results of multiple operation at different times are determined to be unchanged.

- Data consistency. The same field meaning must be the same in different tables, and the name in the specification definition must be used.
- Clear and understandable naming: The table name should be clear, consistent, easy to understand and use.

5.2.4.4 Model Architecture

- Business segment, according to the characteristics and needs of the business, the relatively independent business is divided into different business segments, and the metric or business overlap between different business segments is small. For example, Alibaba e-commerce business includes Taobao, Tmall, B2B and AliExpress (Figure 5.14).
- Specification definition, combined with industry data warehouse construction experience, design a set of data standard naming and management system, specification definition will be applied to model design.
- Model design, based on the theory of dimension modeling and the bus architecture of dimension modeling, constructs consistent dimensions and facts, the implementation of specification definition in the model design stage. At the same time, a set of table specification naming system for model implementation is designed.

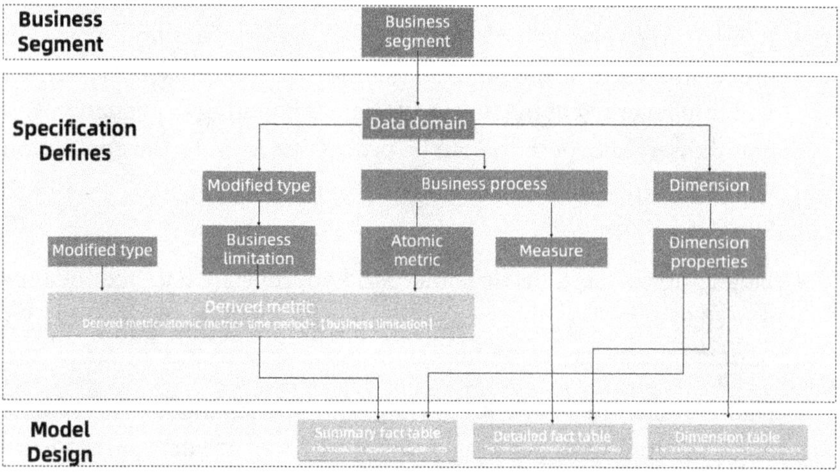

FIGURE 5.14 Data model architecture of Data Middle Office.

5.2.4.5 Data Hierarchy

The data in the Data Middle Office are divided into ODS, CDM store and ADS.

ODS (Operational Data Store): It mainly completes the introduction of structured and semi-structured data such as business system and log to the Data Middle Office, retains the original data of the business system, and is divided into buffer and data service area. The buffer design mainly maintains the consistency with the data source, and ensures that ODS can introduce the source data as it is, without any type conversion and data processing, and provides data service. The service area includes full detail data. Data are the full detail data obtained by type conversion or incremental merging of buffer data, which provides data for CDM and ADS layers.

CDM (common data model layer): It contains most of the data of the whole Data Middle Office, and is the basis of the Data Middle Office, so it is the first thing to ensure the data robustness of this layer. It mainly completes the public data processing and integration, establishes the consistency dimension, and constructs reusable detailed fact table and summary fact table for analysis and statistics. CDM layer is divided into DIM layer, DWD layer and DWS layer, the main functions are as follows:

- DIM (Dimension) is a common dimension layer in CDM layer. Based on the theory of dimension modeling, it constructs enterprise consistent public dimension data.
- DWD (Data Warehouse Detail) is a detail wide table layer in the CDM layer, which is used to store complete and detailed historical data. It is oriented to business process modeling and designed closely around business process. It expresses business process by obtaining measure describing business process, including referenced dimensions and measure related to business process. Its design goal is to provide flexibility and scalability for the subsequent data warehouse model. At the same time, it can directly provide data for the application layer when the DW layer cannot support the requirements. The stability of DWD layer will be affected by business system because of its high coupling with business system.
- DWS (Data Warehouse Summary) is a public summary data layer for storing detailed historical data in CDM layer, and is oriented to

analysis subject modeling. DWS is the core data layer of CDM, and it is the foundation of flexibility and expansibility for application layer.

- ADS (Application Data Service): It provides business or application-oriented data directly, and mainly deals with the architecture of personalized metric data, such as metric data processing without commonality or complexity (such as metric data, ratio data and ranking data). At the same time, in order to realize the demands of data application and data consumption, data assembly oriented to application logic (such as wide table mart, horizontal table to vertical table and trend metric string) is carried out.

5.2.4.6 Business Segment

The division of business segments is based on the use of functions. It is generally recommended to divide the business segments into test segment, development test segment, production segment and extraction segment.

- The DISTILL environment corresponds to the Dataphin business segment LD_ditill: Extraction environment, used in label production.
- The DEV environment corresponds to the Dataphin business section LD_PROD_DEV: Quasi production development test using it, used for model design and development, SQL development, synchronous task creation, etc.
- The PRD environment corresponds to the Dataphin business segment LD_PROD: Real production data environment, used for all kinds of task scheduling, operation and maintenance management.
- Test environment corresponds to Dataphin business segment LD_TEST_DEV and LD_TEST, to familiarize yourself with the product.

5.2.4.7 Data Domain

Data domain refers to a collection of business processes or dimensions that are oriented to business analysis. Among them, business process can be summarized as an indivisible behavior event, such as login, browsing, ordering, payment and receiving of consumer. In order to guarantee the vitality of the whole system, the data domain needs to be abstracted, maintained and updated for a long time, but it is not easy to change. When dividing data domain, it can not only cover all current business requirements, but also be included into existing data domain or expand new data domain without impact when new business enters. The data domain and descriptions of a retail customer Data Middle Office division are as follows (Table 5.1).

TABLE 5.1 Data Domain of a Retail Customer

Data Domain		English Abbreviations	Description
Member domain	member	mbr	Online and offline registered members and potential members information, a variety of basic information data
Commodity domain	Commodity	cod	Data of all products and commodities available for sale both online and offline, including basic information data of categories, brands, SPU, SKU and other relevant commodities
Trade domain	trade	trd	Trade includes online business processes from adding shopping cart to placing order, payment, delivery, refund and return and successful transaction, as well as offline business processes such as scanning code for goods in and out of storage
Marketing domain	marketing	mkt	It contains data deposited by various business processes in offline and online marketing activities
Channel domain	channel	chl	It includes basic data of terminal channels such as cross-industry alliance, e-commerce and stores, as well as business process data such as the creation, maintenance and closure of these terminal channels themselves
Log domain	log	log	All log data recorded by users visiting all platforms including official website, WeChat official account and e-commerce platform, such as GIO behavior data
Public domain	public	pub	Includes the organizational structure, employees, roles and common affairs of the enterprise
Business assistant domain	ba	ba	Including BA, BA training, salary and other data in each channel

5.2.5 Standard Specification Design

5.2.5.1 Common Coding Specification

The common coding specification mainly includes project naming specification, node naming specification, starting point naming specification, temporary table and intermediate table specification, the details are as follows (Table 5.2).

- Project naming specification

  ODS layer project takes ods as suffix, such as tbods; middle layer project takes cdm as suffix, such as tbcdm.

  There are two types of application layer projects: First, bi is used as the suffix for data report, data analysis and other applications, such as tbbi; second, app is used as the suffix for data products and other applications, such as magicapp.

TABLE 5.2 Node Naming Specification

Node or Resource Type	Naming Specification	Example	Note
Virtual node	vt_{virtual node and its meaning}	vt_mtods_start	root node of task
Synchronous importing task node	imp_s_{table name}	imp_s_source	
Synchronous exporting task node	exp_{table name}	exp_{table_name}	If multiple target databases exist, you can add the target database identifier suffix
SQL node	{output table name}	fct_mt_ord_evt_di	
Shell node	sh_{shell name}	sh_small_file_merge_bigfile	
MR node	mr_{shell name}	mr_mt_ilog_parse	
Python node	{shell name}.py	kv_concat.py	
Resource package-jar	{shell name}.jar	partitionBy.jar	
Resource package-python	{shell name}.py	uniq_concat.py	
UDF	{function name}	kv_concat	
Cross-schedule dependencies or checks	chk_{source table}	chk_ads_mt_user	Check whether the task is complete

- Node Naming specification
- Starting node naming specification

  Each project must set up a starting node, named vt_{project_name}_start, which indicates the task starting point of the project.
- Temporary table specification

  Temporary table is a temporary test table. It is a table that can be temporarily used once. It is a table that can be deleted at any time when the data are temporarily saved and then not used.

  Naming specification: tmp_{operation user abbreviation}_{custom business name}.

  Note: The table at the beginning of tmp should not be used for actual use, but only as a test and verification scenario.
- Intermediate table specification

  The intermediate table generally appears in the task and is the table of intermediate data temporarily stored in the task. The scope of the intermediate table is limited to the current task execution process. Once the task is completed, the mission of the intermediate table is completed and can be deleted (it is recommended to keep the intermediate table data for 3 days according to different scenarios, and 32 days for special scenarios for troubleshooting).

  Specification: mid_{table_name}_{numbers start at 0}.

- Where table_name is the target name of a task table. Usually, a task has only one target table. The purpose of adding table name here is to prevent table name conflict during free play and use number instead.

5.2.5.2 *ODS Layer Specification*

ODS layer specification mainly includes three aspects: directory planning, synchronization strategy and development specification, as follows.

1. Directory planning

   The directory planning of synchronization task is as follows:

   - Abbreviation of system name.

- Synchronization task, if there is a case of sub-database and sub-table, add the target table directory and store the synchronization tasks of different database tables in the directory.

  For example, in a retail enterprise, there are three types of business systems in the retail sector: supermarket, department store and shopping center. In the supermarket sector, one set of system for one store, and the system is the same, but the system is vertically split according to the store. First, it will be defined in the specification definition:

  Super Market, SM

  General Merchandise Market, GMM

  Shopping mall, Mall

2. Synchronization strategy

   - Small data scale table: Extraction processing strategy, database direct connection mode, full extraction; storage strategy, full table: Full storage by day, life cycle can be set long period according to business needs (such as 367 days or permanent storage).

   - Large data slowly changing dimension table: Extraction processing strategy, database log parsing mode, incremental extraction to incremental table, and then merge from incremental table to full table. The data volume of some tables is increasing with the development of business. If the full amount of data is synchronized by cycle, the processing efficiency will be affected. In this case, choose to synchronize only the new incremental data each time, then it is combined with the full data obtained in the last synchronization period; storage strategy, the incremental table can be set with a long period (such as 367 days or permanent storage), and the full table can be set with a longer period (such as 183 days or 367 days or 734 days) according to business requirements and storage resources.

3. Development specification

   - Table naming

     Incremental data: {project_name}.s_{source system table name}_delta. Full data: {project_name}.s_{source system table name}.

Increment table by hour: {project_name}.s_{source system table name}_{delta}_{hh}.

Full table by hour: {project}_name}.s_{source system table name}_{hh}.

When the names of the tables synchronized from different source systems to a project conflict, the name of the later table add the dbname of the source system, {project_name}.s_{source system table name}_{dbname}{_**}.

- Field naming

  The default field name is the source system field name.When the field name conflicts with the system keyword, the handling rule is add one '_col' suffix, namely: source field name_col.

- Partition definition

  The data collected every day use ds as the time partition field. If it needs to be split according to the business type, the secondary partition can be set, DS {business type or database name}.

  Partition of hourly collection, ds, hh {business type or database name} is recommended.

  It can be divided into three levels of ds, hh and mi once every 15 minutes.

- Task scheduling

Build virtual nodes for different business systems, named vt_ {business description}_node

For the synchronization task of sub-database and sub-table, build a unified target node to facilitate the later task dependence, the synchronization task of sub-database and sub-table needs to build a unified target node, which is defined by the name of the target table.

5.2.5.3 CDM Layer Specification

CDM layer specification mainly includes three aspects: dimension table design and development specification, fact table design and specification, summary logic table design and specification, the details are as follows.

1. Dimension table design and specification

   - Table naming

     {business segment/project_name}.dim_{business Bu/pub}_{dimension definition}[_{custom named label}], the so-called pub is similar to a business Bu, which can be shared by all Bu, such as time dimension. For example: tbcdm.dim_tb_item(Taobao commodity dimension core table).

- Task naming

  The development rule is that a task has only one target output table, and the task name is defined according to the data table name, such as dim_tb_item.

- Partition definition

  The partition is defined according to the business scenario, the common partition definitions: date (ds), hour (hh) and minute (mi). If necessary, you can design a custom partition.

2. Fact table design and development specification

   - Table naming

     {project_name}.fct_{business BU abbreviation/pub}_{data field abbreviation}_{business process abbreviation}[_{custom table naming label abbreviation}]_{refresh cycle ID}_{single partition incremental or full ID}. Pub indicates that the data include the data of multiple BU, single partition increment or full amount identification: i: represents increment, f: represents full amount.

     For example, tbcdm.fct_tb_trd_ordcrt_trip_di (fact table of Taobao air travel ticket order, daily refresh increment).

   - Task naming

     The development rule is that a task has only one target output table, and the task name is defined according to the data table name, such as fct_tb_trd_ordcrt_trip_di.

   - Partition definition

     The partition is defined according to the business scenario, common partition definition: date(ds), hour(hh), minute(mm). If necessary, you can design a custom partition, note: Different environments have restrictions on the partition level and the maximum number of partitions.

3. Design and development specification of summary logic table

- Table naming

  {project_name}.dws_{business abbreviation/pub}_{data domain abbreviation}_{data granularity abbreviation}[_{custom table naming label abbreviation}]_{abbreviation of statistical time period range}.

  For example: tbcdm.dws_tb_trd_byr_subpay_1d (1-day summary fact table of Taobao buyer's granular transaction staged payment).

- Field naming

  Derived metric naming: {atomic metric}_{time period}_{business qualification}.

  Statistical granularity: Set with the unified specification of dimension.

- Partition definition

  The partition is defined according to the business scenario, common partition definition: date(ds), hour(hh), minute(mi). If necessary, you can design a custom partition, note: Different environments have restrictions on the partition level and the maximum number of partitions.

5.2.5.4 ADS Layer Specification

- Table naming

  {project_name}.ads_{business BU abbreviation/pub}_{abbreviation of analysis topic}_{abbreviation of data granularity}_{abbreviation for custom table naming label}_{time period modifier}.

- Field naming

  The field naming is based on the output of DWS layer's field name. The new ratio type and other user-defined metrics, related naming is involved.

  The partition information is redundantly and physically added to the table field to avoid the exception of partition operation when ADS data table provides external data service

- Partition naming

  Partition can add new sub partition information based on the upper DWS data table according to the demand.

5.3 BIG DATA PLATFORM CONSTRUCTION

Section 5.2 focuses on the data desktop architecture design. This chapter follows the Alibaba Data Middle Office methodology and Alibaba Cloud product practice refers to the Data Middle Office architecture design, refines the detailed design of the Alibaba Data Middle Office, so as to complete the Alibaba Data Middle Office big data platform infrastructure construction. Essentially, the technical base of the data platform is the big data platform, which is a very important part of the data platform. Without the big data platform as the basis, there is no way to build data assets. All kinds of data applications based on data assets have become trees without roots and water without sources, but the data platform is not only the building of the big data platform itself.

5.3.1 Overview of Big Data Platform Construction

Under the framework guidance and constraints of architecture design, the construction of big data platform is the first step of Data Middle Office construction. For an enterprise, the goal of building a big data platform is to enable itself to have the ability of uniform data acquisition, mass data storage, big data distributed computing, data application development and testing, data analysis and display, data-related demands fast response. Refer to Figure 5.15, the big data platform is composed of data storage computing layer, data asset construction management layer and

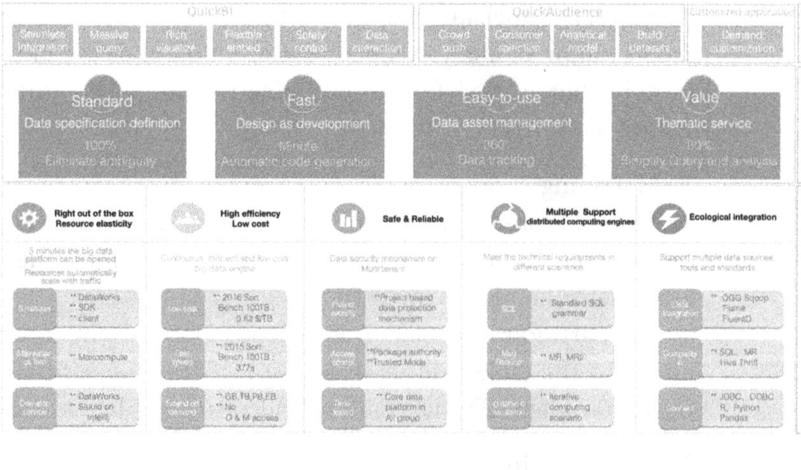

FIGURE 5.15 Basic architecture of big data platform.

data application layer. Each layer carries different functions and scenarios, which are supported by one or more product technology components. For an enterprise whose IT is not the main business, the data storage and computing requirements in the large data field are larger and more complex than traditional business application systems. Without a complete technical team and technical personnel, and without profound technical accumulation, this is a thing that cannot be imagined. A common Hadoop platform can take as little as 6 months to as long as 1–2 years from design to construction, regardless of the overall investment cost, even if the people and organization allow it.

Today on Alibaba Cloud, great changes have taken place, any enterprise, as long as the accumulation of data to allow, you can easily have the required in the form of volume rent cloud big data services, to build a set of their own big data platform, which can take as long as 1–2 weeks to complete. The flexibility of the cloud and the fast ability to provide resources have brought more traditional enterprises closer to emerging technologies on the technological level, promoted the transformation of enterprises to a data-centric governance model, enabled them to collect and manage their own production and operation data more scientifically, fully understand these data, analyze and mine them, and discover their past Business characteristics that have not been paid attention to or found can better guide the production and operation activities of enterprises and even faster business innovation.

If we compare building a big data platform to building a house, the product of the architecture design is the blueprint draft, which is not detailed enough to guide the construction. There are still many technical details, parameters and configuration that need to be further refined by the Data Middle Office architect. The architect will refine the design and explanation of the architecture from five aspects: technology architecture, environment deployment, cloud resource planning, account and permission settings and basic security. The detailed design documents will be output, and the implementation engineer can complete the platform environment by the detailed design documents.

In this process, the requirements for Data Middle Office architects are very high. To be proficient in large data technology components, a complete application architecture and knowledge of cloud products are required, as well as basic network and security skills.

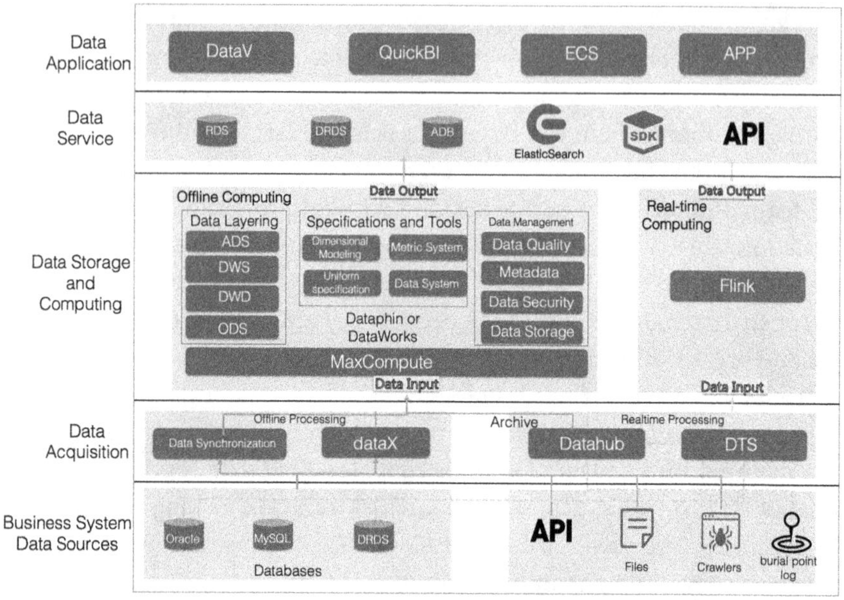

FIGURE 5.16 Data Middle Office technology architecture.

5.3.2 Detailed Technology Architecture

Section 5.3.2 has introduced the content of technology architecture design, which is a classic and very common technology architecture in the construction of Alibaba Data Middle Office. Under different conditions and scenarios, there are differences in the selection of product technical components. Let's start with Figure 5.16, focusing on more detailed analysis from the perspective of product technology selection.

5.3.2.1 Data Acquisition Layer

Data acquisition is also called 'data integration' during the construction of Data Middle Office, that is, the process of data extraction from the business system at the source side to the ODS layer in the Data Middle Office. This layer is the data collection and distribution center of the Data Middle Office and the business system. Its technology selection should not only meet the requirements of business scenarios, but also design and consider for the rich business system data sources.

The most important difference between the demands of upper-level business scenes lies in timeliness. Offline and real-time scenarios currently account for half of the big data field, which can be said to be equal.

Therefore, when data acquisition and access, we must be able to take into account the requirements of these two different timeliness scenarios at the same time.

Offline data collection pays more attention to data throughput and data source compatibility. Alibaba Cloud usually adopts the built-in data integration or synchronization module of big data construction management platform DataWorks or Dataphin. Mainstream commercial databases, open-source databases, some data warehouses (Vertica, HDFS, Hive, etc.) and CSV files can be used for offline data extraction. In some special cases, for example, single-table data with more than 100 million records need to be quickly synchronized offline to the cloud, and the concurrency and maximum bandwidth provided by a single task of data integration or synchronization module are limited. It is recommended to use script to call the built-in DataX tool for more flexible configuration to achieve better data synchronization efficiency.

According to the different data sources, the real-time data acquisition scheme is also different. Database data, such as RDS for MySQL and Oracle for real-time collection, DTS is the preferred data transmission service for real-time analysis and transmission of database logs; Message data, such as Kafka, MQ (Message Queue), real-time computing engine Flink support direct source docking; Application logs and file type data are collected in real-time, and the Loghub of Alibaba Cloud log service SLS is recommended for easy collection. Real-time Data collected through DTS and LogHub are first connected to Alibaba Cloud Data Bus DataHub, which is the streaming data service provided by Alibaba Cloud. It has the function of publish and subscribe streaming data, so that users can easily build streaming data-based analysis and applications. In addition to delivering the incoming real-time data streams to the real-time computing engine Flink, DataHub can also synchronously archive data to the offline computing engine MaxCompute.

The characteristics of the data source itself will also limit or determine the choice of data acquisition scheme. Structured data, semi-structured data and unstructured data are the data sources that the Data Middle Office may collect and summarize. There are many structured data processing methods, which are relatively mature. Semi-structured data such as information and CSV files are also better to be processed directly, while unstructured data such as pictures and videos need to be structured

TABLE 5.3 All Kinds of Data Sources Integration on Cloud

Data Source	Offline	Real Time
Oracle, RDS and other relational databases	Data Integration Data Synchronization	DTS+Datahub Datahub OGG Plugin Datahub Flume Plugin Datahub Canal Plugin
Super large table in relational database	DataX	DTS+Datahub
CSV file	Data Integration Data Synchronization DataX Datahub Fluentd Plugin	Not applicable
Burial point log Application log	Loghub+Datahub Datahub Logstash Plugin	Loghub Datahub Logstash Plugin
MQ, Kafka and other message middleware	Not applicable	Flink

conversion, can be direct identification and data processing by Data Middle Office.

Please refer to Table 5.3 for a summary of the collection tools and solutions of various data sources on Alibaba Cloud.

For some niche data sources, such as PostgreSQL and APIs, data acquisition policies need to be customized. We can use the Kafka connector to collect PostgreSQL data and write incremental data to Kafka and then connect to the DataHub. API data typically require custom development based on the API provider's SDK or subscription specification to be parsed and written to the cloud, typically to MaxCompute. The following is a Python program deployed on Dataphin to capture the thermal map information of personnel activity in the camera area provided by the API:

```
set
# -*- coding: UTF-8 -*-
@resource_reference{"AK"}
@resource_reference{"aliyun_sdk_core"}
@resource_reference{"pycryptodome"}
@resource_reference{"py_reid"}
from odps.tunnel import TableTunnel
from odps import ODPS
import json
import datetime
import os
os.system("mv pycryptodome pycryptodome-3.9.8-cp27-cp27mu-
manylinux1_x86_64.whl")
```

```
os.system("pip install pycryptodome-3.9.8-cp27-cp27mu-
manylinux1_x86_64.whl")
os.system("mv aliyun_sdk_core aliyun-python-sdk-core-2.13.19.
tar.gz")
os.system("pip install aliyun-python-sdk-core-2.13.19.tar.gz")
os.system("mv py_reid aliyun-python-sdk-reid-1.1.7.tar.gz")
os.system("pip install aliyun-python-sdk-reid-1.1.7.tar.gz")
from aliyunsdkcore.client import AcsClient
from aliyunsdkreid.request.v20190928 import
DescribeHeatMapRequest
#Please refer to more interfaces here "https://github.com/
aliyun/aliyun-openapi-python-sdk/tree/master/aliyun-python-sdk-
reid/aliyunsdkreid/request/v20190928"
import configparser
config = configparser.ConfigParser()
config.read('AK')
odps_akid = config.get("AK_INFO","odps_akid")
odps_akpwd = config.get("AK_INFO","odps_akpwd")
reid_akid = config.get("AK_INFO","reid_akid")
reid_akpwd = config.get("AK_INFO","reid_akpwd")

#-----------------------------------part 1 request
API----------------------------
def requestAPI():
    client = AcsClient(reid_akid,reid_akpwd,"cn-beijing")
    request = DescribeHeatMapRequest.DescribeHeatMapRequest()
    o = ODPS(odps_akid,odps_akpwd,'ls_group_
ods',endpoint='http://service.cn.maxcompute.aliyun-inc.com/
api')
    t = o.get_table('ls_group_ods.s_xj_list_emap')
    t2= o.get_table('ls_group_ods.s_xj_list_location')
    with t.open_reader(partition='ds=${bizdate}') as reader:
        for record in reader:
            print("record:%s"%record["emapid"])
            request.set_EmapId(record["emapid"])
            request.set_Date('${date}')
            with t2.open_reader(partition='ds=${bizdate}') as
reader2:
                for record2 in reader2:
                    if record["locationid"] ==
record2["locationid"]:
                        request.set_StoreId(record2["storeid"])
                        response = client.
do_action_with_exception(request)
                        data_list = json.loads(response).
get("HeatMapPoints").get("HeatMapPoint")
                        print(response)
                        print("data_list:%s"%data_list)
                        insertData(data_list,record["emapid"],
record2["storeid"])
```

```
#-----------------------------------part 2   insert 
data-------------------------
def insertData(data_list,emapid,storeid):
    o = ODPS(odps_akid,odps_akpwd,'ls_group_
ods',endpoint='http://service.cn.maxcompute.aliyun-inc.com/
api')
    t = o.get_table('ls_group_ods.s_xj_list_describe_heat_map')
    tunnel = TableTunnel(o)

    upload_session = tunnel.create_upload_session(t.name,
partition_spec='ds='+'${bizdate}')

    with upload_session.open_record_writer(0) as writer:
        for item in data_list:
            record = t.new_record([storeid,emapid,item.
get("X"),item.get("Y"),item.get("Weight")])
            writer.write(record)
    upload_session.commit([0])

if __name__ == "__main__":
    o = ODPS(odps_akid,odps_akpwd,'ls_group_
ods',endpoint='http://service.cn.maxcompute.aliyun-inc.com/
api')
    t = o.get_table('ls_group_ods.s_xj_list_describe_heat_map')
    t.delete_partition('ds='+'${bizdate}', if_exists=True)
    t.create_partition('ds='+'${bizdate}', if_not_exists=True)
    requestAPI()
```

In order to facilitate readers to practice in their work, we have made a brief and concentrated introduction to several products that are often relied on by the data acquisition layer of Alibaba Cloud Data Middle Office.

1. Data Integration Module of DataWorks and Dataphin

   As two mainstream big data construction and management platforms on Alibaba Cloud, Data Integration is a basic capability that DataWorks and Dataphin must provide. In the early days, Data Integration only had offline data synchronization function, focusing on the development of synchronization ability from heterogeneous data sources to MaxCompute on the cloud in complex network environment. With the development of real-time computing technology and application scenarios, these two products have now supported real-time data synchronization.

   I. Offline data synchronization

      Data integration is mainly used for offline (batch) data synchronization. The offline (batch) data channel provides a set of abstract data extraction plug-ins (reader) and data writing plug-ins (writer) by defining the data source and data set of data source and destination, and designs a set of simplified intermediate data transmission format based on this framework, so as to realize data transmission between arbitrary structured and semi-structured data sources. As shown in Figure 5.17:

In addition to data synchronization task, facing the big data application in various industries, data integration will be a lot of appeal, including large amounts of data simply and efficiently configuring synchronization tasks, integrating a variety of heterogeneous data sources, realizing light source data preprocessing, tuning data synchronization tasks (such as fault tolerance, speed limits and concurrency), etc. Therefore, Dataphin also upgrades offline data synchronization to pipeline tasks.

The pipeline task takes multiple types of components and simply drags, configures and assembles them to generate an offline single pipe. The process and transformation components can preprocess the data of the data source (such as cleaning, transformation, field desensitization, calculation, merger, distribution and filtering) to reduce the ETL work after the data integration.

Data integration supports most structured and semi-structured data sources, such as mainstream commercial databases, Oracle, SQLServer, MySQL and DB2; cloud database RDS various engine versions, Redis and TableStore; it also includes unstructured files with structured data, such as OSS, FTP and CSV.

II. Real-time data synchronization

Real-time synchronization synchronizes data changes in the source repository to the target repository in real time. For

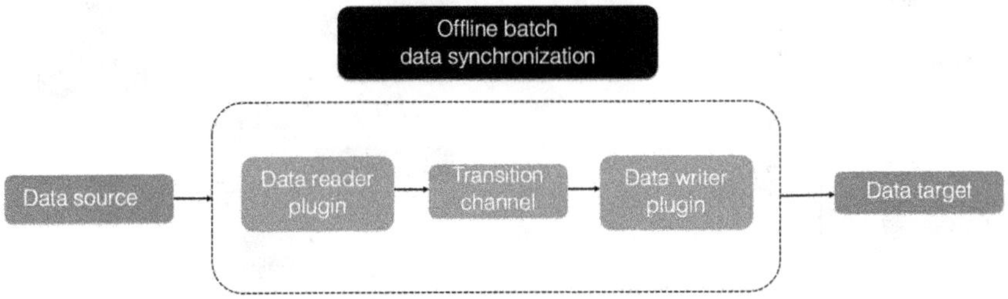

FIGURE 5.17 Basic principle of offline data synchronization.

example, adding data, modifying data and deleting data. If you synchronize with the full amount of historical data and perform real-time synchronization, your source database's data will be synchronized to the target database from the start of the task. Real-time synchronization can keep the target database corresponding to the source database and update it in time.

Real-time synchronization of data integration includes three basic plug-ins: Real-time read, transform and write, which interact with each other through an internally defined intermediate data format. A real-time synchronization task supports multiple transformation plug-ins for data cleaning and multiple write plug-ins for multiple output functions. For some scenarios, a whole DB real-time synchronization solution is supported, allowing developers to synchronize multiple tables in real time at once. Please refer to Figure 5.18:

Real-time data synchronization supports fewer data source inputs than offline data synchronization. It supports MySQL, SQL Server and Alibaba Cloud PolarDB. Other real-time data sources need to be connected to the cloud in more general ways, such as DTS, messaging middleware Kafka, LogHub (Special log access services, this book will not be introduced), or DataHub.

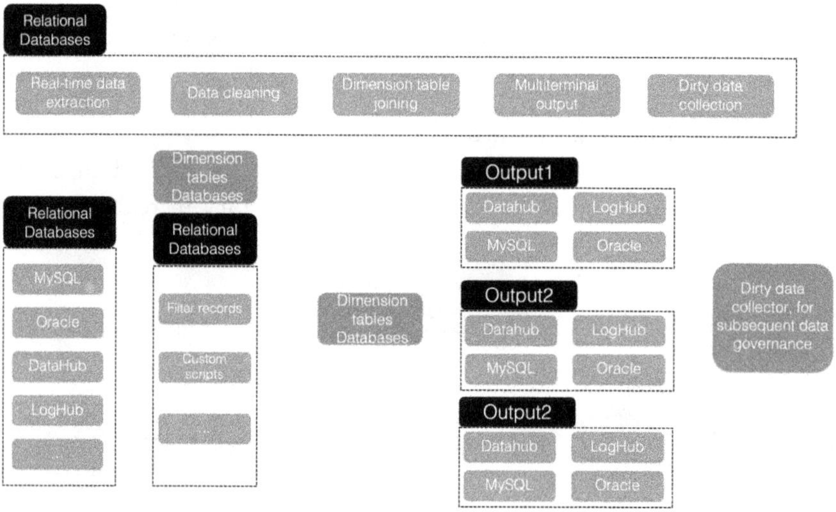

FIGURE 5.18 Fundamentals of real-time data synchronization.

2. DTS

DTS supports data transmission between data sources such as relational databases, NoSQL and big data (OLAP). It is a data transfer service that integrates data migration, data subscription and real-time data synchronization. Data transmission is dedicated to solving the problem of remote, millisecond asynchronous data transmission in public and hybrid cloud scenarios. Its underlying data flow infrastructure is Alibaba's Double 11 Shopping Festival remote multi-active infrastructure, which provides real-time data flow for thousands of downstream applications. It has been running online stably for 6 years. You can easily build secure, scalable and highly available data architectures using data transfer.

In Data Middle Office's scenarios, DTS is mainly used to handle real-time data integration that cannot be covered by DataWorks and Dataphin, such as real-time data acquisition for Oracle databases.

3. Kafka

Apache Kafka builds a high-throughput, scalable and distributed message queue service, which is widely used for log collection, monitoring data aggregation, streaming data processing, online and offline analysis, and is one of the indispensable products in the business platform and big data ecosystem. In real-time data integration scenarios, both open-source version of Kafka and Alibaba Cloud Kafka are supported. Alibaba Cloud provides fully hosted services. Users do not need to deploy operations and maintenance. It is more professional, reliable and safe. It is recommended that users use Alibaba Cloud Kafka.

In the current application architecture, the design of applications and services is more about business flow and loose-coupling architecture through messages. However, if the real-time data integration and DTS tools are not convenient to collect database changes and realize data cloud, it will become a more general real-time data collection scheme to write a program to consume Kafka message and write the message to DataHub.

4. Datahub

Alibaba Cloud Streaming Data Processing Platform DataHub is a platform for streaming data. It provides publishing, subscribing and distributing capabilities for streaming data, allowing you to easily build streaming-based analysis and application. The DataHub service can continuously collect, store and process a large amount of streaming data generated by various mobile devices, applications, web services, sensors, etc. Users can write applications or use the stream computing engine to process streaming data written to the DataHub, such as real-time web access logs, application logs, events and so on, and produce a variety of real-time data processing results such as real-time charts, alarm information, real-time statistics and so on.

The DataHub service is based on Alibaba Cloud's self-developed Apsara OS with high availability, low latency, high scalability and high throughput.

The DataHub service also provides the ability to distribute streaming data to a variety of cloud products and currently supports distribution to MaxCompute (original ODPS), OSS and so on. DataHub can also be seamlessly connected to StreamCompute, Alibaba Cloud flow computing engine, which allows users to easily analyze streaming data using SQL.

5.3.2.2 Data Storage and Computing Layer

1. Data Storage and Computing Engine

   After data acquisition layer's data are integrated to cloud, according to the difference between offline and real-time paths, data enter the technical core of the Data Middle Office, which is the data warehouse. There are two main types of data warehouse design, Lambda and Kappa, which can handle real-time and batch data differently.

   Referring to Figure 5.19, this is a typical Lambda architecture, where offline batch computing and real-time incremental computing are separated, and the results of batch and real-time computing are integrated or not integrated at the data service layer to provide services outside. It requires maintaining two sets of code that run on batch and real-time computing systems, and that produce consistent results, which is often a tricky problem.

   Figure 5.20 is an example of the Kappa architecture, which is different from the core idea of the Kappa architecture. After data collection, it solves the problem of full data processing by improving the stream computing system so that both real-time and batch processes

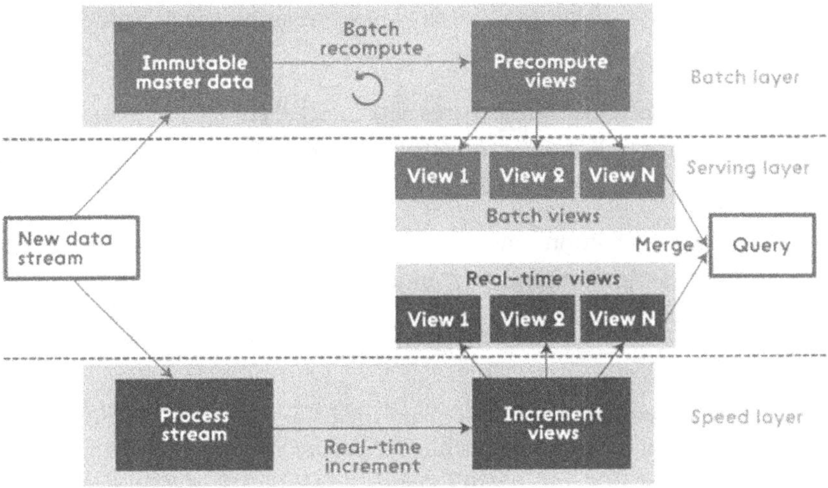

FIGURE 5.19 Lambda architecture.

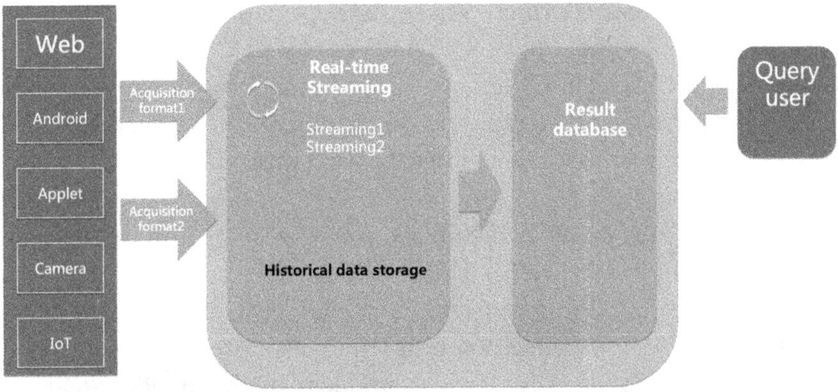

FIGURE 5.20 Kappa architecture.

use the same set of code. In addition, the Kappa architecture considers that historical data will only be recalculated if necessary, and that if recalculation is required, many instances can be started under the Kappa architecture for recalculation.

From the point of view of user development, testing and operation, the Kappa architecture allows developers to face only one framework, which makes development, testing and operation less difficult. This is a very important advantage. However, business needs should take precedence over technological experiences, such as the following scenarios:

- Data initialization requires absolute full storage, temporary storage of the Kappa architecture cannot hold data for such a long time, and rollback costs are too high.
- Data correction (messages out of order, tasks start and stop, transfer and other non-standard operations).
- Restore historical scenes (verification issues).

These scenarios often require backtracking of long periods of historical data, which is a huge challenge to the standard Kappa architecture. Enterprises must build Data Middle Office to do data analysis and mining from the precipitation of past historical data. Therefore, Lambda is also increasingly used in Alibaba's data storage and computing architecture designed for external customers.

MaxCompute (formerly known as ODPS) is usually used for offline storage and computing of Alibaba Data Middle Office, which is a fast and fully managed EB-level data warehouse solution. MaxCompute is dedicated to the storage and calculation of batch structured data, providing solutions to massive data warehouses and analytical modeling services.

In Alibaba Data Middle Office, real-time computing storage and computing are separated. Alibaba Cloud real-time computing Flink version is usually used in the calculation part. It is an enterprise-level, high-performance real-time large data processing system based on Apache Flink. It is officially produced by the founding team of Apache Flink. It has a global unified commercial brand, is fully compatible with open-source Flink API, and provides rich enterprise-level, value-added functions. The results of the calculation need to flow between multiple real-time computing tasks. The channel of flow is actually where the data are stored. The data flow between two real-time computing jobs, we use Kafka messages or Datahub topics most often.

2. Data Asset Construction and Management

Data storage and computing has little business attributes and can only be called a data asset if it can provide direct services to the business, so data need to be built and managed according to the needs of the business. Alibaba Cloud offers two products to help companies build and manage their own data, DataWorks and Dataphin.

DataWorks (original big data development kit) is an important PaaS (Platform-as-a-Service) platform product in Alibaba Cloud. It provides customers with comprehensive product services such as data integration, data development, data maps, data quality and data services, and one-stop development and management interface to help enterprises focus on data value mining and exploration.

Dataphin builds OneData system (OneModel, OneID and OneService) following the big data deposited by Alibaba Group for many years. It integrates products, technologies and methodologies into one, providing customers with one-stop full-link intelligent data construction and management services including data integration, specification definition, data modeling research and development, data extraction, data asset management and data services, which helps enterprises create their own standard unified, capitalized, service-oriented and closed-loop self-optimizing intelligent data system, driving innovation.

How to choose? Dataphin is most closely integrated with Alibaba's Data Middle Office methodology. OneData theory is fully integrated in the product, providing users with configurable and visual dimension modeling, metric processing and label extraction functions. At present, Alibaba Cloud's customers in the new retail industry all choose Dataphin, and some customers in the financial industry also begin to use Dataphin to build Data Middle Office. For customers related to government affairs, due to the relatively weak demand for unified modeling in their business, and more emphasis on data aggregation, centralization, as well as subsequent governance and sharing, DataWorks, which is more universal, has stronger applicability.

5.3.2.3 Data Service Layer

According to OneService requirements, the Data Middle Office must build a unified data service layer, which is not shared with any level of the big data platform data warehouse, that is, the ODS, CDM and ADS layers of data warehouse do not provide services directly to the outside, but need to be exported or encapsulated to provide services. Data services are divided into three main categories:

1. Database tables, which are also the most common type, are imported from the ADS layer of the Data Middle Office by data integration or synchronization tools in reverse to the database. This database is still a data source managed by the big data platform, which in turn provides data query services for insight analysis reports and big data batch queries. RDS for MySQL is the most commonly used data service database. When the query mode is simple, the number of tables queried is less than 3, and the maximum amount of data in a single table is less than 10 million records, the total amount of data is less than 200 GB, using RDS for MySQL is the most cost-effective choice. Cloud-native data warehouse ADB is recommended when the query mode is complex, such as aggregate and window functions, 3 or more table-related queries, tens of millions or more of single-table data and 200 GB or more of total data. ADB is a new generation cloud-native data warehouse that supports high concurrent and low-latency queries, fully compatible with MySQL protocol, and SQL:2003 Grammar standards allow instant multi-dimensional analysis perspectives and business exploration of large amounts of data to quickly build an enterprise cloud data warehouse. Product specifications are optional on demand, balancing cost, capacity and performance for BI query applications. In some special scenarios, such as data applications that need to support fuzzy search and word breaking search, data synchronization to search engines such as Elastic Search is very appropriate. In the history of Data Middle Office external delivery, we also planned to expose the ADS layer hosted by MaxCompute to applications. The application side uses MaxCompute SDK to access data in batches. This design provides convenience, saves storage space of data service layer, but brings potential data security risks. This design idea is no longer used.

2. API interface, which is the future trend, can realize the unified market management of data, effectively reduce the threshold of data opening, while ensuring the security of data opening. API interfaces use a Serverless architecture. Customers only need to focus on the query logic of the API itself. They don't need to care about infrastructure such as the operating environment. Data services will prepare you for computing resources, support flexible scaling and have zero operating and maintenance costs. For a small amount of data

accessed at high frequency, such as a single record accessed through key value, it is very suitable to use API interface to provide synchronous external. The scenario of bulk data access is more asynchronous data reading, which is not suitable for API interface.

3. Message, which is a data service derived from the new requirements of data embedding business enabling business. For example, a shopping mall promotes activities during holidays. In order to better identify the types of consumers, shopping malls label consumers differently through their basic information, shopping information and behavior information, forming different consumer groups. Individual behaviors of different consumers flow to the Data Middle Office continuously in real time. Data Middle Office uses real-time computing power to produce consumers' labels in real time. The same label of consumers will change under the stimulation of promotional activities. At this time, the newly generated label data are produced by messaging middleware and delivered to topic. After consumed by downstream marketing application systems, it can identify the changes of consumer labels in time, and automatically push more suitable promotional activities to consumers in real time according to the rules, so that consumers can have a better activity experience. The messaging middleware is a powerful technical support for Data Middle Office to provide direct value for business.

5.3.2.4 Data Application Layer

Strictly speaking, data applications are not in the scope of Data Middle Office, but a large part of the value of Data Middle Office needs data applications to reveal and transfer. Based on Alibaba's experience in building data applications, the current data applications are mainly divided into three categories:

1. Analysis insight report is different from traditional report. Data Middle Office analysis insight report emphasizes data visualization ability, interactive analysis ability and free exploration and analysis ability of business personnel. QuickBI is a new generation of intelligent BI service platform for cloud users. QuickBI can provide massive data real-time online analysis service, support drag-and-drop operation and rich visualization effect, help customers to easily

complete data analysis, business data exploration, report making and other work. Therefore, we choose QuickBI to implement such applications as analysis insight reports.

2. Digital large screen, analysis insight report data visualization can meet the requirements for business decision-making scenarios, but in some scenarios, a large screen to vividly and real-time display of key metrics. DataV data visualization is a product that uses visual applications to analyze and present large amounts of miscellaneous data. DataV aims to let more people see the charm of data visualization, help non-professional engineers to easily build professional-level visualization applications through graphical interface, to meet customer exhibition, business monitoring, risk warning, geographical information analysis and other business display needs. DataV is often used when data visualization is greater than analysis needs.

3. Data application, report and large screen are the way of using standard products to meet the needs of enterprises to see and use data. In today's increasingly digital development, only standardized products are unable to meet the rich data application needs of enterprise business. For example, digital marketing needs a tool platform based on consumer labels and marketing data, or member operation also needs a label system generated by consumer basic data, behavior data and transaction data. This kind of application relies heavily on data, and needs the Data Middle Office to provide data in real time or offline to drive business development. At this time, we will suggest that customers use cloud resources for customized development. Similar to traditional IT applications, customized development applications are built on the application server ECS, database service RDS, cache service Redis, load balancing service SLB, object storage service OSS and other basic cloud resources.

5.3.3 Environment Deployment

After the detailed design is completed, the next task is to deploy the overall platform environment. Alicloud is an important carrier of the big data platform. In addition, we need to consider the integration of customers' existing environment and cloud environment. Therefore, environment deployment mainly includes hybrid cloud network deployment and Data Middle Office product initialization.

5.3.3.1 Hybrid Cloud Network Deployment

Figure 5.21 is a complex deployment diagram of a hybrid cloud network. It is assumed that the customer does not have any business on the cloud before building the Data Middle Office. His business systems are distributed in IDC owned or rented by many cities in China, such as Beijing, Shanghai, Guangzhou and Hangzhou. Alibaba Cloud Data Middle Office needs to extract data from multiple business systems.

First of all, there are two very important decisive factors to consider how to choose the main region of the Data Middle Office in the cloud. First, if the customer has decided to migrate the company's IT system to the cloud in the future, it is recommended to choose the Alibaba Cloud region of the city where the company's headquarters or information department are located. For example, in Beijing, Alibaba Cloud North China 2 Beijing region or North China 3 Zhangjiakou region can be chosen. If you are in Guangzhou, you can choose from Shenzhen, Heyuan and Guangzhou. If the first condition does not hold, then sort out the city where the business systems needed by the Data Middle Office are mainly concentrated, and the Data Middle Office is also deployed near the corresponding city. This choice is mainly to ensure the network access delay and save bandwidth resources from IDC and office to VPC of Alibaba Cloud. Alibaba Cloud's VPC network across multiple regions also has costs.

Next, we need to consider the IP network segment. Alibaba Cloud provides three large network segments for customers: class A IP address segment 10.0.0/8, class B IP address segment 172.16.0.0/12 and class C IP address segment 192.168.0.0/16. Customers can choose these three types

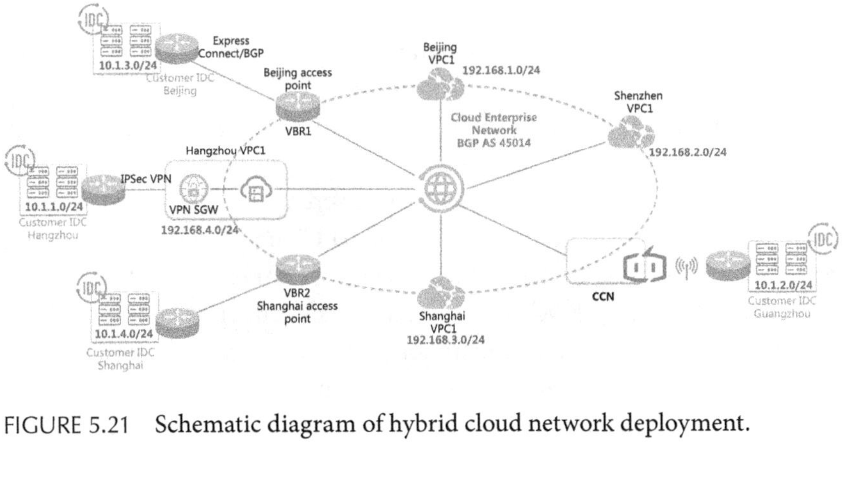

FIGURE 5.21 Schematic diagram of hybrid cloud network deployment.

of addresses at will, but they should ensure that the address segment on the cloud is within the overall IP address planning range of the group, and the IP address conflict between on cloud and off cloud is not allowed. Once the VPC address on cloud is designated, it cannot be modified. Therefore, it should not be too large or too small. Generally, it is appropriate to use a 16–20 bit mask. There are also some best practices for the planning and implementation of VPC that readers need to understand:

- When system is deployed in the same region, the resources needed for product deployment must be sufficient when selecting regions and zones.
- A logical router can be created in VPC, and multiple logical switches can be associated under a logical router.
- For the deployment location of logical switch, we can balance the latency and reliability requirements of different services, and choose to deploy in the same Available Zone or different zones according to the resource quota of the zone.
- VPC deploys security groups to control the access of ECS to the same network segment.
- Cloud firewall is deployed at VPC exit for level 4 and level 7 security protection and access filtering.
- The routing policies of VPC and cloud enterprise network can help to control the access policies between different networks in detail.
- You should deploy EIP, SLB, DNAT gateways and other products, which associate with shared bandwidth packages to maximize bandwidth utilization at VPC exports. When VPC needs to access the Internet, it is recommended to use SNAT to improve security.

Third, how to connect the cloud and the on-premise environment to form a unified hybrid cloud architecture. Alibaba Cloud provides a rich range of product solutions. To solve the problem of connectivity between IDC and VPC on the cloud, you can choose physical dedicated line, VPN or smart access gateway. The comparison of the three schemes is shown in Table 5.4: the physical dedicated line is suitable for group enterprises and needs stable and broad bandwidth, such as 1 GB or more; VPN is more

TABLE 5.4 Comparison of Network Interconnection Schemes on Cloud and On-Premise

Item	Express Connect	Smart Access Gateway (SAG)	VPN Gateway
Link	Lease line	Hybrid link (lease line+Internet)	Internet
Cost	High	Medium	Low
Quality	High	Medium	Low
Delivery cycle	Long(moths)	Medium (hardware version, weeks; software version, minutes)	Short (minutes)
Business scenario	Enterprise migrates to cloud	Enterprise migrates to cloud, interconnect of branches	Enterprise migrates to cloud

flexible than the physical dedicated line, but it can also be satisfied for enterprise applications with 100 m bandwidth; smart access gateway is more suitable for retail enterprises, such as the scene that multiple stores need to quickly access to the cloud.

When the number of VPCs exceeds two, the advantages of cloud enterprise network will be highlighted. As long as all VPCs need to be interwoven are added to the cloud enterprise network instance, users do not need additional configuration. The network realizes automatic forwarding and learning of multi-node and multi-level routing through the controller, so as to realize fast routing of the whole network convergence. Cloud enterprise network also has high availability and network redundancy. There are multiple groups of independent redundant links between any two points in the whole network. Based on the above advantages, cloud enterprise network is now fully replacing the high-speed channel.

Through the appropriate product mix, users can quickly build a network of hybrid cloud and distributed business system, which is used to support the data transmission needs of Data Middle Office.

Finally, it is the deployment of application and system. As an IT system, the big data platform follows the standard IT system environment design specifications, and it is recommended to deploy four environments, namely, DEV, integrated test environment (SIT), UAT environment and PRD. The PRD environment is physically isolated from other environments, which can be realized by VPC. The DEV, integration test environment and user acceptance environment can be deployed in different virtual switches to achieve logical isolation under the same VPC.

5.3.3.2 Initialization of Data Middle Office Products

The overall technical solution of Data Middle Office consists of a series of products, among which the core ones are: intelligent data construction and management platform Dataphin, data visualization analysis tool QuickBI and consumer operation intelligent user growth tool QuickAudience.

1. Dataphin

   - Set the super account access key (AK) to ensure that the main account of Dataphin can have sufficient authority to manage the ordinary account.
   - Configure the computing engine. Offline computing engine supports Alibaba Cloud MaxCompute or AnalyticDB for PostgreSQL, and real-time computing engine generally adopts Alibaba Cloud real-time computing Flink.
   - New business sectors and projects can be divided and configured by referring to the architecture design of the Data Middle Office.
   - Add users, Dataphin supports importing accounts from Alibaba Cloud RAM system. Only when added as a Dataphin user, RAM accounts added can access Dataphin.
   - User privilege management, after a user is added into Dataphin, it does not have project space permissions by default. It needs super administrator or project space administrator to add them into the project space. Refer to the accounts and permissions part of Section 5.3.4 for an introduction to Dataphin's multi-user roles.

2. QuickBI

   - Create a new administrator account, similar to Dataphin.
   - Add users, you should operate in the organization management function. The new users can be other main accounts of Alibaba Cloud or RAM accounts. New users can be divided into organization administrators and ordinary users according to their roles.
   - Divide user groups and workspaces, and group users according to a certain organizational structure. At the same time, delimit workspaces and assign users to workspaces. Different users have

different member types. For an introduction to QuickBI multi-member types, see the accounts and permissions part of Section 5.3.4.

3. QuickAudience

- Create a new administrator account, similar to Dataphin and QuickBI.
- Add users, you should operate in the member management interface of the organization. The new users can be other Alibaba Cloud main accounts, also can be RAM account. New users can be divided into organization administrators and ordinary users according to their roles.
- Divide users into groups and group them according to a certain organizational structure for better management.

5.3.4 Security Solution Deployment

5.3.4.1 Security System of Data Middle Office

The author believes that the security of Data Middle Office can be realized through product, technology and system. First of all, the data security of the Data Middle Office should be based on Alibaba Cloud's cloud platform and cloud products. Through the security capabilities of the cloud platform itself and cloud products, the data security can be guaranteed from the basic platform level. On the basis of cloud capability, we can guarantee the security of data assets in the whole stage before, during and after the event by means of system and technology.

The construction of Data Middle Office on Alibaba cloud platform will use a large number of Alibaba Cloud public services or products. As a cloud service provider, Alibaba Cloud public cloud platform has passed the level 3 protection certification of China Ministry of public security, and Alibaba Financial cloud has even passed the level 4 protection certification of China Ministry of public security. Each product has its own security function module to ensure the user's data security and system security. Readers can learn more about it on Alibaba Cloud's official website. This book does not elaborate on it.

Besides platform and product, how to guarantee the security of data platform from the system specification and technical scheme? The author is engaged in three stages: before, during and after.

1. Advance work

Nip in the bud, cure in no disease and security work are important in advance. We start from two aspects of management system and technology implementation:

I. Management system

Before the data is really touched by the builders and users, the data security work should be carried out through the way of management system.

- Security organization construction: A complete security organization should include decision-making level, management level, implementation level, grass-roots level and audit supervision level.
- Personnel security training: Personnel who may be involved in data need to continuously improve their data security ability, and cultivate their data management ability, data operation ability, data technology ability and data compliance ability.
- Daily security management regulations: Through the formulation of security management regulations, data security requirements are put forward for all involved personnel. Stakeholders need to clearly understand what to do and why not to do, so as to use data safely and normatively.
- Confidentiality agreement: Confidentiality agreement is used to restrict the behavior of personnel who can directly or indirectly touch data. In case of violation of relevant provisions, corresponding punishment shall be given according to this confidentiality agreement.
- Data application approval process: The use of data is not a simple process customization, and the appropriate data service delivery mode needs to be selected according to different scenarios. In addition, the application and approval process of data under different security levels may also be different. In the system, we need to determine the data application approval process in different scenarios, so that the data can be used safely and reasonably.

- Operation and maintenance guarantee system: It includes not only data platform and system operation and maintenance guarantee system, but also data task operation and maintenance guarantee system.

II. Technology implementation

- Data asset sorting: Data asset detail sorting refers to sorting out the scope of all data assets to form a data asset list. According to the list of data assets, data security policies can be specified and real-time management can be promoted.

- Data classification: For the existing data, determine the sensitivity level of the data and define the sensitivity level of the data, so as to provide the basis for the subsequent data security work. In government and financial institutions, data sensitivity markers are usually divided into four categories: level 0 (unclassified), level 1 (confidential), level 2 (sensitive) and level 3 (highly sensitive). It is suggested that the label security of the personnel in the Data Middle Office should be set to level 0 by default, and the default level of the data should be set to level 0. Ordinary data authorization is completed through the authorization system. For sensitive data, the sensitive level of the data is set according to the division. When the data security level is higher than the personnel security level, the personnel cannot access the modified data. It is necessary to adjust the personnel security level after applying for approval, or to make the data desensitized and get the low sensitive data before authorizing the personnel to use it.

- Account authority design: In order to effectively sort out the data authority, it is convenient to register and manage the authority and authorization of priority data, and it is also convenient to stop loss and trace the source of data problems in case of problems in subsequent data. In the next section, we will focus on the account permissions of several important products in the Data Middle Office.

- Specification of data construction: One of the important reasons for the safety problems is that the production operation is not standardized. If there is no agreed and standard practice and everyone implements it according to their own understanding, the result may be uncontrollable and unsafe. It is necessary to define the data construction stage, including data cloud, data development, data modeling and other stages of data specification, and strictly follow the specification in the actual construction stage, so as to better avoid the security risks in the process of data construction.

- Data acquisition security: Mainly including data source authentication security, data transmission security. Before the internal and external data sources are collected into the Data Middle Office, a variety of technical means and strategies are needed to ensure the data security and reliability. There are data transmission steps for the data source's data entering and exporting to the cloud platform. It is necessary to ensure the data security of data transmission, such as using HTTPS to encrypt the link.

- Data storage security: MaxCompute is the precipitation of Alibaba Cloud's underlying data storage and computing technology, and can undertake the summary storage and computing tasks of PB-level data. The underlying uses of mature encryption technology, key management and multiple backup recovery technology are to improve the high availability and security of data storage.

- Data computing security: There are many scenarios in the process of data computing in the Data Middle Office, such as number of offline data warehouse layered with sensitive data encryption, data access across the project space is authorization mechanism and data mining environment isolation, etc.

- Data using security: Data need to reflect the value by providing external services, and the corresponding data sharing process needs to be standardized according to different data use methods in the use process. At present, there are three main ways to use data: the most common data API, data table and data batch export. Different ways are applicable to different security measures.

- Data destruction security: Alibaba Cloud has established a security management system for the whole life cycle of equipment (including receiving, saving, placement, maintenance, transfer, reuse or scrap). The access control and operation monitoring of equipment are strictly managed, and the equipment maintenance and inventory are carried out regularly. Alibaba Cloud establishes a data security erasure process on discarded media. Before disposing of data assets, check whether the media containing sensitive data and legitimate authorized software has been erased, degaussed or bent, and cannot be recovered by forensic tools. When some hard copy materials are no longer needed due to business or legal reasons, they should be physically destroyed, or the damage certificate of the third party of data processing should be obtained to ensure that the data cannot be reconstructed.
- Data application security: The downstream of the Data Middle Office – data application is also the security work that must be concerned about. The correct use of various security capabilities on the cloud platform also affects the security degree of data application. When applying for cloud platform resources, we need to use cloud resources reasonably according to best practices. Before the trial run/launch of the cloud platform, it is also necessary to verify whether these best practices have been met according to the best practices again, so as to ensure the data security on the cloud.

2. Work in the process

When a security incident occurs or a major event requires convoy protection service, it is necessary to establish a mechanism for the discovery, handling and upgrading of security incidents as well as technical contingency plans to reduce the losses caused by security incidents.

I. Management specification

- Safeguard norms: In the big promotion, holidays or other important time points, the business system focuses on the security during the guarantee period, to avoid the occurrence of hacker attacks, tampering with data and then causing adverse effects of security incidents, we need to do a good job in the safeguard norms from the aspects of network security, host security, application security, security inspection system, etc.

- Emergency support system: In order to ensure the system security and prevent the occurrence of emergency events such as deliberate attack, destruction of network system and data security, the emergency handling mechanism scheme is used when encountering major security problems and failures, including the emergency plan organization system, system failure classification, failure handling procedures and emergency handling processes.

II. Technology implementation

- Safety clue discovery: How to find and identify security clues is the first problem to be solved in the work. Security problems may be discovered and reported by people after security problems occur. For example, when someone discovers some security vulnerabilities or leaked data at work, they will be reported. In this case, we need to ensure that the reporting and flow process of security problems is smooth. But Data Middle Office can also find some possible data security problems from the daily logs of users.

- Stop Loss in Emergency: In case of an emergency data security risk, you can turn on data protection by setting SET ProjectProtection=TRUE on MaxCompute to prevent data from being exported and potentially further compromised. If the data export prohibition function is used from the beginning, then you can modify the Policy file to disable all data export functions as appropriate.

- Handling of problem accounts: First, we need to define which account has problems. If it is a security risk or event located through the log, we can locate it through the user account field in the log. After locating the problem account, the owner of the corresponding account needs to be informed to disable the existing AK (Application Key)/SK (Security Key) and generate a new AK/SK.

3. Afterwards work

   Afterwards work, it mainly refers to the audit, check, problem review and continuous improvement after data development and use.

   I. Management specification

      - Problem review: After the occurrence of data security incidents, it is necessary to review the problem after the event, pay attention to the cause, process and solution process of the incident, so as to form a security problem, and then deeply study the background and causes of the problem, learn from experience and lessons and optimize the work specification and technology.

      - Continuous improvement: With the emergence of new data security risks, the continuous innovation of information security technology and the continuous expansion of data security applications, the work of data security also needs continuous improvement. The data security team needs to be more active in the construction of security and stability, and helps other new business teams to quickly reuse relevant experience to form normalization and continuous improvement.

   II. Technology implementation

      - Behavior audit: First of all, we need to record the behavior of the whole life cycle of data use, and form behavior log and operation log as the basis of audit. On the one hand, it is based on the log to determine whether there are high-risk operations, and carry out targeted investigation; on the other hand, it is based on the log to determine whether there are system exceptions, such as system exception alarm or error, or abnormal collapse. Data risks are identified through various data security models, and high-risk operations are timely reminded through business system push, large monitor screen and other ways.

5.3.4.2 Account and Authority

The Data Middle Office project will involve many departments and personnel of internal and external technology and business. The account and authority management of these personnel are very important. We focus on the user permission system of three core products: Dataphin, QuickBI and QuickAudience.

1. Dataphin

   Referring to Figure 5.22, Dataphin accounts have two levels: super administrator and other accounts. When Dataphin is deployed, a super administrator account will be created, and all subsequent accounts will be added and authorized by the super administrator. Each account can be granted different roles in different project spaces, including project administrator, data development, analyst, data operation and visitor.

   With super administrator, the permissions of the six roles are different in different function modules and scenarios of Dataphin. In order to make it easier for readers to understand the role allocation principle, please refer to Figure 5.22: super administrator is only used for the overall operation and maintenance and management of Dataphin, and is not recommended for the development and testing of project space. In the project space of the DEV, ordinary data developers should be granted the role of 'data development', data analysts should be granted the role of 'analyst', while other people can be granted the role of 'visitor' if they want to simply view the configuration and task information and are not allowed to view the data content; in the project space of the PRD environment, there are only

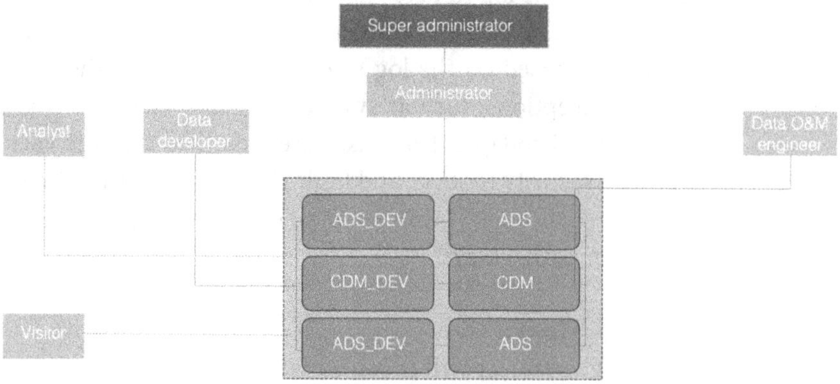

FIGURE 5.22 Dataphin account system.

'administrator' and 'data operation and maintenance'. Two types of roles, to ensure the separation of data task development and release online responsibilities.

2. QuickBI

Referring to Figure 5.23, QuickBI account and permission management are more complex and flexible than Dataphin. At the organizational level, there are only two organizational roles: organization administrator and ordinary user. The administrator is usually the technical director or the administrator authorized by the technical director, and the developer has all the permissions of the space, but for the modification and deletion of objects, only the content created by himself can be operated, and the developer is generally the report developer.

Users can also perform centralized management through user groups, and the menu permissions and row-level permissions of QuickBI report portal can be granted to user groups. In addition, the user can define the member label when creating, and the row-level authority management of report data can be realized through the member label.

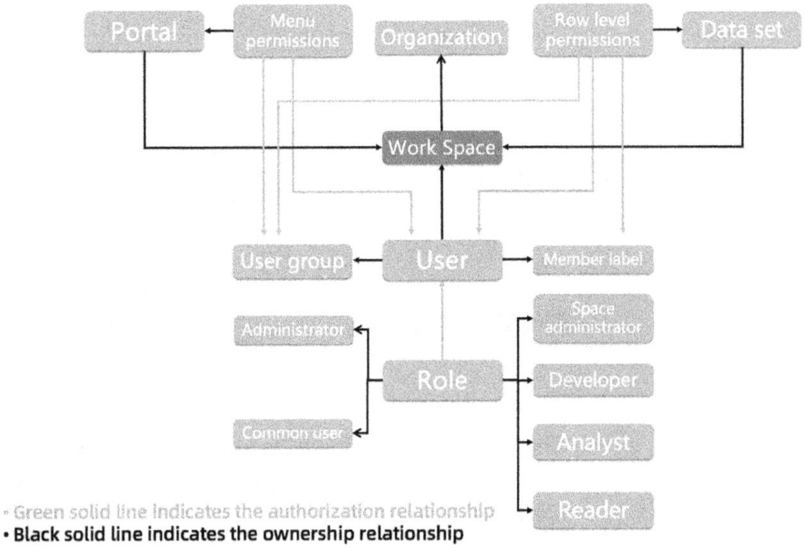

FIGURE 5.23 QuickBI account system.

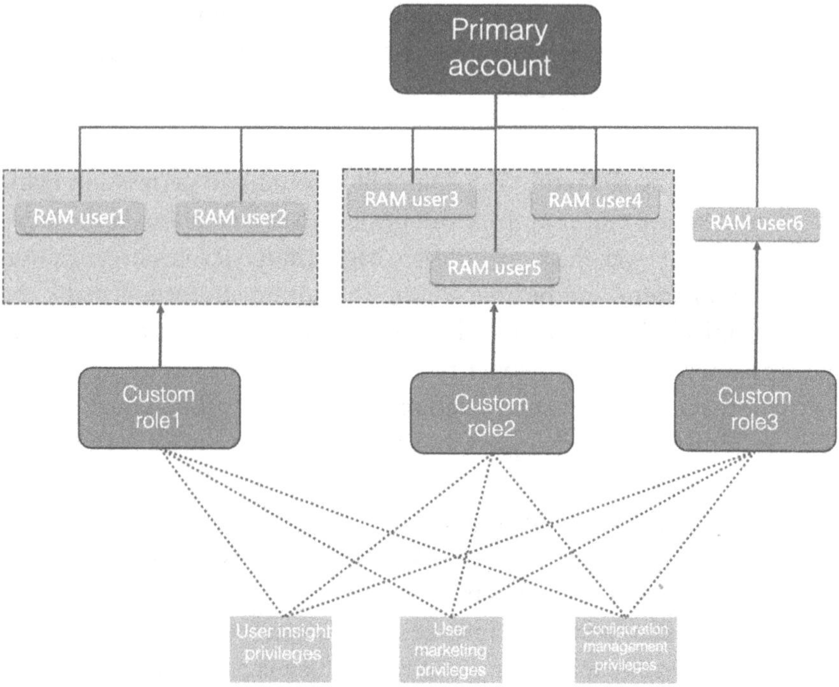

FIGURE 5.24 QuickAudience account system.

3. QuickAudience

   Referring to Figure 5.24, QuickAudience's account permission system is closer to the classic account management system design, that is, the account supports grouping, and various user-defined roles can be set. The user-defined roles package several system permissions, and the account and account grouping can be granted different roles.

5.4 DATA ASSET CONSTRUCTION

5.4.1 The Data Asset Definition

Data asset refers to data resources, such as documents and electronic data, that are owned or controlled by the enterprise and can bring future economic benefits to the enterprise. In an enterprise, not all data constitute a data asset, and a data asset is a data resource that generates value for the enterprise – From *Data Asset Management Practice White Paper 4.0.*

5.4.2 Data Asset Acquisition

5.4.2.1 Overview of Data Asset Acquisition

Data asset acquisition in Alibaba's platform is often referred to as data integration. One of the features of data integration is integration, to bring together different sources of data and different forms of data, so synchronizing various data sources to a data warehouse from different business systems is the beginning of everything. The data integration mainly selects the appropriate cloud technology solution for different data source types and business scenarios, synchronize data to the ODS layer of the data middle platform.

There are many types of data in an enterprise's business system. There is structured data from relationship databases such as MySQL, Oracle, DB2, SQLServer and OceanBase more. There is also unstructured data from non-relational databases, for example, HBase and MongoDB. These type of data are usually stored in a database table. There is also structured or unstructured data from the file system, such as Alibaba Cloud Object Storage OSS and File Storage NAS. These types of data are usually stored as files. In the process of Data Middle Office construction, data integration is for different data sources, data volume and network architecture to choose the appropriate integration technology solutions for data access implementation. The purpose of data integration is to provide raw data for the central ODS layer, to consolidate the calculation, insight and mining potential information of the data, to realize the value of big data and to empower business and create value.

5.4.2.2 Data Integration Process

The data integration is divided into two parts: the design of the integration scheme and the implementation of the data integration. The data integration scheme design is based on the data source list, data catalog, data dictionary data source combing, data integration scheme confirmation, synchronization strategy confirmation on the final output details of the data on the cloud acquisition list.

Data integration implementation is based on the data acquisition inventory into the development stage, divided into table creation, inventory implementation, stock implementation, incremental implementation, incremental data verification and other stages.

5.4.2.3 Data Integration Design

Data integration design is a comprehensive consideration of the customer's business system data source type, data volume, synchronization time, network planning, resource costs and security requirements to choose the appropriate integration solution.

1. Data source combing

   According to the list of data sources produced in the system and data research stage, the combing class of the data source type of the business system is carried out, and the data source types of the total scenarios such as relationship type, NoSQL database, traditional number warehouse platform, SAAS platform, data file and log file, at the same time pay attention to whether there is real-time data access scene and data source and the network mode of the central platform as the basis for the selection of data integration.

2. Introduction to the integration tool

   According to the timeliness of data, data integration is divided into offline and real-time two categories, the corresponding synchronization method of bulk loading and real-time acquisition of two categories of tools (Table 5.5).

   I. If the real-time requirements are not high, then the use of bulk loading class tools such as DataX, timed to complete the bulk data loading can be. For example, offline number of warehouses, offline BI data analysis and other scenarios.
   II. If real-time requirements are high, and data generation requires immediate ability to see the results of the analysis, use real-time acquisition tools such as DTS, OGG, Logstash and Fluentd.

      Dataphin's data access function is based on DataX integration, can meet the data synchronization needs in most of the scenarios, the following is mainly Dataphin to introduce the data access function for data synchronization, for data sources with special time requirements or not satisfied can refer to the synchronization scenarios described.

TABLE 5.5 Comparison of Data Integration Tools

Product Name	Function	Usage Scenarios
DataX	Open-source heterogeneous data source offline synchronization tools are dedicated to achieving stable and efficient data synchronization between heterogeneous data sources including relationship databases (MySQL, Oracle, etc.), HDFS, Hive, ODPS, HBase, FTP, and more.	Offline data synchronization
DTS	Supports data transfer between RDBMS, NoSQL, OLAP and many other data sources. It provides data migration, real-time data subscription and real-time data synchronization and other data transmission methods.	Live online sync
Datahub	Streaming Data are processing platforms that provide publishing, subscription, and distribution capabilities for streaming data, with high availability, low latency, high scalability, and high throughput.	Real-time calculations
Real-time Compute	Alibaba Cloud Real-time Compute is a one-stop, high-performance real-time big data processing platform built on Apache Flink.	Real-time calculations

3. Data synchronization scene introduction

   I. Database data integration

      - Hybrid cloud database data integration

        Data are written to MaxCompute from a directly connected database through data access and cycle scheduling in the cloud (Table 5.6).

      - Local data stream push

        Build an on-premises DataX server cluster and task scheduling platform in IDC to push data to MaxCompute on the cloud by configuring DataX synchronization tasks (Table 5.7).

TABLE 5.6 Summary of Database Data Integration

The applicable scenario	Database types need to be within Dataphin support The environment is self-controllable, the security requirements are high, and there is a certain amount of bandwidth cost support
The network link	Self-built IDC, cloud connected via VPN/private line Build a hybrid cloud
Sync policy	Actively initiate synchronization jobs from the cloud, connect to self-built IDC databases and write to the cloud cluster
Safety	Network environment security, hybrid cloud in the same virtual lanyon The network is closed to the outside world
Performance	On the synchronization node, enable the simultaneous decimation and compression strategy Greatly increases the extraction speed
Operations	The operation and operation are simple, the scheduling operation is unified in the cloud configuration and management

TABLE 5.7 Summary of Local Data Stream Push

The applicable scenario	Database security isolation is demanding and does not provide database connection information or permissions, and data pushes are initiated by the data source holder.
The network link	Self-built IDC, the cloud is connected via a public network or private line.
Sync policy	Export data streams from the on-premises database and proactively write to the cloud. The data do not land in the middle.
Safety	Security levels are low by setting whitelists for security control.
Performance	On DataX servers, enabling methods such as the same extraction strategy and data compression can increase the speed.
Operations	Operations are complex. Self-built IDC scheduling jobs and cloud job scheduling are independent of each other, need to be processed in the cloud, need additional scheduling server resources.

- The file interface

  In self-built IDC, database data are converted to csv or txt format, compressed files and target bits are scheduled by the offline dispatch server to sync data to FTP on the cloud, and Dataphin synchronizes data to MaxCompute by checking the flag bits regularly (Table 5.8).

TABLE 5.8 Summary of File Interface Data Integration

The applicable scenario	Environmental isolation is demanding and needs to be transited through a file form, suitable for cross-sectoral, third-party data exchange scenarios.
The network link	Self-built IDC, the cloud through the public network connection, data based on the public network transmission.
Sync policy	Export data files from on-premises data and push them to the cloud FTP interface/OSS storage. Dataphin jobs are synchronized to the MaxCompute cluster.
Safety	The database is isolated from the intranet and is not made public.
Performance	Speed is guaranteed. Compress files to save network bandwidth. In the Dataphin synchronization task, enable the synth processing mechanism to speed up synchronization.
Operations	Operations are complex. The scheduling jobs in the self-built IDC are independent of each other and need to be processed in the cloud, which requires additional server resources.

II. Data file integration

- Hybrid cloud data files integration

  Convert data files to csv or txt formats in self-built IDC, passing compressed file and flag bit information.

  The offline dispatch server regularly syncs the data to the offline FTP server, and Dataphin syncs the data to MaxCompute by regularly checking the flag bit (Table 5.9).

TABLE 5.9 Summary of Hybrid Cloud Data Files Integration

The applicable scenario	The environment is self-controllable, the security requirements are high, and there is a certain amount of bandwidth cost support
The network link	Self-built IDC, the cloud through VPN/line connection, to build a hybrid cloud
Sync policy	Proactively initiate synchronization jobs from the cloud to pull data from the self-built IDC room to the MaxCompute cluster
Safety	Network environment security, hybrid cloud in the same virtual LAN, the network closed to the outside
Performance	The file is compressed on the file interface machine and the network link performance is good
Operations	Low operation and maintenance costs, scheduling operations in the cloud unified configuration and management

- On-premises data files push to the cloud

  Convert data files to csv or txt formats in self-built IDC, synchronize compressed file and flag bit information to the cloud FTP server at regular times through an offline dispatch server, and Dataphin syncs data to MaxCompute by regularly checking the flag bit.

III. Cloud on several warehouse platforms

When the amount of data is small, the HDFS file is read directly to MaxCompute via data access and cycle scheduling in the cloud. Stock data or historical data can be considered by using the MaxCompute Migration Assist (Table 5.10).

IV. API data integration

Custom API development through interface documentation or convention planning, API deployment and scheduling, reading data from the source interface and writing it to MaxCompute (Table 5.11).

V. Unstructured data integration

Online and offline IDC or on-cloud ECS synchronizes unstructured data into OSS storage via OSSimport, converting unstructured data into structural data and syncing it to MaxCompute with knowledge maps and sound pattern recognition (Table 5.12).

TABLE 5.10 Summary of Warehouse Integration

The network link	Self-built IDC, the cloud through VPN/line connection, to build a hybrid cloud
Sync policy	Actively initiate synchronization jobs from the cloud, connect to self-built HDFS file systems and write to cloud clusters
Safety	Network environment security, hybrid cloud in the same virtual LAN, the network closed to the outside
Performance	By compressing HDFS files to save network bandwidth, conse00 extraction in the cloud guarantees decimation speed
Operations	The operation and operation is simple, the scheduling operation is unified in the cloud configuration and management

TABLE 5.11 Summary of API Data Integration

The applicable scenario	Docking with third-party systems or packages, data synchronization cannot be directly connected to the database due to technical or security isolation.
The network link	Connect via a public network or private line.
Sync policy	Customize API data readers and deploy them in the cloud Dataphin scheduling system. The cloud schedules wake-up jobs, gets data from the interface and writes to the cluster. Data as far as possible do not land.
Safety	Secure your data with whitelisting and AK verification.
Performance	Subject to the app layer access policy.
Operations	The operation and operation is simple, the scheduling operation is unified in the cloud configuration and management.

TABLE 5.12 Summary of Unstructured Data Integration

The applicable scenario	Unstructured data requires a structured transformation
The network link	Connect via a public network or private line
Sync policy	Configure one-time or incremental cycle tasks to synchronize data to OSS Security with OSS Security With whitelisting and AK verification for data security
Performance	Subject to the app layer access policy
Operations	The operation and operation is simple, the scheduling operation is unified in the cloud configuration and management

VI. Real-time data integration

- The MySQL database is integrated in real time

  Write MySQL real-time data to Datahub with DTS data synchronization, and regularly archive data to MaxCompute or combine real-time computing for data online processing (Figure 5.25).

  Oracle real-time data are written to Datahub via the DTS data subscription feature or the Datahub OGG plugin, and Oracle real-time data synchronization is used to periodically archive data to MaxCompute or online processing in conjunction with real-time computing (Figure 5.26).

  If real-time data are needed for archival analytics, it can be subscribed to through DTS data, based on Kafka connector.

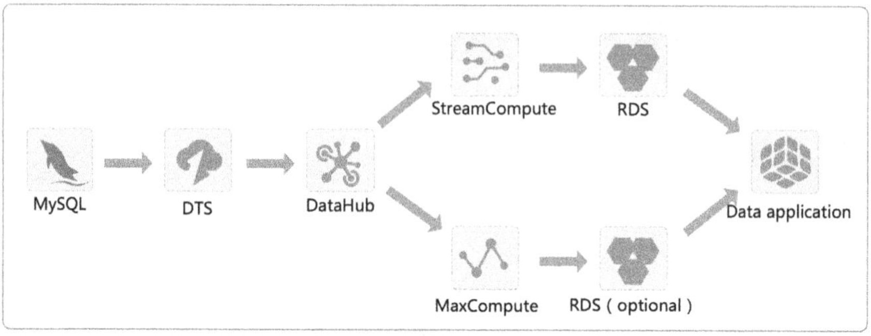

FIGURE 5.25 Real-time data integration of MySQL.

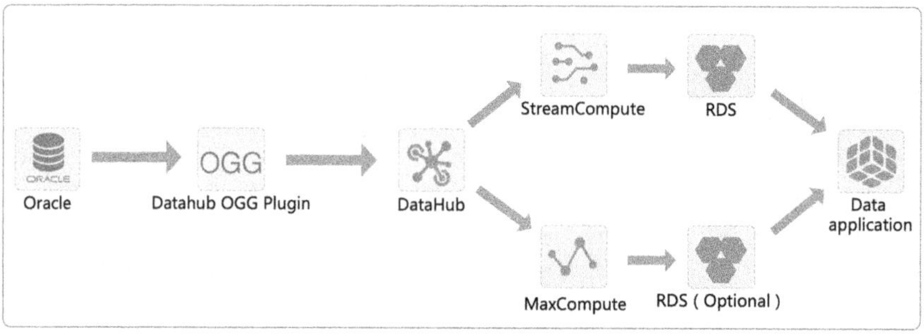

FIGURE 5.26 Real-time data integration of Oracle.

When you post consumer data to Datahub, you can consume it with data connector, downstream to Flink and use Datahub's data archiving capabilities to synchronize data to MaxCompute.

Without real-time data archiving, you can use kafka connetor directly to deliver consumer data to real-time computing for online processing without Datahub.

- Log integrated on cloud in real time

  Write real-time data from log files to Datahub through Datahub's log plug-ins Logstash and Fluentd plug-ins, and regularly archive data to MaxCompute or combine real-time computing for data online processing using Datahub (Figure 5.27).

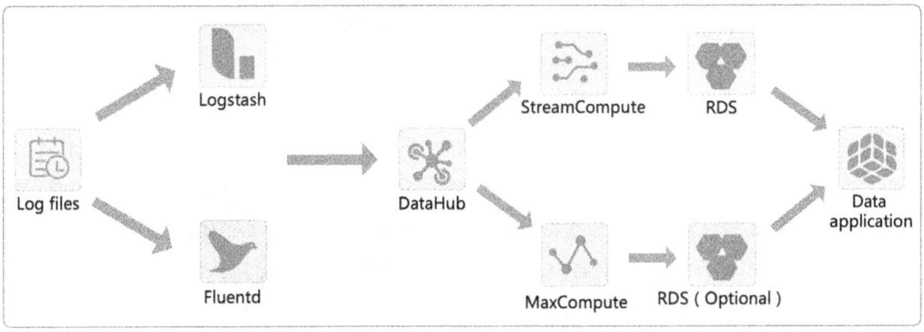

FIGURE 5.27 Real-time integration of log data.

4. Sync policy selection of data integration

Data synchronization strategy is mainly according to the size of the business system table data volume, synchronization number, database role and special restrictions of the table to make appropriate synchronization policy selection, in order to ensure the timeliness, accuracy and stability of data synchronization.

A full-volume synchronization policy means that there is no need to filter the data when the data are integrated, each synchronization is a full amount of data, and the choice of a full-volume synchronization policy can be considered as follows:

- Full synchronization may be considered for dimensional or fact tables with a full amount of data less than 1,000,000 rows and daily increments of no more than 10,000 rows.
- If the table with the extract source as a read-only library can be relaxed to 1,000 W in moderation, a table with a large amount of data is not recommended to go full synchronization, because in the case of multiple simultaneous synchronization, a large number of full-table scans can easily lead to IO pressure on the database and cause failure events.
- Full synchronization strategies can also be considered as appropriate for non-incremental fields and for tables with physical deletions to ensure data consistency.

Incremental policy refers to the synchronization of the time of data processing is not enough for the business system to be too large to meet the time of data processing by means of full synchronization, and to select each time the increase of new changes is synchronized, and the data are synchronized, which is applicable to the fact table and log table (Table 5.13).

The main purpose of data collection list is to sort out the list of tables that need to be synchronized, and to sort out the synchronization strategy, synchronization conditions, synchronization fields and scheduling time of the corresponding tables. At the same time, as the problem tracking table on the cloud records the cloud's exception list to facilitate problem follow-up and backtracking (Table 5.14).

5.4.2.4 Implementation of Data Integration

1. Create a table

   The first step in the integration implementation is to create incremental and full scales in MaxCompute, which can be generated procedurally and manually. The field name is as consistent as possible with the source system, as described in the ODS layer table design specification:

   I. Incremental table structure (Table 5.15)

   II. The full-scale structure (Table 5.16)

   Rule description:

   Addition and deletion identification, business data change time, incremental data bulk write time, these three new fields

TABLE 5.13 Data Synchronization Policies

Scene	Synchronization Policies
Small data scale (dimensional table: Table with less than 10 million rows of data) or no incremental field, with physically deleted tables	Full synchronization, full daily extraction of source data, written to the daily/monthly full partition table
Big Data Scale (Fact Table: Data volume greater than 10 million rows)	Incremental synchronization, incremental data extracted into the daily incremental partition table and the previous day's daily full-scale partition for merge, written to today's full-scale partition
The log table (no update action class)	Incremental synchronization, incremental data are taken to the daily delta table and stored in increments by day

TABLE 5.14 Data Integration List

Work Class	Content of Work	Description
Integration table list sorting	The synchronization method is confirmed	Depending on the type of data source to confirm data access, whether existing tools are supported, specific support scenarios can refer to the cloud synchronization scenario on the data.
	Synchronize policy confirmations	Depending on the size of the table data, whether there are incremental fields, and whether there are physical deletions, the policy for confirming the data integration is full or incremental synchronization. You can refer to the 2.33 cloud sync policy selection on your data.
	The synchronization cycle is confirmed	Depending on the requirements of the task calculation cycle, set the synchronization period, which is generally sufficient for T1.
	Incremental condition confirmation	Confirm the name and type of the incremental field, which is generally recommended for creating or updating time fields, while checking the source database for indexes.
	Synchronize task run window confirmation	Depending on the business system architecture, load conditions and network bandwidth size, it is recommended to avoid peak business hours, while startup time is recommended after 0:30 to avoid data loss due to data synchronization delays.

TABLE 5.15 Incremental Table Structure Field Description

Source Library Table Structure	Add or Delete Logo	Time at Which Business Data Changes	Incremental Data Bulk Write Time	Partition Key
col_a,col_b	busi_flag	busi_date	etl_time	ds

TABLE 5.16 Full-Scale Structure Field Description

The Source Library Table Structure	Add or Delete the Logo	The Time at Which Business Data Changes	Incremental Data Bulk Write Time	The Partition Key
col_a,col_b	busi_flag	busi_date	etl_time	ds

are mainly for the convenience of directly from the data level to troubleshoot data problems, in addition to the first full collection, daily data write value, depending on whether the business library has this field, or log acquisition method (such as DTS), otherwise these three fields are empty every day (Table 5.17).

TABLE 5.17 ODS Layer Table Naming Rules Description

	Incremental Table	Full Scale
Table naming rules	{project_name}.s_{source database name}_{source table name}_delta	{project_name}.s_{source database name}_{source table name}
Field naming rules	The field uses the source system field name by default Processing rules when a field name conflicts with a system keyword: Add a '_col' suffix, that is, the source field name is _col	The field uses the source system field name by default Processing rules when a field name conflicts with a system keyword: Add a '_col' suffix, that is, the source field name is _col
The field type	To prevent data accuracy loss, all fields in the delta table use the string type	Refer to the field type mapping rules from SQLServer/MySQL to MaxCompute below
Add or delete the logo	U is used uniformly when there is no distinction A – First full data; D – Delete; U – Update; I – Insertion	U is used uniformly when there is no distinction A – First full data; D – Delete; U – Update; I – Insertion
The time at which business data changes	19000101 – First full data; Not empty – incremental data, consistent with the source library bar value, or generated by log acquisition tools such as DTS Empty value – The source library has not been acquired	19000101 – First full data; Not empty – incremental data, consistent with source library bar values, or generated by log acquisition tools (for example, DTS) Empty value – The source library has not been acquired
Incremental data bulk stocking time	Not empty – Data are library write time	Not empty – Data are library write time
The partition key	One partition per day; the first initial incoming key value is '19000101'	One partition per day
Life cycle	Default: Never expired	Default: 30 days, lifecycle 30

III. Table field type mapping design rules

- Incremental table

  To prevent data accuracy loss, all fields in the delta table use the string type, which is created based on the source library map.

- Full scale

  If the data type has no special needs, it is recommended to use MaxCompute's older version of the data type, follow up the source data volume type is different, select the corresponding mapping transformation specific reference MaxCompute data type mapping table:

  https://help.aliyun.com/document_detail/54081.html?spm=a2c4g.11186623.6.736.329a1ce714PLbh

2. Implementation of stock data

   The stock data integration is to synchronize the existing historical data in the business system to the cloud, for the one-time task of data initialization, the implementation process mainly ensures that the target library delta/full-scale structure conforms to the specification, inserts the synchronized full amount of data into the delta table and permanently retains the initialized data for subsequent data problem troubleshooting and data pollution recovery.

   Stock data migration process:

- Incremental table data synchronization: This step completes the initialization of incremental tables by synchronizing historical data from the business system table (T1) to the incremental table (T1_Delta) through a data synchronization task.

- Full-scale data synchronization: This step completes the initialization of the full scale by synchronizing the data from the delta table to the full scale through a code task.

3. Stock data verification

   After the inventory data migration is complete, a data validation of the stock data of the full scale is required, and the validation dimensions can be referred to as follows (Table 5.18).

TABLE 5.18 Data Check Rules

Whether Must	The Test Item	Test the Method	Description
Is	The amount of data is compared to	Data comparison	Data scales need to be kept reasonably volatile
Is	The full text is compared	Data comparison	For scenarios where the expectation should be exactly the same, such as the report adding/modifying field logic, a full-text ratio should be made to ensure the accuracy of the original field data
Is	The number of fields is compared to	Data comparison	Check to see if the field is missing
Whether	Field Enumeration value distribution ratio	Data comparison	For log class data that is more commonly used, compare the count of the field top 100
Whether	Fields sum, max, min, avg ratio	Data comparison	Metric field comparison, applicable to data revision, data reconstruction scenario
Whether	The field distinct is compared to	Data comparison	Dimension field comparison, which applies to verifying the consistency of fields such as primary keys

4. Incremental data implementation

In bulk data synchronization, some tables have more and more data as the business grows, which can affect processing efficiency if synchronized in full on a cycle-by-cycle scale. In this case, you can choose to synchronize only the incremental data for the new changes at a time, and then close in with the full amount of data obtained in the previous synchronization cycle to obtain the latest version of the full data.

In traditional data consolidation scenarios, most of the consolidation techniques use the merge method (update plus insert): The current popular big data platform basically does not support update operations, now the more recommended way is full external connection (full outer join) and data full override reload (insert overwrite), that is, daily scheduling, the day's incremental data and the previous day's full amount of data to do a full external connection, reload the latest full data. The following is a common way to synchronize incremental data:

I. Non-flowing transaction table data synchronization method, use for data update tables

   - Incremental table data synchronization: This step synchronizes the new data for T1 in the business system table (T1) to the partition ds day2 of the incremental table (T1_Delta) by filtering data synchronization task.

   - Full-scale data synchronization: This step follows the code task to make the data in the delta table (T1_Del ta) ds day2 with the data in the full scale (T1_Full) ds dayl, and to write the latest full amount data todssday2 in (T1_Full).

II. Incremental synchronization of log tables for tables with no data updates

   - Incremental table data synchronization: This step synchronizes the new data from T1 to partition ds of the incremental table (T1_Del ta) by filtering the criteria in the business system table (T1) through the data synchronization task.

   - Full-scale data synchronization: This step, through a code task, writes the data in the delta table (T1_Del ta) in dss day2 todssday2 in (T1_Full).

III. Incremental data synchronization for small data tables

   - Incremental table data synchronization: This step synchronizes the new data for T1 in the business system table (T1) to the partition ds day2 of the incremental table (T1_Delta) by filtering the data synchronization task.

   - Full-scale data synchronization: This step follows the code task to make the data in the delta table (T1_Del ta) ds day2 with the data in the full scale (T1_Full) ds dayl, and to write the latest full amount data todssday2 in (T1_Full).

5. Incremental data validation

   After the incremental data migration is completed, a data verification of the incremental data of the full scale is required to verify that the dimension can refer to the inventory data verification method, focusing on the completion of the incremental data, such as whether the modification data are effective and whether the incremental data are composited.

6. Description of the data-dependent inspection mechanism

   For some special data sources cannot confirm the arrival time and data accuracy of the data, it is necessary to increase the status and readiness of the data check task before the synchronization task starts. If the direct connection database is not allowed in the government industry for data synchronization, the owner is required to export the data as a data file and verification file and upload it to a transit, after the data synchronization by a third-party system. For such cases, the data synchronization task needs to be initiated before the data synchronization is started by adding the data state checker steps to check the integrity and correctness of the data transfer.

5.4.3 Data Asset Building (Figure 5.28)

5.4.3.1 Dimension Table Design

The steps for dimension design are summarized below:

1. The specification of dimensional design

   Data source of data warehouse is a large number of decentralized business system data, these data into the data warehouse need to be

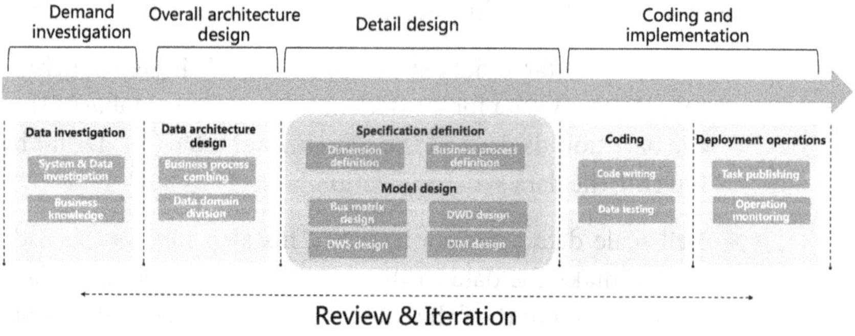

FIGURE 5.28 OneModel modeling process.

standardized and standardized integration, convenient for downstream analysis and use. The specifications are as follows:

I. The unification of naming specifications, table names, field names, etc. For example, a member ID is cst_id or member_id.

II. The uniformity of the field type.

III. Uniformity of common code and code values. For example, the currency uniform use of RMB, gender unity with 1/0 for men and women, or unified with male/female for men and women.

2. The point of note for dimension property design

Determining dimension properties is the key to dimension design, and the following dimension attributes determine the points of note:

I. Enrich the dimension properties as much as possible

To provide a good basis for downstream data statistics, analysis and exploration, we need to enrich the dimension attributes as much as possible. For example, the core members of the retail industry and commodity dimensions have nearly a hundred dimensions.

II. Give as many meaningful text descriptions as possible

In dimensional modeling, general encoding and text exist at the same time, such as the item ID and product name in the product dimension, store ID and store name. IDs are typically used for associations between different tables, and names are used for report labels.

III. Distinguish numerical attributes from facts

Numerical fields, as fact measures or dimension attributes, are used primarily for the general purpose of reference fields. Typically used to query constraints or group statistics, it is used as a dimension attribute, and typically to participate in the calculation of measures, as a fact. Numeric fields are more likely to exist as dimension attributes if they are discrete values (which can be enumerated), such as store IDs, and more likely to exist as factual measures, but not absolutely, to refer to the purpose of the field.

IV. Try to precipitate common dimension attributes

Some dimension properties need to be obtained through more complex logical processing, either through multi-table associations, or by mixing different fields of a single table, or parsed by a field in a single table, so that common dimension properties need to be precipitated as much as possible. For example, the member registration channel is associated with the member master form and the member registration behavior record form, and can then be grouped through the member registration channel. Whether e-commerce goods are online is an important constraint, processing logic is that the commodity status is 1 and the shelf time is less than or equal to the current time.

There are two advantages to precipitate common dimensional attributes, one is to facilitate the use of downstream, reduce complexity, and the other is to avoid the inconsistent of different logics in downstream analysis.

3. Best practices of dimension table design

The detailed design of dimension tables mainly involves the integration and splitting of dimensions.

I. Dimensions integration

Several ways dimensions are integrated are as follows:

- How the master-from table is merged. Place fields from both tables or multiple tables in the main table (basic information) and dependent information in their respective tables. For the primary key of the main table, it is generated by the 'source primary key plus system distinguishing flag' composite primary key. This is currently relatively rare.
- Direct merge. Vertically consolidated, different source tables contain the same data set, but store different information. For example, the membership basic information table, the member extension information table are all members-related information table, can be integrated into the membership dimension model, rich member dimension table properties. Common information and personality information are placed in the same table, which is a good way to fit a high-fit

table, and if the re-happenability is low, there are a large number of empty values in the field.

- Do not merge. If there is a large difference between the source table structure and the primary key, it is stored separately in the number of warehouses.

II. Dimensions split

When splitting dimensions, there are two main considerations:

- Depending on the business irrelevance of dimension attributes, dimension properties that are less relevant are split into multiple physical table stores. For example, 1688 commodity dimension and Taobao commodity dimension, the two BU business development, can be divided into two commodity dimension table for storage.

- For dimension tables with too many dimension properties and too many sources, you can do the appropriate split.

  Some dimension properties have an earlier source table output time, while some dimension properties have a late source table output (for example, some label information that needs to be worked out), or some dimension properties are hot, frequently used, some dimension properties are less hot, and some dimension properties are stable and some dimension attributes change frequently. In view of these situations, we can design the main dimension, the main dimension table storage stability, early output time, high heat properties, from the dimension table storage changes quickly, output time late, low heat dimension properties.

  For example, commodity dimensional table, according to the actual situation can be divided into commodity main dimensional table and commodity expansion dimensional table. The commodity main dimensional table stores properties with early output time and high heat, and the commodity expansion dimensional table stores label information that produces later.

  Whether it is better to have richer dimension attributes or to split different dimensions appropriately requires comprehensive consideration based on actual data and business conditions.

5.4.3.2 Detailed Fact Sheet Design

Alibaba data common layer design concept follows the idea of dimension modeling (Kimball dimension modeling), dimension design is mainly based on dimensional modeling theory, based on dimension data model bus architecture, to build consistent dimensions and facts.

1. Proposals for the design of a detailed fact table

   Based on the experience of some fact table design practices, some of the design recommendations summarized are as follows:

   I. Integrate business processes that are as dispersed across business systems as possible

      In the business systems of an enterprise, the same business process may involve different data sources, such as membership coupons, there may be a table in the membership system to record it, at the same time WeChat platform may have WeChat coupons records also, so abstract the business process of receiving coupons as much as possible, generate a coupon fact table.

   II. Generate metrics in the fact table that are easy to use for downstream analysis

      For example, a report requires statistical access duration, which is only recorded in the business system record table, while a new generation of a measure field can be made in the fact table called access duration, access duration, access end time–access start time.

   III. The definition of the granularity of the fact table is recommended

      In the process of fact table design, the finer the granularity is defined, the better, and it is recommended to start with the lowest level of atomic granularity, which provides maximum flexibility and can better support the user's needs at all levels of detail. For example, a typical trading fact table design contains a trade master fact table and a trade detail fact table, which contains atomic granularity flexibility.

   IV. Fact measurement should be consistent

      Units measured in the same fact table should be consistent. For example, the amount of member consumption, preferential amount, etc., the use of a unified unit of measurement, unified into a dollar or points, the number of points also need to be consistent.

V. Handles the null value of the fact table

In the actual project, there may be empty values in the fields of the fact table due to data source data quality problems, and for the processing of empty values, some fields are not allowed to be empty and need to be fed back to the data source to be completely resolved or filled with values in consultation with the business.

- The foreign key of the fact table cannot have an empty value because it involves association with the dimensional table.
- The field of the measure cannot be empty, or the results of the statistical analysis will be incorrect.
- Fields that involve filtering conditions may cause more than, less than, equal, etc. to take effect, and it is recommended to negotiate fill values with the business.
- Fields are not used for statistical analysis and can be temporarily allowed to be empty.

2. Detailed fact table design best practices

The design of the fact table is mainly divided into transaction fact table and periodic snapshot fact table, the cumulative cycle snapshot fact table is less applied, not involved here, there is a need to supplement later.

I. Transaction fact table

A transactional fact table is also a fact table that expresses one or more business processes at the same time. Therefore, the incremental table transaction fact table is mainly divided into single-transaction fact table and multi-transaction fact table. The so-called transaction is also the business process mentioned earlier, such as the order is a business process, payment is also a business process. To illustrate more intuitively, the three most common business processes used in e-commerce transactions – ordering, paying and confirming receipts – illustrate the fact tables for both types of transactions. If it is a single-transaction fact table, then the fact table expresses a business process, either under order or paid, and a record expresses only one business process information. A multi-transaction fact table, on the other hand, means that a table contains multiple business processes, and a record represents information about one or more business processes (Table 5.19).

II. The periodic snapshot fact table

Compared to the transaction fact table, the periodic snapshot fact table is more difficult to understand, the following focus on the characteristics and types of the periodic snapshot fact table.

- Cycle snapshot fact table characteristics

  Statistics are measures of the interval period, such as history to date, natural year to date, quarter to date and so on.

  The periodic snapshot table has no concept of granularity and is replaced by a combination of period-state measures,

TABLE 5.19 Comparison of Single-Transaction Fact Tables with Multi-Transaction Fact Tables

	A Single-Transaction Fact Table	A Multi-Transaction Fact Table
Business processes	One	Multiple
Granularity	It's irrelevant to each other	Same granularity
Dimension	It's irrelevant to each other	Consistent
Fact	Take only the facts of the current business process	Keeping the facts in multiple business processes, which are not in the current business process, needs to be zeroed
Redundant dimensions	Multiple business processes require redundancy multiple times	Different business processes require only redundancy once
Level of understanding	Easy to understand and not confusing	Difficult to understand and needs to be qualified by labels
Calculate storage costs	More, each business process needs to be computed stored once	Fewer, different business processes merge together, reduce the amount of storage computing, but there are a lot of zero values for measures that are not current business processes

such as the historical total order total fact table is a sparse table and the periodic snapshot table is a dense table. Sparse table refers to only the day only the operation occurred will have a record, such as the following orders, payments and so on. Dense table refers to the day no action will also have a record, such as for the member's history so far the amount of consumption, regardless of whether the member has expected to spend on the same day, will give the member a record line.

– Single-dimensional cycle snapshot fact table

  A snapshot fact table for a single dimension, such as a member history-to-date cycle snapshot fact table, includes historical consumption amounts, purchases and so on.

– The mixed dimension cycle snapshot fact table

  In contrast to a single dimension, a blended dimension is sampled for multiple dimensions only during the daily sampling cycle. Typical is Taobao buyer seller history so far cycle snapshot fact table, sampling cycle is every day, the dimension is the buyer plus seller, reflecting the different buyers for different buyers of the order amount, payment amount, number of orders and so on.

– Take a full snapshot of the fact table

  For the first two cycle snapshot fact tables, often obtained through transaction fact table processing, such as member history-to-date cycle snapshot fact table, is through the member transaction fact table processing. In addition to the first two periodic snapshot fact tables, there is a special snapshot fact table that can be called a full snapshot fact table.

  For the data that the state has been changing, use the full snapshot table to count the latest state to date, such as order evaluation, good and bad evaluation will change every day, using the full snapshot table, that is, the daily full update. This type of snapshot fact table uses more, and if the amount of data on the business data table is not large or does not identify the timestamp increment, it is synchronized in full every day, or it can be designed as a full snapshot fact table.

5.4.3.3 Summary Fact Table Design

Based on Alibaba's OneData methodology best practices, the aggregate data layer is modeled on the subject object of analysis, and a common granularity summary is built based on the metrics needs of the upper-level applications and products. A table of summary data layers typically corresponds to a statistical granularity (dimension or combination of dimensions) and several derived metrics under that granularity. A fact table, called a summary fact table, obtained by summarizing detail data, belongs to the summary data layer (DWS). By accessing the summary fact table, the user can reduce the calculation of the query, respond quickly to the user's query and also facilitate the different users to access the detailed data to bring inconsistent results, such as Ali Group buyers in the last day of the transaction summary table and the most recent week seller transaction summary table, are included in the summary fact table.

1. The basic principles of the summary fact table

   I. Consistency

      The source of the summary table is a detail-granular fact table, so you need to provide query results that are consistent with the query granularity fact table.

   II. Avoid a single-table design

      Don't store different levels of summary data in different rows in the same table, for example, if some rows in the summary fact table hold transaction volumes summarized by day, and some rows store transaction volumes summarized by month, this can be misused by consumers. It is more efficient to store two columns of transactions by day and monthly summary, and then distinguish them by column name or comment.

   III. The granularity of the summary can vary

      Summary does not need to be as granular as the original schedule of facts, it only needs to care about the dimensions of the query. For example, the dimensions involved in the order are goods, buyers, sellers and regions (provinces, cities, etc.), so that you can summarize 1 day's transaction volume by commodity,

1 week's transaction volume by buyer or 1 month's transaction volume by seller and region.

IV. Data commonality

The biggest purpose of a summary fact table is to be used by a third party, and this summary fact table is often provided to downstream tasks to use such a third party when it is used without having to obtain and repeat a large number of calculations in the detail fact table each time. Summary fact tables provide an improvement in query performance, but they also increase the number of layers of CDM and add complexity, so whether or not they are common is also an important criterion for establishing summary fact tables.

V. Does not span the data domain

A summary fact table is made up of a detailed fact table, so a summary fact table does not cross facts or data domains, and whether it crosses a data domain is also a criterion for dividing whether the table belongs to the CDM layer or the ADS layer.

VI. Distinguish between statistical cycles

You should also be able to describe the statistical period of the data on the name of the table, referring to naming conventions. For example, '_1d' is for last day, '_td' is for by that date and '_nd' is for the recent N days.

2. Basic steps for summarizing fact table design

According to the core formula of the metric, the derived metric consists of atomic metric, business qualification, statistical period and statistical granularity, and in Dataphin, there are two sources of statistical metrics that make up the summary fact table, as follows (Figure 5.29):

- Through Dataphin specification modeling, when the system generates derived metrics, it automatically converges according to the same statistical granularity.
- Common physical table fields created by non-derived metric creation are aggregated into the corresponding summary table in the same way that sql is written according to the principle of the primary key.

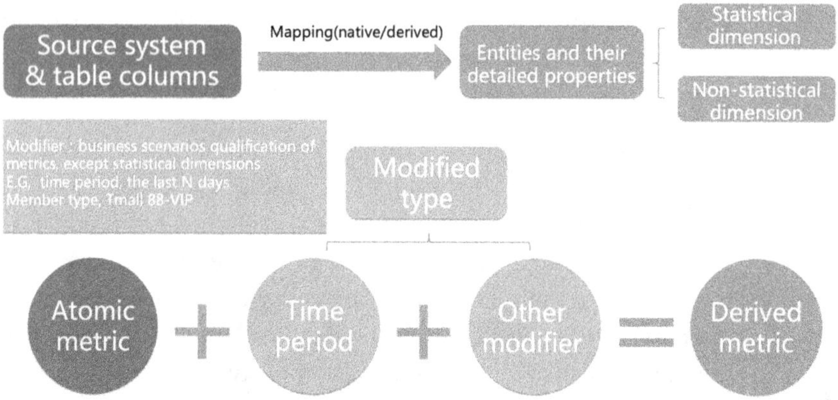

FIGURE 5.29 The core formula of the metric.

5.4.3.4 OneID Build

'OneID' is the natural person's identity code. OneID is similar to a real-life ID number, for each natural person in the Internet world is given a stable virtual identity ID – OneID through algorithms, and identifies all kinds of identity IDs and tool IDs owned by natural persons, such as account ID: taobao_id and alipay_id, also such as device ID: IMEI and IDFA. OneID aggregates IDs from various systems and fields of natural person origin, removes IDs that do not belong to natural persons, highlights active IDs, eliminates dead IDs, etc., thus helping businesses such as commercial marketing, crowd selection and label aggregation.

The cloud on OneID targets the actual needs of customers for easy porting, efficient deployment, stable performance and diverse needs.

1. OneID terms

   I. ID

      ID means information that theoretically belongs to only one natural person at the same time and within the scope of the study. ID is the basic concept of OneID, in real life everywhere ID figure, such as device ID: mobile phone138XXX, license plate number京A123456, another example is the enterprise's registration ID: Taobao ID and Alipay ID; of course, it also includes the inherent identity ID of a natural person, such as an identity card number and a passport number.

II. Unique identity ID

Unique identity ID: This ID type, the same natural person can only have 1. Core ID is a classification of ID, and OneID divides it into core ID and non-core ID according to the characteristics of ID. Core ID refers to a natural person who has only one such ID at the same time and within the scope of the study. The above-mentioned Identity card number, passport ID, etc. are the core ID. In real life, there are many instances of core IDs, such as identity card number, passport ID, certain special IDs (for example, research scope, source system or other system issued to them the only ID, such as a company's membership ID), but we usually recognize that the more important ID, such as Taobao ID and Alipay ID does not belong to the core ID.

III. Non-unique identity ID

As compared to the concept of core id, a natural person can have multiple IDs of the same kind of non-core ID. For example, in life, a natural person can have more than one mobile phone number, Taobao account number, Alipay account number, license plate number and so on. Non-core IDs do not limit the number, but their limits belong uniquely. This means that although these IDs can be owned by the same individual, their owners can only have one. There may be a lot of controversy here, such as the couple share a mobile phone number, a look at the mobile phone number belongs to the husband and wife. In fact, the above situation, OneID calculation will eventually belong to the phone number of one of them, because this common usually have a tendency, OneID calculation will be based on who uses more, who recently used frequently, who used it to make more important registration and other information to determine attribution.

IV. ID relationship pair

The relationship of ID forms a three-way relationship between two IDs with related relationships, that is, ID1, ID2, SCORE, and SCORE reflects the true degree of relationship between the two IDs. The range of values is 0–1. It uses improved data in Alibaba OneID cosine value. The methods for forming relationship pairs have a direct acquisition method (obtained from logs and dimensional tables), and indirect acquisition methods (based on auxiliary information) to process new relationship data.

V. OneID

OneID is Alibaba's uniform identity for a natural person's identity ID; a natural person enjoys a OneID issued by a OneID system, a long string of numbers and letters; usually it has 32 or 48 bits. It is a virtual ID, not a realistic meaning, but it is a natural person identity ID in OneID technology. The real ID of a natural person in real life is aggregated in it, thus forming an OneID with an ID group, which we usually refer to, both the 48-bit virtual ID of the OneID itself, or the ID owned by the natural person, which needs to be distinguished according to the scene.

VI. Personal attributes

Refers to some personal attributes and information that a natural person can describe in addition to ID. This information includes name, gender, date of birth, nationality, place of origin and even graduation school, major. These auxiliary information do not meet the uniqueness requirements of ID, but there is still a certain degree of identification of natural persons, it will be one of the information sources of OneID calculation.

2. OneID technology route

I. OneID Research

The purpose of the ID category and quantity survey is to get an overall understanding of the customer's ID status, to collect the types and sources of IDs to be concerned about and to collect the number of systems distributed across the systems. This step is important for ID source data entry and testing.

- Basic information research

  To understand the category and number of overall characteristics of the customer ID, the survey table for the overall characteristics of the ID is below (Table 5.20).

- ID Details Research

  After understanding the basic characteristics of ID, we need to start counting and combing the detailed characteristics of ID, and develop an 'ID detail feature table', which

TABLE 5.20 Sample of Survey Table for ID Characteristics

No	ID Keytype	System	Id Level	Id Num	Id Total Num
1	gr_member_id	CMS	primary_id	1472394	1472394
2	phone	Opera	secondary_id	13234	36709
3		F&B		23475	
…	…	…	…	…	…

mainly includes ID name, source table name and whether it is a fact table.

- ID relationship to research

  After completing a detailed study of the ID and auxiliary information in Table 5.21, it is not difficult to find out the ID relationships that are available in each main table. For statistical convenience and viewing convenience, we add a column to the table ID detail feature table as a way to obtain relationship pairs (Table 5.22).

II. The ID relationship gets

- The fact table gets it directly: Ids that appear in fact tables can form ID relationship pairs if they contain more than one ID type.

- Dimension table gets directly: Relationship pairs can also be obtained from dimension tables. Because the dimension table information is relatively complete, some ID information customers indicate it at registration, but the actual number of times used is very small.

- Algorithm acquisition

  ID relationship pairs obtained directly from a record are the most effective method. However, in real life, several IDs appear in a recorded situation is not all cases, still a large part of the ID is separate in the facts. The idea of using algorithm to construct relationship pairs is to collect the personal information brought by ID, the similarity of the two computational information, if the calculated value is greater than the set threshold of judgment conditions, can be set to the new ID relationship pair.

TABLE 5.21 Sample of ID Detail Feature

No	Table	System	KeyType	Fact	Fields	Level	Calculation	Special	Info Type	Example
1	s_Member	ERP	Passport	N	Passport no	2	Passport_no is not null	—	ID	ER436438
			Idcard		Idcard	1	—	—	ID	3424233
			Credit_card_id		credit	1	Join credit_detail_d on credit=cid	Concat(naeid, credit)	ID	7374376492864

TABLE 5.22 Sample of ID Relationship Pairs

No	Table	System	KeyType	…	Special	Info	Example	Id_pairs
…	…	…	Passport Idcard Credit_card_id	…	…	…	…	Idcard-passport idcard-credit_card_id…

III. ID relationship pair calculation

The calculation of ID relation pair mainly includes the following processes:

- Daily relations pair

  Calculate the daily occurrence frequency of ID.

- Cumulative ID calculation

  Calculates the total number and density distribution of IDs over a cumulative period of time.

- Day-to-day relationship pair calculations

  Calculate the frequency of daily relations pairs, cumulative ID relations pair calculations, cumulative ID relationship pairs for a period of time, and correlation.

- Correlation calculation

  The ID relationship uses ID space distance calculation for correlation.

- Attenuation

  Frequency attenuation is commonly used in exponential decay and linear attenuation. Exponential attenuation: If the customer's OneID needs to pay special attention to the timeliness of the ID, it can be used exponential decay, such as most Internet companies ID information change faster, if an ID is not in the study time range, it can usually be considered dead.Linear decay and traditional enterprise ID is less sensitive to time, such as an international travel group, customer ID appeared in the group time span is very large, perhaps a certain ID has been more than 1 year or even many years. In such a scenario, linear attenuation can be used, and Figure 5.30 is two parameter adjustments for exponential attenuation and an example of linear attenuation.

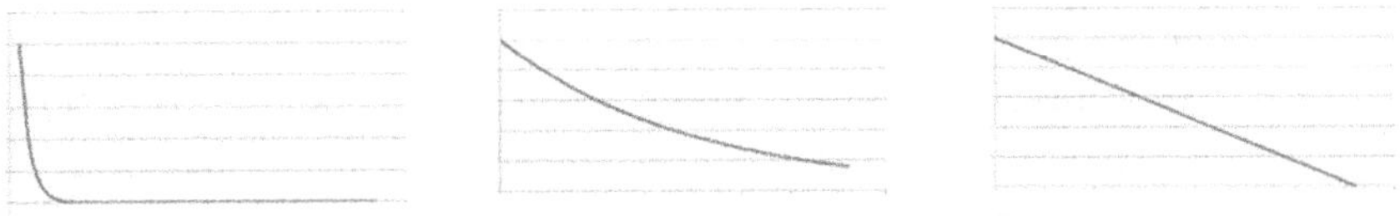

FIGURE 5.30 Different attenuation function.

- Correlation calculation

  The relevance of the fact table relationship pair, calculated using spatial distance.

IV. Diagram is connected

Make a graph connectivity calculation for a related ID. Forms a connecting weight. When the ID cluster exceeds 100,000 IDs (thresholds can float) processing:

- Tool Class Processing ID (or Cheat ID)

  With the same ID, the association ID is cut to a super-thousand pair, and all related relationships are cut.

- TOP class processing ID

  The same ID and another ID of the same type, the relationship to the super N pair, then topN above are pruned, generally 100.

- Legitimate class processing ID

  If there is a problem with the details of the Id content, the ID is discarded directly and the relationship is cut. ID detail recognition is generally available in a normal way.

- A giant cluster of unknown reasons

  Limits the transmission of information about super-jumbo cluster part type ID. When none of the above three methods can contain the giant cluster, the connectivity limit of the giant cluster is required (Figure 5.31).

V. Graph iteration

The ID relationship uses ID space distance calculation for correlation.

A diagram iteration is a natural person identification process, the main basis for the diagram iteration consists of the following:

Figure calculation starting point: Multiple unique identity ID entities.

Figure calculation end point: Natural person split ends, OneID inverted table formed.

The figure is calculated by ID type (unique identity ID and non-unique identity ID), personal attributes (gender, age,

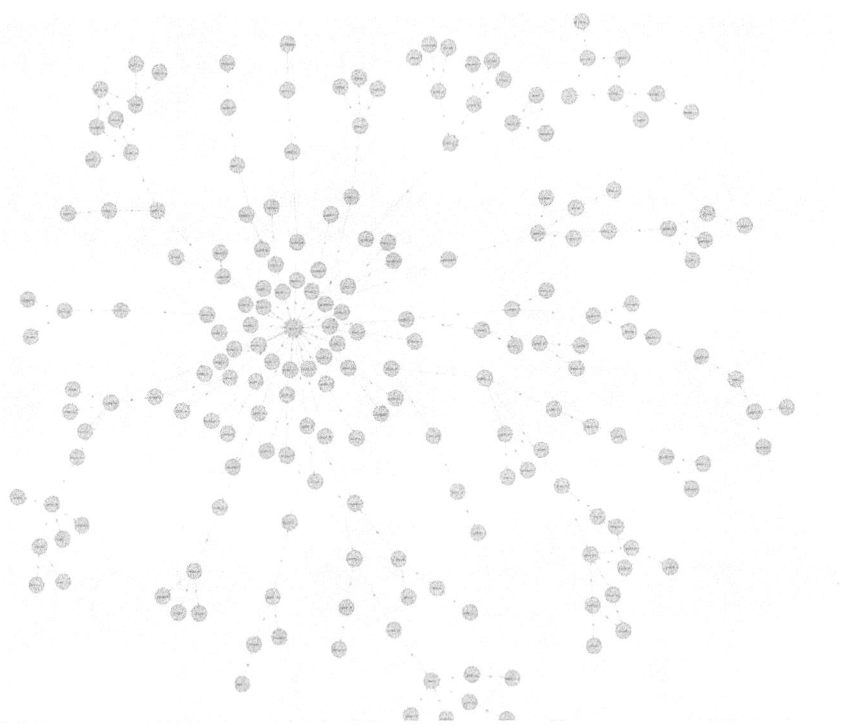

FIGURE 5.31 Giant cluster.

gender, etc.). The process of splitting mixed natural-to-human hybrid relationships, ID relationship importance sorting.

Figure calculation output: OneID inverted table results (Figure 5.32).

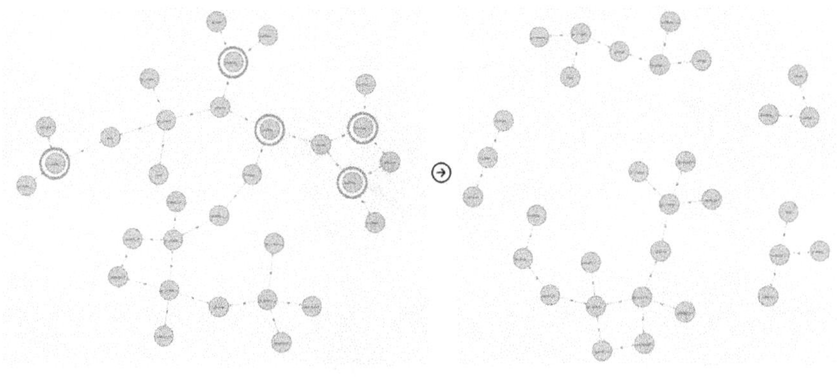

FIGURE 5.32 OneID Graph iteration results of inverted table output.

3. OneID inverted table

The OneID inverted table is the final output of OneID. The structure of the OneID inverted table is shown below. It is a secondary partition, and OneID is generally used as follows (Table 5.23):

I. Query global information based on an ID

key_id and keytype can be approximated as union primary keys, which locks the Information of the OneID.

II. Query all the natural person's information based on OneID

Lock natural person queries directly based on OneID, that is, based on ds and OneID queries.

III. ID sorting based on intimacy relationship (Figure 5.33)

TABLE 5.23 One Record of OneID Inverted Table

key_id	OneID	email	credit	passport	phone	guest_id	keytype	Ds
28632@ xx.com	passport# 194100203 153153103 149108154 149107107 198102156 150	28632@ xx.com :1.0	0xe412 c37ad 9e4ec: 0.532	SGP#0 xdd35 f19db 8911c 26:0.832	0x4688 ca9963 cea11f 9ef35c: 0.8382	3590760: 0.737|21 3124:0.12	email	2018 1208

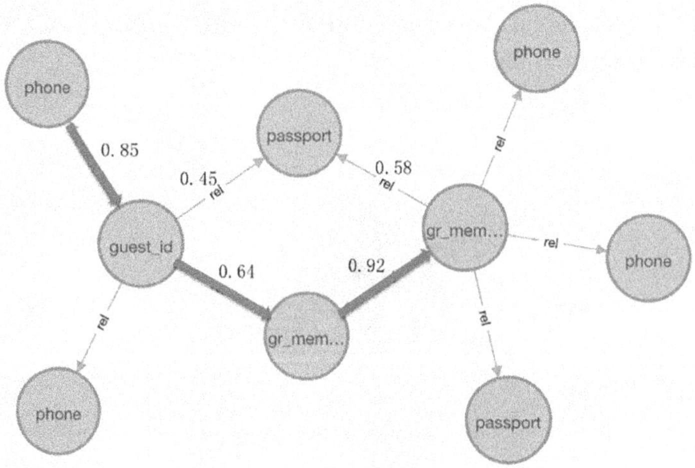

FIGURE 5.33 ID intimacy relationship.

4. OneID inheritance

   OneID inheritance is inherited from OneID's value, the 48-bit virtual ID number. Every natural person has a unique OneID number in the OneID inverted table. OneID inheritance, on the other hand, refers to the same natural person, and if it is not first seen in OneID inverted rows, the previously existing OneID is chosen to be followed. There are many benefits to keep OneID unchanged, and when OneID is used downstream, it is often through OneID's key_id to find the original person, not an ID.

5.4.4 Data Asset Management

There are three important directions in the field of data asset management: asset analysis, asset governance and asset application, and the need for technical research and practice based on these three directions, the process, experience, standards and norms and other products, and ultimately constitute a unified data asset management platform.

5.4.4.1 Asset Analysis

The asset analysis includes two parts: asset inventory and asset evaluation. Asset inventory is designed to give data users a better understanding of the data, which can be understood and reasoned through knowledge maps or to build an enterprise asset catalog, while asset assessment evaluates the activity and input–output ratio of assets.

The asset analysis consists of the following three parts:

- The asset analysis object.
- Take enterprise-wide big data as the object of asset analysis.
- Multi-dimensional data asset analysis system.

Based on asset analysis object, grass-roots metadata, user behavior log, data knowledge map as the material, through the comprehensive human brain and machine learning algorithm is a means to fully understand the data asset content, complete all kinds of data asset analysis, understand the data content.

The user collaborates and establishes the data confidence mechanism, which realizes the multi-dimensional data asset analysis system, which complements the data content understanding and data confidence mechanism.

Asset Analysis to be as product-like as possible, focus on the following:

- Based on the multi-dimensional data asset analysis system, asset inventory, asset evaluation and asset exploration are carried out behind the technology side and products that the user cannot see, so as to output easy-to-read and understandable asset reports to the user.
- Provide asset navigation services to facilitate users to find the data they want and their details in a variety of ways.
- Provide asset analysis services for specific topics, such as core asset analysis and user-defined asset analysis.
- Provides easy-to-use configuration management that facilitates asset analysis and productization, such as data category configuration management, data asset labeling and more.

5.4.4.2 Asset Governance

Asset governance includes governance in computing, storage, quality, model, security, cost and other fields, and forms an effective intelligent governance closed loop, which precipitates governance methodology as tool product output.

Asset governance consists of the following two parts:

1. Closed-loop system of asset governance

   Establish a closed-loop system of asset governance, including status analysis, problem diagnosis, governance optimization and effect feedback.

   The content of each link is enriched and perfected, and the problem diagnosis includes not only the computational storage resource diagnosis, but also the field diagnosis of data quality and data security.

2. Asset governance multi-dimensional output

   Asset governance is committed to opening up governance to the outside world. Output through dimensions such as standard output, custom products, capability output and building collaboration mechanisms.

5.4.4.3 Application of Assets

Asset applications enable end-to-end access over a full link, evaluate application input–output ratios and perform secure inspection controls.

Asset application mainly refers to asset application all-link system, through full-link data tracking, data from acquisition to data processing to data application, to achieve end-to-end access.

Asset application productization. Provide application analytics products around end users, driven by the nature of data assets. Includes:

- Full-link 'blood' relationship, clearly show the data of the context.
- Full-link protection: To let users know the various safeguards and problems, and why asset applications can be stable and healthy operation.
- Access analysis: A comprehensive analysis of the data applied to the product and scenario access.
- ROI assessment: Indicates to the user the input and output of the current product or scenario-based application.

5.4.4.4 Asset Management Tools

Through asset analysis, asset governance and asset applications, we strive to move big data from a cost center to an asset center, with businesses committed to data asset building and management. This from two perspectives to carry out platform practice: First, research and development perspective, in the data research and development link for data security, data quality, data standards, data assets construction, pay attention to task production nodes, production monitoring, in order to provide continuous and stable data processing results to provide external services. For the user base is mainly research and development. This platform practice is Dataphin hosting and output, and the second is the big data management and business unit perspective, which requires a business perspective and management perspective on data assets, they can be seen, checked, managed, estimated and used, answering the cost value and management questions of data assets, and using the lowest threshold to make data available, accessed and used by business people. The following focuses on the domain-wide data asset operations platform.

After the enterprise completes the asset construction from the research and development link, the following challenges must be faced in the management and interoperability of data external services, and this series of challenges from the big data management department are urgently needed to free up the human resources through the platform way, in order to build a good data application ecology:

- A large number of mentions, check the demand.
- A steady stream of asset construction needs.
- Cost and value assessment of data assets.
- The security management of data assets, the integrity and traceability of asset definitions.
- The service convenience of data assets: Self-service can be completed with the lowest threshold.

Another cloud output of data in Taiwan products: The global data asset operation platform is positioned to build a manageable, verifiable, controllable, accessible and estimable data asset service center centered on core data assets, supports the circulation of data assets within the enterprise, enhance the awareness of data use, assist enterprise digital-driven production model transformation. It is a shared tier product that links production platform Dataphin and data consumption scenarios.

At the same time, from the perspective of enterprise data asset management, the assets of the Data Middle Office and Taiwan are only a part of the enterprise data assets, and whether from the development perspective of the Data Middle Office or the whole domain management perspective, it is required to manage the metadata of the business system.

1. Data asset management section

    - Standard asset introduction process: A systematic approach to asset management links from introduction to standardization to classification to catalog planning. Asset introduction should first solve the problem of collection and adaption at data source, which needs to adapt to multi-type data sources. In addition, due to the different metadata collection conditions of different sources, it is necessary to design a standard layer to isolate the impact of version upgrade, so as to provide stable support of the standard model upward.

- Standardized process after the introduction of assets: The standardization here mainly refers to the shelf classification of assets, the maintenance of asset definitions and descriptions, the attribution of assets, the marking of the business labels of assets. And from the security management level for the asset visible and consumer objects for centralized control.
- Catalog planning after the standardized process: From the perspective of business classification of data assets, standing in the enterprise to view the actual situation of data asset classification for catalog planning.

2. Data Asset Services Section

The production of data assets is for use, before use needs to be available, available, after use needs to be assessed, the platform in a systematic way to solve the business perspective dimension of the asset presentation and query, and through blood relations, asset preview and other ways to help users fully understand and locate the data assets used before use.

After positioning the data assets, the next thing to solve is the service problem, the asset platform by docking the service market tools online docking, such as the numbering tools and analysis tools. From the usage process, the user can directly launch a consumption request for the positioned asset, and after approval, the asset platform creates a dataset on the tool side. Asset platform to achieve the consumer side of the rights control based on the realization of users from the one-stop link to see through, the process also liberated the IT department a large number of check and take demand.

3. The value assessment part of the data assets

The problem to be solved by value evaluation is to help managers assess the costs and benefits of on-shelf assets, thus complementing off-shelf decision-making and asset use assessment analysis. In terms of cost, assets are measured in terms of storage and calculation, and in terms of revenue, they are analyzed mainly from the dimensions of service invocation times, service output departments, etc., to assess the extent and breadth of use of Data Middle Office assets across the division. We'll cover the value of your data assets in more detail in the next section.

5.4.5 Data Asset Value

What is the value of data? 'The true value of data lies in the application, and data without application value are just data'.

Then how to evaluate the value of data, you can establish a suitable data value evaluation model based on data demand analysis, which mainly includes data cost and benefit evaluation methods, evaluation metrics, etc., and supports the dynamic update of data value evaluation methods and various metrics.

Cost method, income method and market method – the three traditional valuation methods have their applicability when applied to data asset valuation, but they all have certain limitations. At present, no mature data asset valuation method has been formed. Gradually exploring and advancing the value analysis of data assets in specific fields or specific cases may be a feasible way to continuously deepen this research in the future (Table 5.24).

TABLE 5.24 Metrics of Data Value Evaluation

Metrics Classification	Metrics Name	Metrics Description
Unit cost D=acquisition cost c1+integration cost c2+maintenance cost c3	acquisition cost c1	The cost of obtaining data
	integration cost c2	The cost of integrating data
	maintenance cost c3	Storage equipment and security investment, the cost of maintaining data
Intrinsic value coefficient a=data quality a1×data size a2×data sharing a3	data quality a1	Data quality, including data completeness and accuracy
	data size a2	How much data size
	data sharing a3	The business system involved in the data and its complexity
Use value coefficient B	Use efficiency b	Data usage, including the number of users and user evaluation
Market value coefficient D=market demand d1×market supply d2×competitor pricing d3	market demand d1	Market demand for this data
	market supply d2	The supply of the data in the market
	competitor pricing d3	The pricing of similar data by competitors in the market

Data asset value evaluation methods, cost method, income method, and market method – the traditional three evaluation methods have their own advantages and limitations when applied to data asset valuation. At present, no mature data asset valuation method has been formed. Then we use the cost method as the basis to integrate other methods to calculate the value of the data, and the relevant calculation formula is as follows:

Data value V = unit cost C (yuan/TB) × data volume N (TB) × intrinsic value coefficient A × (use value coefficient B + market value coefficient D)

The value of data assets has received more and more attention, but the current research on data asset evaluation is still in the early stage, and the above evaluation methods and implementation also need to be gradually improved. The complexity of data assets also makes the methods borrowed from intangible asset evaluation inadequate. Which method is most suitable for the value evaluation and final pricing of data assets, as well as its executable, still needs to be continuously explored.

5.5 INTELLIGENT DATA APPLICATIONS

5.5.1 Data Applications Overview

5.5.1.1 Why Are Data Applications Essential

In the practice of enterprise Data Middle Office, architecture design, platform construction and capitalization all fall within the category of platform construction. The achievements of platform data assets construction are a standardized modeling and extraction designed data assets system, which is built based on 'Three One' methodology of Data Middle Office. But if there is no further construction in the application layer and it will not be used in business process, the data assets system will be just like 'gold mines' lying in mines, not being mined conveniently, understandable, friendly and business scenario based for applications, and that will greatly limit the full use of middle-office assets by enterprises. At the same time, data should be made into a pool of 'living water', and any data assets should be transferred to the front-end business scenarios for use, observation of the effect and continuous tuning, so that the value of data will be continuously polished and shining.

5.5.1.2 Definition and Category of Data Applications

The applications of Data Middle Office are user-oriented. Learn more about the pain points of data usage and combine data assets and technologies (such as real-time technique and algorithm technique) to construct application system that can be suitable for users in different roles or in different business scenarios, and its specific carrier of the system is data products. Figure 5.34 shows the top-level design of Alibaba's Data Middle Office, and the product is an important part that occupies a corner of the triangle. Data product is a kind of product that solidifies data, data analysis and decision logic into an application product and provides data value for business decision.

Data applications are oriented to different users, and the categories are as follows:

- Visualized large screen and fixed chart, which are usually used by decision-makers, external customers or users who cannot analyze data and only need to get data results, and its main characteristic is to display the results directly without showing analysis logic and thread of the design.
- Interactive data analysis products, such as common BI tools, can drag and drop freely to generate analysis statements and analyze business problems for users who need to analyze data. They are usually used by data operators to discover problems, position issues and find the

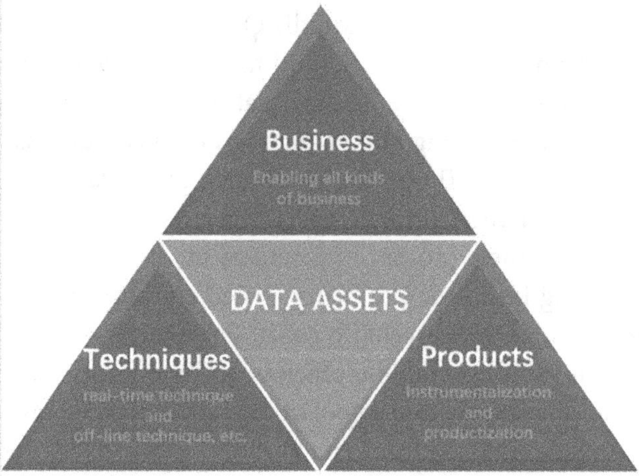

FIGURE 5.34 The Top-Level design of Alibaba's Data Middle Office.

causes through business judgments and data analysis results, so as to complete business optimization.

- Intelligent decision data products are a kind of data products that automatically execute decisions and actions according to the results of data analysis and algorithm technique. For example, the risk control decision engine of the financial industry, credit and anti-fraud strategies directly intervene in business processes through the judgment results of various rules.
- Mixed data products are comprehensive products that combined with specific business application scenarios, usually include both interactive analysis and decisions of algorithm technique, and feedback data application results in the form of fix statements.

5.5.2 Planning and Practice of Data Applications

Enterprises often need to consider the overall goal and the decomposition of the implementation phase when building Data Middle Office, so that they can better concentrate resources to balance the current situation and goals, as well as the short- and long-term conflicts. In the practice of enterprise Data Middle Office, we found that many enterprises expect to decision by intelligence instead of manual before their data assets have been deposited, and they believed that building Data Middle Office can solve all problems. But the accumulation of data by enterprises does not happen overnight, the term 'asset' of data assets requires planning and operation.

5.5.2.1 Analysis of Business Status

Data applications should fit users and business. User-fitting means to understand the ability of data analysis and data application of those who need data in the enterprise, and must avoid 'pull-out' design. Especially in the first phase of Data Middle Office construction, appropriate data applications should be considered to discover the value of enterprise data assets that like water sources need to be tapped and made to flow, so that all employees of the enterprise can use the results of Data Middle Office most widely in the first phase of the project. This is very important and essential for expanding a second or a third phases of the project.

In the first phase of the project, usually the main pain points of data faced by enterprises are inconsistent definitions of metrics, low data quality, insufficient data sources, etc. Therefore, the first phase is also the 'credit construction' phase, and the data provided by Data Middle Office must be credible and usable, and if the data source can be expanded, it will add additional values.

The first grip for business status analysis is user analysis. Pay attention to the following points.

1. Role analysis: Data requirements form high-level, middle-level or grass-roots-level users are all need to be considered, but the granularity of the analysis that they need is different. The high levels need 'wide' and 'less' data. What they want are results, and it is best to supply them with external competitors and market data to help them make decision references. The form of expression is tailored according to the senior management's own professional background and preferences. For example, some executives like to view detailed data, and some just like to view the results for performance tracking and problem discovery. The middle levels need the result data within the scope of responsibility, and also need the analysis process to help them finding the causes of the problems and the optimization points. The grass-roots levels need a lot of detailed data to view and need tools or products that can empower business with data.

2. Analyze the pain points of using data: For example, the high levels will be confusing when they view the same metric as before but get different results. Then we should unify the metric caliber and solidify a fix statement, so that the seniors can view the data in one place. The grass-roots levels who do precision marketing need to think the data logic and extract the demand to IT department by themselves. So here needs a product, which can directly select labels on the interface to choose and launch users.

3. Ability level of user data analysis: At the application level, whether the empowerment of data is to provide wide tables to write SQL queries or package them into interface operations, to reduce the programming threshold of data analysis, requires a considerable understanding of the data analysis capabilities of the internal personnel of the enterprise. Even in terms of not only analytical tool capabilities, but also analytical thinking. If it is relatively weak, then it is necessary to provide classic analysis statements to solidify analysis ideas to empower front-line staff.

The second grip for business status analysis is the design of the overall digital transformation phase of the enterprise. While focusing on the future, we must respect the current phase. First of all, our vision and planning for the future can temporarily let go of the constraints of the status quo to discover and locate the target. In our project, Alibaba's experts and corporate personnel work together to imagine how the application of data can empower business in various business sectors, including the form, purpose, effect to achieve, etc. Then return to the status quo to find the constraints of real conditions, enumerate the preconditions for implementation and determine the phased goals.

5.5.2.2 Planning of Application Products

Application product planning is essential to give the product matrix by roles and stages, and do over-design in practice, and enumerate the preconditions for realization. The stages that usually need to be followed are online business, digital operation and intelligent decision-making. In a project, we imagined a good product-marketing sandbox. This product uses data and algorithm to support decision-makers deployment strategies in marketing activities. For example, according to budget and marketing goals, the system can perform optimal decision-making simulations to give strategic suggestions on channel placement, product placement, crowd placement, etc. However, the company itself has not yet completed the online business phase, and there is no marketing management system to initiate and store activity data online, so it is impossible to jump to the third stage at once. In planning, we often find that the Data Middle Office project has the effect of 'checking for omissions and filling deficiencies' in informatization capacity building.

The entire planning process is as follows (Figure 5.35).

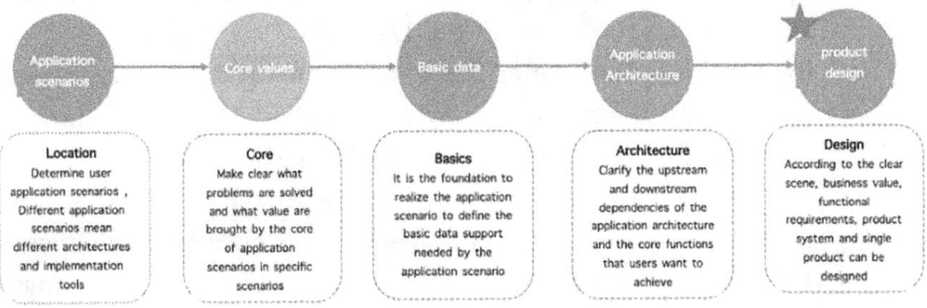

FIGURE 5.35 The entire planning process of Data Middle Office.

5.5.3 Practice of Industry Data Applications

5.5.3.1 General Data Applications

General data applications refer to data applications that are built on the Data Middle Office and do not have specific industry attributes. And its purpose is to manage, display and use data assets conveniently. The application functions are generally used in multiple industries, or 80% of the functions can be generally used in multiple industries. This book introduces two types of application practices.

1. Data assets platform

   The purpose of this application is to open the data assets of the Data Middle Office to the entire company's business departments in an understandable and low-threshold manner, and at the same time serves as a platform for the big data department to manage and operate data assets. It is an indispensable asset operation platform after the Data Middle Office launched. In the same way, we start from the user role to dismantle the functions and goals of the platform.

   - The first type of roles: Enterprise manager

     Enterprise executives need to understand the total amount, types, cost and service output invocation of their own company's data assets to judge the degree of support for the entire company's businesses in the Data Middle Office. At the same time, from the perspective of digital transformation, the scale, degree of standardization and use of data assets in the enterprise data reflect the progress of this process to a certain extent.

   - The second type of roles: big data department

     After the Data Middle Office goes online, we must consider operational issues. First, how to make the entire company widely used, so that the assets of the Data Middle Office become 'living water'? Second, after the whole company understands, how to deal with the demand blowout situation? The third is how to calculate the cost and benefit of the Data Middle Office.

     There are two very important solutions to the first question, one is to let the whole company know what the Data Middle Office is with the lowest threshold? What does it contain? What changes can it bring to their business? Second, it must be implanted into one or two business scenarios through valuable data applications.

It can be a very small function, but it must make real changes to the business. Therefore, this platform must carry the functions of asset maps and asset panoramas. The valuable data applications always contain strong business attributes and are not presented in general applications.

For the second question, it is necessary to distinguish between demand types. Generally speaking, the first blowout must be access demand and various report demands. Therefore, the platform must include a self-service access function. It is a very natural process from viewing to accessing. Of course, we hope that it will be presented in one platform.

For the third question, it is necessary to find the metrics for calculating the cost and benefit of the Data Middle Office. The cost should be clear, and it can be calculated by computing and storage resource consumption. For value calculation, many companies are keen to develop value evaluation models. If the results calculated by this model are interpretable, it can be widely recognized by the whole company, and can be implemented (such as performance evaluation based on this value), then it is successful. But if the model just become a score or metric, and put it on the page only, then it is better to directly calculate the number of asset calls.

- The third type of roles: business department

  The business department here refers to the rest of the business and functional departments except the IT and Big Data department. Most staff do not know about it because it is impossible to participate in the Data Middle Office construction project. And they usually have a lot of data requirements, but they don't know what the Data Middle Office has and what it can provide. At the same time, there is a high probability that they cannot write query statements, and need to constantly find IT or big data departments to raise various requirements. Therefore, functions such as asset maps, asset panoramas and self-service access are also needed.

2. User portrait and circle platform

A very important category of assets in Data Middle Office are tags. There are many entities of tags, such as customers, stores, merchandise, pilots and employees. The tags are used to describe the characteristics of entities from different dimensions to facilitate the user has a good understanding of the analyzed entity objects. At the same time, the tags are designed to allow users to perform refined operations. For example, to wake up sleeping users, you can circle the people who have not purchased in the past 6 months and people who have purchased in the past year and have high value. Three labels are used in this scenario. This combination relationship can be flexibly used based on the operational ideas of front-line operators. The design of the label system is not introduced in this chapter. This chapter mainly focuses on the asset application function. We still start from the user role to explain.

- The first type of roles: big data department

  The asset objects of this platform are the labels, which are opened to business departments from the application point of view. Therefore, there must be a label management function in it to solve the problem that how to monitor the label effect and provide the highest quality label for users to use. At the same time, the Data Middle Office should include the optimization functions to solve the problems that how to perform offline management for labels that do not perform well? How to provide and manage user-created tags or combined tags? How to present a complete and clear label catalog to users?

  Therefore, this application must first create a label catalog according to different label entities, which is convenient for users to query flexibly and clearly. The second is the monitoring effect. In addition to the number of uses and the number of queries, the effect of the label can also be observed before and after the commercial use of the label. This type of effect is often applicable to algorithmic labels. For example, the label of high probability purchase crowd, which is an algorithm label with supervised learning, then in the algorithm training, there can be metrics such as accuracy and recall to evaluate the credibility of the algorithm, and the commercial effect after the label is released. It is necessary to compare activities by group to compare the promotion situation, including promotion of conversion rate and promotion of GMV. Third, according to the effect, the administrator can carry out label and shelf management. This management can be at the application level or related to the scheduling task at the platform level. Finally, it is necessary to provide users with convenient functions to create combined tags and upload custom tags.

- The second type of roles: business department

  In order to be used by business department, the first problem to be solved is to have a labels catalog, and the second is that users can flexibly choose the labels to interpret the portrait of the entity object, and the next is to do interface operations that can be selected by a series of drag-and-drop or point-and-click operations. Of course, you must be able to see the effect in the end.

5.3.3.2 Data Application of Industry

Several categories of data products have been mentioned before in this book. The product carriers of visual large screen and interactive data analysis are similar and slightly different, and the content presented depends on different businesses. This section mainly introduces some practices in the retail industry, while other industries introduce the most industry-characteristic application practices to readers.

1. The aviation industry

- Pilot supply and demand matching platform

  Before introducing the application, first describe the business background.Why is there a supply and demand match for pilots? Pilots are very important and valuable personnel assets of airlines, and their education and training are related to the flight safety. Therefore there is always a contradiction between supply and demand of pilots. On the one hand, the strong demand based on the civil aviation market, airlines continues to introduce new aircraft and new routes, on the other hand, the cultivation of the pilot to graduate from aviation school into the airlines will continue in the airlines training base in training, from the narrow body machine to wide-body aircraft, from the co-pilot to captain, every advanced technology level need to accomplish a certain number of flight hours and training qualifications, not happen overnight. Long-term planning and short- or medium-term matching are required between all supply and demand. The supply and demand face two levels of matching, one is the match of people, one is the match of flight hours. In order for a person to be matched to an aircraft type, the number of flight hours is based on the number of available flight hours of a person, which is related to the qualification level. At the same time, the number of unavailable flight hours caused by rest, illness and shift lockdown due to management functions shall be removed.

The pain points and problems that the platform needs to solve are as follows: First, based on the 'aggregation' of data in the Middle Office, the data previously scattered in various business systems, offline and 'expert experience' are presented on one platform to improve efficiency, including aviation network plan, aircraft introduction plan and experienced man–machine ratio; second, the platform can make flexible calculation according to long-term planning and short-term adjustment, refine the modification requirements of each model, and select the corresponding candidates according to the operation experience to determine the modification requirements; and the third is to further match the flight hours refined to the finest granularity according to the route, base, qualification level, etc. to determine the difference plan.

2. Financial industry

- Transaction anti-fraud risk control decision engine

  Transactions related to Internet finance include payment, wallet, marketing activities and so on. Every business needs to develop anti-fraud strategies in several scenarios to conduct real-time screening and return relevant decision results to the business system, such as pass, reject and manual review. The application design of the risk control decision engine can be abstractly reduced to the invocation of scene strategy based on the business system. The decision engine returns the result to the business system according to the real-time parameters passed by the business system and the pre-configured strategy of each scene. Of course, there are several key points that need to be defined.

The first is scenario, this is the basis for the operation of the whole decision-making engine and the key to communication with the business system. This data function will be implanted into the operation process of the business system. Once the risk control decision-making results output end need immediate execution engine, of course, if hang up for risk control engine's own reason or during large presses. Based on the relevant risk control and management system of the company and on the premise of ensuring the key metrics of risk control, demotion can be done technically. Scenario can be viewed as risk control decisions in business processes, such as login, reset/change of password, setup of non-secret payment, order payment, etc. Each link requires different input parameters of the business system and exit parameters of the decision engine. It should be noted that the reason represented by the return code of the business system may not be the real risk control reason based on security policy considerations.

The second is the configuration of rules. The growth of business will bring about an increase in rules. As the saying goes, 'While the priest climbs a post, the devil climbs ten'. However, the more confusion brought to the risk control decision engine is how to solve the conflicts of many rules (the risk control strategy has too many personnel, and it is difficult for human to evaluate which logic is conflicting or duplicated), the ranking of rules priority and the effectiveness of rules. The conflict/repetition problem requires the decision engine to analyze and judge the possibility of conflict according to the configuration of the policy and submit it to the human decision. Priority issues involve the definition of sorts, and the platform needs to agree on a set of ordering rules for policy personnel to follow; the most important thing is that the effectiveness of rules must be tracked, from the hit rate, accuracy and other metrics to observe and remove useless rules.

Rules can also be set in the 'dual path' mode, namely online and bypass. The online rule means that the system runs the results in real time and feeds back to the business system, while the bypass is usually when the strategy staff wants to test the effect of another set, and only the results run on the system are not fed back to the business system.

Because the risk control engine is an application on the data platform, a large amount of historical data can be used to verify the effect of rules when they are online, so as to ensure the effect of rules more effectively.The third, case disposition. Manual disposal is a very important command in anti-fraud transactions. As for the suspended state when the business system returns, the decision engine will uniformly present such cases in the case disposal module and distribute them to the case handler for review. This role will retrieve a large amount of detailed data and user data within the scope of authority for judgment.

The fourth, list library. In rule settings, many scenarios refer directly to the list for blocking or releasing. There are many types of risk control list, such as political list, white list, black list and gray list. Some of the data comes from external data, such as the list of political leaders, while others are the list libraries accumulated by the platform.

3. Retail industry

- Visualize large screen applications

  The visual large screen is generally used in the scene of the cockpit and corporate external publicity. The front is for the overall design of corporate executives' overview of the operating conditions of the enterprise, while the latter is displayed based on the data screened in the enterprise's own publicity. This book mainly introduces the cockpit.

  The cockpit is to provide the global data viewing function for the leadership, and to support them to quickly understand the overall business achievement and operation situation. At the same time, combined with the existing market data of users, the horizontal comparison can be made to evaluate their own development trend and industry development direction, and provide reference and guidance for the formulation of future planning and the setting of their own goals.

  The key elements and steps of cockpit implementation are shown in Figure 5.36.

  Usually, if it is based on the cockpit of the retail industry, you can refer to the structure scope in Figure 5.37. Of course, the specific scope should be determined according to the project.

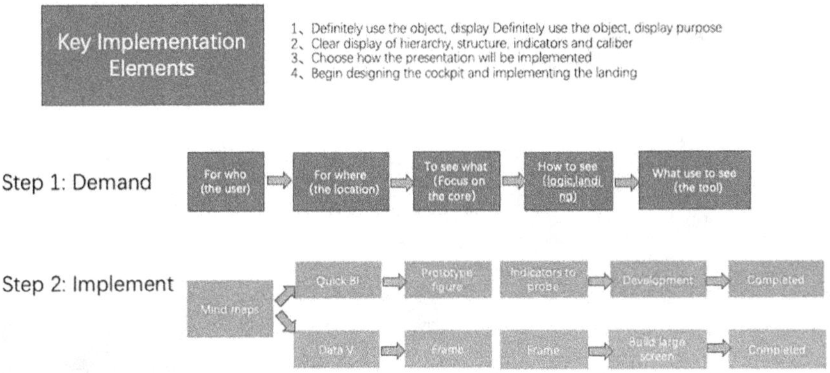

FIGURE 5.36 Cockpit implementation key elements and steps.

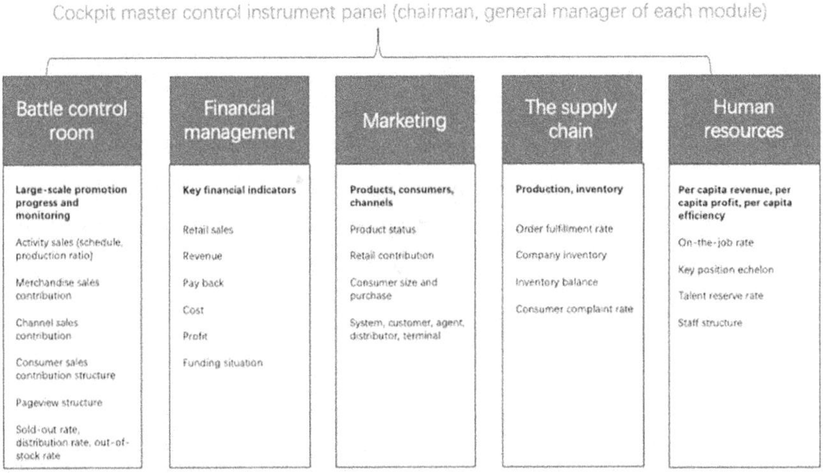

FIGURE 5.37 Functional structure of cockpit instrument panel.

Figure 5.38 takes the combat command room as an example to show the large screen effect.

- Interactive data analysis applications

  Interactive data analysis will depend on the BI tools complete, consumers in the retail industry the scenario, the data of China project practice is divided into business domain-oriented transverse analysis theme, such as consumer, merchandise, stores, guides, marketing, consumer rights and interests and so on,

FIGURE 5.38 Large screen in combat command room.

also can do the theme of the longitudinal characteristics of oriented business scenario theme analysis. Horizontal means that multiple business analysis scenarios can be supported, such as consumer novelty, re-purchase, wake-up call, optimal reduction of stores and crowd characteristics of commodities, which can be screened, searched, drilling down and linkage in the set of consumer analysis Kanban. Vertical means starting with a business scenario, spanning multiple business domains and screening analysis modules for users and combining analysis links to solve the analysis needs of this business scenario.

For interactive analysis applications, the presentation of business analysis logic is crucial. In addition to business understanding, user thinking and landing thinking are also required, as explained in Figure 5.39.

Figure 5.40 is a typical retail industry insight system by business domain on the project.

4. Beauty makeup industry

- Text mining

  Beauty makeup industry for market changes, particularly sensitive to the voice of the customer, thus guide for customer evaluation, customer service call, text mining, such as customer's

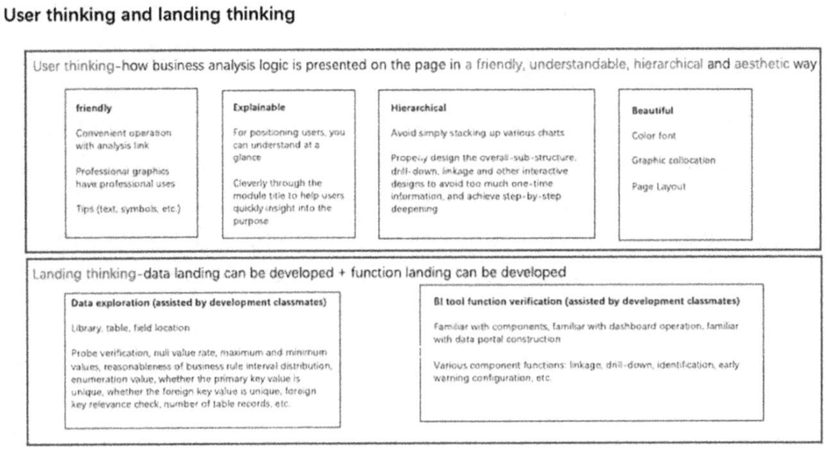

User thinking and landing thinking
User thinking-how business analysis logic is presented on the page in a friendly, understandable, hierarchical and aesthetic way
friendly
Convenient operation with analysis link
Professional graphics have professional uses
Tips (text, symbols, etc.)
Explainable
For positioning users, you can understand at a glance
Cleverly through the module title to help users quickly insight into the purpose
Hierarchical
Avoid simply stacking up various charts
Properly design the overall-sub-structure, drill-down, linkage and other interactive designs to avoid too much one-time information, and achieve step-by-step deepening
Beautiful
Color font
Graphic collocation
Page Layout
Landing thinking-data landing can be developed + function landing can be developed
Data exploration (assisted by development classmates)
Library, table, field location
Probe verification, null value rate, maximum and minimum values, reasonableness of business rule interval distribution, enumeration value, whether the primary key value is unique, whether the foreign key value is unique, foreign key relevance check, number of table records, etc.
BI tool function verification (assisted by development classmates)
Familiar with components, familiar with dashboard operation, familiar with data portal construction
Various component functions: linkage, drill-down, identification, early warning configuration, etc.

FIGURE 5.39 User thinking and landing thinking.

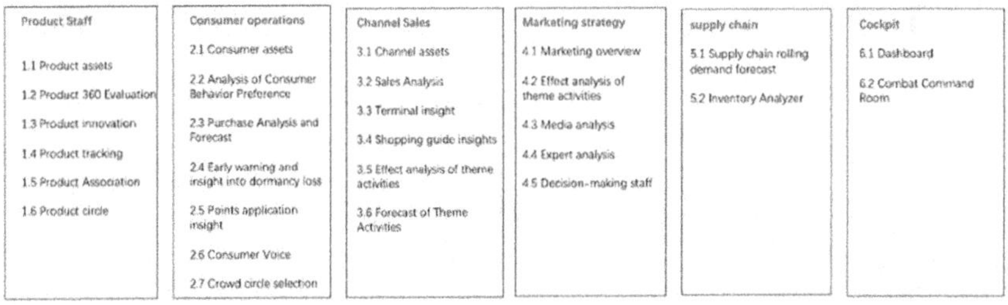

Product Staff
1.1 Product assets
1.2 Product 360 Evaluation
1.3 Product innovation
1.4 Product tracking
1.5 Product Association
1.6 Product circle
Consumer operations
2.1 Consumer assets
2.2 Analysis of Consumer Behavior Preference
2.3 Purchase Analysis and Forecast
2.4 Early warning and insight into dormancy loss
2.5 Points application insight
2.6 Consumer Voice
2.7 Crowd circle selection
Channel Sales
3.1 Channel assets
3.2 Sales Analysis
3.3 Terminal insight
3.4 Shopping guide insights
3.5 Effect analysis of theme activities
3.6 Forecast of Theme Activities
Marketing strategy
4.1 Marketing overview
4.2 Effect analysis of theme activities
4.3 Media analysis
4.4 Expert analysis
4.5 Decision-making staff
supply chain
5.1 Supply chain rolling demand forecast
5.2 Inventory Analyzer
Cockpit
6.1 Dashboard
6.2 Combat Command Room

FIGURE 5.40 Retail industry business domain insight system.

dialogue is especially important, and by cleaning, extraction of text mining is a text file, effective, high value, understandable, novel characteristics of digital information, use the information to organize the process of service application scenario. Text mining is a branch of artificial intelligence research, which is a multidisciplinary field.

It involves the processing of unstructured data, knowledge graph, graph calculation, neural network, clustering and so on.Alibaba Data Middle Office has accumulated a large number of successful cases of text mining, among which commodity evaluation is particularly common. We can introduce the application of text mining through the following four scenarios of Alibaba Data Middle Office.

- Competitor Customer Voice Analysis (Figure 5.41)

  Through the analysis of the customer voice of the competing products, a lot of real information about the competitors can be obtained. For example, in this case, the special effect of the competing products is moisturizing and hydrating, and the taste is good. The attraction of the activity is the gift, and the main customer group is mothers, mothers and housewives. In this way, we can design targeted market strategies and improve our competitiveness.

- Customer voice analysis of its own products (Figure 5.42)

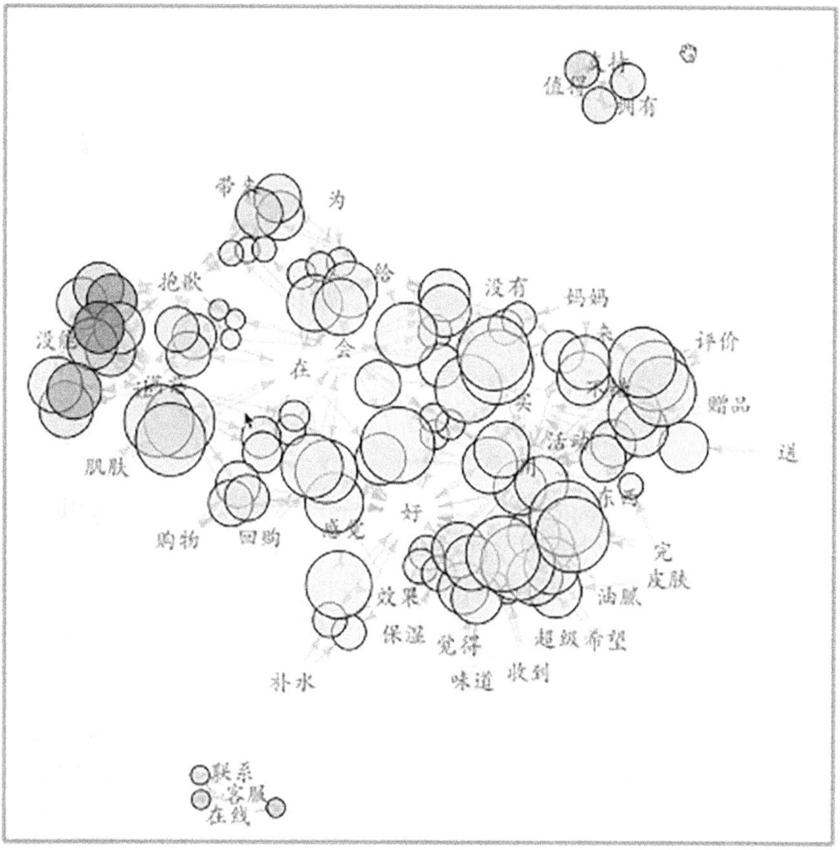

FIGURE 5.41 Competitor customer voice data mining.

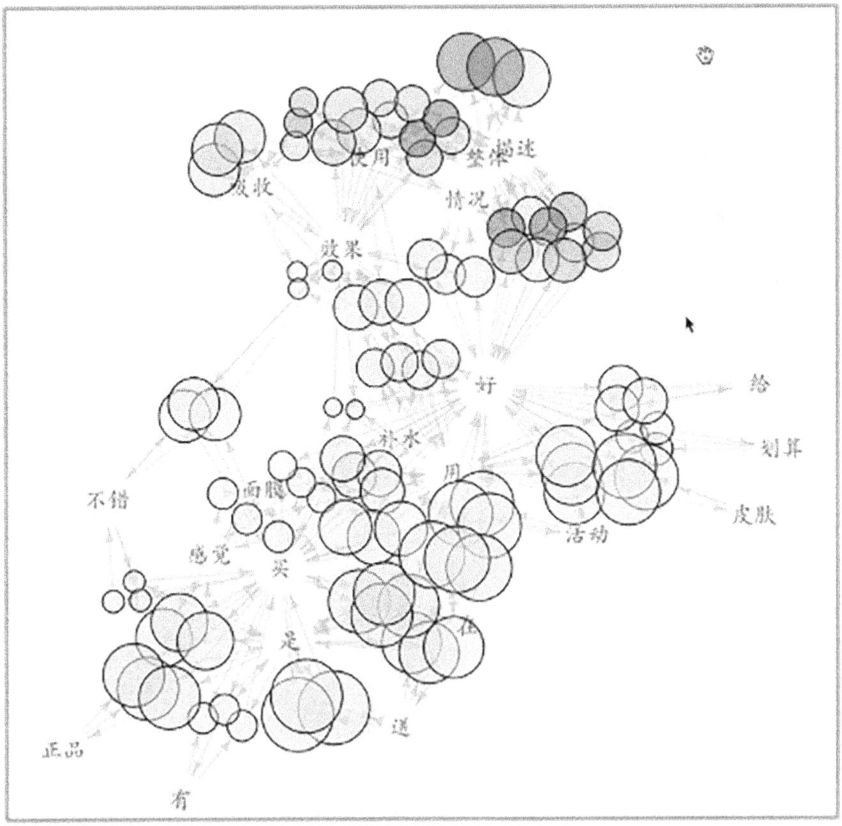

FIGURE 5.42 Customer voice of its own products data mining.

At the same time, we can also through their own customer voice analysis, assist us in product positioning, planning strategy, etc.

- Voice analysis of negative customers (Figure 5.43)

  In addition, through the emotional model, we can screen out the complaining factors of negative customers and find the pain points of our products. For example, in this case, we identify that the customer is not satisfied with the points exchange. Through the effective identification of negative emotions, products can be improved in a timely manner and customer experience can be enhanced.

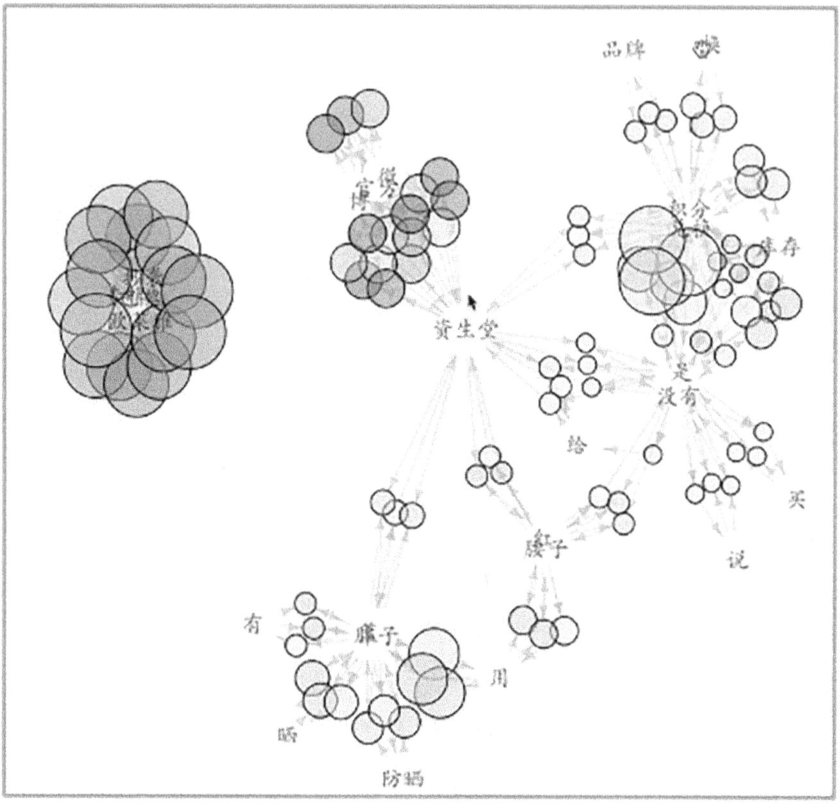

FIGURE 5.43 Negative customer voice data mining.

- Establish customer evaluation system (Figure 5.44)

  Finally, we through Alibaba Data Middle Office establish a perfect customer evaluation system, using text mining technology and China union, real-time quantitative monitoring the granularity of product or public opinion of the market, insight, analysis, and combined with the business strategy in a timely manner the ground improvement, so as to achieve rapid response to the market system of closed-loop operation.

5. Clothing industry

- Sales Forecast

  The clothing industry is particularly sensitive to inventory, so it is necessary to study how to accurately forecast future sales. Forecasting model is a common supervised model, which

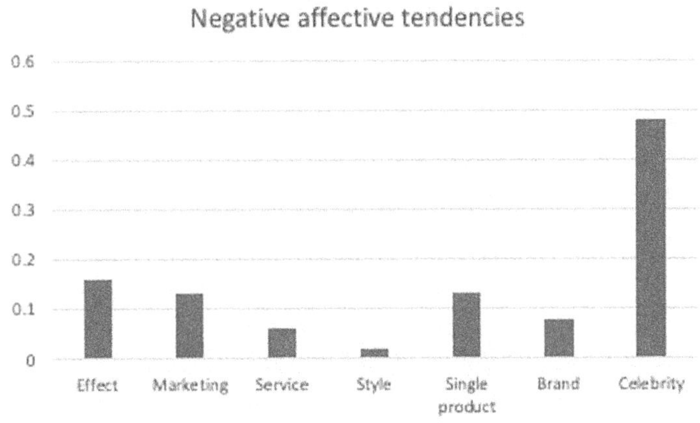

FIGURE 5.44 A sample of customer evaluation system.

is often used in precision marketing and financial planning. Precipitation data through Alibaba Data Middle Office, a large number of consumers painting history, historical sales data, sales promotion information buy stores as well as the ability to model, weather and geographical information, after cleaning, denoising, the training sample data, filter characteristics, build model, evaluation model and the master steps, for merchants to provide online sales forecast data, to achieve inventory optimization, cost reduction and efficiency.

5.6 TECHNOLOGY PRACTICE OF DATA MIDDLE OFFICE

5.6.1 Data Governance

With the digital transformation of enterprises, they often encounter the problems of how to manage data assets, how to view data assets and how to quickly find and use data. Alibaba Data Middle Office will well solve the problem of enterprise data governance.

The pain points of enterprise data assets management are: unknowable, uncontrollable, undesirable and unconnected. Here, agnostic means that the enterprise assets are not clear, the business personnel do not know where the data are, the technical personnel do not know the business meaning of the data, and the enterprise managers do not know how to use it; uncontrollable means that the quality of the enterprise data is not high, and the quality and scale are not controllable; inadvisable means that the business personnel need to collect and obtain the data, and the process

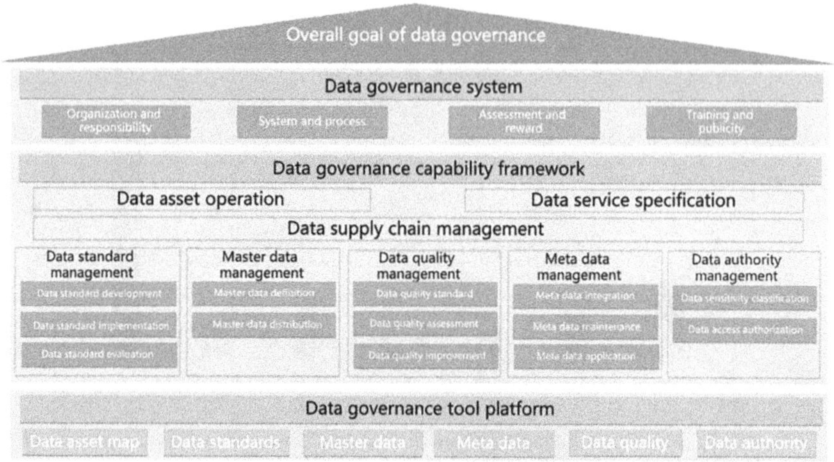

FIGURE 5.45 Overall architecture of data governance.

is difficult, the cycle is long, the link is blocked, and the data service is not convenient. It can effectively support the business, but unconnected means that the data and business knowledge cannot be linked, and it is difficult to generate value. In the practice of Data Middle Office, we gradually summed up a set of effective methods of data governance to solve the above problems (Figure 5.45).

5.6.1.1 Data Asset Operation Governance

Through the asset operation of data, a unified view and management screen is built based on the data platform to improve the pain points of unclear data assets, make the data clearly visible to enterprise managers and departments of consumption data, deepen the cross collision of data and mining value.

5.6.1.2 Data Service Specification Governance

Relying on data assets operation, combined with Data Middle Office to provide unified data services.

5.6.1.3 Data Supply Chain Management Governance

Through the analysis of the whole data supply chain, the optimization and governance, including data generation, data integration, data analysis and data consumption four links. In the process of data generation, we ensure the quality of data source through data quality tools, ensure the data definition through data standardization management, master data management and metadata management, and ensure the efficiency, speed and automation of data collection and cleaning through the data Mid-office. In the process of data integration, all kinds of data sources can be converged and connected in the data Mid-office, so that structured, unstructured, real-time, offline and other data can be managed conveniently and quickly. In data analysis, the data Mid-office can combine with Bi and python to provide data analysis, model building and other functions. In the aspect of data consumption, the data Mid-office provides a unified data service interface. By dredging the whole data asset supply chain, the efficiency of enterprise data link can be improved.

5.6.1.4 Data Standard Management

Establish enterprise-level data standards and specifications throughout the whole process of data generation and consumption to ensure the consistency of business and technical understanding.

- Data definition and format: Mainly formulate data definition and format standards to ensure the consistency of business description, classification and semantics of data. Generally speaking, the formulation method is to negotiate and formulate the agreed data standards from the aspects of foundation, sharing, uniqueness, business priority, cross department or system use, scope of use, data cycle, etc.
- Data acquisition and verification: Mainly formulate standards and specifications for data acquisition and verification to ensure that the data are transferred in a correct and reliable way and keep the data continuously applicable. The development method is generally based on the analysis of the production, circulation and use process of existing data, analyzing and solving the problems that may be encountered in the implementation of data standards, and forming executable data creation, storage and use process specifications.
- Data system and process: Mainly formulate system and process standards, ensure that data and process have rules to follow, so as to solve conflicts in time, and define how to implement, manage and measure data and process. Generally, it needs to be combined with

the data management organization structure and process system to define the management organization structure and process system of data standards, so as to realize the effective management of standard update, standard conflict and other issues.

5.6.1.5 Data Quality Management

Based on data quality management tools, a virtuous circle of discovery-evaluation-improvement is gradually formed. Through the configuration of data quality rules, automatic and intelligent management of data quality is realized. Data quality rules are generally designed from four dimensions: integrity, accuracy, consistency and timeliness.

5.6.1.6 Master Data Management

Master data are the core data of an enterprise, and almost all businesses are carried out around these business objects, so it is necessary to ensure the consistency of them in various systems.

5.6.1.7 Meta Data Management

Through the enterprise-level metadata sorting, we can effectively link up the business and technical aspects of data, unify all staff's understanding of data, improve the metadata management system, realize the functions of metadata monitoring, change management, version management and knowledge management, and comprehensively transform from metadata management to metadata drive.

5.6.1.8 Data Authority Management

Data authority management is to formulate a set of reasonable and scientific data classification standards for the data used by enterprises, and establish a set of data use authorization mechanism according to the classification standards, so as to meet the requirements of business for data use. The goal is to establish a complete set of data sensitivity classification standards for the data in the business data analysis environment according to its sensitivity and importance, so as to ensure the number of businesses according to the compliance and convenience of use. First of all, according to the data characteristics of the enterprise and the sensitivity and importance of the data, we need to grade the data, which can be multi-dimensional and cross formed fine-grained hierarchical management system to meet the needs of different scenarios. The second is authority management, which establishes a data leakage prevention system from three dimensions of personnel, process and technology to comprehensively protect sensitive data from illegal use, storage and transmission. Through training and other ways, the staff can improve the user's awareness of data rights and the ability to control risk behaviors; the process is based on clarifying the control requirements for data leakage prevention in the enterprise-level data authorization strategy, establishing institutionalized behavior norms and corresponding processes, incorporating the awareness of sensitive data protection into the enterprise culture, and reducing the risk of data leakage, including hierarchical process and sensitive data management according to the evaluation process, sensitive data isolation process, access permission approval process, compliance control process, risk assessment process and so on. In technology, based on the identification of sensitive data, the data leakage prevention technology control means combining active defense and passive monitoring is adopted to protect and monitor the sensitive data.

5.6.2 Real-Time Data Middle Office

From the generation of data to the formation of data assets, data assets eventually feedback and enable the business. In this process, the value of data itself decreases rapidly over time, and time is the biggest enemy of data assets. With the development of business and technology, the demand for real-time computing is exploding. With the introduction of real-time computing power, the value of the Data Middle Office is multiplied in timeliness. Timeliness here includes real-time data, real-time analysis, how to make the results quickly available to production systems, how to make business self-help rapid analysis, how to quickly respond to business changes and so on. Only by doing these aspects well can we give full play to the value of data and make the Data Middle Office serve the business better.

At the enterprise level, the real-time computing technology is still difficult to be effectively applied. The reason mainly lies in the high technical threshold, the difficulty in controlling the development and maintenance costs, and the lack of mature product solutions. In the process of building data desktop for enterprise customers, Alibaba Cloud GTS summarizes a set of real-time data desktop programs based on Alibaba Cloud technology product components. As a real-time computing framework for streaming data, it can effectively reduce the delay of full-link data streams, real-time computing logic and spread computing costs, ultimately effectively meet the business needs of real-time processing of large data, and help enterprises take the lead in fierce competition.

5.6.2.1 Overall Architecture of Real-Time Data Middle Office

The real-time Data Middle Office is a part of the Data Middle Office, so the overall architecture is basically consistent with the overall architecture of the Data Middle Office. The difference between the two is still the core part of the technology – the data warehouse. In this particular scenario, the data are always flowing and the calculation is carried out in real time (Figure 5.46).

The real-time data warehouse is mainly divided into three layers: DWD, DWS and ADS. The ODS layer in the offline data warehouse is removed, and strictly speaking, it is weakened. At present, data transmission between each layer of real-time data warehouse is carried out through Datahub or Kafka, data processing and calculation are carried out through Flink.

ODS (Operational Data Store) layer: In the hierarchical division of a real-time data warehouse, this layer is relatively incomprehensible because data are not persisted after being collected in real-time, but transmitted through streaming data channel Datahub or message middleware Kafka. Essentially, topic of Datahub or Kafka is the ODS layer of real-time data warehouse. Unlike traditional offline data warehouses, it is a real-time flow.

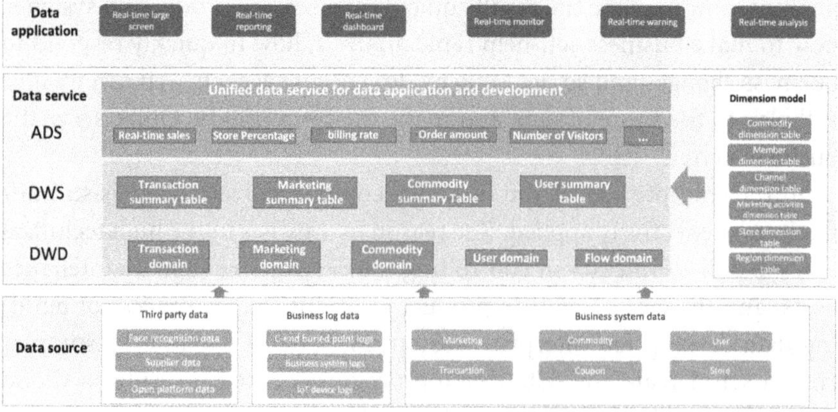

FIGURE 5.46 Overall architecture of real-time Data Middle Office.

DWD (Data Warehouse Detail) layer: Real-time computing engine subscribes to business data message queues, and then combines data cleaning, multiple data source join, streaming data and offline dimension information to associate all dimension attributes of some same granularity business systems and dimension tables, increasing data usability and reusability to obtain final real-time detail data. This part of the data has two branches, one that falls directly into the ADS for real-time detailed queries and one that is sent to a message queue or Datahub for DWS computing.

DWS (Data warehouse summary) layer: Build a public summary layer with the concept of data domain and business domain. Unlike offline data warehouses, the summary layer is divided into mild summary layer and high summary layer, which are produced simultaneously. The light summary layer wishes to reuse the common data warehouse layer ideas as much as possible and produce some standardized, multi-business common basic multi-dimensional detail metrics. The light summary layer can be written into the ADS layer's analytical database for front-end product complex OLAP Query scenarios to meet the needs of self-service analysis and output reports; high summary layer data are written to RDS or some other NoSQL database, like Hbase, provide simple query applications for real-time data, such as real-time large screen.

ADS (Application Data Service) layer: The ADS layer is not built one-to-one according to the needs, but is designed to meet different needs to form a unified metric library to support more demand scenarios quickly. ADS layer exists as database tables in real-time computing scenarios. They can provide services directly to the outside world, or they can be API encapsulated through OneService architecture to provide unified data services through APIs.

5.6.2.2 Technology Architecture of Real-Time Data Middle Office

Real-time Data Middle Office technically solves three major problems: fast collection of streaming data, fast real-time processing calculation of data and real-time data analysis and query. The architecture of real-time Data Middle Office technology is also designed and built around these three major issues (Figure 5.47).

The core technical component of real-time computing is the computing engine. Storm, Spark Streaming, SAMZA, Apache Flink and other real-time computing engines are available on the market. After a long

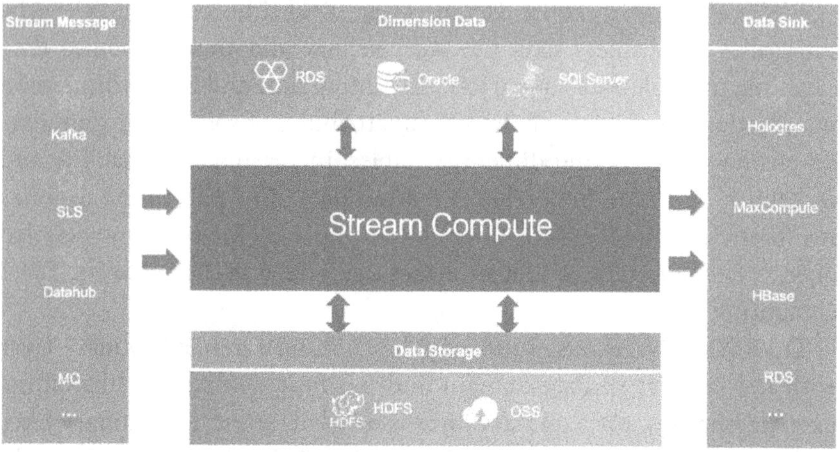

FIGURE 5.47 Real-time Data Middle Office technology architecture.

period of internal practice comparison, Alibaba Group has identified Flink as the only real-time computing engine. Flink achieves a high throughput, low-latency and high-performance real-time streaming computing framework by implementing the Google Dataflow streaming computing model. At the same time, Flink supports highly fault-tolerant state management, which prevents states from being lost due to system anomalies during computation. Flink periodically maintains state persistence through distributed snapshot technology checkpoints, enabling correct results to be calculated even when the system is down or in abnormal situations.

Alibaba acquired Flink commerce company Ververica in 2019 and made internal improvements to open-source Flink to create Alibaba's version of Flink, known as Blink. In the same year, Alibaba also contributed Blink to the open-source community. Now Blink has become the computing engine of choice for real-time computing of Alibaba Cloud.

Blink supports a wide range of source and sink data. For structured data such as RDS, Oracle and SQL Server, Blink can connect directly to the data sources and synchronize data to Blink for calculation in real time. This product can also access data streams such as Kafka, SLS, Datahub and MQ in real time to realize fast calculation of business data in Blink (including cleaning, data processing and light summary). The product selection and scenario analysis are described in Sections 5.3.2 and 5.4.3 of this book, and we do not expand here.

Real-time calculation results can be stored in different sink storage according to different business scenarios. Real-time analysis can choose Hologres, real-time reports and large screens can choose RDS, real-time large data concurrent queries can choose HBase, real-time message push can choose MQ, data backup can choose different scenarios such as MaxCompute.

5.6.2.3 Real-Time Data Middle Office Resource Evaluation Model

Compared with offline computing, real-time computing is bound to increase the consumption of IT resources in order to ensure timeliness. For a better ROI of enterprises, we briefly introduce the resource evaluation of Alibaba Cloud real-time computing Blink, which can be used as a reference for readers to make resource planning for real-time Data Middle Office.

1. Customer Business Assessment

   The resource unit of Alibaba Cloud real-time computing is CU, namely 'Computing Unit'. For simple services, such as single-stream filtering, string transformation and other operations, 1CU can process 10,000 records per second; for complex operations, such as SQL of join, window functions and group by syntax, 1CU can process 1,000–5,000 data records per second.

   Note: RPS given by customer business evaluation in traditional industries is often low and sometimes not of reference value.

2. Estimated capacity planning model

   Capacity planning formula (structured data): total number of nodes = total number of records per second/(1,000 RPS (1CU processing capacity per second) * machine Core number * 0.75) + 3 * masters (Note: a certain amount of buffer should be reserved according to the situation, and the calculation result of this formula is the minimum configuration in the case of parameter simulation). The minimum 6 nodes (3 Mater + 3 Worker nodes) is recommended, and the actual available computing resources of the customer should not be less than 50% of the total resources.

3. Example of capacity planning

   I. Business evaluation

      Real-time calculation of a customer mainly includes scenes related to coupons, orders and users. Real-time request of the coupons work every second estimate maximum concurrency of 500 RPS (Request Per Second), real-time request of the relevant work every second order estimate maximum is 3,000 RPS, concurrent users of large screen real-time request of the relevant work every second estimate maximum concurrency of 1,000 RPS, customer business processing is relatively complex, contains join and group by operation.

II. Capacity planning

- Customer concurrent RPS: 500+3,000+1,000=4,500 RPS
- Single CU processing capacity: 1,000 # RPS business complexity is relatively complex, containing join and group by operation, 1 CU=1/4 g core CPU Memory
- Customers use models of single node performance: 4 Core CPU, 16 g Mem
- Estimate number of worker nodes: 4,500/(1,000 * 4 *0.75) = 1.5 Nodes #0.75 is the percentage of available resources other than those required by the system
- Number of master nodes: 3 nodes # By default, 3 Master nodes are expected to manage less than 1,000 Worker nodes
- Total number of resources: 1.5 nodes (Worker node) * 2 (development and production environment)+3 nodes (Master node)=6 nodes

5.6.2.4 Real-Time Data Middle Office Application Scenarios

1. Real-time business monitoring

Traditionally, aggregating data into an offline computing engine can no longer meet the fast-growing business needs by generating data reports in N+1 days. The online characteristics of the Internet also push business needs to real-time. It is becoming more and more common to adjust strategies according to current customer behavior at any time. Traditional data warehouses begin to transform to real-time data warehouses. The real-time Data Middle Office uses data visualization tools QuickBI and DataV to form real-time insight analysis capabilities for sales, marketing, member operations and real-time executive cockpits for enterprises (Figure 5.48).

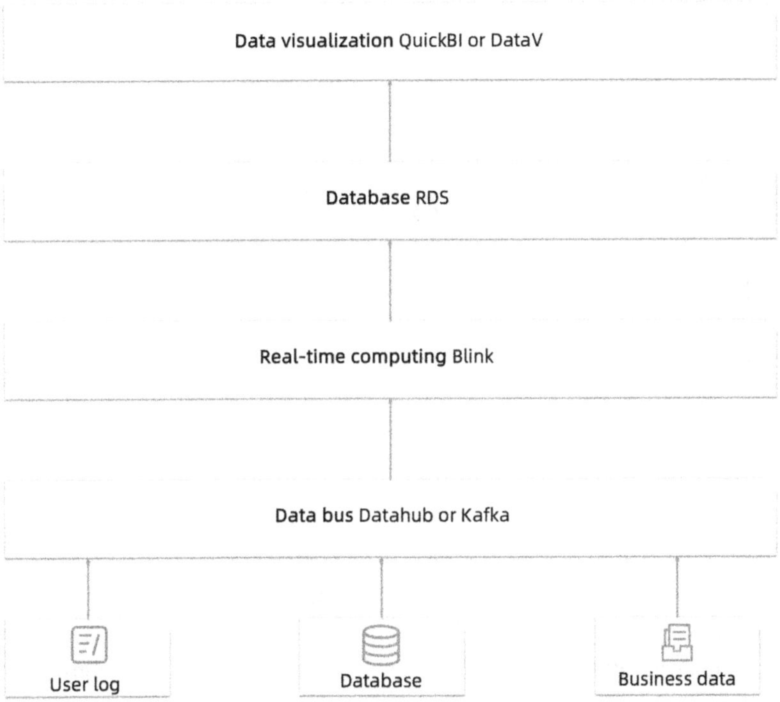

FIGURE 5.48 Real-time business monitoring architecture diagram.

2. Real-time risk control

In the financial industry, risk control is the core function of the business, which has a very wide range of applications, including content risk control, financial risk control, marketing risk control and other types. With the advent of the Internet era, a large amount of data, access and requests are generated. New business requirements prompt the traditional risk control system to transform to the big data scenario. If risk control cannot be done in real time, customer business will cause huge losses.

Under the architecture of Figure 5.49, users' real-time events can invoke the rule engine synchronously for real-time feedback, and further model analysis can be done by sending messages asynchronously to the real-time computing engine to provide data support and feedback for evaluation, penalty and monitoring systems.

The risk control business system based on the real-time Data Middle Office has two characteristic advantages: first, strong real-time performance, high throughput, low delay and millisecond

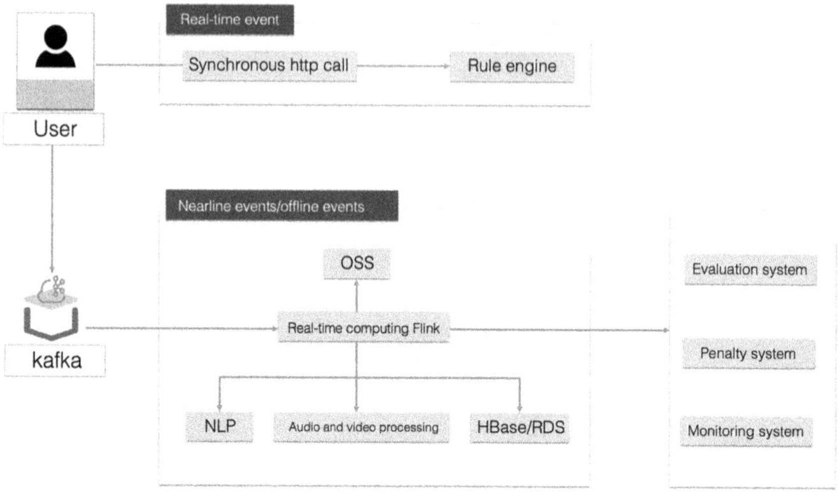

FIGURE 5.49 Real-time risk control architecture diagram.

alarm interception under the big data scene; second, unique event rule support, real-time computing Flink's unique CEP syntax naturally supports rule setting, which perfectly fits the risk control scenario.

3. Real-time machine learning

Machine learning has been applied in more and more industries as an important scenario for large data, but traditional algorithms mainly focus on training and providing predictions using static models and historical data. How to accurately and individually predict users' short-term behavior becomes a new problem for dynamic decision-making (Figure 5.50).

Based on the consistency of real-time data collection and computing engine with real-time data warehouse architecture, real-time computing Flink/Blink version can spit out the cleaned and standardized data stream to the machine learning system. The training results of machine learning are used to support real-time user portraits, real-time association analysis and other business scenarios, which greatly accelerate the timeliness of machine learning.

Real-time machine learning provides customers with two important values: first, operational refinement. It supports the rapid portrait of millions of customers, accurate positioning of user characteristics,

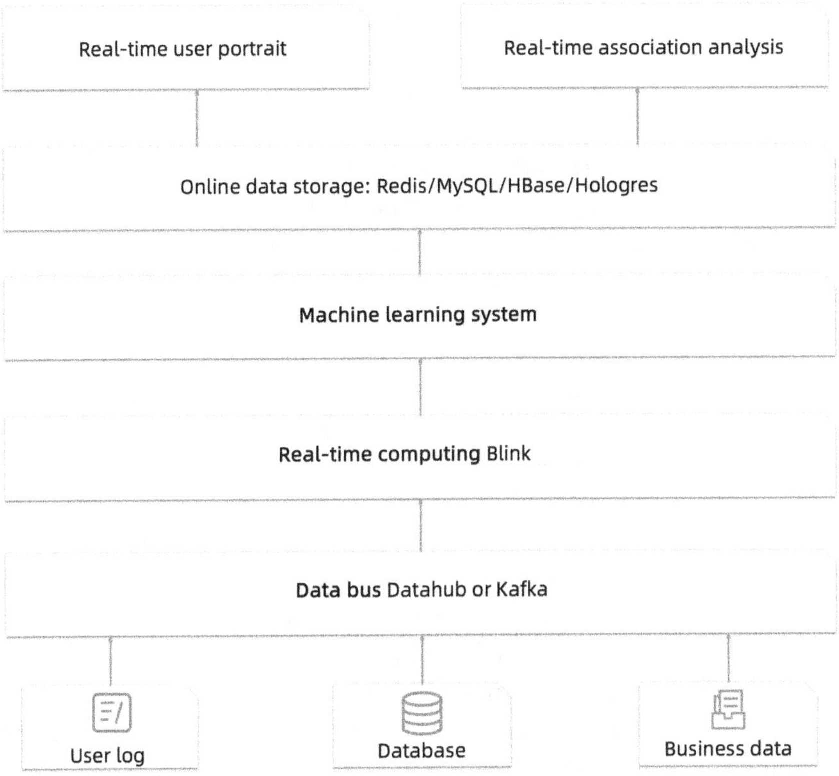

FIGURE 5.50 Real-time machine learning architecture diagram.

personalized operation to improve customer turnover and retention rate. Second, low threshold. Seamless integration of real-time computing learning algorithm platform, one-stop to complete the whole process of data development, model training, scene prediction.

5.6.2.5 Development Trend of Real-Time Data Middle Office

1. Dual engines of real-time Data Middle Office

   First of all, Flink or Blink does not fully represent real-time computing. The understanding of real-time is a process of development, from the earliest Lambda architecture, to the Kappa architecture proposed later, and then to the dual engine architecture of stream computing and interactive analysis.

   Furthermore, the key to real-time data-enabling business is how to involve the business side rather than let the platform side complete it alone. Only business staffs can deeply understand what business needs when business development is returned from real-time platform personnel. Finally, how to make real-time system respond to changes quickly and shorten TTM (Time to Market) time becomes more and more important. Traditional real-time development of an application is often an order of magnitude slower than offline data warehouse development. How to achieve the same or even higher level of development timeliness and offline data warehouse is a growing demand. Providing direct service for business is also a proposition that Data Middle Office pays special attention to.

The dual engine architecture of stream computing and interactive analysis is a more ideal architecture, which takes the interactive analysis with real-time storage as the center to build the whole real-time link. In this architecture, stream computing is responsible for basic data, while interactive analysis engine is the center. This engine must have its own storage. Through the collaborative optimization of computing storage, it can achieve high write RPS, high query QPS and low query latency. In this way, real-time analysis and on-demand analysis can be realized in batch mode, and business changes can be responded quickly. Business parties use their familiar development tools and development language (SQL), just like traditional BI, to develop real-time applications based on big data. Through interactive analysis, they can achieve the optimal experience by using the query as you go. Thus, the big data real-time data warehouse experience is aligned with the traditional OLAP database experience, and maximizes link and user experience simplification.

2. Stream and batch in one

The technical concept of stream and batch in one was first proposed in 2015. Its original intention was to enable developers to use the same set of interfaces to implement stream and batch calculations of big data, thereby ensuring the consistency of processing and results. Subsequently, big data vendors/frameworks such as Spark, Flink and Beam have come up with their own solutions. Although the implementation methods are different, to some extent, the idea of stream and batch in one has been widely recognized in the industry.

The evolution of architecture is essentially in the direction of batch streaming in one, to enable users to complete real-time computing in the most natural and cost-effective way. Second, the full-link real-time and SQL trend is very clear. On the one hand, more and more businesses need real time, which in many cases refers to minute-level delays and acceptable cost of implementation; on the other hand, there is a clear trend to express all aspects of computing in SQL, preferably standard SQL.

However, in order to truly go from the theory to the implementation of batch integration, especially in the large-scale implementation of core data business scenes of enterprises, it is often faced with the dual challenges of technology and business, so the actual production landing cases are rare. Alibaba Group brought this idea to the ground on its 'Double 11 Shopping Festival', and the corresponding application scenario is promotion marketing analysis. As the core data product used to serve decision-making and guide operation during the promotion period, marketing activity analysis products cover the analysis of the pre-activity, mid-activity and post-activity links, which need to meet the different requirements of data timeliness and data flexibility of different role staffs in different stages.

How to quickly respond to the frequently changing business demands and more efficiently deal with the data problems during the period are becoming more and more important. The upgraded new generation of marketing activity analysis architecture needs to meet the following characteristics:

- The real-time and offline data warehouse can unify the data model, data storage and data access interface, and truly achieve the flow and batch in one.
- More powerful data storehouse is needed, which can not only satisfy the concurrent writing query of massive data, but also satisfy the timely query function of business.
- Simplify the existing product logic of structure, to reduce the complexity of product realization.

After a long time of calling and trying, we finally chose Dataphin based on the integrated real-time computing engine, Hologres based on the delivery of real-time analysis engine and FBI based on Alibaba's internal visual analysis tool to realize the Framework Reconstruction and upgrading of Tmall marketing activity analysis:

- Through the upgrade of the flow and batch in one architecture, the flow and batch SQL get unified on logic and computing engine level. First, unified logical table encapsulation, one logical table is mapped to two physical tables, namely real-time DWD and offline DWD. Data calculation codes are all developed based on this logical table. Second, the stream and batch in one development platform Dataphin supports the code development of logical tables, the personalized configuration of stream–batch computing modes, as well as different scheduling strategies, forming a convenient integration of development, operation and maintenance. Finally, the unified storage layer based on OneData specification is not only the unified model specification, but also the unified storage, achieving a seamless connection.
- The data storage and query are unified by Hologres. Hologres's real-time query capability provides strong support for core business modules such as live broadcast, pre-sale, additional purchase and traffic monitoring; in the face of complex and changeable marketing scenarios, Hologres's OLAP real-time query capability can well support.
- By using the ability of FBI products, we can reduce the construction cost, meet the high flexibility of business and meet the needs of different roles for reports. Aiming at the underlying data of batch flow integration defined by marketing activities, in order to meet the flexibility of user analysis of real-time data, real-time comparison and hour comparison, the FBI abstracts a set of standard data model of real-time and offline integration. After creating the model, it can realize the accurate comparison of real-time data.

Both computing and analysis engines and streaming and batch in one are Alibaba's active attempts and explorations in the field of real-time computing. Alibaba's big data team believes that the integration of streaming and batch in the storage layer and the evolution from streaming and batch to 'Data lake and data warehouse in one' are all possible directions. Alibaba Cloud will continue to integrate the internal polished technology architecture and platforms, such as Dataphin, QuickBI, Flink and Hologres, and export them to the cloud to serve more external users, so as to help external customers better carry out digital transformation.

5.6.3 Data Desensitization

For enterprises, there is a certain degree of sensitive privacy data. Such data have the need for privacy desensitization, usually with the following basic demands:

- Data desensitization, you cannot directly view the original value.
- The desensitized data can be reversed to obtain the original value or part of the original value, which can be used to participate in business calculations. For example, analyze the age and place of birth according to the ID card; analyze the country according to the passport number.
- The desensitized data should be distinguished as much as possible and can participate in calculations such as OneID.

The data desensitization function can realize the data desensitization capability between data source products in the cloud environment. Through static desensitization, users can realize scenarios where production data are desensitized and transferred to the development and test environment for analysis and use, avoiding the risk of data leakage caused by direct access to sensitive data. The module logic architecture diagram is as follows (Figure 5.51).

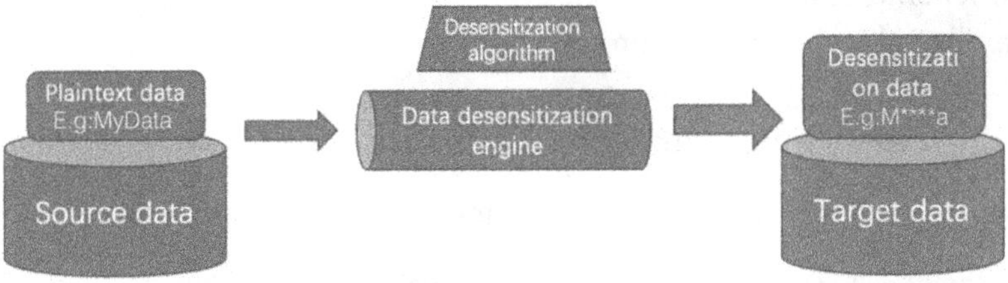

FIGURE 5.51 Data desensitization technology scheme.

Supports the desensitization transfer capability of heterogeneous data sources. Supported data source types include structured data sources RDS, MaxCompute and ADS, and structured data stored in unstructured data sources OSS (table type files csv, xls and contains Txt of structured content). Provides atomic-level desensitization capabilities based on table dimensions, realizes the management of desensitization tasks through task configuration, and supports three modes of manual triggering, timing triggering and combined triggering, providing enterprises with flexible desensitization options. At the desensitization algorithm level, it supports 6 categories and a total of 30 common desensitization algorithms, including hashing, masking, replacement, transformation, encryption and shuffling, which can meet the desensitization needs of various business scenarios.

In the process of data desensitization, the data column in the source structured data will be first identified for sensitive information, and the results will be stored in the sensitive data identification and classification management for use by the desensitization module. When performing desensitization, the user first selects the source data table that needs to be desensitized, and then automatically provides the corresponding desensitization algorithm pre-configuration option based on the sensitive data recognition result of the table, and the user can directly use it or perform the desensitization based on business needs. Edit the sensitive algorithm and customize the parameters. After completing the algorithm configuration, select the target data table to be written, define the task trigger mode and the conflict judgment between data rows and columns, and finally complete the task configuration.

5.6.3.1 Data Desensitization Scheme

There are three kinds of desensitization schemes: irreversible desensitization scheme, reversible desensitization scheme and source-dependent system reversible desensitization scheme.

1. Irreversible desensitization scheme
   - Architecture of irreversible desensitization (Figure 5.52)
   - Irreversible desensitization algorithm (Table 5.25)
2. Reversible desensitization scheme

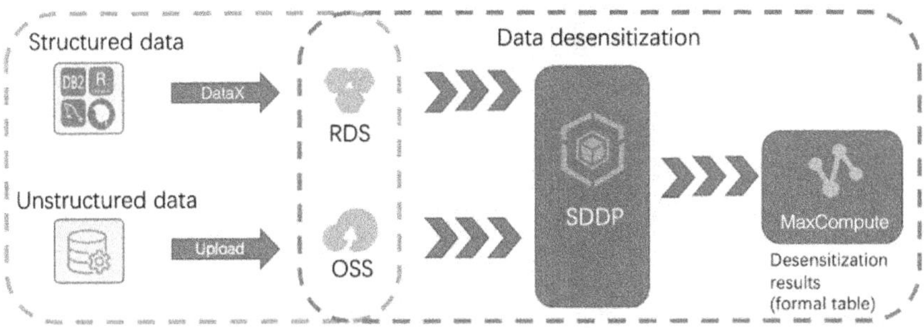

FIGURE 5.52 Irreversible desensitization.

I. Reversible desensitization architecture (Figure 5.53)

II. Inverse encryption algorithm (Table 5.26)

III. Key security mechanism

Data encryption and desensitization can be done through the Alibaba Cloud data encryption product SDDP. The secure retention of keys is achieved through the following mechanisms:

- Project permission management

  The encryption and decryption results are stored in an independent MaxCompute project space, and the authority management is extremely strict. Only a few people have the authority to join the project.

- JAR package secure hosting

  The JAR that implements the encryption and decryption logic will be stored in MaxCompute as a resource. Only the accesskey and accessid of the project space can complete the modification of the JAR package file.

- JAR package code obfuscation

  The code of encryption and decryption logic will be obfuscated to further increase the difficulty of cracking.

- Secret key strength

  Reversible encryption will use AES symmetric encryption algorithm, and the key strength is in line with China Southern Airlines' internal security regulations.

TABLE 5.25 Irreversible Desensitization Algorithms

Serial Number	Data Type	Desensitization Algorithm	Sample after Desensitization
1	ID card	First replace the last 4 digits as xxxx, the original value is encrypted after md5, AES symmetrically encrypted and then converted to hexadecimal	Data before AES encryption 43070319870101xxxx (replace the last 4 digits with xxxx) +e10adc3949ba59abbe56e057f20f883e (the original value is encrypted after hash) AES symmetrically encrypted and then converted to hexadecimal 49ba59abbe56e057f20f88349ba59abbe56e057f20f88349ba59abbe56e057f20f88349ba59abbe56e057f088357
2	passport	First replace the last 4 digits as xxxx, the original value is encrypted after md5, AES symmetrically encrypted and then converted to hexadecimal	Data before AES encryption 59046xxxx (replace the last 4 digits with xxxx) +e10adc3949ba59abbe56e057f20f883e (the original value is encrypted after the hash) AES symmetrically encrypted and converted to hexadecimal 49ba59abbe56e057f20f88349ba59abbe56e057f20f88349ba59abbe56e057e057f20883e
3	Other personal identification documents such as military card	First replace the last 4 digits as xxxx, the original value is encrypted after md5, AES symmetrically encrypted and then converted to hexadecimal	Data before AES encryption 5943xxxx (replace the last 4 digits as xxxx) +e10adc3949ba59abbe56e057f20f883e (the original value is encrypted after hash) AES symmetrically encrypted and converted to hexadecimal 49ba59abbe56e057f20f88349ba59abbe56e057f20f88349ba59abbe56e057e057f59abbe56

(*Continued*)

TABLE 5.25 (*Continued*) Irreversible Desensitization Algorithms

Serial Number	Data Type	Desensitization Algorithm	Sample after Desensitization
4	mailbox	Replace the last 4 digits of the mailbox prefix as ****, the original value is encrypted after md5	shifu****@alibaba-inc.com+f5a173175cb930c20791bccd4dc26 88d
5	name	Keep the first word and replace other characters with *, the original value is encrypted after md5	xiang**+15baf6ee9a5dd8ae8235b0d67340734e
6	mobile phone number	First replace the last 4 digits to 0000, the original value is encrypted after md5, AES symmetrically encrypted and then converted to hexadecimal	Data before AES symmetric encryption: 13810470000+2d19830b04e48c6714d71ae5f8242027 AES after symmetric encryption to hexadecimal 30b04e48c6714d71ae5f824230b04e48c6714d71ae5f8242
7	Landline number	First replace the last 4 digits to 0000, the original value is encrypted after md5, AES symmetrically encrypted and then converted to hexadecimal	Data before AES symmetric encryption: 01047650000+2d19830b04e48c6714d71ae5f8242027 AES after symmetric encryption to hexadecimal 30b04e48c6714d71ae5f824230b04e48c6714d71ae5f8242
8	Address (country, province, city)	No desensitization	Hangzhou, Zhejiang
9	Address (street, house number)	Convert to hexadecimal after AES symmetric encryption	9e9947ba34bfb3842482af8fccf08da1
10	Bank card number	Replace the last 6 digits with 000000, AES symmetrically encrypted and converted to hexadecimal, the original value is encrypted after md5	Data before AES encryption 6226097632300000 (the last 6 digits are 000000) +e10adc3949ba59abbe56e057f20f883e (the original value is encrypted after hash) AES symmetric encryption is converted to hexadecimal ba59abbe56e0ba59abbe56e0ba59abbe56e0ba59abbe56e0

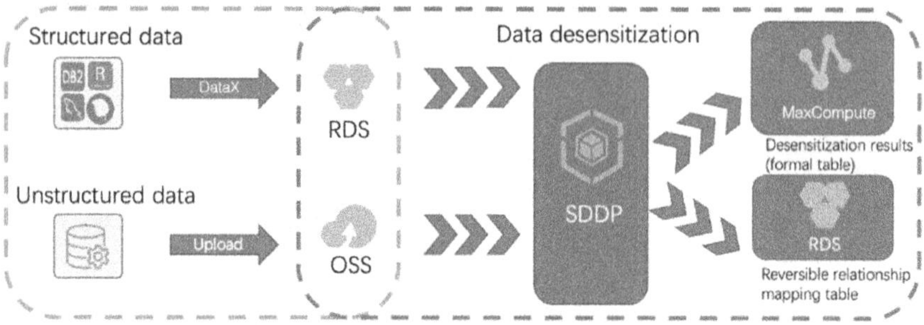

FIGURE 5.53 Reversible data desensitization.

TABLE 5.26 Inverse Encryption Algorithms

Serial Number	Data Type	Desensitization Algorithm	Sample after Desensitization
1	ID card	Convert to hexadecimal after AES symmetric encryption	After AES symmetric encryption, convert to hexadecimal 49ba59ab be56e057f20f88349ba59abbe56 e057f20f88349ba59abbe56e057 f49ba59abbe56e057f20f883
2	passport	Convert to hexadecimal after AES symmetric encryption	After AES symmetric encryption, convert to hexadecimal 49ba59ab be56e057f20f88349ba59abbe56e 057f20f88349ba59abbe56e057f 49ba59abbe56e057f20f883
3	Other personal identification documents such as military card	Convert to hexadecimal after AES symmetric encryption	After AES symmetric encryption, convert to hexadecimal 49ba59ab be56e057f20f88349ba59abbe56e 057f20f88349ba59abbe56e057f 49ba59abbe56e057f20f883
4	mailbox	Convert to hexadecimal after AES symmetric encryption	Same as above
5	name	Convert to hexadecimal after AES symmetric encryption	Same as above
6	mobile phone number	Convert to hexadecimal after AES symmetric encryption	Same as above

(*Continued*)

TABLE 5.26 (*Continued*) Inverse Encryption Algorithms

Serial Number	Data Type	Desensitization Algorithm	Sample after Desensitization
7	Landline number	Convert to hexadecimal after AES symmetric encryption	Same as above
8	Address (country, province, city)	No desensitization	Original value
9	Address (street, house number)	Convert to hexadecimal after AES symmetric encryption	Same as above
10	Bank card number	Convert to hexadecimal after AES symmetric encryption	Same as above

3 Reversible desensitization scheme of dependent source system

I. Architecture diagram (Figure 5.54)

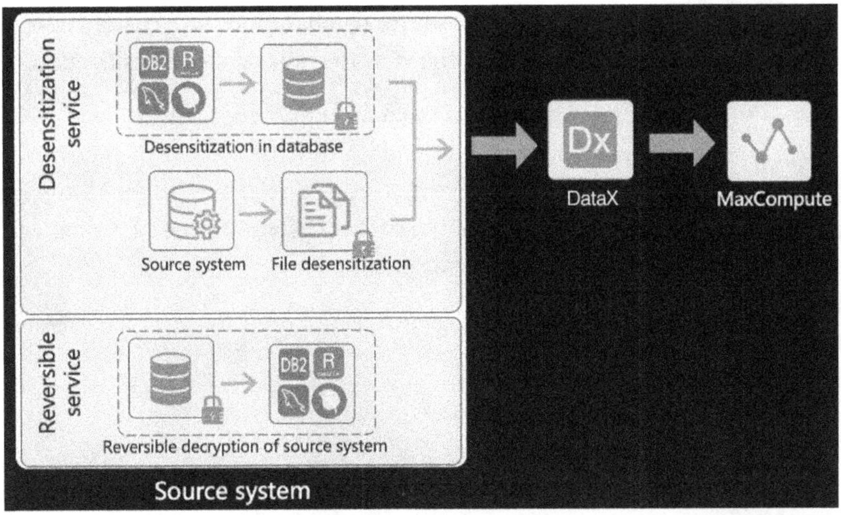

FIGURE 5.54 Architecture diagram of a reversible desensitization scheme for source-dependent systems.

5.6.3.2 Cloud Desensitization of Database

In the process of uploading structured database data to the cloud, data desensitization can be carried out in multiple stages according to different needs.

1. Cloud desensitization
   I. Key points of the scheme:
      - The data are uploaded to the cloud intact and temporarily stored in a temporary table.
      - Call the SDDP service desensitization module to process the temporary table and write the desensitized data to the formal table. This solution makes full use of the advantages of cloud native to ensure data security and control. This method is recommended.

   II. Architecture diagram (Figure 5.55)
2. Source database desensitization
   I. Key points of the scheme
      - Add a new view in the source database and call a custom desensitization function in the view to realize data desensitization.
      - Authorize the view to the data export account.

        This solution achieves data desensitization in the source database, and the security level is relatively high. However, some modification of the source system is needed, which will bring system modification costs.

   II. Architecture diagram (Figure 5.56)

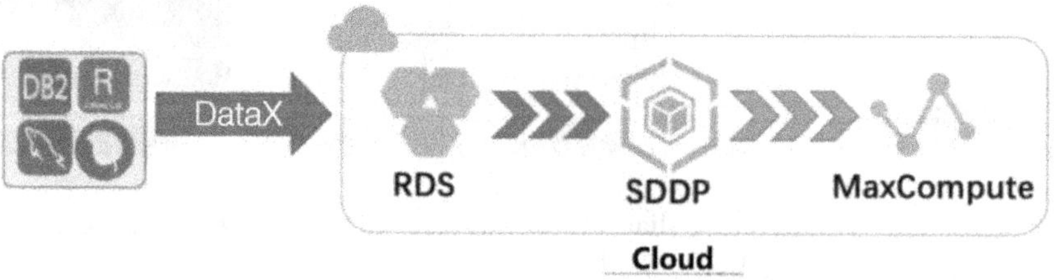

FIGURE 5.55 Desensitization on cloud architecture diagram.

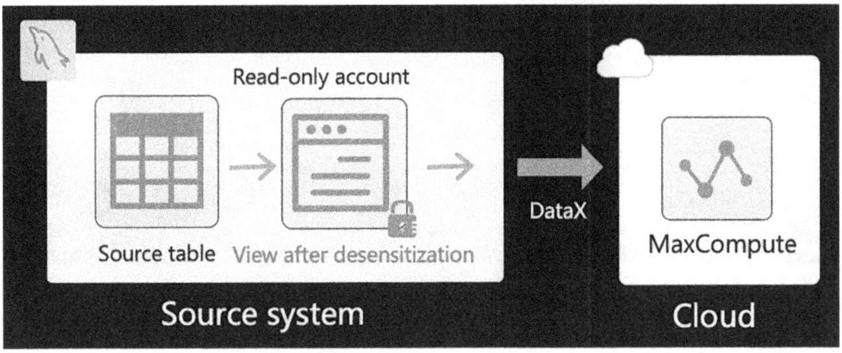

FIGURE 5.56 Source database desensitization architecture diagram.

3. Desensitization during extraction

   I. Key points of the scheme

      - When DataX reads data from the database, call system functions to achieve desensitization.

      - This solution realizes desensitization in DataX, and does not change the source library. However, the desensitization algorithm and secret key will be exposed in the script, and there is a certain risk of leakage.

   II. Architecture diagram (Figure 5.57)

4. Desensitization during conversion

   I. Key points of the scheme

      DataX supports data conversion and can call custom JAR packages to achieve desensitization. This solution realizes

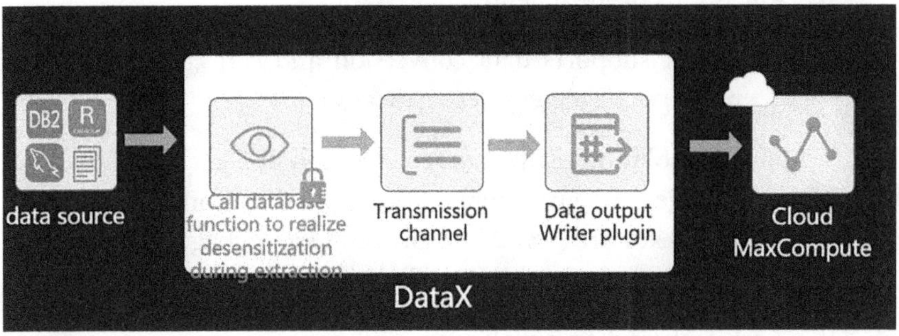

FIGURE 5.57 Desensitization architecture diagram during extraction.

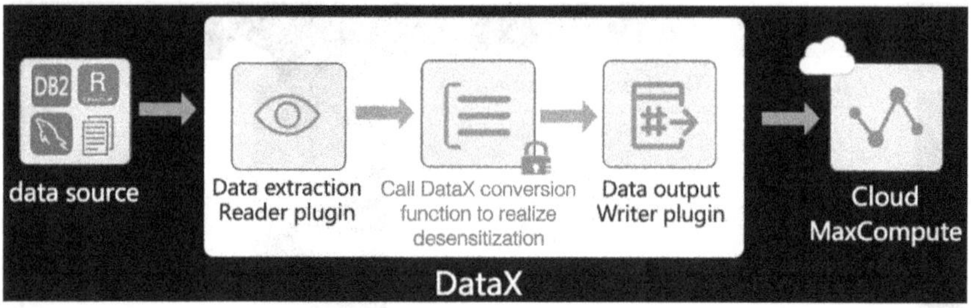

FIGURE 5.58 Desensitization architecture diagram during transformation.

desensitization in DataX, and does not change the source library. The desensitization algorithm and secret key are stored in a custom JAR package, and the security is relatively guaranteed.

II. Architecture diagram (Figure 5.58)

5.6.3.3 Cloud Desensitization of Data Files

If you upload to the cloud as a data file, it is recommended to save the file through OSS, and then use SDDP for desensitization. If there is highly confidential information, in principle, the data file producer is required to desensitize the data before uploading. The specific desensitization method depends on the respective data file production system. This content is ignored in this article.

1. Cloud desensitization

   Refer to the cloud scenario of the database, so I won't go into details.

2. Desensitization during conversion

   I. Key points of the scheme

      - DataX supports data conversion and can call custom JAR packages to achieve desensitization.
      - This solution realizes desensitization in DataX, and does not change the source library. The desensitization algorithm and secret key are stored in a custom JAR package, and the security is relatively guaranteed.

   II. Architecture diagram (Figure 5.59)

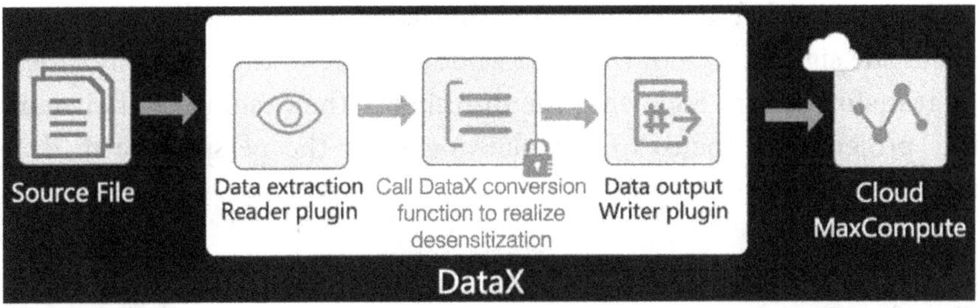

FIGURE 5.59 Data integration of data files desensitizing architecture.

5.7 ALIBABA DATA MIDDLE OFFICE PROJECT CASES

5.7.1 Case of Beauty Brand Industry

One of the top 10 brands in the beauty industry in China used to rely on high-recognition brands and the expansion of e-commerce channels, with annual sales reaching billions of RMB. On the road of digital transformation, customers hope that the construction of digital capabilities will empower business and bring higher and more precise marketing effects. In the implementation of the entire Data Middle Office project, the company's senior executives proposed that the basis of the Data Middle Office is technology, and the value lies in the ability to productize in the Data Middle Office.

At the beginning of the project, the customer's chairman had high hopes for the project: 'The foundation of the Data Middle Office is technology, and the value lies in the ability to realize data productization in the Data Middle Office'. The company must establish a 'Data Middle Office organization', which will be the best and important department that reflects the value of strategy and technology. Data Middle Office will have two types of products, technical products and commercial data application products... Data Middle Office will have several technical products in the future, but the commercial data applications products are the valuation reflected that we really want to realize. 'Create ideals from pain point', which must become the thinking orientation of the Data Middle Office project team. 'Bold assumptions, believe in the possibility and make breakthroughs with all efforts', which must become the values of the Data Middle Office project team. 'The top level of the customer is right'. The cognition of the Data Middle Office is very clear about the significance of the Data Middle Office to the company's digital transformation. Following the thinking of the customer's chairman, this project case focuses on explaining from the perspective of data application.

5.7.1.1 *Implementation Methodology*

Therefore, in the project implementation methodology, we innovated the way from business map to product map, and finally carried out the results of data asset construction through incubation in product form. The business map refers to the construction of a data asset system based on the perspective of business blueprints by segmenting business themes based on the business pain points and data application goals of the enterprise. The product map undertakes the business blueprint design. From the user's perspective, based on the specific scenarios of the data application, land and assemble data assets and upper-layer applications, so that it can help users solve the data-enabling application requirements of specific business scenarios (Figure 5.60).

The conversion from the business map to the product map actually includes three layers, as shown in Figure 5.61.

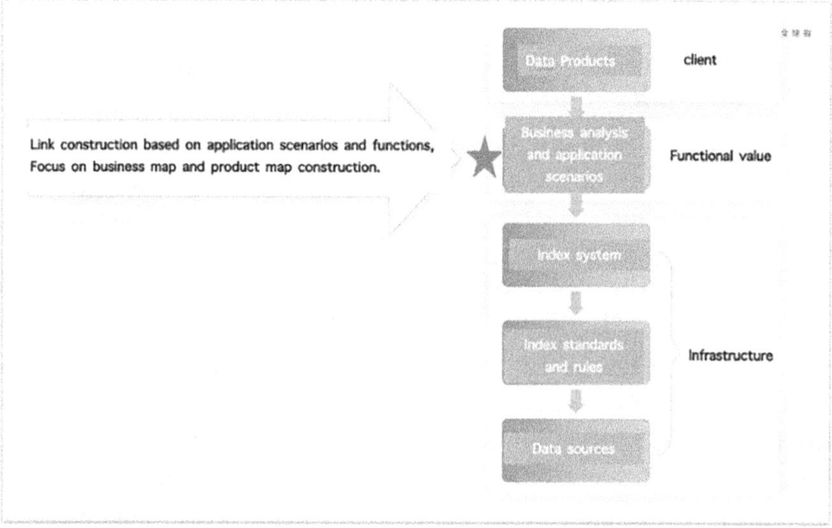

FIGURE 5.60 Data Middle Office implementation methodology.

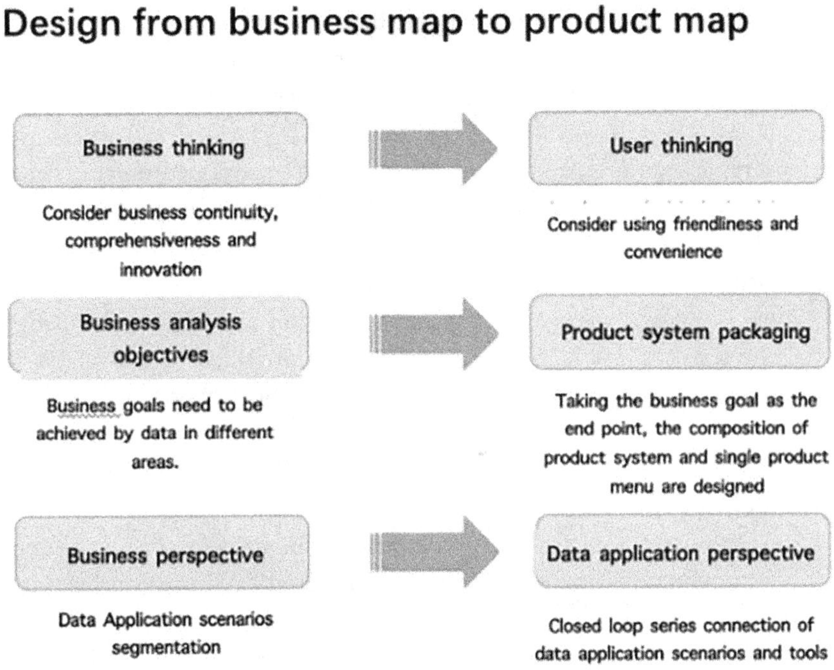
Design from business map to product map
Business thinking
User thinking
Consider business continuity, comprehensiveness and innovation
Consider using friendliness and convenience
Business analysis objectives
Product system packaging
Business goals need to be achieved by data in different areas.
Taking the business goal as the end point, the composition of product system and single product menu are designed
Business perspective
Data application perspective
Data Application scenarios segmentation
Closed loop series connection of data application scenarios and tools

FIGURE 5.61 **Design ideas from business drawing to product drawing.**

5.7.1.2 Packaging and Construction of Product System

The product system of the group's Data Middle Office is built from six aspects: products, users, sales, marketing, cockpit and supply chain. Each product has clear users. The design of each menu and topic solves how to use the capabilities of Data Middle Office's data assets to empower specific business operations or decisions from the perspectives of insight and crowd marketing.

After the establishment of the product system, from the later operation and the presentation of the brand image of the company Data Middle Office, the project also packaged a portal, opened up with the company's DingTalk account system, and realized scan code login.

5.7.1.3 Intelligent Algorithm Blessing

In the product system, some business scenarios, such as consumer churn prediction and early warning, purchase prediction and consumer evaluation, will use the algorithms in implementation. And following we'll introduce the main algorithms and application scenarios in several projects.

- Conversion loss prediction algorithm

  The whole architecture of the conversion loss prediction algorithm is shown below. The original data source is processed in Dataphin to obtain the relevant data of the ODS or ADS layer. The features and sample data required by the algorithm are processed on this basis, and DataWorks reads it through MaxCompute after processing the physical table in Dataphin to perform corresponding characteristic sample processing. After the processing is completed, feature engineering, model training and prediction are processed through the features and algorithms components provided by PAI platform to obtain the prediction results of the algorithm. After obtaining the algorithm results, Dataphin provides the algorithm prediction results to the downstream report (QuickBI) by reading the physical table (Figure 5.62).

  Purchase forecast/loss prediction is a binary classification problem, which can be trained by GBDT (Gradient Boosting Decision Tree) model. The algorithm result will output the probability value in the form of a label and display on the insight page. When it is warned, the users can circle and release the crowd in the crowd selection menu.

- Sales forecast algorithm

  In the previous data exploration, it was found that the sales data mainly fluctuate with the promotion activities, and the modeling needs to consider the impact of the big promotion. The complete framework of the model is shown in Figure 5.63. The whole modeling ideas of supermarket channel and e-commerce channel are basically the same, but there will be some differences in the feature

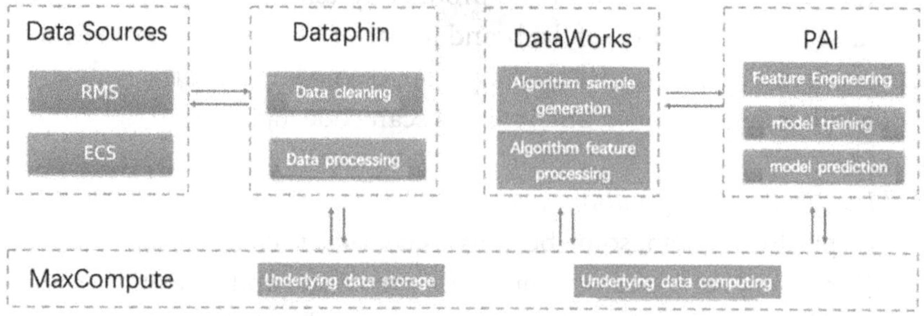

FIGURE 5.62 Data Middle Office technology and data architecture.

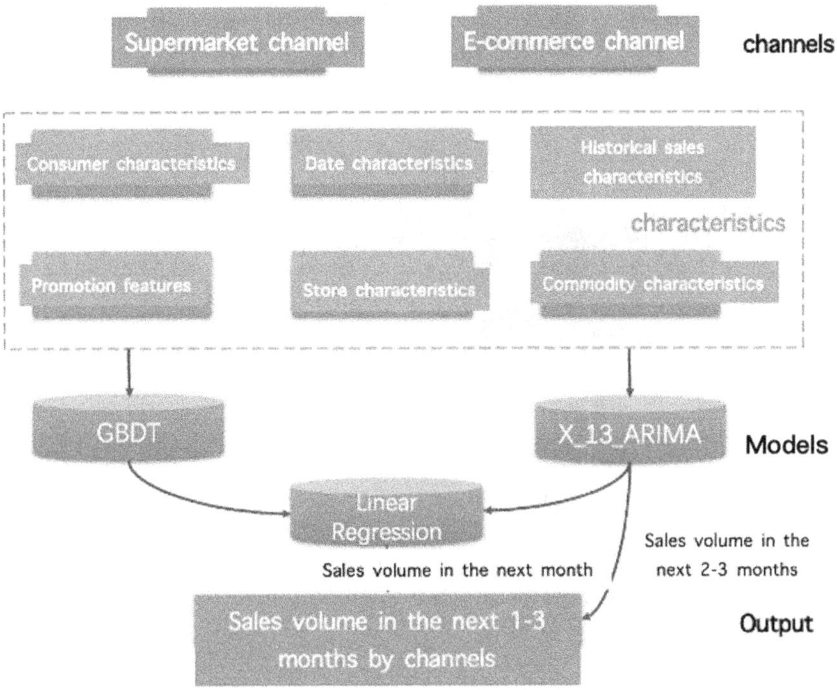

FIGURE 5.63 Sales forecast algorithm architecture.

construction of the regression model. The whole idea of building the model is that the time series model uses historical data to predict monthly sales volume for the next 1–3 months, and the regression model predicts the sales volume for the next month, and finally, the sales volume forecast results for the next month trained by the regression and time series model will be integrated into a final forecast value, and the sales volume forecast value in the next 2–3 months is directly given by the time series model.

5.7.1.4 Technical Solutions

1. Overall architecture

   The overall solution has the following features (Figure 5.64):

- Multi-terminal data collection, group data horizontal access.
- Data standard modeling, solidify standard uniform caliber and eliminate ambiguity 100%.

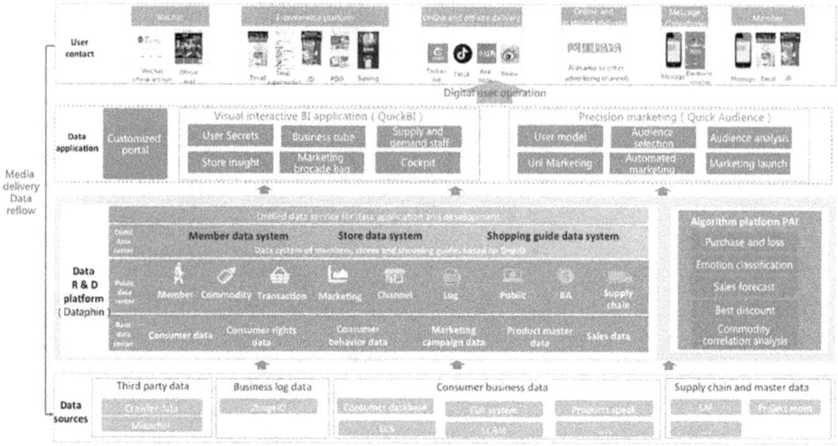

FIGURE 5.64 Data Middle Office overall architecture.

- Integrating algorithm into data application combined with machine learning PAI.
- More electric touch, closer to users.
- Integration improves the scalability and imagination of data Mid-office.

2. Portal Solution

With the advent of the era of big data, enterprises pay more and more attention to data, the application forms of data are more and more diversified, and more and more products are involved in the application. The multi-entry of multi-product makes the user experience unfriendly, the multi-account of multi-product increases the complexity of user use, and the multi-style of multi-product cannot be well integrated with the corporate culture. In this context, portal emerges as the times require. Portal integrates multiple data application products into a unified portal, unified authority and unified style, and can customize the unique functions of the enterprise to improve the data application scenarios. In order to meet the needs of this scenario, we need to integrate QuickBI, QuickAudience and DataV pages into one portal, which is in line with the enterprise culture; and we need to unify the permissions of QuickBI, QuickAudience and DataV through DingTalk, so as to realize single sign on (Figure 5.65).

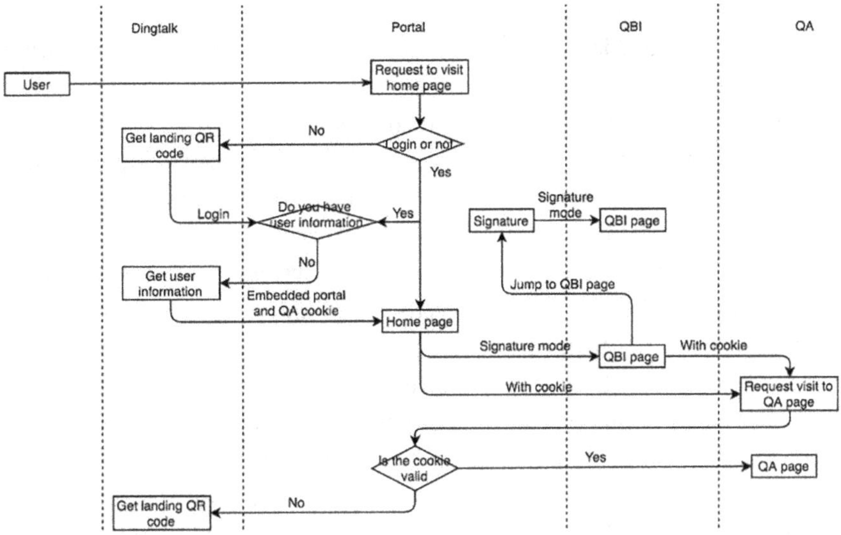

Remarks:

- The login mode of signature is the same as that of portal, and the validity of the original login link is 1 minute
- The login mode of cookie is the same as that of portal
- QA access can only be authorized through the portal, not logged in alone
- Qbi accounts are synchronized in qbi, and permissions are configured in qbl
- QA accounts are synchronized in the portal, and permissions are configured in QA

FIGURE 5.65 Privilege integration scheme.

5.7.2 Case of the Easyhome Group

5.7.2.1 Project Background

After 20 years of development, the group has more than 400 stores in 220 cities in China, with a turnover of nearly 100 billion RMB in recent years. However, this group's main business is still the traditional mode of home furnishing stores and stalls leasing. Like most traditional enterprises, there are common problems of data 'not clear', 'not well managed' and 'not well used'. The wave of digitalization has a huge impact on traditional enterprises. Data will become the core resource of future enterprises. The transformation to digitalization has become the consensus of excellent traditional enterprises. Since 2016, Easyhome has put forward the strategic planning of 'transforming from property management to big data-driven', which has been ahead of most traditional enterprises. Alibaba began cooperation with this client from the new retail domain in 2018, Alibaba Cloud Intelligent GTS started construction of Data Middle Office project for the client in 2019, this is a series of measures to speed up the pace of the client's digital transformation.

5.7.2.2 Solutions of Alibaba Cloud

Data Middle Office is a tool for customers to realize business integration and resource sharing. It helps businesses transform from property space integration to in-depth business integration, and helps the transformation from 'big household' to 'big consumption'. In the early stage, Alibaba Cloud and this customer reached an agreement on the overall goal of the project: to build 'a set of role system', that is, to support customer's multi-role online collaborative data platform system; support 'four types of main business', namely data-driven refined marketing platform, investment operation platform, franchise platform and property management platform; improve 'two levels', internal management level and external service level; and finally build 'one ecological role data-enabling platform', to form the group's intelligent business ecological network.

Furthermore, it is necessary to realize three transformations: realize 'contact point digitization' through the construction of consumer, shopping guide and brand contact system, realize ' core business online' through the construction of marketing, investment promotion, chain franchise and property management business system dominated by the original offline and realize 'operation management digitization' through the Data Middle Office's global business monitoring ability. From the Data Middle Office point of view, the biggest significance of the first two 'transformation' is to realize the continuous precipitation of C-end and B-end data to the Data Middle Office, and at the same time to carry the processed data results, so as to guide and enable the business.

1. Overall architecture

   With the Data Middle Office as the center, the overall architecture is shown in Figure 5.66. Middle Data Office collected from the store market order management system, POS system, CRM system, card & coupon system, customer service system and other business system of structured data sources, and to collect AMAP LBS data, stores passenger flow data, the national and regional economic operation data and industry data that provided by the third party, there will be some structure processing on these semi-structure data after data integration to cloud.

   In the Data Middle Office, three data centers are built, including vertical data center, public data center and extraction data center. Vertical data center stores synchronized data collected from data

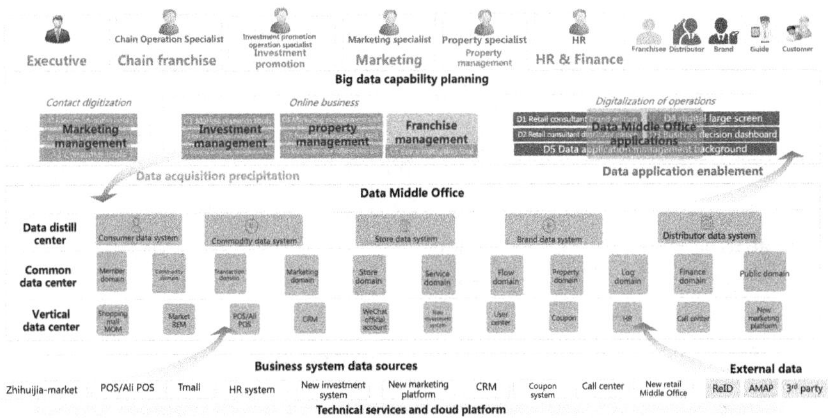

FIGURE 5.66 The overall architecture of Eayshome's Data Middle Office.

sources. The public data center divides the data domain of the vertical data center, designs the dimension model according to the modeling idea of Data Middle Office specification and outputs the core data model of data domain such as member domain, transaction domain and marketing activity domain. These models produce abundant atomic and derived metrics after the definition of metric specifications. The extraction data center undertakes the task of data label processing and extraction, and forms five data systems around the output labels of consumers, commodities, stores, brand manufacturers and dealers.

Through the unified OneService system, the metrics and labels produced by the Data Middle Office can provide services for four types of business systems and four types of data applications. Among them, the marketing management system can realize 'contact point digitization', the investment management, property management and franchise management can realize 'business online', and the data application can realize 'operation digitization'. When the business system calls the data service, it also continuously deposits the data assets to the Data Middle Office, strengthens the accuracy, breadth and intelligence of the data, and forms an iterative and rising cycle system.

2. Data asset construction

We sort out about 1,000 core business tables from nearly 10,000 tables of the business systems of the Easyhome Group. Based on the metric label requirements of this project, we finally implemented more than 600 tables data integration. Half of the 600 tables realized dimension modeling through OneModel methodology, and generated logical fact tables and logical dimension tables in Dataphin. Around these fact tables and dimension tables, more than 90 atomic metrics are defined, and about 1000 derived metrics are produced by atomic metrics plus modifiers such as analysis dimension, business limit and time cycle. In response to the two application scenarios of customer portrait and hot marketing of marketing categories, we also built 45 consumer labels and 25 commodity labels. All the above are the data asset construction results of the offline computing part of this project.

At the same time, we also introduced Alibaba Cloud real-time computing blink to support store order sales and new membership scenarios. Referring to Figure 5.67, the two scenarios use the same set of technology architecture: there are two ways to collect data: one is that the program consumes ActiveMQ topic to write the Datahub in real time, and then the Datahub is written to the real-time computing engine blink; the other is that DTS parses the DRDS database binlog and writes it directly to the real-time computing engine blink. We introduce the concept of real-time data warehouse in blink. Data access is in ODS layer. After data cleaning and normalization in ODS layer, it is transmitted to CDM layer real-time computing task through Datahub again. CDM layer task performs correlation

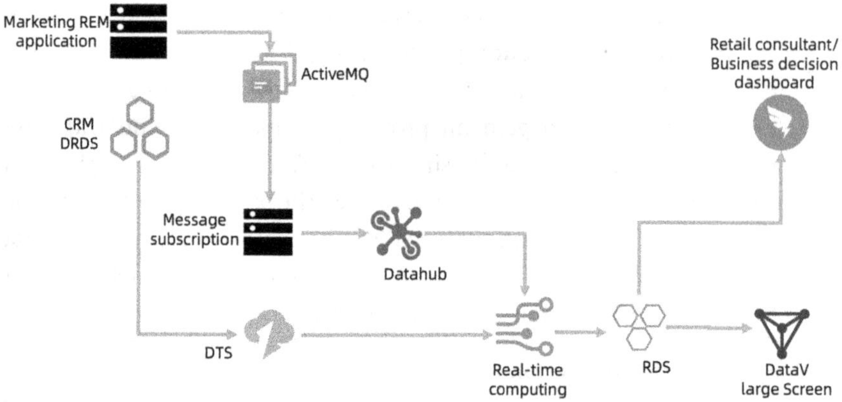

FIGURE 5.67 The real-time computing solution of Easyhome Group.

calculation on data flow of fact table and dimension table, and produces light summary fact table. Some of the metrics that can be shared have been in the light summary fact table. The result data flow is still sent through the next level ADS layer blink task of Datahub. Finally, the data task in ADS layer further processes and calculates the personalized data metrics required by different data applications. This is the first time that Alibaba Cloud intelligent GTS implements a real-time computing solution when it delivers Data Middle Office outside.

3. Data application construction

   Obviously, the construction of business systems such as investment promotion system and marketing system helps customers' daily operation and management actions to better realize online management. From the perspective of Data Middle Office, we are more concerned about data application:

   - Retail advisor brand edition and dealer Edition: These two business analysis tools for two types of ecological partners with different roles of Easyhome refer to the product ideas of Alibaba business advisor, and provide functions such as customer collection in store, surrounding potential customers amplification, site selection analysis, competition analysis, transformation effect analysis, marketing effect analysis, product distribution and product ranking monitoring, which can be used for business analysis. Customers can get insight to the operation and sales data at any time, so the merchants can establish intelligent decision-making capabilities such as customer insight, business decision-making and risk assessment.
   - Business decision dashboard: A data application that is concerned and used by staffs at all levels of operation and management of Easyhome. Data are analyzed from the themes of sales, shop stalls, investment attraction, chain stores, brands, marketing and so on.
   - Digital big screen: This is a series of data visualization screens designed and built for the senior management of Easyhome and the need to build social influence, including sales, factory, investment, chain, brand and member.

Careful readers will have noticed that the multiple themes of data application are similar, and that the metrics presented and analyzed are actually quite repetitive or similar. Thanks to Alibaba Data Middle Office OneModel unified metric specification and Dataphin product tools, we fully thickened the data public layer, continuously deposited public metrics in the public layer, realized the reuse of public metrics, which not only ensured the unified diameter of metrics, but also avoided the waste of workload caused by repeated construction.

What is more worth mentioning is that Alibaba Cloud Intelligence began to implement the strategy of integrating cloud and DingTalk from 2020. We also resolutely carried out the strategy in the customer project of Easyhome. All applications of the Data Middle Office are accessed through DingTalk as the entrance, and all people can easily access these applications on their mobile phones. We believe that lightweight, simple and clean mobile insight analytics dashboard could be an important trend for BI in the future.

5.7.2.3 Customer Value

We view the project delivery results from the perspective of customers (Figure 5.68). The Data Middle Office gathers the core business data of 368 stores and 7.21 million members of the group. Data Middle Office's

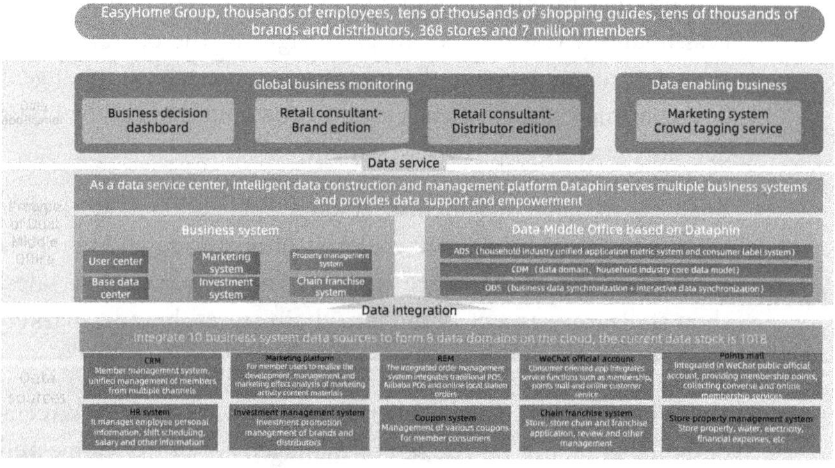

FIGURE 5.68 Project delivery results.

business empowerment covers thousands of leaders and employees at all levels of the group, the management of tens of thousands of brands and dealers, and tens of thousands of shopping guides. In response, Li Xuanxuan, vice president of Easyhome Group, highly commented: 'the management decision board solves the problem that managers at different levels know the business under his jurisdiction as they know it. The decision-making digitization of Easyhome has been refined to the smallest operating unit. Don't underestimate the digital achievement now, all the product tools of this project have entered the most painful point of business. In most enterprises of above scale, you ask managers what happened in the past week, day, hour in their business field. Most of them don't know'.

We will make a simple summary of the value of the project, summed up in three points: quality, efficiency and empowerment.

1. Quality improvement: The pace control and quality control in the construction process of Data Middle Office project are relatively reasonable. Although the impact of the epidemic of COVID-19 from December 2019 to May 2020 was experienced, the project team used DingTalk video conference tools combined with remote daily meeting weekly meeting and other project management means to ensure the successful completion of the delivery target. The construction of the project has realized the online, mobile and refined management of the core data assets of Easyhome. Help Easyhome to realize the decision-making of operation management, provide data support, digital decision-making touch to the smallest business unit. This is also an important means and way for enterprises to do better in operation and management when the demographic dividend gradually disappears and the market tends to be saturated. We believe that the essence of 'digital transformation' is to use technology means to improve the management ability and efficiency of enterprises, to achieve the change of management dimension by refining the granularity of management and improving the accuracy of management, and then to seek business innovation.

2. Efficiency improvement: The big data computing capability of the Data Middle Office helps customers greatly improve the output baseline of all reports. The monthly report is advanced from the next

month to the daily output of the current month; the daily output baseline is advanced from 8:00 p.m. to 8:00 a.m.; with the help of real-time computing, the business insight timeliness is improved from t+1 day to second-level*** real time. Relying on data products and operation, business personnel and technical personnel deduce the problems of business, application and data itself from data metrics, realizing the consistency of business analysis among stores, branches and groups of Easyhome Group. With the triple support of platform, version management specification and Data Middle Office agile development capability, the development cycle of business requirements has been upgraded from monthly to weekly. The data applications kept iterative speed once a week, and the average bug solving time is less than 1 day.

3. Empowerment: First of all, the project enabled the marketing and investment promotion department to gain a grasp of data analysis and support the daily chain, investment promotion and management work; Second, the project enabled partners to quickly and accurately understand the operation of the industry, the brand itself or their stores; Third, the project enabled the new retail team of Easyhome, this team developed and launched a set of data application operation monitoring screen based on the methodology and technical system of Data Middle Office. These phenomena and changes mean that it is possible for the team to transform from the traditional cost department to the Revenue Department and realize the value of data. Finally, we have enabled the ISV partners on the Alibaba side of the project to help them form a localized team familiar with the methodology and product technology of Alibaba Data Middle Office, which has enriched the Alibaba data platform delivery ecosystem.

5.7.3 Case of D Province Commercial Group

5.7.3.1 Customer Background

The commercial group (Hereinafter referred to as 'D Commercial Group') is a provincial state-owned enterprise formed in the 1990s by the government departments. It is now highly focused on the retail industry and health industry. After several years of development, the group now has several listed companies, universities, national research and development platforms, academician workstations, national well-known trademarks, hundreds of subordinate companies, the total assets of more than 100 billion Yuan. Its retail industry covers new retail, e-commerce, consumer finance, vocational education and other fields. It has hundreds of large retail stores such as department stores, supermarkets, shopping centers, home furnishing specialized stores, auto 4S stores and chain supermarkets in the province and the whole country.

With the continuous growth of the domestic residents' consumer market, the domestic supermarket department store industry continues to expand. However, due to the traditional and extensive business model, the industry is facing multiple challenges in the post epidemic era, such as the loss of offline consumer groups, overstock of inventory, capital turnover and strategic transformation. At the same time, the passenger flow and sales of retail physical stores are greatly affected by the epidemic situation, and the traditional mode is not suitable due to the huge changes in the environment. The retail industry is in urgent need of digital transformation.

In recent years, the D Commercial Group established a leading group to promote digital transformation, and established the intelligent technology center to carry out the construction work related to digital transformation. The 'D Commercial Group Data Middle Office and digital application' project was completed and put into operation in July 2020. In the next step, relying on the construction of Data Middle Office, we will build three capabilities of 'data global convergence, data capitalization and data service', continue to carry out industry empowerment, promote digital transformation upgrading and adjustment and build a digital new D Commercial Group with data empowerment, innovation guidance and collaborative ecology, so as to better serve the people.

5.7.3.2 Solution of Alibaba Cloud

Through the Alibaba Cloud Data Middle Office products and construction implementation methodology, build the data platform system and data analysis service system of D Commercial Group, and realize the enterprise level data standardization, data capitalization and data service of this group, which can be divided into the following three aspects:

- Build high-level blueprint design of big data to provide D Commercial Group with big data capability blueprint and implementation route in the next 3 years.

- Build the group common date model, realize the unification of data construction and management, and form a standard reusable data asset system.
- Create business-oriented data application construction to provide data support for business decision-making.

The following will start a brief introduction from the above two aspects:

1. High-level blueprint design of big data

   Taking the Data Middle Office of department stores and shopping centers in the retail sector of D Commercial Group as an example, the panoramic planning of the Data Middle Office is as follows (Figure 5.69).

   I. Market and commercial real estate brand value: Based on the survey data of precipitation city business circle, analyze the customer group situation of market business circle. Establish data-driven, optimize brand marketing strategy and enhance the brand viscosity of commercial real estate. Analysis of online commercial real estate brand promotion, including online reviews, interactive data, public sentiment data and third-party platform channel consumer evaluation data, optimizes brand marketing strategy (Figure 5.70).

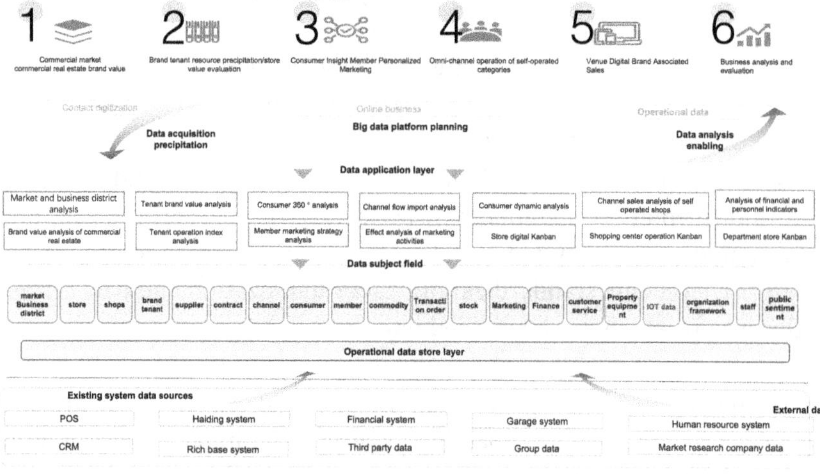

FIGURE 5.69 Data Middle Office panoramic planning.

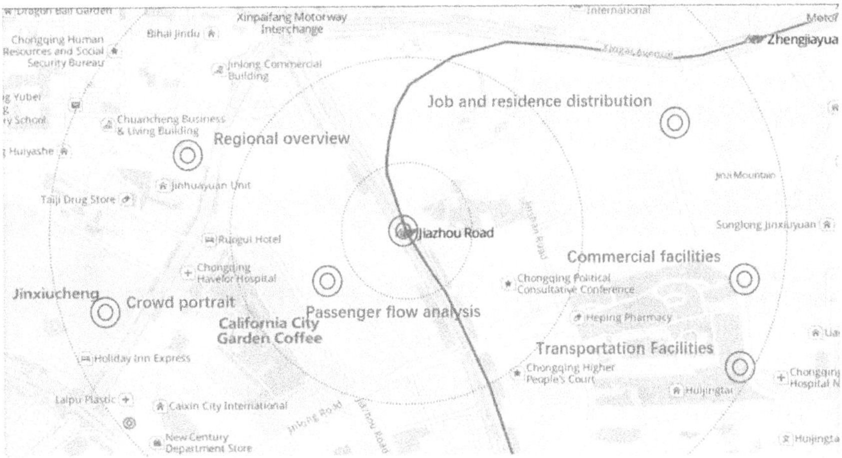

FIGURE 5.70 Urban business district data analysis.

II. Brand tenant resource precipitation and store value evaluation: Tenant brand resource pool, precipitation intention and cooperation brand resources establish hierarchical tenant brand rating mechanism, provide the basis for tenant brand cooperation strategy (Figure 5.71).

III. Consumer insight and member personalized marketing: Based on consumer behavior, consumption, preference and other data, multi-dimensional and multi-level classification constitutes a

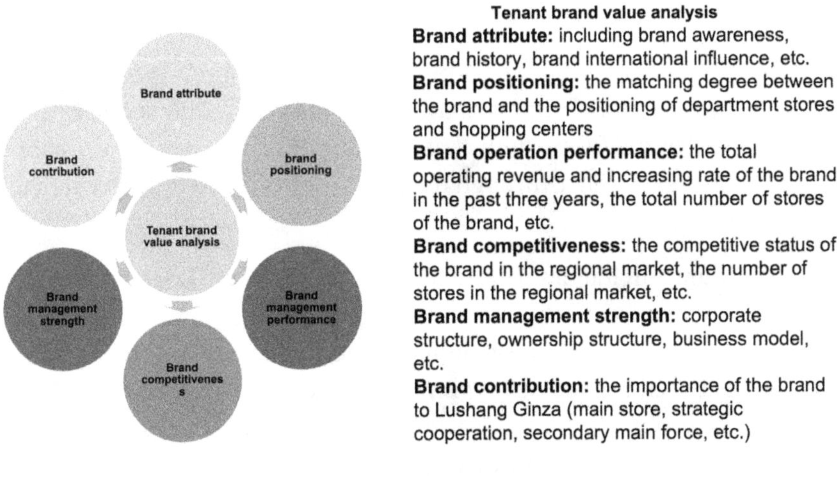

Brand attribute
brand positioning
Brand management performance
Brand competitivenes s
Brand management strength
Brand contribution
Tenant brand value analysis
Tenant brand value analysis
Brand attribute: including brand awareness, brand history, brand international influence, etc.
Brand positioning: the matching degree between the brand and the positioning of department stores and shopping centers
Brand operation performance: the total operating revenue and increasing rate of the brand in the past three years, the total number of stores of the brand, etc.
Brand competitiveness: the competitive status of the brand in the regional market, the number of stores in the regional market, etc.
Brand management strength: corporate structure, ownership structure, business model, etc.
Brand contribution: the importance of the brand to Lushang Ginza (main store, strategic cooperation, secondary main force, etc.)

FIGURE 5.71 Tenant brand value analysis.

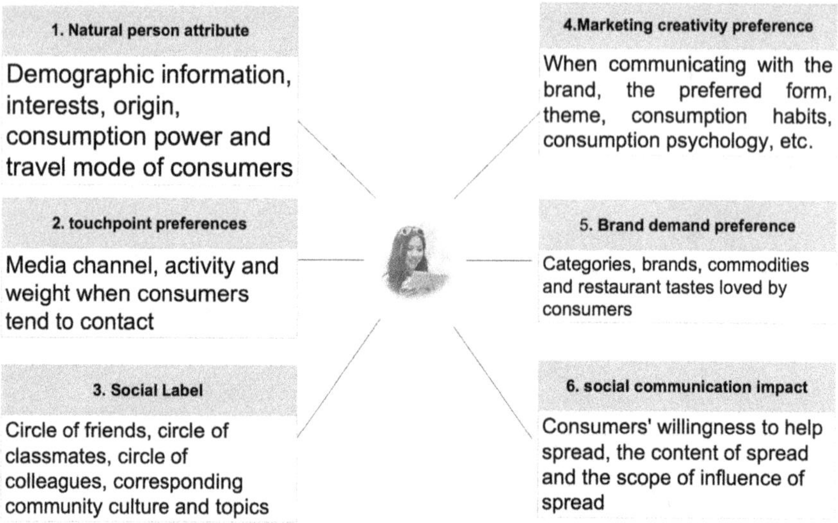

1. Natural person attribute
Demographic information, interests, origin, consumption power and travel mode of consumers
2. touchpoint preferences
Media channel, activity and weight when consumers tend to contact
3. Social Label
Circle of friends, circle of classmates, circle of colleagues, corresponding community culture and topics
4.Marketing creativity preference
When communicating with the brand, the preferred form, theme, consumption habits, consumption psychology, etc.
5. Brand demand preference
Categories, brands, commodities and restaurant tastes loved by consumers
6. social communication impact
Consumers' willingness to help spread, the content of spread and the scope of influence of spread

FIGURE 5.72 Customer insight.

data label system, such as basic information, social attributes, behavior information (which can be divided into online behavior and offline behavior) and preference information (which can be divided into life preference and consumption preference). The labeled customer information is displayed in the same view, forming a 360° customer view. Based on the insight of consumer data, make personalized marketing strategy (Figure 5.72).

IV. Self-operated brand Omni channel operation: Open up members' digital identities in different channels, build member rights center and form Omni channel seamless consumption and shopping experience through the design of front-end different contact scenes.

V. Venue digital brand-related sales: Based on the sales data of consumers in the venue and the order of consumption and shopping, identify the customers' moving line, and identify the relationship between brands through moving line and shopping analysis, so as to play a guiding role in developing related sales in different industries in the future (Figure 5.73).

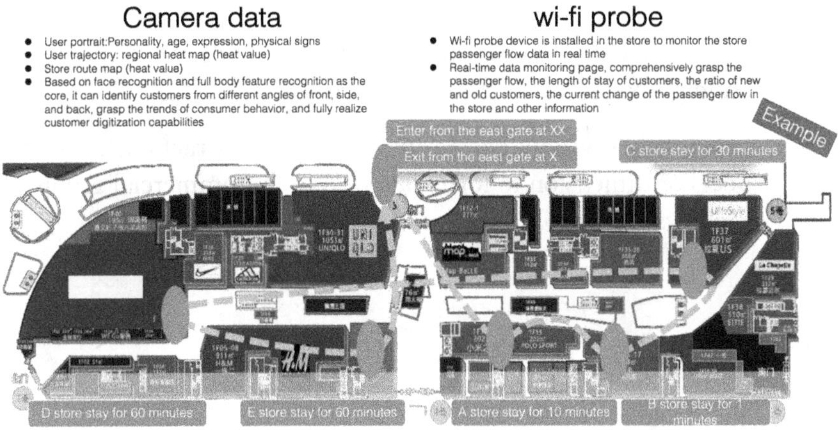
Camera data
User portrait:Personality, age, expression, physical signs
User trajectory: regional heat map (heat value)
Store route map (heat value)
Based on face recognition and full body feature recognition as the core, it can identify customers from different angles of front, side, and back, grasp the trends of consumer behavior, and fully realize customer digitization capabilities
wi-fi probe
Wi-fi probe device is installed in the store to monitor the store passenger flow data in real time
Real-time data monitoring page, comprehensively grasp the passenger flow, the length of stay of customers, the ratio of new and old customers, the current change of the passenger flow in the store and other information
Example
Enter from the east gate at XX
Exit from the east gate at X
C store stay for 30 minutes
D store stay for 60 minutes
E store stay for 60 minutes
A store stay for 10 minutes
B store stay for 1 minutes

FIGURE 5.73 Site digitization.

VI. Business metrics business analysis and evaluation: Business metrics, including shopping malls, department stores, headquarters, brands, shop assistants and other dimensions, analyze scenarios such as store passenger flow analysis, energy consumption analysis, customer complaint analysis, brand consumer portrait, marketing activity effect analysis and commodity portrait analysis through the analysis and evaluation of business metrics, provide the basis for business decision-making.

2. Group data public layer construction

Based on the construction experience of OneModel in Alibaba Data Middle Office, and based on the business requirements of Data Middle Office construction and data asset inventory information input of D Commercial Group, the construction of group CDM is completed, which is mainly divided into the following steps:

I. Business process analysis: Business process is an inseparable behavior event. Combined with business demand analysis and research, referring to business planning and design documents and business operation (development, design, change, etc.) related documents, comprehensively analyzes the source system and business management system involved in data warehouse, and then defines the business process, part of business processes and processes are as follows (Figure 5.74).

II. Data field division: For business analysis, abstract business process or dimension to form data field of D Commercial group, as follows (Figure 5.75).

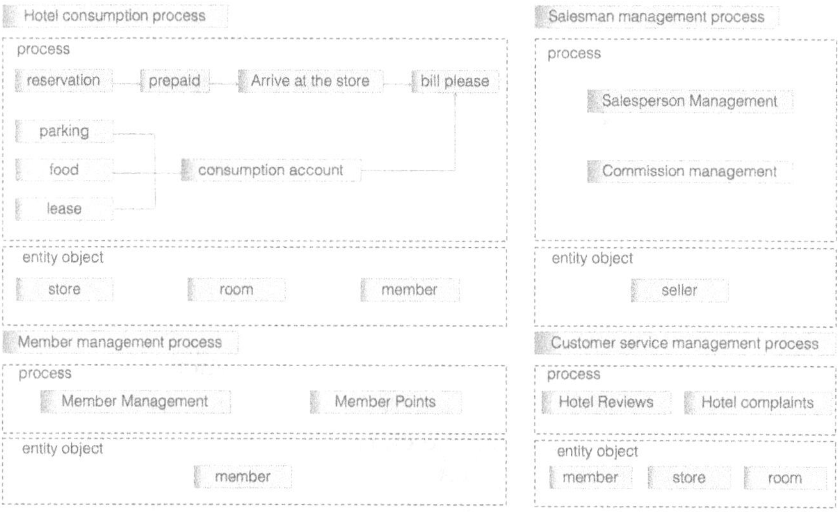

FIGURE 5.74 Business process analysis.

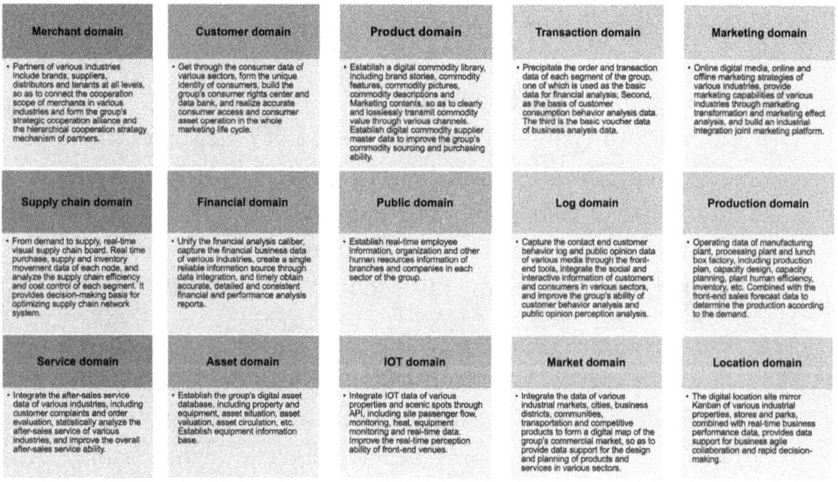

FIGURE 5.75 Data domain division.

Business process			Consistency dimension						
Data domain	Business process	Name	Store	Commodity	Customer	Supplier	Channel	Logistics	Payment method
Transaction	Supermarket transaction order	trd_cs_ord_evt	√	√	√	√			
	Department store transaction order	trd_bh_ord_evt	√	√	√	√			
	Supermarket payment flow	trd_cs_pay_evt	√		√				√
	Department store payment flow	trd_bh_pay_evt	√		√				√
	Hotel order	trd_jd_ord	√	√	√		√		
	Self owned payment channel payment order	trd_jr_pay_order	√	√	√	√	√		√
	Household transaction order details	trd_jj_order_evt	√	√	√				
	Shopping center transaction order	trd_gwzx_ord_evt	√	√	√				√
	Add to cart	trd_add_to_cart		√	√	√			
	Order management	trd_manag		√	√	√		√	√
	Order distribution management	trd_distb_manag	√	√	√	√		√	

FIGURE 5.76 Bus matrix of transaction data domain.

III. Bus matrix construction: Combining business process, dimension and data field, the bus matrix is constructed based on dimension (Figure 5.76).

IV. Dimension table design: Dimension is an entity that actually exists and does not exist due to events. It is constructed based on the principle and method of dimension design.

V. Design of fact table: Based on the principles of integrity, consistency and additivity, complete the design of fact table.

VI. Definition of metric specification: Based on the definition of metric specification, complete the design of atomic metric, business limit, statistical cycle and statistical granularity, and then complete the definition of DWS layer-derived metric.

5.7.3.3 Customer Value

The first phase of Data Middle Office construction project of D Commercial Group has been completed, reaching the phased goal:

- Data asset construction: Based on the data asset inventory of more than 40 existing business systems, 600 + data tables have been put into the cloud. Based on this, the global Data Middle Office of the data group has been constructed to realize the standardization and online of the group's data assets, provide the group's data center (PaaS) platform capability and provide the group's data resources for various industries.
- Data-enabling business: Through intelligent technology center and various industrial groups to establish a joint project team mechanism, build data platform front-end applications, such as retail store

operation analysis and consumer portrait, to enable store digital operation decision-making; through the integration of smart applications in big data domain, upgrade commodity circulation and sales process, and improve store operation efficiency.

- Service ecology construction: Connect with various industries to create an agile digital creation mechanism, and reach a consensus with the industry on gradually promoting the construction of digital service scene by stages.
- Data asset operation: On the basis of consolidating the base of data platform, build the core competence of D Commercial Group intelligent technology center, empower various industries, serve ecological enterprises through the operation of data asset products and promote the transformation of intelligent technology center from cost center to profit center.

CHAPTER 6

AIoT on Cloud

Shun Shunhou and Yang Peng

6.1 INTRODUCTION TO AIoT

Under the trend of the global economic environment, enterprises have made intelligent strategic transformations and business upgrades, accelerated the supply-side structural reforms for intelligent interconnection and opened up a new blue ocean in the intelligent era through the new supply of intelligent interconnection. As the upgrade and evolution direction of the mobile Internet, smart Internet is becoming an important driving force for economic development, transformation and upgrading and social progress.

In the past decade of development, more and more industries and enterprises have established their vision of becoming digital native enterprises. After years of exploration, more and more industries and applications have combined artificial intelligence (AI) and Internet of Things (IoT) to serve the digital transformation of enterprises. With the technical characteristics of Artificial Intelligence Internet of Things (AIoT), AIoT will play five key roles in digital transformation: one is the prerequisite for the digitization of products and services; the second is the means of connecting and interacting to obtain data; the third is the tool for analyzing and mining value; the fourth is the implementation of management control. Way; fifth is the cornerstone of business model change.

6.1.1 AIoT Definition

The Intelligent Internet of Things (AIoT) is a concept that emerged in 2018. It refers to the system collecting all kinds of information in real time

DOI: 10.1201/9781003272953-6

through various information sensors, and intelligently analyzing data through machine learning at the terminal device, edge or cloud center, including positioning, comparison, Forecasting and scheduling.

At the technical level, AI enables the IoT to acquire perception and recognition capabilities, and the IoT provides data for training algorithms for AI.

At the commercial level, the two work together on the real economy to promote industrial upgrading and experience optimization. In terms of specific types, there are three main categories: intelligent networking equipment with perception/interaction capabilities, equipment asset management through machine learning methods and systematic solutions with networking equipment and AI capabilities. From the perspective of collaboration, it mainly solves the problems of perception intelligence, analysis intelligence and control/execution intelligence.

There are more and more industry applications that combine AI with IoT. AIoT has become the best channel for the intelligent upgrade of major traditional industries, and it will also become an inevitable trend in the development of the IoT.

6.1.1.1 Internet of Things

The IoT is the 'Internet of Things Connected'. It is an extended and expanded network based on the Internet. It combines various information sensing devices with the Internet to form a huge network that can be realized at any time, Any place, the interconnection of people, machines and things.

The IoT is an important part of the new generation of information technology. The IT industry is also called: Pan-interconnection, which means that things are connected and everything is connected. Therefore, 'The Internet of Things is the Internet of Things'. This has two meanings: first, the core and foundation of the IoT is still the Internet, which is an extended and expanded network based on the Internet; second, its user end extends and extends to any item and item to carry out information exchange and communication. Therefore, the definition of the IoT is a kind of network to connect any item to the Internet through information sensing equipment and according to an agreed protocol for information exchange and communication, so as to realize the intelligent identification, positioning, tracking, monitoring and management of items.

The concept of the IoT first appeared in the book *The Road to the Future* by Bill Gates in 1995. In *The Road to the Future*, Bill Gates has mentioned the concept of the IoT, but it was limited by wireless networks, hardware and sensing equipment at the time. The development of China has not attracted the attention of the world.

In 1998, the Massachusetts Institute of Technology in the United States creatively put forward the idea of the 'Internet of Things' called the EPC system at that time.

In 1999, the United States Auto-ID first proposed the concept of 'Internet of Things', which was mainly based on item coding, radio frequency identification technology (RFID) and the Internet. In the past in China, the IoT was called a sensor network. The Chinese Academy of Sciences started research on sensor networks as early as 1999, and has achieved some scientific research results and established some applicable sensor networks. In the same year, the International Conference on Mobile Computing and Networks held in the United States proposed that 'the sensor network is another development opportunity facing mankind in the next century'.

In 2003, the US 'Technology Review' proposed that sensor network technology will be the top 10 technology that will change people's lives in the future.

On November 17, 2005, at the World Summit on the Information Society (WSIS) in Tunisia, the International Telecommunication Union (ITU) released the 'ITU Internet Report 2005: Internet of Things', which formally put forward the concept of 'Internet of Things'. The report pointed out that the ubiquitous 'Internet of Things' communication era is coming, and all objects in the world, from tires to toothbrushes, from houses to paper towels, can be actively exchanged through the Internet. RFID, sensor technology, nanotechnology and intelligent embedded technology will be more widely used.

6.1.1.2 AI+IoT

With the development of emerging technologies such as AI, cloud computing and the IoT, some technologies that complement each other in function are inevitably combined.

AI began at the Dartmouth Conference in 1956, and the concept of the IoT has been proposed for more than 20 years. AIoT stands for Artificial Intelligence Internet of Things, which is a collective term that combines AI and IoT technologies.

Artificial Intelligence is referred to as AI. Refers to the process of imitating people's thinking and consciousness by machines realized through computer technology and application systems. Mainly include language recognition and image recognition AI is AI, which is based on artificially trained language, behavior, etc., to make machines produce human-like thinking and behaviors such as self, consciousness and thinking.

The core concept of IoT is the interconnection of everything, that is, connecting machines, people, things, etc., using information technologies such as local area networks or the Internet through sensors and controllers.

AIoT is the integration of AI and IoT technology. First, collect a large amount of information among machines, people and things through the IoT, and store this information in the cloud. Second, combined with AI, that is, the machine can think like a human, and continuously train and optimize the machine. Finally, the information input by the user is combined with the cloud information collected by the IoT to make it a process of feedback to the user through big data analysis, that is, the AIIoT. In theory, it should be a powerful symbiosis system. The IoT provides AI with data, and then the AI feeds back tangible instructions and improvements to the IoT.

The combination of two powerful technologies, AI and IoT, brings a variety of applications and flexibility. The advantage of AIoT is that it can promote and provide viable options. The IoT can provide information about devices, but by adding machine learning algorithms, organizations can predict decisions and results, making it possible for the IoT to make decisions on its own in the future.

6.1.2 AIoT Industry

The IoT industry refers to a collection that is engaged in or mainly engaged in the production or service activities of the IoT, including two major categories of IoT manufacturing and IoT service industries, which mainly include manufacturing, medical, energy and public utilities, agricultural technology, and wisdom City, financial services, smart furniture and other industries.

6.1.2.1 System Architecture

The architecture of the IoT is divided into four levels from bottom to top: perception layer, network layer, platform layer and application layer (Figure 6.1).

According to these four levels, the industrial chain of the IoT can be roughly divided into eight major links: Chip providers, sensor suppliers,

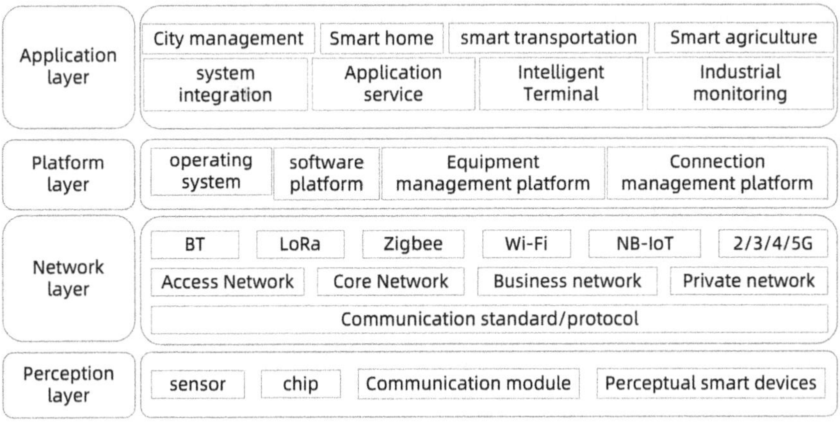

FIGURE 6.1 IoT system architecture.

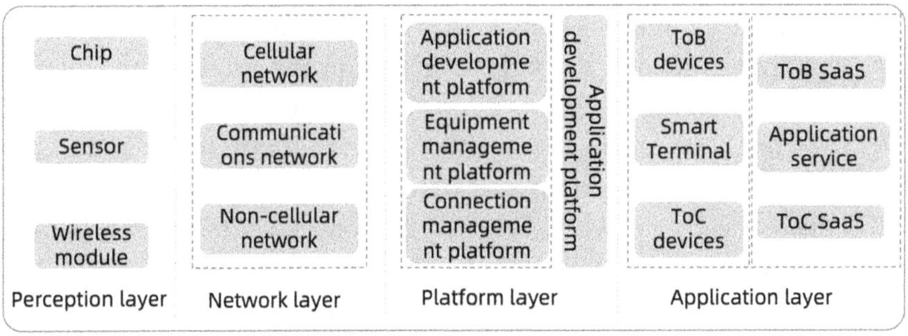

FIGURE 6.2 Eight major links of the IoT industry chain.

wireless module (including antenna) manufacturers, network operators (including SIM card vendors), platform service providers, system and software developers, smart hardware manufacturers, system integration and application service providers (Figure 6.2).

6.1.2.2 Industry Chain

In 1999, Kevin Ashton first proposed the term 'Internet of Things'. In 2005, he formally proposed the concept of 'Internet of Things' in the article 'ITU Internet Report 2005: Internet of Things', and the world started Related construction of the IoT. China started relatively late, and it only began its incubation period in 2009, focusing on the research and development of related technologies, and began to apply in the public management and service market; 2013–2015, China's IoT market as a whole entered the introduction period, system integration companies in good results have been achieved at this stage, and the scale of the industry has reached about 1 trillion. From 2016 to 2020, IoT-related products will begin to be used on a large scale at the enterprise, household and personal levels. Innovative services continue to grow, and the entire industry's technology The further improvement and comprehensive popularization of the standard system indicates that the industry has entered a growth period; after 2020, the IoT industry has begun to enter a mature period, with all-round integration of people, things and service networks. Related technologies have begun to be applied in depth, and the market size is initially expected to reach 100 billion yuan.

1. The chip is the 'brain' of the IoT, and a semiconductor chip with low power consumption and high reliability is one of the essential key components in almost all links of the IoT. According to the different chip functions, the chips required in the IoT industry include chips integrated in sensors and wireless modules to achieve specific functions, as well as system chips embedded in terminal devices to provide 'brain' functions-embedded micro. The processor is generally in the form of MCU/SoC (Mirco Control Unit/System on Chip).

   At present, in the field of the IoT, there are a large number of chip manufacturers, a wide variety of chips and obvious differences in personalization. However, the chip field is still dominated by international giants such as Qualcomm, TI (Texas Instruments) and ARM (Advanced RISC Machines). Although there are many domestic chip companies, most of the key technologies are imported from abroad, which directly leads to the lack of profitability of many chip companies and it is difficult to occupy market share.

2. The sensor is the 'five senses' of the IoT, essentially a detection device, which is a device used to collect various types of information and convert it into a specific signal. It can collect identity, movement status, geographic location, posture, pressure, temperature, humidity, light, sound, smell and other information. Sensors in a broad sense include traditional sensitive components, RFID, bar codes, barcodes, two-dimensional codes, radars, cameras, card readers and infrared sensor components.

   The sensor industry has a long history and is currently dominated by several leading companies in the United States, Japan and

Germany. About 70% of the sensor market in my country is occupied by foreign-funded enterprises, and the market share of domestic enterprises in my country is relatively small.

3. The wireless module is the key equipment for the IoT to access the network and locate. Wireless modules can be divided into two categories: communication modules and positioning modules. Common LAN technologies include WiFi, Bluetooth and ZigBee. Common WAN technologies mainly include 2/3/4G, NB-IoT, and unlicensed bands that work in licensed frequency bands, such as LoRa, SigFox and other technologies. Different communications correspond to different technologies. Communication modules NB-IoT, LoRa, and SigFox belong to low-power wide area network technologies, which have the characteristics of wide coverage, low cost and low power consumption, and are specifically developed for the application scenarios of the IoT.

   In addition, in a broad sense, there are smart terminal antennas related to wireless modules, including mobile terminal antennas and GNSS positioning antennas. At present, in terms of wireless modules, foreign companies still dominate. Domestic manufacturers are also relatively mature and can provide complete products and solutions.

4. The network is the channel of the IoT and the most mature link in the current IoT industry chain. Broadly speaking, the network of the IoT refers to the integrated network formed by various communication networks and the Internet, including cellular networks, local ad hoc networks and private networks, so it involves communication equipment, communication networks (access networks, core networks, etc.), Business, SIM manufacturing, etc.

   Considering that the IoT can largely reuse existing telecom operator networks (wired broadband networks, 2/3/4G mobile networks, etc.), and domestic basic telecom operators have monopoly characteristics, which are currently the most important promoter for the development of domestic IoT. So in this link we will focus on the three major telecom operators and are closely related to them, and will benefit from closely related to them, and will benefit from being closely related to them, and will benefit from being closely related to, and will benefit from. The growth of cellular IoT terminals is the SIM card manufacturer.

5. The platform is the basis for effective management of the IoT. As an important link of equipment convergence, application services and data analysis, the IoT platform must not only implement the 'management, control and operation' of the terminal downwards, but also provide platform as a service (PaaS) services for application development, service provision and system integration upwards. According to different platform functions, it can be divided into the following three types:

   - Device management platform: It is mainly used for remote monitoring, system upgrades, software level, troubleshooting, life cycle management and other functions of the IoT devices. All device data can be stored in the cloud.
   - Connection management platform: Used to ensure the stability of terminal networking channels, the management of network resource usage, tariff management and bill management. Package change, number/address resource management.
   - Application development platform: Mainly provides IoT developers with application development tools, back-end technical support services, middleware, business logic engine, API, interactive interface, etc. In addition, it also provides a highly scalable database, real-time data processing, intelligent prediction offline Data analysis, data visualization display applications, etc., allow developers to quickly develop, deploy and manage without considering the underlying details, thereby reducing costs in a short time.

   As far as platform-level enterprises are concerned, foreign manufacturers include Jasper and Wylessy. There are mainly three types of domestic IoT platform companies. The first is the three major telecom operators, which mainly start with the construction of connection management platforms; the second is Internet giants such as BAT, JD.com and Huawei, which use their traditional advantages to build Equipment management and application development platforms; the third is platform vendors in their respective segments, such as Yitong Century, Shanghai Qingke and Xiaomi.

6. The system and software can make the IoT devices operate effectively. The system and software of the IoT generally include operating

systems and application software. Among them, the operating system (OS) is a program that manages and controls the hardware and software resources of the IoT. It is similar to the IOS and Android of smart phones. It is the most basic system software that runs directly on the 'bare metal'. Other application software is available. It can run normally under the support of the operating system.

Currently, IT giants, such as Google, Microsoft, Apple and Ali, are mainly releasing IoT operating systems. As the IoT is still in its infancy, application software development is mainly concentrated in areas with strong versatility such as the Internet of Vehicles, smart homes and terminal security.

7. Smart hardware is the bearer terminal of the IoT, which refers to equipment that integrates sensor devices and communication functions, can connect to the IoT and implement specific functions or services. If divided according to the customer-oriented purchasers, it can be divided into To B and To C:

    - To Class B: Including meters (smart water meters, smart gas meters, smart electricity meters, industrial monitoring and testing instruments, etc.), vehicle front-mounted (vehicles), industrial equipment and public service monitoring equipment.
    - To C category: Mainly refers to consumer electronics, such as wearable devices and smart homes.

In view of the extremely rich application scenarios of the IoT, there are many types of terminals, we will only list some To B type, market demand and relatively concentrated manufacturers of this type of terminal manufacturers.

8. System integration and application services are important links in the deployment, implementation and application of the IoT. The so-called system integration is the process of verifying and integrating multiple products and technologies into a complete solution based on the requirements of a complex information system or subsystem. The current mainstream system integration practices include equipment system integration and application system integration.

    Based on this, we believe that the system integration of the IoT is generally for large customers or vertical industries, such as government departments, urban investment companies, real estate companies and industrial manufacturing companies, and often provides comprehensive solutions in the form of main. Facing the complex application environment of the IoT and equipment in many different fields, system integrators can help customers solve various equipment and subsystems' interfaces, protocols, system platforms, application software, etc. and subsystems, building environment, construction cooperation, Organize management and staffing related issues to ensure that customers get the most appropriate solutions.

6.2 THE DEVELOPMENT OF AIoT

In recent years, as a high-tech industry, the IoT has led to wave after wave of global technological revolutions. It is currently the focus of attention of countries, enterprises and academia, and its development has risen to a strategic level. All countries are trying to make achievements in the new round of global technological revolution set off by the IoT in order to enhance the country's comprehensive strength and prestige. Whether the IoT technology can lead is crucial to the development of the IoT in various countries. The United States included the IoT in the 'Six Key Technologies Potentially Affecting U.S. Interests in 2025'. The 2009 'Economic Recovery and Reinvestment Act' made the IoT a key strategy to escape the financial crisis; the European Commission formulated the 'The European Union's Internet of Things Action Plan' to lead the development of the world's IoT; South Korea, Japan and other countries have also formulated corresponding IoT strategies to strive to seize opportunities for IoT technology and industrial development.

As a new industry, the IoT in my country has developed very fast. The development of IoT technology plays a very important role in promoting the development of various industries in my country. At the end of the last century, the Chinese Academy of Sciences organized a team to conduct research on the IoT technology, and invested a lot of money. It has achieved certain results in many aspects such as microsensors and mobile base stations. At present, it is in the supply of materials, technical support and network support. A very complete system has been established. At the same time, my country has strengthened its cooperation with other developed countries in the IoT technology, and my country also has an important influence on the formulation of international standards for the IoT. My country has applied for related intellectual property rights and technical patents in the fields of technology, network and other related IoT technology. Based on the existing IoT technology and research theory as the foundation, it has promoted the long-term development of my country's IoT technology.

As the Chinese government pays more attention to the development of the IoT technology and continues to promote the research and development of the IoT technology, in a relatively short period of time, my country has gradually made the IoT evolved from a relatively unfamiliar technology term to today's impact on all aspects of people's lives. Industrial entities.

At present, my country's IoT is still facing some development bottlenecks, but with the joint efforts of national strategy, industrial policy and industry professionals, my country's IoT will accelerate integration and innovation and empower the real economy. The development of the IoT industry shows the following trends:

First, the global IoT has entered a stage where the acceleration of industry implementation and network supervision and rectification are equally important. The global IoT equipment continues to be deployed on a large scale, with the number of connections exceeding 11 billion; the module and chip market has a strong momentum, the trend of platform concentration is obvious, and the investment in the industrial field is becoming more active. Affected by the new crown pneumonia epidemic, the market size growth is expected to be lowered, but the overall upward trend remains unchanged.

Major economies have accelerated their network and security layouts. The United States has successively introduced multiple bills to emphasize 5G international leadership and focus on IoT innovation and security; the European Union has issued a strategy to consolidate IoT data foundations and take multiple measures to improve cybersecurity risk control capabilities; Japan has established new The establishment of IoT terminal defense countermeasures and South Korea speeds up the 6G research and development layout, and continues to increase capital investment in IoT-related fields.

Second, the scale of my country's IoT industry has grown faster than expected, with outstanding results in network construction and application promotion. Driven by national strategies such as network powers and new infrastructure, my country has accelerated the promotion of IPv6, NB-IoT, 5G and other network construction. The number of mobile IoT connections has exceeded 1.2 billion, and device connections account for more than 60% of the world's total. Consumer Internet of Things The Industrial Internet of Things has gradually begun large-scale applications, and breakthroughs have been made in the development of 5G and Internet of Vehicles. Data show that the scale of the industry in 2019 exceeded 1.5 trillion yuan, which has exceeded the expected planned value.

The third is to increase the layout of leading enterprises, and the development of 5G network construction and edge computing will deepen the application of the two-wheel drive IoT. Since 2019, leading companies such as Huawei, Alibaba and Haier have each focused on increasing their layout, leading venture capital institutions have been investing actively and the average financing amount in the IoT field has increased. With the acceleration of the commercialization of 5G and the large-scale deployment of NB-IOT, the integration and innovation of the IoT, AI, and big data have accelerated. At the same time, the increase in device connections has driven the growth of edge computing demand, and application scenarios such as the Internet of Vehicles, industrial Internet and smart medical have further deepened.

6.2.1 Policies and Regulations

The IoT is an information carrier such as the Internet and traditional telecommunications networks, which enables all ordinary objects that can perform independent functions to achieve interconnection. In recent years, China's IoT industry has developed rapidly, and the scale of the industry has grown rapidly. The rapid development of my country's IoT industry mainly benefits from the support of relevant national policies.

National policy for the IoT industry in 2013: The country began to propose a special development strategy for the IoT, and the IoT industry follows the strategic deployment of the industry's development formulated by the government.

National policy for the IoT industry in 2017: The Ministry of Industry and Information Technology issued the 'Thirteenth Five-Year Plan for the Internet of Things' to clarify the development goals of the IoT.

2018 National Policy for the IoT Industry: The Ministry of Industry and Information Technology issued the 'Internet of Things Security White Paper', which elaborated on the current status of Internet of Things security and the development direction of protection strategies.

National Policy for the IOT in 2019: At the two sessions just held this year, Guo Yonghong, a deputy to the National People's Congress, suggested improving the regulatory system to protect user privacy. It can be seen that people from all walks of life are paying attention to the security of the IoT. How to protect the IoT is the next stage. The government's focus on this issue is expected to invest more financial support for the development of IoT security.

In recent years, relevant national departments and local governments have issued a series of relevant policies for the IoT industry. In order to better understand relevant policies in various regions, the '2019 China Internet of Things Related Policies Collection' provides reference for the majority of industry professionals.

6.2.1.1 National Policies

1. In February 2013, the State Council issued the 'Guiding Opinions of the State Council on Promoting the Orderly and Healthy Development of the Internet of Things': By 2015, it will be used in industry, agriculture, energy conservation and environmental protection, commerce and trade, transportation energy, public security, social undertakings, urban management. Realize the pilot and demonstration applications of the IoT in areas such as safety production and national defense construction. The level of large-scale applications in some areas has been significantly improved, and a number of IoT application service companies have been cultivated.

   With the goal of mastering the principles to achieve breakthrough technological innovation, grasping the direction of technological development, focusing on the urgent needs of applications and industries, clarifying the development focus, strengthening the research and development and industrialization of low-cost, low-power, high-precision, high-reliability, and intelligent sensors, and focus on Break through basic common technologies such as core chips, software and instrumentation of the IoT accelerate the research and development and innovation of key technologies such as sensor networks, smart terminals, big data processing, intelligent analysis and service integration, and promote the IoT and the new generation of mobile communications, cloud computing, and offline the integration and development of technologies such as the Internet and satellite communications.

Make full use of and integrate existing innovative resources to form a batch of IoT technology research and development laboratories, engineering centers, and enterprise technology centers, promote cooperation between application units and related technology, product and service providers, strengthen collaborative research, and break through industrial development bottlenecks.

2. In September 2013, the National Development and Reform Commission and several ministries and commissions issued the 'Special Action Plan for the Development of the Internet of Things (2013–2015)': The plan includes top-level design, standard formulation, technology research and development, application promotion, industrial support, business model, and security 10 special action plans for protection, government support, laws and regulations and talent training. Each special plan sets out the overall goals that the IoT industry will achieve in 2015 from their respective perspectives.

3. In January 2017, the National Development and Reform Commission issued the 'Thirteenth Five-Year Development Plan for the Internet of Things': By 2020, an internationally competitive IoT industry system will be basically formed, including the overall system of perception manufacturing, network transmission and intelligent information services. The scale of the industry exceeded 1.5 trillion yuan, and the proportion of intelligent information services increased significantly. Promote the planning and layout of IoT perception facilities, and the number of public network M2M connections exceeded 1.7 billion. The research and development level and innovation capabilities of the IoT technology have been significantly improved, a standard system that adapts to industrial development has initially taken shape, the scale of IoT applications has continued to expand, and a ubiquitous and secure IoT system has basically taken shape.

   Build 10 distinctive industrial clusters, cultivate and develop about 200 key enterprises with an output value of more than 1 billion yuan, as well as a group of 'specialized, special and new' small- and medium-sized enterprises and innovation carriers, and build a group of broad coverage and strong support A public service platform to build an internationally competitive industrial system.

4. In June 2017, the Ministry of Industry and Information Technology issued the 'Notice on Comprehensively Promoting the Construction and Development of Mobile Internet of Things (NB-IoT)'

   By 2020, the NB-IoT network will achieve universal coverage across the country, and achieve deep coverage for application scenarios such as indoors, traffic road networks and underground pipe networks. The scale of base stations will reach 1.5 million. Strengthen the capacity building of the IoT platform, support the access of a large number of terminals and improve the big data operation capability.

5. In December 2018, the Ministry of Industry and Information Technology issued the 'Internet of Vehicles (Intelligent Connected Vehicles) Industry Development Action Plan'

   By 2020, a breakthrough will be achieved in the cross-industry integration of the Internet of Vehicles (Intelligent Connected Vehicles) industry, intelligent connected vehicles with high-level autonomous driving functions will achieve scale applications in specific scenarios, a comprehensive Internet of Vehicles application system will be basically constructed, and user penetration will be greatly increased. The level of smart road infrastructure has been significantly improved, the policies, regulations, standards and safety guarantee systems that adapt to industrial development have been initially established, and an open, integrated, innovative and developed industrial ecology has basically taken shape to meet the diverse, individualized and continuously upgraded consumer needs of the people.

6.2.1.2 Local Policies

1. Shanghai: Shanghai's implementation opinions on promoting the 'Internet plus' action

   Among them, Internet+R&D and design, virtual production, collaborative manufacturing, supply chain, smart terminals, energy, finance, e-commerce, commerce, culture and entertainment, modern agriculture, new business forms, new models, crowd-creation space, transportation, health, education, Tourism, Smart Home, Public Security, Urban Infrastructure and 21 Special Actions for E-government.

2. Zhejiang Province: The 13th Five-Year Development Plan for the Internet of Things Industry in Zhejiang Province

   Among them, the proposed development goals are: By 2020, to form 1 IoT enterprise leader with a main business income exceeding 80 billion yuan, more than 5 leading enterprises with more than 10 billion yuan, and more than 50 key enterprises with more than one billion yuan, agglomeration of industrial chains. With more than 500 core enterprises, it has become a high ground for the development of the IoT industry with strong international influence.

   By 2020, the IoT industry system integrating R&D and manufacturing, system integration, demonstration applications and standard promotion will be relatively complete. The capabilities of key core technology research and development, system integration and service provision will be comprehensively improved. The scale of the IoT industry, innovation capabilities and the application level is leading in the country. The national IoT industry center has basically been established, and a global industrial center has been formed in the fields of digital security. The main business income of the province's IoT industry has exceeded 500 billion yuan.

   There are 100 IoT companies that have established enterprise technology centers at or above the provincial level to support the development of universities and research institutes in the field of IoT, overcome a number of key technologies for the development of the IoT and produce a number of innovative results of the IoT. The province's IoT enterprises have led to participate in the formulation of 30 international standards, more than 100 national standards, and more than 500 high-tech enterprises. Hangzhou National Independent Innovation Demonstration Zone has become a world-class IoT technology innovation cluster.

3. Fujian: Eight measures on accelerating the development of the IoT industry

   Put forward eight measures to focus on development priorities, strengthen innovation support, build a characteristic platform, support leading development, support entrepreneurial innovation, expand the national market, accelerate the cultivation of talents and increase fiscal and tax support.

Policy rewards:

- If the total investment exceeds 15 million yuan, the provincial Internet Economy New Guidance Fund will give a maximum subsidy of not more than 5 million yuan.
- Implementing major special projects for the core technology of the IoT industry, the Provincial Department of Science and Technology will give up to 5 million yuan for each item, and the acceptance results are the country's leading, and the province's new Internet economy guidance funds will be given an additional 2 million yuan.
- Promote Fuzhou and industry leaders to jointly build a narrowband IoT open laboratory, and the provincial Internet economy will provide 30% subsidies, and the subsidy funds will not exceed 10 million yuan.

4. Guangdong: Guangdong Internet of Things Development Plan (2013–2020)

   The plan states:

   I. Accelerate the basic design and construction of the IoT, mainly by improving communications and accelerating the construction of public support platforms to improve the design of the IoT.

   II. To promote the integration and innovation of the IoT technology and the development of industrialization, the national level is highly focused on this development. Three deployments are proposed in the plan:

      - Speed up technological integration and innovation.
      - Develop core industries and heart-shaped liquids.
      - Optimize industrial layout.

   III. Promote the creation of our world-class smart city cluster, which is mainly to create smart homes, smart environmental protection and promote the development of smart cities in eastern Guangdong.

IV. Promote the application of the IoT in the fields of production and business services.

V. Promote the application of the IoT in the field of social services. Two aspects are mainly mentioned here, one is government public services, and the other is the application of social and people's livelihood.

5. Guangxi: Guangxi's implementation plan for promoting the orderly and healthy development of the IoT

   Development goals: To realize the wide application of the IoT in various fields of economy and society in our district. The research and development capabilities of related technologies for IoT applications have been significantly improved. The IoT industrial system with strong competitiveness, reasonable spatial layout, and a certain industrial scale has become an important force to promote the intelligent and sustainable development of Guangxi's economy and society.

   The IoT has become an important strategic fulcrum for driving social innovation and development, accelerating the transformation of economic development power and accelerating economic transformation and upgrading.

6.2.2 Development of Standards

Although the formulation of standards for the IoT in my country is in its infancy, it is developing rapidly. The IoT standardization organizations have been established one after another, and the number of standards development has been increasing year by year. The details are as follows:

1. Basic standards

   The National Internet of Things Basic Working Group established an 'overall project team' to develop my country's IoT terminology, architecture and IoT test and evaluation system standards.

2. Perception standards

   My country has formulated more than 500 technical standards directly related to sensors in the instrumentation and sensitive device industries; it has established a basically complete RFID standard system that can provide support for my country's RFID industry, and has completed RFID basic technical standards and major industry standards. Application standards and other work.

Developed sensor interface standards, defined data collection signal interfaces and data interfaces; formulated other public document frameworks, data exchange formats, performance testing and other standards for biometric identification; formulated audio, image, multimedia and hypermedia information coding standards.

3. Service support standards

   The SOA standard working group carries out the standardization (revision) work in the fields of SOA, Web services, cloud computing technology and middleware; the basic standards working group of the IoT develops a series of standards for the IoT information sharing and exchange, collaborative information processing, and perception object information fusion research on models; the big data working group will carry out my country's big data standardization work, and the cloud computing work group will carry out my country's cloud computing standardization work.

4. Business application standards

   In the field of public safety, in March 2011, my country established a working group for the application of standards for the IoT in the public safety industry, and listed the standardization project as a part of the state-supported public safety national IoT demonstration project.

   In the field of health care, in 2014, the Health and Family Planning Commission applied for the establishment of a working group on medical and health IoT application standards, and promoted the formulation of 11 medical and health IoT standards including the 'Medical and Health IoT Application System Architecture and General Technical Requirements'.

   In the field of intelligent transportation, my country has established a working group for application standards in the field of IoT transportation to carry out (general technical requirements for vehicle remote service systems) and other related standardization work of transportation IoT; also established 'digital TV receiving equipment and home network platform interface'. The Standards' working group, the 'Resource Sharing and Collaborative Service Standards Working Group' and the 'Home Network Standards Working Group' carry out relevant standardization work.

In the field of intelligent manufacturing, the Ministry of Industry and Information Technology and the National Standards Committee jointly issued the 'Guidelines for the Construction of the National Intelligent Manufacturing Standard System' on December 29, 2015.

In the fields of agriculture, forestry and environmental protection, my country has established the clothing industry IoT application standard working group, the forestry IoT application standard working group and the environmental protection IoT application standard working group, and launched the forestry IoT in agriculture, forestry and environmental protection IoT terms. Related standardization work.

5. Common technical standards

   The 'Identification Project Team' is set up under the National Internet of Things Basic Working Group to develop basic technical standards for my country's IoT coding and identification; 'National Internet of Things Security Project Team' to develop basic technical standards for my country's IoT security; 'Electronic Label Standard Working Group'. The purpose is to establish China's RFID standards and promote the development of China's RFID industry.

6. Formulation of future standards

   Standard formulation and industry promotion is a long-term work. There is no final standard. It will continue to be formulated, revised and improved with the development of the IoT industry. The IoT standards are based on the development of IoT technology and industry. The IoT involves a wide range of technical fields, diverse application fields and a huge industry. These characteristics determine the arduous task of formulating IoT standards. At the same time, competition among different countries and regional alliances is fierce in standard setting, which makes collaboration and coordination between different standards more difficult. However, with the advancement of the IoT technology and the development of the IoT industry, standards in the field of IoT will definitely develop in a unified and coordinated direction.

6.2.3 Industry Application

The new generation of information technology is pushing human society into a new era. Technologies such as the IoT, AI, cloud computing and big data empower various industries, which in turn derive business models such as smart cities, smart manufacturing, smart transportation and new retail, and create. The IoT has always been regarded as the third wave of the development of the world's information industry after computers and the Internet. It has benefited from the maturity and improvement of AI technologies, products and applications in recent years. The AIoT combined with the IoT gives the IoT more intelligent advantages, enabling the real Internet of Everything to be realized. More and more companies use the IoT to connect all physical devices to collect a large amount of information and data, and use deep learning to refine intelligent decision-making, ultimately allowing the entire industry to maximize commercial value.

All walks of life are talking about digital transformation, but transformation and upgrading are long way. Many industries are looking at the dividends of AIoT and cannot keep up. Some industries have already taken the lead in the forefront of this intelligent track. From industrial manufacturing to medical, transportation, retail and other fields, automation and informatization are needed. For example, the use of robots in the medical industry to perform surgery is not only to achieve higher accuracy, but also to avoid bacterial infections, because robots create more Clean space, and these trends have created development opportunities for the IoT industry. In 2019, 87% of healthcare institutions will adopt IoT technology for healthcare institutions and IoT smart pills, smart home care, personal healthcare tubes, electronic health records, management of sensitive data and overall higher levels of patient care. In short, the possibilities are endless. This improvement can be applied to many vertical and horizontal industries.

1. Energy and public utilities

   The IoT can connect more energy sources with increasing demand, and it can also integrate renewable energy sources to achieve clean power generation. Through the IoT, people can understand equipment information in almost real time, thereby reducing the impact of temporary power outages.

2. Transportation and distribution

   The IoT takes trajectory tracking to a whole new level. It enables all parameters in the delivery process to be recorded, not only the location, but also temperature, humidity, vibration, tilt and so on.

This comprehensive monitoring of drivers and equipment greatly improves safety. Especially making driverless cars closer to reality.

3. Agricultural science and technology

    The IoT helps realize precision agriculture. It helps to make sowing, irrigation and fertilizer use more accurate. The IoT can monitor soil quality, wind speed and sunlight, which allows farmers to know what level their crops have grown. At the same time, the use of the IoT in agricultural production can also save resources, reduce costs and reduce the impact on the environment.

    For example, sensor data can be used to suggest how much water is more suitable for irrigation. These forecasted irrigation suggestions save water and electricity resources, can prevent crop diseases, reduce costs and improve crop quality.

    For another example, the IoT can also help farmers manage their farms. Through the use of applications and sensors, farmers can collect, store and track farm data, including temperature, air quality, energy supply and feed use, so that various businesses on the farm can be remotely observed and managed.

4. Smart City

    The complete physical form of a smart city may not yet appear. But smart cities are under construction. The IoT can be used to reduce energy use, manage traffic and increase citizen safety. The IoT can help urban residents, who account for half of the global population, make their lives easier, cleaner, safer and more enjoyable.

5. Retail

    The IoT has been changing the retail industry. It can make the in-store shopping experience more personalized. In addition to consumers no longer having to worry about getting lost in the mall, the IoT can also recommend you ‘may like’ channels to you through your purchase history behavior.

6. Financial services

    In a data-driven global financial environment, the IoT helps to improve intelligence, reduce risks and provide a better digital experience. It can be used to calculate insurance costs, for accurate credit analysis, to personalize retail banking experience and provide customized new products.

7. Smart life

The scope of reference is very wide, and applications such as smart home, smart real estate and smart communities can be summarized in the field of smart life. Through research, we learned that the application of IoT platforms in smart life scenarios is mainly divided into two stages:

The first is the equipment IoT stage. In the early days, many start-ups made a lot of investment in smart hardware. Although the smart home industry did not achieve the expected success at the time from the commercial level, many companies even transformed or died within 3 years, but the remaining companies have already taken the opportunity. The access capability from the underlying hardware to the cloud has been opened up, accumulating abundant customer resources.

The second is the data operation stage. The device IoT in the field of smart life is much simpler than in the field of industrial manufacturing. Whether it is docking with large and small home appliances/lighting/smart products in the home, access control/cameras/parking barriers/security systems in the community, most platform companies think. The problem of the IoT in this field has been solved. At least the vision of single product intelligence and linkage intelligence has been realized. The next stage is to open up the interconnection between various devices and systems, and think about how to provide data for customers, especially operators. Level of service to enhance the value of products to customers.

As the roles of all parties have joined the field, the four identities including property, real estate developer, owner and business will all be connected, and manufacturers will output a complete set of smart residential systems for this purpose, providing high-quality smart life service. However, due to the wide scope of the industry and the limitations of corporate genes, players in the smart life field usually choose to prioritize investment in their areas of expertise.

The IoT is a key driving force for the digital transformation of multiple industries. After decades of exploration, more and more industries and applications have combined AI and IoT. AIoT has become a major traditional industry including manufacturing, transportation, construction, energy, electricity, education, logistics, and retail. The only way to intelligent upgrade. The impact of AIoT can be measured through a wide range of application cases. This technology has the ability to span enterprise and personal use, which marks its long history; moreover, AIoT will continue to expand to various industries and expand its applications and product portfolio.

6.2.4 Application and Development of Technology

As a concept proposed for more than 20 years, the IoT has achieved various breakthroughs in technology, including front-end sensing technology to promote smart devices to obtain data, communication technology responsible for data transmission, and big data technology. The IoT is undergoing upgrading from basic equipment such as hardware and sensors to software platforms and vertical industries. At the same time, as hardware costs drop, cloud computing, big data, AI and industry scenarios are combined, as well as 5G and NB-IoT network technology evolution. The current stage of the industrial ecology is on the eve of the outbreak. The IoT platform will be the most important infrastructure for how to support the secure and stable connection of massive devices to the cloud, as well as provide smooth and reliable data channels and device management capabilities.

In the construction of communication technology, 5G is the latest generation of cellular mobile communication technology, and it is also an extension after 4G (LTE-A, WiMAX), 3G (UMTS, LTE) and 2G (GSM) systems. The main advantage of the 5G network is that the data transmission rate is much higher than the previous cellular network, up to 10 Gbit/s, faster than the current wired Internet, and 100 times faster than the previous 4G LTE cellular network. Another advantage is the lower network latency (faster response time), less than 1 ms, compared to 30–70 ms for 4G. Due to faster data transmission, 5G networks will not only provide services for mobile phones, but will also become a general home and office network provider, competing with wired network providers. Previous cellular networks provided low-data-rate Internet access for mobile phones, but a cell phone tower cannot economically provide enough bandwidth as a general Internet provider for home computers. At present, my country has built more than 130,000 5G base stations. In order to speed up the progress of 5G 'new infrastructure' construction, on March 12, 2020, China Telecom announced that it will complete the construction of 250,000 5G base stations nationwide in the third quarter of 2020 with China Unicom. China Tower also stated that as of early March 2020, China Tower has built more than 200,000 5G base stations, and plans to deploy 500,000 in 2020. In addition, China Mobile has fully completed the first phase of the 5G project and realized 5G commercial use in 50 cities. On March 6, 2020, China Mobile officially launched the centralized procurement of 5G Phase II wireless network main equipment in 2020. A total of 28 provinces, autonomous regions and municipalities have released centralized procurement, and the demand for a total of 232,143 5G base stations. Strive to reach 300,000 5G base stations by the end of 2020, and ensure that 5G commercial services will be provided in all prefecture-level and above cities across the country by 2020.

With the commercialization of 5G technology, the later core planning of IoT enterprises will no longer focus on the underlying hardware or communication technology, but on software platforms and vertical industry applications, and deeply explore the commercial value brought by the IoT. At that time, the IoT platform will become the infrastructure of various applications, and become a large-scale, low-cost and easy-to-use software and service to meet the needs of IoT solutions in the context of the Internet of Everything.

6.2.5 Competitive Landscape

Recently, Gartner, an internationally renowned research organization, released the latest report on the competitive landscape of global IoT platforms, and Alibaba Cloud has become the 10 most competitive companies. Gartner report shows that Alibaba Cloud has a strong performance in the IoT market, providing tens of billions of IoT device connectivity and tens of thousands of industry solutions, covering smart communities, parks, cities, industry, agriculture and other fields.

In this latest 'Competitive Landscape: Internet of Things Platform Suppliers' annual survey report, Gartner analyzed hundreds of technology companies and compared them from the perspectives and dimensions of product landscape, technological advantages and ecological layout, and finally selected 10 enterprises: Alibaba Cloud, Amazon AWS, Microsoft Azure IoT, Automation Intellect, AspenTech, CloudPlugs, Samsung SDS, Toshiba and Tuya Smart.

Not surprisingly, the 3A that occupies the first echelon in the cloud computing market – Amazon AWS, Microsoft Azure and Alibaba Cloud also appear in this pattern and occupy a strong position. Amazon AWS focuses on the industrial IoT field, Microsoft Azure focuses on manufacturing, energy, construction, etc., while Alibaba Cloud focuses on solving complex cross-domain applications in the fields of cities, transportation, industry and medical care.

3A is transferring the layout and technical advantages of cloud computing to the field of IoT. If cloud computing is compared to the heart of an adult, the IoT is a neural network, AI is equivalent to the brain and data are the flowing blood, so that the entire body can operate efficiently. This is why cloud computing giants have deployed the IoT.

In 2018, Alibaba Cloud announced that IoT is the group's new main track after e-commerce, finance, logistics and cloud computing. Alibaba Cloud plans to connect 100 edge-to-end collaborative computing in the next 5 years.

The manufacturing industry is the first bridgehead that the IoT must break through. In Guangdong Province, the most developed manufacturing industry in China, thousands of small- and medium-sized enterprises have already connected to the Alibaba Cloud Industrial IoT platform to realize the vertical connection of the machine IoT and the horizontal connection of the industrial chain, and the integration of enterprise orders, procurement, production and storage. The logistics is fully digitalized, the average digital transformation cost has been reduced by 80%, and the production efficiency has increased by 30%.

Gartner pointed out in the report, 'The future of the Internet of Things is driven by vertical market segmentation, and needs to focus on applications in the segmented areas… But most suppliers are still in the rough technology stage, so technology companies need to prioritize the Internet of Things. Only through the in-depth strategy can we seize the opportunity in the future market.'

At present, Alibaba Cloud has launched four major IoT platforms, including cities, industry, life and parks, and provides more than 20,000 IoT solutions and applications with partners, covering complex interactive applications in cities, transportation, industry, life and aviation Scenes.

6.3 AIoT PROJECT CONSTRUCTION PROCESS

Based on the in-depth practice of AIoT in multiple industry customers, we summarized the AIoT mid-station construction process as shown in Figure 6.3. It mainly includes four stages of demand & plan research, plan design, project facilities and service online promotion and operation.

As the AIoT project is very closely related to the equipment system, this chapter will specifically focus on the IoT construction process.

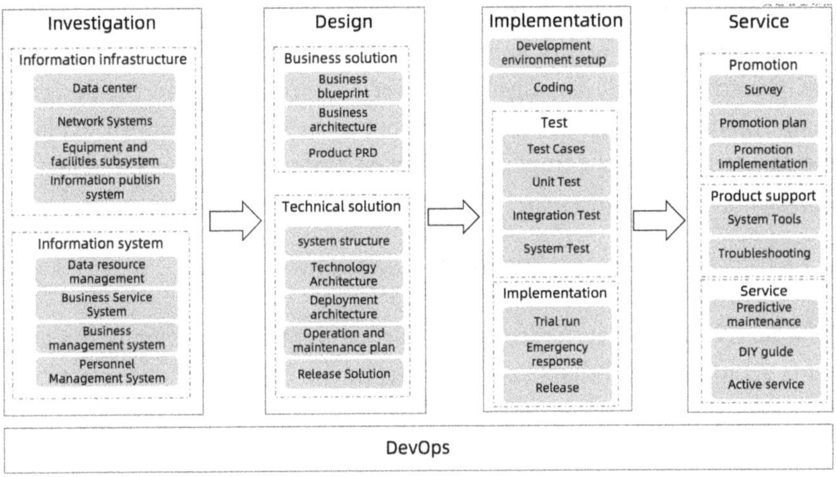

FIGURE 6.3 The construction process of the Internet of Things project.

6.3.1 Status Quo Investigation

First, conduct a demand survey on the relevant business departments of the project construction, and after sorting out and analyzing the demand obtained from the survey, compile the system demand specification. Submit the prepared requirements specifications to the business department for review. After the review is passed, the system requirements specifications will be signed and confirmed as the basis for the next step.

Because the AIoT project involves a wide range of areas, at least automation technology, embedded technology, communication technology, technology and so on. The complexity of the technology further shows that the system investigated in the AIoT project is very complex, and each subsystem adopts its own construction method, the construction party is not unified, and the system construction is relatively fragmented. It mainly involves the investigation of two types of systems. One is the investigation of information infrastructure, which involves the construction of Data Middle Office computer rooms, network systems, equipment and facilities subsystems and information release subsystems; the second is informationization Business & management system research, including data asset management, business service system, management system, personnel management and financial management.

In addition to the control of the software part in the AIoT project, the most complicated part is mainly the control of the hardware part. Among them, due to the particularity of the AIoT project and the construction of equipment and facility subsystems, there is no uniform standard in the industry, and the management is very complicated, and the design is Traditional weak current transformation, so this chapter mainly explains the construction of the implementation part of the equipment and facility subsystems in the infrastructure of information construction.

6.3.1.1 Investigation and Report on the Transformation of Weak Current System

In order to ensure that the scope of subsequent implementation of the project is accurate, controllable and feasible, in the early stage of project construction, on-site investigations and business combing were carried out for the implementation of AIoT projects, and the investigation and evaluation report was formed. The main points of the investigation are as follows:

1. Material collection

    Collect project plans, weak current intelligent system drawings and other drawings to understand the equipment distribution location, number of points and network topology; the project party fills in the survey form to understand the current status of the project's intelligent system (such as equipment brand, model and quantity) and the project Operational data.

    Obtain the contact information of the docking person of the intelligent system manufacturer, communicate and obtain the agreement, interface documents, product introduction documents, understand the software and hardware capabilities of the existing equipment, and facilitate the subsequent assessment of whether the equipment meets the business needs.

2. Site investigation

    On-site verification of equipment brand, model, quantity, location, network topology, modification/connection conditions, etc.

    Communicate with intelligent system users, operation and maintenance personnel and project operation management personnel to understand business needs.

    Evaluate whether the existing equipment hardware equipment meets the business needs, and determine the best solution among several ways of using docking systems, installing sensors, adding points and directly replacing equipment.

Evaluate whether the basic network topology of the existing project meets the transformation needs, and determine whether the wired and wireless solutions are used for network transformation implementation.

Assess project renovation risks, such as equipment replacement and ground excavation in the passage area, water and power cuts, and large equipment entering and exiting the site, which affect the normal operation of the project. The construction needs to be implemented in strict accordance with the construction plan, and the project party shall be notified 3–5 days in advance to notify the enterprise, and the cross-reconstruction shall minimize the impact.

3. Preparation of research report

   Current status interpretation and construction planning: Sort out the functional requirements for the transformation and upgrading of various intelligent systems, and confirm the construction plan for the scene operation requirements of the business departments; sort out the brands, models, existing functions, interface methods, whether to meet the docking needs, project usage, current status of project network topology, etc.

6.3.1.2 Weak Current Transformation and Upgrading Plan

Reconstruction list of weak current equipment, systems, pipelines, etc.: After confirming the business requirements, transformation methods and weak current implementation drawings, output the transformation list according to the engineering quantity, and list the selection specifications, quantities, brands and other information of equipment materials in a subsystem. Materials such as systems and pipelines meet the following requirements:

- Requirements for construction drawings.
- Technical requirements of engineering specifications.
- Requirements of standard engineering specifications.
- The latest requirements of government departments for construction projects.
- Weak current equipment purchase list: After confirming the business requirements, transformation methods and weak current implementation drawings, output the purchase list according to the

engineering quantity, and list the recommended selection specifications, quantities and brands of equipment materials by system.

- Software and hardware interface protocol list: According to the project transformation method, communicate with the manufacturer to obtain the software and hardware interface protocol and manage the docking progress and the protocol list.

Blueprint of weak current reconstruction construction design:

- Clearly embody the weak electronic system and realize its functions by means of system diagrams, software interfaces, function lists, etc.
- Clearly reflect the technical metrics and technical types of key equipment by means of equipment material tables and system diagrams.
- Clearly reflect the number of points, layout and location of weak current wells, control room, common area trunking specifications, reserved holes, buried pipelines, etc. of each subsystem with a plan view.

Workload and budget evaluation:

- Compile a bill of quantities for bidding: Compile a bill of quantities including software and hardware integration in accordance with the program design specifications.
- Detailed requirements for software and hardware docking: Provide detailed docking instructions for software and hardware docking including process framework, development guidelines, development kit demo, development examples and functional requirements.
- Pipeline calculation sheet: According to project planning and drawing design, process standards, accounting for different pipeline consumption and costs.
- Functional budget: Draw up a construction budget according to different construction modules.
- Technical specifications: Prepare technical specifications of different intelligent modules according to business needs and construction planning.

- List of main equipment and materials: According to project planning and drawing design, process standards, sort out the list of main equipment and materials.

Assist in the completion of the bid evaluation work for the weak current renovation:

- Reply to bidding and return bid question paper.
- Review the technical content and evaluation report of bidding.
- Cooperate with technical services in the construction phase of weak current reconstruction.
- Technical clarification during the construction phase.
- Drawing review.
- Suggestions on equipment product technology selection.
- Review of design changes.
- Suggestions on construction technology.
- Cooperate with technical services at the completion stage of weak current renovation.
- Opinions on review of completed drawings.
- Participate in the completion acceptance.
- Facilities risk assessment and improvement suggestions.
- Facility operation and maintenance design and optimization suggestions.

6.3.2 Program Design

Based on the research results of the previous stage of business and information system construction, conduct in-depth analysis of the needs of informatization construction infrastructure upgrades, informatization business system construction upgrade management requirements, and current business system functions, and carry out business architecture, application architecture, and data architecture, Technology architecture design and other content design.

1. AIoT overall architecture design

   The general overall architecture of AIoT is designed in the form of 'equipment+edge gateway+AIoT platform+business platform + application & IOC'. In the overall technology architecture, the edge gateway and AIoT platform are built based on Alibaba Cloud's IoT platform architecture and support a large number of intelligent equipment access; the IOC application layer of the project is based on the mature AIoT intelligent field service center and the microservice architecture model for secondary development. Fully designed and implemented in terms of decoupling, openness, security and high availability of the architecture.

   The technology architecture design is designed based on the principle of decoupling. The edge gateway, IoT platform, business center and application & IOC four-tier core architecture on the project side and each component in the architecture support flexible deployment across networks, and all levels support. Data interaction is carried out through the interface to meet the future continuous optimization, iteration and flexible expansion needs of new business scenarios for each component.

   The edge side adopts Alibaba Cloud's comprehensive edge gateway management and application publishing capabilities, and developers can develop new access devices and subsystems by themselves. With Alibaba Cloud's centralized publishing capabilities for edge applications, it can quickly deploy newly released docking programs and edge applications in the entire system. At the same time, it adopts the complete Ali IoT edge service framework, and has pre-integrated the cloud function between the edge gateway and the IoT platform. Developers only need to complete the driver development work of device docking and the docking of subsystems on the edge side. There is no need to pay attention to the specific details of the cloud on the device, nor to develop the cloud function on the device.

2. System deployment architecture

   It is recommended that the classic AIoT platform deployment architecture is a highly reliable solution deployment architecture of local edge application deployment+cloud SaaS. Cloud software as a service (SaaS) adopts a cluster deployment method to support high availability (service redundancy, mutual backup, no single point of failure), and it is also convenient for future horizontal expansion of various service units.

Each service component in cloud SaaS can be deployed on the same server or on a separate server. The deployment structure is as follows:

- Deploy the connection service and session management service together, with multiple machines for mutual backup.
- Application Programming Interface (API) services, management consoles and business sub-services are also deployed in a multi-machine highly available manner.
- The cache service is also deployed with the API service in a multi-machine high-availability method.
- The database adopts multiple separate machines and is deployed in a cluster.

- For the high-availability deployment solution of edge servers, it is recommended to use the property management all-in-one machine cluster deployment plan (2 or more) to ensure that after a property management all-in-one machine fails, the business is not interrupted or can be restored in a short time.

3. Business Middleware architecture design

   The business middleware adopts microservice solutions to allow system services to have flexible loose coupling capabilities, making development teams easy to develop; through microservice solutions, the system can be service-oriented, interface-oriented and modularized. Let each business system become a microservice, realize the intercommunication between different services, and let the SaaS architecture have good openness, through the way of API to organize core business assets and open them to applications; through the access control of services, Security audit and other functions to ensure service security, improve service levels and expand cooperation space.

   The microservice architecture has good openness, and the intercommunication between different services can be realized through API and HTTP interface calls. With this feature, we can open up the service capabilities of the business center and have a good docking capability with the front end.

4. Technical solutions design

Before implementation, it will conduct integration and business scenario plan design and output summary design report according to the research situation and business scenario requirements; and conduct detailed design of the networking, IP planning, system specifications, interface implementation, etc. of each component in the solution, and output details Design report.

I. The summary design report mainly contains the following contents:

   - Project demand analysis: Project background, scale, objective overview, system overview.
   - Overall system architecture: Platform architecture, functional architecture.
   - Description of the functional scenarios of each business module.
   - Operating environment: Software platform, hardware platform.
   - Interface design.
   - Description of safety, reliability and other solutions.
   - Service descriptions such as training and maintenance.

II. The detailed design report mainly includes the following contents:

   - Hardware configuration/virtual machine resource allocation, physical/logical device networking diagram & table (including local/peer-end IP, port, VLAN and protocol).
   - Main data flow direction (to IP+port level).
   - Firewall policy.
   - The username/password involved.
   - Description of related external systems (such as GIS/NTP/SMS/mail server and IP/port/user name/password/expiration date description of each subsystem).

6.3.3 Delivery Implementation

The entire process of software development will follow a unified coding standard, which mainly clarifies and regulates code naming, code writing format, programming skills, performance, etc. The development server of this project is mainly based on the J2EE technology architecture, so the software development coding should mainly refer to the Java program coding standard. The following will not elaborate on the Java program coding standards.

After the system is developed, testers will test system functions and system performance. During the testing process, corresponding documents such as 'Test Plan' and 'Test Report' will be generated. The 'Test Report' will be submitted to the owner and the owner will. The party shall check and confirm the content of the 'Test Report' to verify whether the submitted product meets the requirements of the test specification, and then formally submit the installation package and corresponding documents to the owner.

1. System development integration

   Complete the development, integration, docking, online and personalized customization of each terminal/subsystem, each business system and each component in the solution (function/UI, etc.)

   - Development and integration of docking protocols between edge nodes and various subsystems

     Complete the docking protocol and data interaction development with various equipment and facility subsystems. Deploy different numbers of edge nodes according to the actual situation of each project. As the subsystems with the same function have different manufacturers and versions in different projects, some pilot projects will conduct specific investigations to determine the transformation plan after the contract is signed, complete the relevant docking development according to the transformation plan, and obtain relevant data through the interface. At the same time, complete the network connection with each project subsystem, docking and commissioning, trial operation and business launch.

   - Development and integration of edge nodes and AIoT platform

     Refer to the standard interface or SDK provided by the AIoT platform to complete the docking protocol and data interaction development between the edge node and the AIoT platform. Complete the network connection, docking and commissioning, trial operation and business launch of the project edge node and the AIoT platform.

- AIoT platform integration

  Complete the network connection, docking and commissioning, trial operation and business launch of the AIoT platform and the edge nodes of each project, and realize the registration, access and management of edge nodes and subsystems on the AIoT platform. Complete the network connection, docking and commissioning, trial operation and business launch between the IoT platform and the project, and realize the uplink and downlink data interaction between the IoT platform and the project.

2. System trial operation

   The trial operation of the application system is actually a continuation of the test, checking the stability and applicability of the system. The conversion of the new and old systems requires managers to enter the new system from the traditional management system. How to enable users of different levels to adapt to the new system as soon as possible is a problem to be solved in trial operation management. At the same time, changes in the internal and external environment, shortcomings in operation, in order for the project to perform its due role in an expected period of time, in a good operating state, there must be a unified trial operation management.

- Before the trial operation of the project, the project team shall formulate a trial operation plan, user manual and system operation and maintenance manual, as well as necessary training materials, and submit them to the owner and business users for approval.

- During the trial operation, special personnel will be provided to escort the smooth implementation of the system trial operation. The system problems exposed during the trial operation need to be recorded by a dedicated person, and the recorded problems should be assigned to the corresponding developers for tracking and modification, and the recorded system problems and modifications will be reported to the owner every day.

After the trial operation, the project team is responsible for improving the user manual, operation and maintenance manual and other materials.

3. Formal operation and acceptance of the system

   - The system is officially running
     The system enters the system operation and maintenance period after the system is officially operated. The work of the system operation and maintenance period mainly includes: daily system monitoring and maintenance, application system upgrade and update, fault recovery, fault analysis and processing and system backup.

   - System acceptance
     According to the requirements of the business side, prepare the acceptance materials, submit the acceptance application and the business user will organize the acceptance work.
     In addition to the requirements of the business user, the acceptance of the application system should also comply with national laws, administrative regulations and relevant mandatory standards and industry standards. On this basis, the acceptance materials are prepared and the acceptance application is submitted to the business party, who will organize the acceptance work.

6.3.4 Service Operation

6.3.4.1 Survey and Research

In order to ensure that the scope of project implementation is accurate, controllable and feasible, Alibaba Cloud AIoT conducts project site survey and research before the system goes online.

1. Research on project basic equipment information:

   - Basic project information: Construction area, number of buildings, delivery period, etc.

   - Project management information: Number of property engineers, frequency of inspections, operation and maintenance overview.

   - Collection of project materials: Floor plans, weak current diagrams, on-site photos.

- Project transformation conditions: Whether water and electricity can be cut off, and whether it supports replacement.
- The degree of project informatization, etc.

2. Research on project equipment information:

- Equipment categories: Power supply and distribution, water supply and drainage, fire fighting, elevators, etc.
- Equipment information: Geographic location, reconstruction object, reconstruction quantity.
- Network overview, node information, bandwidth, system network topology.

6.3.4.2 Operation and Maintenance System

Alibaba Cloud AIoT provides owners with enterprise-level operation and maintenance service guarantee support and training, including platform software release, changes, service status monitoring, service disaster tolerance and data backup, etc., as well as routine service inspections and fault emergency handling.

1. Design review

   In the product development and design review stages, the relevant mechanisms from the perspective of operation and maintenance are fully considered to make the service meet the high-availability requirements of operation and maintenance access.

2. Service management

   Responsible for formulating online business upgrade changes and rollback plans, and implementing changes. Grasp the services in charge, the relationship between the services and the resources on which the services depend. Be able to find defects in the service, report them in time and promote their solutions. Formulate service stability metrics and access standards, while constantly improving and optimizing the functions and efficiency of procedures and systems, and improving the quality of operation. Improve monitoring content and improve alarm accuracy. When the online service fails, respond as soon as possible, report the known online failure according to the process and execute it according to the plan, and organize the related personnel to jointly troubleshoot the unknown failure.

3. Resource management

   Manage the server assets of each service, sort out server resource status, data center distribution, network dedicated lines and bandwidth, use server resources reasonably and allocate servers with different configurations according to the needs of different services to ensure full utilization of server resources.

4. Routine inspection

   Develop routine service inspection points and continue to improve them. According to the established service checkpoints, the service is regularly checked. The problems found in the investigation process shall be tracked in time to eliminate possible hidden dangers.

5. Plan management

   Develop routine service inspection points and continue to improve them. According to the established service checkpoints, the service is regularly checked. The problems found during the investigation shall be investigated in time to eliminate possible hidden dangers. Develop routine service inspection points and continue to improve them. According to the established service checkpoints, the service is regularly checked. The problems found during the investigation shall be investigated in time to eliminate possible hidden dangers.

6. Data backup

   Develop routine service inspection points and continue to improve them. According to the established service checkpoints, the service is regularly checked. The problems found in the investigation process shall be investigated in time to eliminate possible hidden dangers.

6.3.4.3 Edge Computing Operation and Maintenance

1. Connection diagnosis

   When Link IoT Edge cannot connect to the cloud, you can use the lectl tool and use the diagnose command to make a preliminary diagnosis.

2. Management log

   You can use the lectl tool to package logs or temporarily change the behavior of the log service to facilitate uploading of logs or temporary troubleshooting when problems occur. You can package logs and manage log services through commands.

3. Management configuration

   Configuration management mainly implements operations such as obtaining configuration, device configuration, removing configuration, obtaining matching keys and forced writing through the config subcommand.

4. Management equipment

   The device-related information of Link IoT Edge can be managed through the device subcommand. The current device status information can be viewed through the lectl device show command.

5. Management function calculation

   The function calculation on Link IoT Edge can be managed through lectl. Including display information, deploy function, remove function, reset deployment, call function and restart function.

6. Management flow data analysis

   You can start streaming data tasks, stop streaming data tasks and verify SQL.

7. Remote operation and maintenance

   Remote service access includes remote connection (SSH protocol), remote file management (SFTP protocol) and remote access to other network services based on TCP protocol.

6.3.4.4 Training

In order to ensure that the project equipment can be installed and operated normally, the business system can be launched normally, the project results can be better understood and used by the project property management personnel, the operation difficulty and the lowest operating cost after the project are completed and put into production need to provide corresponding technical training. To train relevant personnel and units.

1. The training objects are mainly related technical personnel and business personnel of the project property management party.
2. The training content provided should include but not limited to the following:

   - Phased results of the project consistent with the project implementation schedule.

- The project includes basic knowledge, working principles, performance technical metrics, system parameter configuration and management and maintenance methods of equipment, materials, products or systems, and requires the original training of equipment manufacturers.
- Training on the use of business systems, such as parking lot management and personnel traffic management business systems. In the field of security applications, it provides training on the setting of security alarm business rules and work order handling.

6.3.5 Delivery Value of AIoT

The industry AIoT platform is also facing many challenges while welcoming huge opportunities. The market demand is constantly changing. This requires AIoT to quickly provide the solutions required by the project. The development of a complete IoT solution may require products in Constant polishing in the project. With the vigorous development of the AIoT market in recent years, the Ali AIoT platform has solved the following problems for industry customers:

1. Decoupling applications and hardware

   Because of the diversity and complexity of connection protocols and control systems, the problem of difficult device access is reflected in many sub-sectors. AIoT realizes a software–hardware separation architecture through a unified access standard, and successfully realizes the application decoupling from hardware devices.

2. Unified integration of decentralized self-made subsystems

   Before the large-scale market promotion of AIoT projects, there are already many third-party systems in the market, such as parking subsystems, access control subsystems, EBA subsystems and security subsystems in the real estate and park industries. What AIoT platform manufacturers can do is to connect with the original system. The degree of openness of the original system to the AIoT platform will directly affect the platform's security, response speed and other metrics. If customers want to overthrow the old and adopt the idea of removing subsystems to build a new platform, customers need great determination; Alibaba Cloud AIoT platform helps customers to manage each subsystem in a unified manner through powerful system integration capabilities, assists customers in completing the construction goals of unified management and control of decentralized subsystems in the business, and improves the efficiency of system management and control.

3. Achieved the standardization of AIoT products

   Because most of the customer's IoT requirements are different, and a set of software cannot adapt to all requirements, most platform vendors have to face many different levels of customization and modification, which is very difficult for some platform companies that are not well-standardized. High pressure will lead to higher costs, longer cycles and even affect customer satisfaction. Alibaba Cloud AIoT continues to accumulate services in the industry and at the same time establish equipment access standards through layered architecture design, and implement it through the object model mechanism. The standardization of equipment management functions and services, and the standardization of connection protocols simplify the difficulty of equipment access diversity and complexity.

4. Improved product competitiveness

   It is very difficult for the IoT technology to move from the laboratory to the product and from the product to the market. At present, most companies are stuck in pushing technology to the product stage, but they are not going very smoothly on the road to marketization. The reasons are for example, platform companies' insufficient understanding of the industry, insufficient platform neutrality and lack of benchmark cases. Wait. Alibaba Cloud AIoT co-creates scenarios, products and solutions with customers in vertical industries such as smart parks, smart cultural tourism and smart communities, and greatly enhances the market-oriented competitiveness of AIoT products by creating benchmarks in the AIoT segment.

5. Establish industry standards

   The standards of different customers in different industries are different. For example, energy saving and emission reduction, cost reduction and efficiency increase, and how much can be reduced to make money, so only energy industry companies propose to develop standards to measure product value and promote customers' understanding of platform value and awareness. Through the implementation of the project, Alibaba Cloud and customers jointly create industry standards and improve the digital construction goals of the industry's operation management and control.

In the project delivery and implementation, Alibaba Cloud AIoT starts from the top-level design architecture according to the characteristics of the IoT project, plans the access requirements of various intelligent facilities and systems as a whole and integrates and connects to the AIoT platform, thereby realizing equipment data standardization and service model standardization, and can develop various standardized applications based on the AIoT platform, form a rich AIoT service ecosystem and realize the effective landing of AIoT digital applications.

Alibaba Cloud AIoT relies on its own technical strength and the co-creation of customers in the industry, and has designed the following reference principles for the intelligent engineering implemented by AIoT projects to share with the industry to promote the healthy development of the AIoT market:

1. Unified architecture: In combination with the construction planning of the building space, the intelligent engineering infrastructure construction is included in the overall design category, and the overall planning is carried out based on the principles of 'integrated planning and integration, high-quality construction', and 'uniform equipment standard and standardized access'. Relying on the standard protocol supported by the AIoT platform, it integrates intelligent facilities such as security monitoring, video intercom, personnel passage, vehicle passage and access control management to quickly complete the infrastructure construction of the digital part of the AIoT project.

2. Normative: According to the relevant software/hardware used by the intelligent hardware equipment used in the AIoT project, it needs to comply with relevant industry standards, covering mainstream IoT architecture, chip and device types, adopting mainstream framework design systems, and using mainstream database development.

3. Security: Based on the smart devices and system services used in the AIoT project, a comprehensive multi-level security mechanism is designed, including device identity authentication, security operation center, trusted execution environment and trusted service management.

4. Scalability: In the design and selection of system software and hardware, the scalability is fully considered. The system structure is easy to expand and supports unified operation and maintenance to adapt to the larger task load that may appear in the future. The hardware platform is upgradeable, shielding the brand differences of similar equipment to ensure a more stable operation and maintenance environment and consistency of user experience, as well as lower operation and maintenance costs.

6.4 ALIBABA CLOUD AIoT MIDDLEWARE

A. **Alibaba Cloud AIoT development history**

From the Internet of Everything to the Intelligent Connection of Everything, it is inseparable from 'ubiquitous computing'. Alibaba Cloud IoT has deployed collaborative computing on the cloud side. Based on the IoT enabling platform, joint developers, chip module manufacturers and industry partners have completed in-depth layouts in the four major areas of city, life, automobile and manufacturing. The '1234' strategy of one cloud, two terminals, three types of partners and four major areas. Its development process is as follows:

In 2014, the Smart Life Division was established and the IoT began to be deployed.

In April 2017, Alibaba Cloud IoT Division announced the establishment.

In June 2017, Alibaba Cloud and more than 200 IoT industry chain companies established the IoT Standard Alliance – ICA Alliance to promote national and international standards.

On October 12, 2017, Alibaba Cloud officially released the Alibaba Cloud Link IoT platform at the Hangzhou Yunqi Conference, providing three major infrastructures including the IoT cloud integration enabling platform, the IoT market and the ICA Global Standards Alliance. Intelligent networking in the three major areas of life, industry and cities.

On January 9–12, 2018, Alibaba Cloud IoT appeared at the CES show in the United States, showing shared bicycles and AliOS-based module chips.

On March 28, 2018, Alibaba Cloud announced its strategic investment in edge computing and launched its first edge computing product Link Edge.

On March 28, 2018, Alibaba announced its full entry into IoT, which is the new main track after cloud computing.

In September 2019, the new AIoT operating system AliOS Things 3.0, AIoT activation center, and urban AIoT platform 2.0 were launched.

At the Yunqi Conference·Shenzhen Summit held in 2018, Hu Xiaoming, Senior Vice President of Alibaba Group and then President of Alibaba Cloud, officially announced that Alibaba will fully enter the field of IoT, positioning itself as a builder of IoT infrastructure and providing IoT Connect platforms, AI capabilities, and realize cloud-side-end integrated collaborative computing, and output to society. In 2019, Alibaba Cloud announced the upgrade to Alibaba Cloud Intelligence, taking 'cloudification of IT technical facilities, Internetization of core technologies, and digitalization and intelligence of applications' as its new strategy, and plans to make all Alibaba technologies available through Alibaba Cloud. Open output will help lower the threshold for digital transformation in all walks of life. Alibaba Cloud itself will insist on being 'integrated' and do its best at cloud and data intelligence technology links, making it a part of partner technology. The industry also focuses on new retail, Digital government and new finance. On the whole, Ali's AIoT is based on IoT, AI is ubiquitously reflected in various solutions and AIoT-related data intelligence technology capabilities can be empowered to various industries through partners.

B. **Alibaba Cloud AIoT development goals**

From the development history of Alibaba Cloud IoT, it can be seen that as early as 2008, Alibaba formulated a 'cloud computing' and 'big data' strategy, established a cloud service platform based on Feitian operating system, and used it to grasp. With the wave of cloud computing development, it has become the largest cloud computing vendor in China and even ranks among the best in the world. The success of Alibaba's cloud computing business also laid the foundation for the future development of Alibaba's IoT. In April 2017, Alibaba Cloud IoT Division announced the establishment. And set a clear vision and development goals for himself, as shown in Figure 6.4.

Although many companies have put forward their own IoT strategies, due to their own positioning and different industries, when

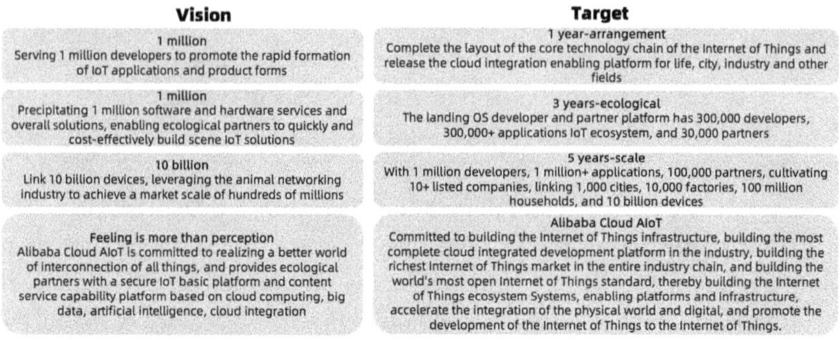

Vision
Target
1 million
Serving 1 million developers to promote the rapid formation of IoT applications and product forms
1 year-arrangement
Complete the layout of the core technology chain of the Internet of Things and release the cloud integration enabling platform for life, city, industry and other fields
1 million
Precipitating 1 million software and hardware services and overall solutions, enabling ecological partners to quickly and cost-effectively build scene IoT solutions
3 years-ecological
The landing OS developer and partner platform has 300,000 developers, 300,000+ applications IoT ecosystem, and 30,000 partners
10 billion
Link 10 billion devices, leveraging the animal networking industry to achieve a market scale of hundreds of millions
5 years-scale
With 1 million developers, 1 million+ applications, 100,000 partners, cultivating 10+ listed companies, linking 1,000 cities, 10,000 factories, 100 million households, and 10 billion devices
Feeling is more than perception
Alibaba Cloud AIoT is committed to realizing a better world of interconnection of all things, and provides ecological partners with a secure IoT basic platform and content service capability platform based on cloud computing, big data, artificial intelligence, cloud integration
Alibaba Cloud AIoT
Committed to building the Internet of Things infrastructure, building the most complete cloud integrated development platform in the industry, building the richest Internet of Things market in the entire industry chain, and building the world's most open Internet of Things standard, thereby building the Internet of Things ecosystem Systems, enabling platforms and infrastructure, accelerate the integration of the physical world and digital, and promote the development of the Internet of Things to the Internet of Things.

FIGURE 6.4 Alibaba Cloud AIoT's vision and goals.

looking at the IoT, the solutions and measures implemented are not the same, and it can even be said to be varied.

In Alibaba's view, the past 20 years are the era of the 'Internet of People', and the next 20 years are the era of the 'Internet of Things'. The former is to digitize human activities through e-commerce, social networking, cultural and entertainment activities, which gave birth to today's booming Internet market. The latter is to digitize the entire physical world, from forests, rivers to roads, cars, everything, all-encompassing, all-inclusive, the so-called 'interconnection of all things', the ultimate goal is to achieve 'things' and 'things'. The interaction between 'things' and 'people' will have a far greater impact than the 'Internet of People'.

In this process, Alibaba Cloud IoT has also established a very clear position for itself, that is, Alibaba Cloud IoT will be the builder of the IoT infrastructure. Its core value lies in solving three major problems:

First, provide an open and inclusive IoT connection platform;

Second, provide powerful AI capabilities;

Third, realize the integrated collaborative computing of cloud and end.

Take the construction of smart cities as an example. In order to achieve this goal, in October 2017, Alibaba Cloud launched the Alibaba Cloud City Link City IoT platform, providing a set of cloud, edge and end smart city solutions by building local IoT. The platform accelerates the access of urban components, unifies data standards and improves urban governance efficiency, people's livelihood experience and comprehensive coordination. Corresponding to the construction of the IoT in different cities, different platforms have been customized and given new names. For example, the IoT platform supporting Wuxi Hongshan Town is named Feifeng Platform.

C. **Alibaba Cloud AIoT product matrix**

The IoT is a very fragmented market. It is difficult to make the industry through only a few manufacturers or industries. If we want to turn value into an application model, we need more partners and companies to truly realize the ecological value. In doing so, each link in the ecological chain can truly find its own positioning and commercial value in the IoT market. Alibaba has built an AIoT product system and delivery system through deepening industry exploration and ability precipitation; Alibaba Cloud AIoT is a unified open platform for ecological partners based on basic IoT platform capabilities, industry application capabilities and ecological application capabilities, to work with ecological partners to create a safer, smarter and more open IoT application ecosystem.

Alibaba AIoT builds IoT infrastructure from all levels of 'end-side-management-cloud'. The product matrix often includes IoT operating systems, IoT communication IoT Hub, edge computing platforms, IoT development platforms, security services and IoT. Contents such as the networked market constantly improve the ecosystem, and realize the intelligent integration of people, things and clouds in the digital world (Figure 6.5).

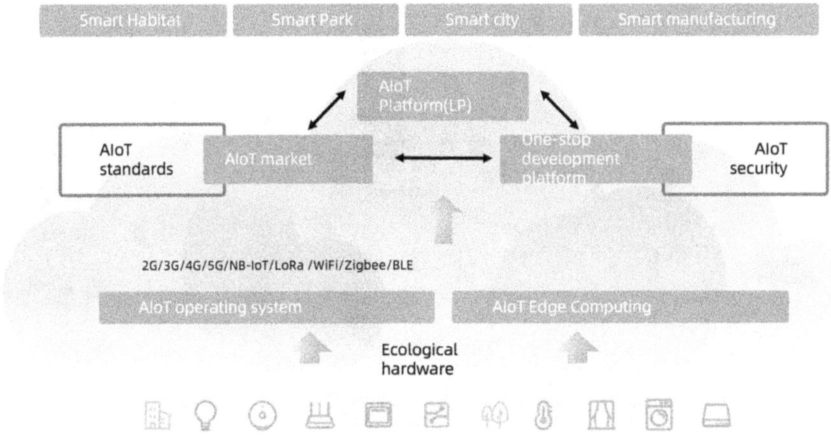

FIGURE 6.5 Alibaba Cloud AIoT product matrix.

6.4.1 One-Stop Development Platform IoT Studio

IoT Studio (formerly Link Develop) is a productivity tool provided by Alibaba Cloud for IoT scenarios and is part of the Alibaba Cloud IoT platform. It can cover various core application scenarios of the IoT industry, help you complete the development of equipment, services and applications efficiently and economically, and accelerate the construction of IoT SaaS. The IoT development service provides a series of convenient IoT development tools such as mobile visualization development, Web visualization development, service development and device development, and solves the problem of long development links, complex technology stacks, high coordination costs, and difficulty in program transplantation in the field of IoT development. The problem of redefining IoT application development.

6.4.1.1 Features of IoT Studio

Based on the completion of the device-side development, it can connect to the basic IoT services, and then use the service development, web visual development and mobile development capabilities provided by IoT Studio to develop IoT applications cost-effectively (Figure 6.6).

1. Visualized construction

   IoT Studio provides visual construction capabilities. You can quickly complete the development of web pages, mobile applications and API services related to device data monitoring through drag and drop and configuration operations. You can focus on your core

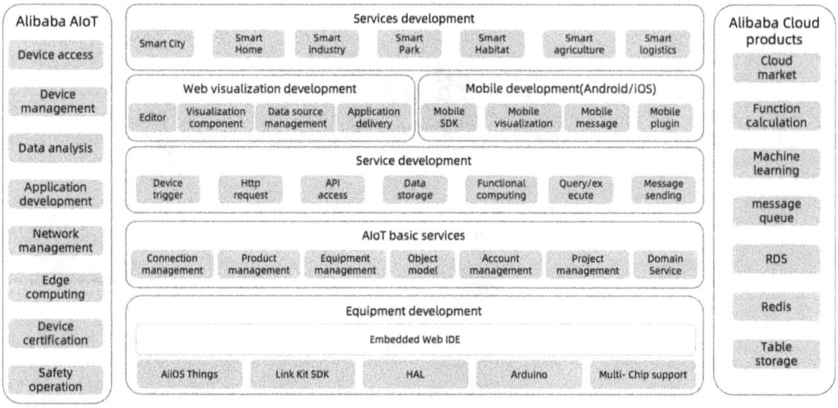

FIGURE 6.6 IoT Studio architecture diagram.

business, get away from the tedious details of traditional development and effectively improve development efficiency.

2. Seamless integration with equipment management

   Device-related attributes, services, events and other data can be directly obtained from the IoT platform device access and management module. IoT Studio and the IoT platform are seamlessly connected, which greatly reduces the workload of IoT development.

3. Abundant development resources

   Each development platform has a large number of components and rich APIs. As the product is iteratively upgraded, the component library will become more abundant, and IoT Studio will help you improve development efficiency.

4. No deployment required

   Using IoT Studio, after the application service is developed, it is directly hosted in the cloud and supports direct preview and use. It can be delivered and used without deployment, eliminating the trouble of purchasing additional servers and other products.

6.4.1.2 Advantages of IoT Studio

IoT application development is the only way for IoT enterprises to go to the cloud, but it is very difficult to achieve this because there are many pain points in IoT application development:

1. High cost: IoT application development is usually not the core business of equipment manufacturers or solution providers, but with the gradual increase in software labor and outsourcing costs, this fee increases year by year, and even the cost of IoT application development of some enterprises is relatively low. The profit is equal.
2. Link length: IoT application development involves embedded development, protocol integration, data development, service development, front-end development and other links. To build a complete IoT application requires multiple roles to cooperate with each other and experience 23 links can be completed.
3. Strong customization: What's more terrifying is that most IoT applications need to be customized. For example, in the construction

industry, each building is different, and each factory is different, and on-demand construction can only be implemented in a project manner.

The following is a comparison of traditional development and IoT Studio-based development (Table 6.1):

6.4.2 AIoT Cloud Platform

Alibaba Cloud IoT Platform is an open platform dedicated to IoT and intelligent hardware cloud services. The platform relies on Alibaba Cloud's powerful platform and technical support to provide industry users with network access, data storage, real-time data processing and display, as well as application development and application operation services for industry users.

The platform minimizes the technical threshold of enterprise smart device networking through foolish tools, continuously enhanced SDK and API service capabilities, reduces R&D costs, improves product production speed and helps companies build their own IoT industry applications and better connections, Serve the ultimate consumer.

IoT IoT cloud+terminal integrated management platform, realizes platform management functions such as CMP (connection management), DMP (device management platform), AEP (application enable platform), BAP (business analysis platform), and realizes AI-enabled interaction All things are connected.

Based on an instantiated deployment architecture, the Alibaba IoT platform provides device access, device management, monitoring operation and maintenance, data flow, data management, processing and analysis and other IoT building capabilities, helping companies to have more complete production materials during digital transformation. In the IoT platform, every enterprise needs to help the digital upgrade of the industry (Figure 6.7).

6.4.2.1 IoT Hub

IoT Hub helps devices connect to Alibaba Cloud IoT platform services and is a data channel for secure communication between devices and the cloud. IoT Hub supports PUB/SUB (publish/subscribe) and RRPC (Reverse Remote Procedure Call) two communication methods, where PUB/SUB is message routing based on Topic. IoT Hub has the following features:

TABLE 6.1 Comparison of Traditional Development and IoT Studio-Based Development

	Traditional Development	Based on IoT Studio
One-stop shop	Traditional development requires the cooperation of multiple roles such as embedded development, protocol development, data development, back-end development and front-end development.	Provide a variety of development tools to complete the complete IoT SaaS construction on one platform, including server/Web application/H5 mobile development, etc.
Low cost	You need to purchase servers, CDNs, databases and upload resources for each deployment cycle, which may cause errors.	Provides a fully managed application service, available immediately after development is completed, no need to purchase servers, databases or deployment services.
Easy to customize	It is necessary to purchase servers to build a load-balanced distributed architecture, and it takes a lot of manpower and material resources to develop a complete set of 'access+computing+storage' IoT systems.	By visually building panels/services, every developer can become a full-stack engineer of the Internet of Things. It is as simple as using PPT to complete the visual application in 10 minutes.
Performance	It is extremely difficult to implement the scalability architecture by itself, and it is extremely difficult to schedule the server, load balancing and other infrastructure from the device granularity.	It has the long-term connection capability of hundreds of millions of devices and the ability of millions of concurrency, and the architecture supports horizontal expansion.
Stability	You need to discover the downtime and complete the migration by yourself, and the service will be interrupted during the migration. Stability cannot be guaranteed.	Service availability is as high as 99.9999%. Decentralization, no single point of dependence. Has multiple data center support.
Safety	Additional development and deployment of various security measures are required, and ensuring the security of device data is a great challenge.	Provide multiple protections to ensure device cloud security: 1. Equipment certification guarantees equipment safety and uniqueness. 2. Transmission encryption to ensure that data are not tampered with. 3. Cloud shield escort and permission verification to ensure cloud security.

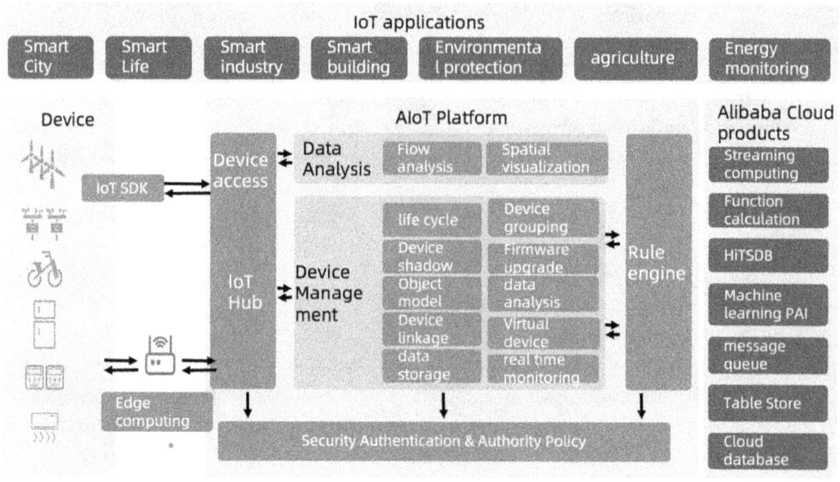

FIGURE 6.7 Architecture diagram of the Internet of Things Platform (LP).

- High-performance expansion: Supports linear dynamic expansion, which can support the simultaneous connection of 1 billion devices.
- Full-link encryption: The entire communication link is encrypted with RSA algorithm and AES (Advanced Encryption Standard) to ensure the security of data transmission.
- Messages arrive in real time: After the device and the IoT Hub successfully establish a data channel, a long connection will be maintained between the two to reduce handshake time and ensure that messages arrive in real time.
- Support for data transparent transmission: IoT Hub supports the transmission of data to its own server in the form of binary transparent transmission, without saving device data, thus ensuring the security and control of data.
- Support multiple communication modes: IoT Hub supports RRPC and PUB/SUB communication modes to meet your needs in different scenarios.
- Support multiple device access protocols: Support devices to use CoAP, MQTT and HTTPS protocols to access the IoT platform.
- In addition to supporting a single device access method, Alibaba Cloud IoT hub also supports the access of gateway devices.

6.4.2.2 *Equipment Connection*

The Connect Management Platform (CMP) is one of the most basic platforms of the IoT platform. It is mainly responsible for supporting the secure and efficient interconnection between massive and multi-protocol devices, as well as between the cloud and the devices. At present, the connection management platform supports a wealth of access protocols, such as MQTT, CoAP, HTTP, HTTP2, LWM2M and other protocols. It also supports network standards such as LoRa, NB-IoT, and provides device authentication, connection management, message routing and protocol analysis. And other basic capabilities and rich expansion capabilities.

First, Equipment certification

1. Basic certification

   The IoT platform uses ProductKey to identify products, Device Name to identify devices and device certificates (ProductKey, DeviceName and DeviceSecret) to verify device legitimacy. Before the device is connected to the IoT platform through the protocol, it needs to report product and device information according to different authentication methods. After the authentication is passed, it can be connected to the IoT platform. For different use environments, the IoT platform provides multiple authentication schemes.

   The IoT platform currently provides three authentication schemes, namely:

   - One device, one secret: Each device burns its own device certificate (ProductKey, DeviceName and DeviceSecret).
   - One type and one secret: Burn the same product certificate (ProductKey and ProductSecret) under the same product.
   - Sub-device dynamic registration: After the gateway connects to the cloud, the sub-device obtains the DeviceSecret through dynamic registration.

   The three solutions have their own advantages in ease of use and safety. You can choose flexibly according to the required safety level of the equipment and the actual production line conditions.

2. Pre-certification

   Pre-authentication means that the device is first authenticated from the unified center, and then connected after obtaining the access address assigned by the cloud. The typical scenario is that the device is distributed in various places after shipment, which can realize functions such as access routing and cross-region nearby connections.

Solutions for other functions of device connection are as follows:

- Device connection: The IoT platform supports the connection of a large number of devices, and the device and the IoT platform can conduct stable and reliable two-way communication.
- Device access: Provide device-side SDK, drivers, software packages, etc. to help different devices and gateways efficiently access the IoT platform. Open source code for devices on multiple platforms provide cross-platform migration guidance, and empower enterprises to access devices based on multiple platforms.
- Device transmission: asymmetric mode: The data transmission channel between TLS, DTLS devices and the IoT platform is asymmetrically encrypted to ensure that the data cannot be tampered with; MQTT-TCP non-encrypted mode: The traditional data channel between the device and the cloud supports MQTT-TCP connection.
- Data routing: The device communicates with the IoT platform and needs to communicate with the specified Topic. The platform regulates the Topic to facilitate management; supports cross-account authorization for devices to realize the transfer of device permissions; provides APIs to configure Topic routing relationships to realize M2M scenarios.
- Multi-protocol support: Support devices to connect to the cloud through OPC-UA, Modbus, BACnet, SNMP and other protocols; it not only meets the real-time requirements of long connections, but also meets the low power consumption requirements of short connections.
- Two-way communication: Two-way communication between the device and the IoT platform, supporting device reporting data, supporting cloud control devices; the platform can broadcast communication to a batch of devices; MQTT (Pub/Sub) asynchronous communication model, supporting devices reporting

data; MQTT (RPC/RRPC) synchronous communication model, suitable for business scenarios that need to wait for a reply.

- Network heterogeneous: Provide 2G/3G/4G/5G, NB-IoT, WiFi, LAN and other different network equipment access capabilities to solve the needs of enterprise heterogeneous network equipment access management.

6.4.2.3 Equipment Management

Relying on the massive device connection management function of the IoT platform, the edge devices and services in the edge devices are encapsulated into edge instances, and edge instances are used to carry edge computing functions. In the edge instances, sub-device configuration, management and device attributes are allowed.

The IoT platform provides complete device life cycle management functions, supporting device registration, function definition, data analysis, online debugging, remote configuration, firmware upgrade, remote maintenance, real-time monitoring, group management, device deletion and other functions. Provide device models to simplify application development.

1. Equipment grouping

   The IoT platform provides device grouping functions, which can manage devices across products through device grouping, among which:

   - A group can contain up to 100 first-level sub-groups.
   - Grouping only supports three levels of nesting, namely grouping> sub-grouping>sub-subgrouping.
   - A sub-group can only belong to one parent group.
   - The nested relationship of the group cannot be modified after being created, but can only be deleted and recreated.
   - When there are sub-groups under the group, the group cannot be deleted directly. The parent group can be deleted only after all the child groups are deleted.
   - When searching for groups, support fuzzy search for group names, including searching in group list and sub-group list.

2. Device shadow

The IoT platform provides a device shadow function for caching device status. When the device is online, you can directly obtain cloud instructions; after the device is offline, you can actively pull cloud instructions when you go online again.

The device shadow is a JSON document used to store the reported status of the device and the expected status of the application.

Each device has one and only one device shadow. The device can obtain and set the device shadow through MQTT to synchronize the status. The synchronization can be shadow synchronization to the device or device synchronization to the shadow.

3. Object model

Object model refers to the digitized entity in the physical space and the data model of the entity constructed in the cloud to describe the function of the entity.

Thing Specification Language is a file in JSON format. It is an entity in the physical space, such as the digital representation of sensors, electrical lighting, and large and small home appliances in the cloud. It describes what the entity is, what it can do and what information it can provide from the three dimensions of attributes, services and events. The definition of these three dimensions of the object model completes the definition of product functions.

The object model divides the product function types into three categories: attributes, services and events.

4. Data analysis

The standard data format defined by the IoT platform is Alink JSON. However, devices with low configuration and limited resources or requirements for network traffic are not suitable for directly constructing JSON data to communicate with the IoT platform, and can transparently transmit the original data to the IoT platform. The IoT platform provides a data analysis function, which can convert data between the device-defined format and JSON format according to the submitted script.

Currently supports the analysis of two types of data:

- Custom Topic uplink data, that is, the custom format data payload that the device reports to the cloud through the custom Topic parses into JSON format.

- Topic data of uplink and downlink object models, that is, the custom format object model data reported by the device to the cloud are parsed into Alink JSON format, and the Alink JSON format data sent by the cloud are parsed into a device-defined format.

5. Data storage

   The IoT platform supports devices to upload files to the IoT platform server for storage via HTTP/2 streaming channel. After the device uploads files, you can perform management operations such as downloading and deleting on the IoT platform console.

6. Online debugging

   The IoT platform provides virtual device functions. The virtual device simulates the real device to establish a connection with the IoT platform, and reports simulated attributes and event data; the simulated cloud pushes attribute settings and service call instructions to the virtual device. Users can complete application development and debugging based on the data of the virtual device.

   Use restrictions:

- The minimum time interval for continuous push is 1 second.
- Push up to 1,000 messages continuously.
- The push button can be used to push debugging information up to 100 times a day.
- If the data format of the device is transparent transmission/customization, input the Base64-encoded string of binary data, the length does not exceed 4,096 characters.
- When the real device is online or the device is disabled, the virtual device cannot be started. After the real device goes online, the virtual device will automatically go offline.

7. OTA upgrade

   The IoT platform provides OTA upgrade and management services. First, ensure that the device supports OTA service, then upload a new upgrade package on the console, and push the OTA upgrade message to the device, and the device can be upgraded online.

8. Remote configuration

   Using the remote configuration function, you can remotely update the device's system parameters, network parameters and other configuration information online without restarting the device or interrupting the operation of the device.

6.4.2.4 Rule Engine

The rule engine provides data flow and scene linkage functions. Simple rules can be configured to seamlessly transfer device data to other devices to realize device linkage; or transfer to other cloud products to obtain more services such as storage and computing.

- Support forwarding the data in the platform to another topic, realize message flow between devices, interconnection and intercommunication, configure rules to realize communication between devices and quickly realize M2M scenarios.
- Support forwarding data to message service and message queue to ensure the stability and reliability of application device data.
- Support forwarding data to table storage, and provide a joint solution of device data collection + structured storage.
- Support forwarding data to cloud database, and provide a joint solution of device data collection + relational database storage.
- Support forwarding data to the data collector, providing a joint solution of equipment data collection + big data calculation.
- Support forwarding data to time-series spatio-temporal database and provide a joint solution of equipment data collection + time-series data storage.
- Support forwarding data to function calculation, and provide a joint solution of device data collection + event calculation.

6.4.2.5 Data Analysis Management

IoT data analysis provides equipment intelligent analysis services, and the entire link covers equipment data generation, management (storage), cleaning, analysis and visualization. IoT data analysis mainly includes four parts: data collection, data cleaning, data analysis and data display.

After the device generates data, the data management module in the IoT data analysis stores IoT device data and management business data, and supports cross-domain analysis of data to clean the data. After data cleaning, it can be further analyzed through the data development module to quickly build development tasks to process equipment data. Finally, the data analysis results can be displayed in the spatial visualization module, or the streaming data analysis module can be used to transfer the data flow to other cloud products or edge display.

1. Data analysis

   The IoT platform provides data analysis functions for the IoT, supports the development of the IoT platform and provides intelligent analysis services for equipment. The full link covers equipment data generation, management (storage), cleaning, analysis and visualization, which can effectively reduce the data analysis threshold. Provide basic support for IoT application development.

   The IoT platform supports real-time data analysis, provides tools for real-time analysis of streaming data on the cloud and can build custom real-time data analysis and calculation tasks. Real-time data analysis tasks include component orchestration tasks and SQL tasks to process, filter or divert data.

   Component orchestration tasks: By dragging and dropping components, the relationship between device data input and output is established, and data can be filtered by setting attributes, and data can also be split.

   Basic components: Device input, data source output, data filtering, aggregation calculation and dimension table association can be used to configure basic data cleaning and data aggregation flow calculation rules.

   Advanced components: Anomaly detection, support device data anomaly detection.

   SQL type tasks: Write business logic based on StreamSQL, define a variety of data processing functions and operators in SQL, save them as tasks and process data.

2. Data management

   The data analysis service of the IoT platform supports one-click configuration of IoT device data storage and business data management, and supports cross-domain analysis of IoT device data and business data.

IoT data storage: Users can create and formulate data storage cycles for subsequent data analysis.

Other data sources: Support the use of other data source functions (original data source configuration), configure database accounts and analyze the correlation between business data on RDS and operating data generated by the IoT, for use in stream computing and data development to link with business data.

3. Data development

The IoT platform supports interactive query services on the cloud for data development, without the need for data preprocessing, and directly uses standard SQL statements to analyze data on the device.

Data development supports the following functions:

- Create data development tasks and support SQL analysis.
- Data sheet of commonly used devices on the open IoT platform.
- Support RDS service data sheet for user configuration.
- Grammar verification checks whether it complies with grammar execution rules.
- SQL statement analysis query results support CSV download.
- Table management: Manage user data in system data tables (platform system tables, platform equipment data tables) and other data sources.
- Function list: List the functions supported by IoT data analysis and function descriptions. Functions need to be used in development tasks. You can view function information through the function list.
- Data development supports the analysis of device data in the following three data sources: System data of the IoT platform, device data defined based on product capabilities and business data authorized by users.
- API service development: The data tasks in data development (data tasks developed using SQL) can be encapsulated into

APIs, which are convenient for developers to call. It can directly respond to device-side requests or be used for server-side data docking. Supports two methods of generating API services: Generating API through data tasks and developing API services under the API service catalog.

6.4.2.6 *Video and AI Algorithm*

Video Edge Intelligent Service (LinkVisual) is a video service product that provides video stream access to the cloud, cloud storage, distribution and video AI functions; it is designed to help video equipment manufacturers, solution providers and service providers to quickly add stock or build new cameras. The device is connected to the cloud. In addition, it also provides a wealth of video algorithms, as well as cloud-side collaboration (algorithm cloud training, cloud delivery, edge computing inference) video intelligent services (Figure 6.8).

LinkVisual products provide users with end-to-end complete video capabilities. For users to solve the problem that video equipment is connected at the edge through the Alink protocol, the middleware of the Alibaba Cloud basic platform is integrated in the cloud to provide high-quality video distribution and storage capabilities, and the App SDK for industry ISV users can be realized on the client side. Live broadcast, on-demand, picture management, news subscription and other functions.

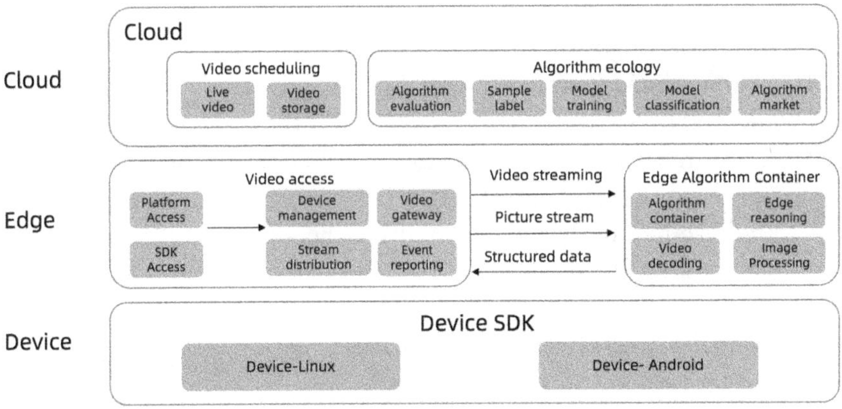

FIGURE 6.8 Link Visual architecture diagram.

1. Video management capabilities:
   - Video equipment connection: Support GB/T28181, ONVIF protocol to connect directly to the cloud, or connect to the cloud through an edge computing server, support the definition of integrated private protocol SDK, complete data upload to the cloud, SDK can be provided to upload device data to the cloud.
   - Encoding: Support H264 or H265 format encoding.
   - Device control and management: Support PTZ, management and other functions provided by national standards or private protocols; support MQTT command issuance to complete PTZ/OTA/alarm events/message linkage/recording plan settings.
   - Video distribution: Support external rtmp, rtsp, hls, http-flv video stream output.
   - Video playback: The playback terminal supports Android, IOS, PC, WEB playback, and the player supports videojs, ckplayer and other browsers, and supports H5 browsers to play hls.
   - Cloud API: Support live broadcast, video playback, local video list, cloud video list, set video plan, read video plan, cloud storage, event linkage, alarm capture, PTZ control, stream switching, heartbeat keep alive, authentication Right, export in the form of API.
   - Device access: Support device access, triple management and device configuration.
2. AI analysis algorithm

   The Ali AIoT IoT platform has human-related, vehicle-related and event-related AI algorithms, and has the ability to continuously iterate more AI and algorithms. It supports the deployment and operation of algorithms in different forms in the cloud, edge servers and terminal devices. Collaborative scheduling. Meet the real-time synchronization of data generated by edge applications to the cloud platform. The following example illustrates two algorithms implemented in the smart campus scenario:
   - Face recognition

     It can automatically analyze facial information, identify the facial features of personnel and strengthen personnel management.

The face recognition algorithm supports automatic face capture and recognition of passing pedestrians. The system enters the personal database photos, and after the system recognizes the face, the system captures and compares and distinguishes the identities of the persons in the face database and those outside the database.

People in the face database can also be identified. The algorithm supports pushing the recognized face picture, capture time, personnel identity and other information into the system, and automatically judges whether to alarm according to the alarm conditions set in the system.

- Restricted area monitoring (intelligent arming)

  Forbidden zone monitoring can be set in some key areas, and an alarm will be generated immediately if someone breaks in. The area intrusion algorithm is used to automatically monitor whether there is abnormal intrusion within the set range. The algorithm supports the identification of whether humans and various types of vehicles (tricycles, trucks, cars, bicycles, etc.) have invaded. This algorithm can support users to draw electronic fences by themselves and monitor any area within the coverage of the camera.

  In the algorithm task management, add a new task, select the area intrusion algorithm, select a camera device for monitoring, set the algorithm detection frequency (used for the algorithm detection frequency), algorithm execution time period (used to set the algorithm detection time Segment, you can customize the time range of the check), click OK to complete the task creation.

  In the area intrusion algorithm task, select the label area to customize the drawing of the algorithm monitoring area. The drawn area supports rectangles and arbitrary polygons, and can be drawn arbitrarily according to the actual monitoring scene. At the same time, the algorithm confidence level can be set. The confidence level mainly controls the confidence threshold value of the algorithm detection. The higher the value, the fewer times the algorithm detects, but the more reliable it is.

  The algorithm pushes the recognition result, recognition time, and recognized captured pictures into the system. The system uses its own early warning rules to make judgments, and finally transmits the information that needs early warning through the device terminal or other means to achieve monitoring, protection, warning and processing.

6.4.2.7 Security Capabilities

The IoT platform provides multiple protections to effectively ensure the security of equipment and cloud data.

1. Identity authentication

   - Provide chip-level security storage solution (ID^2) and device key security management mechanism to prevent device keys from being cracked. The security level is very high.
   - Provide a one-device-one-secure device authentication mechanism to reduce the security risk of equipment being compromised. It is suitable for the ability to pre-assign device certificates (ProductKey, DeviceName and DeviceSecret) in batches, and burn the device certificate information into the chip of each device. The security level is high.
   - Provide one type one secret device authentication mechanism. The device burns in product certificates (ProductKey and ProductSecret), and dynamically obtains device certificates (including ProductKey, DeviceName and DeviceSecret) during authentication. It is suitable for situations where the equipment certificate cannot be burned into each equipment during mass production. The security level is normal.
   - Provide device authentication mechanism of X.509 certificate, support devices directly connected based on MQTT protocol to use X.509 certificate for authentication. The security level is very high.

2. Communication security

   - Support TLS (MQTT\HTTP), DTLS (CoAP) data transmission channel to ensure the confidentiality and integrity of data, suitable for equipment with sufficient hardware resources and not very sensitive to power consumption. The security level is high.
   - Support device authority management mechanism to ensure secure communication between devices and the cloud.

- Support the isolation of device-level communication resources (Topic, etc.) to prevent problems such as device unauthorized access.

6.4.3 Edge Computing

Edge computing refers to an open platform that integrates core capabilities of network, computing, storage and applications on the edge of the network close to the source of things or data, and provides edge intelligent services nearby to meet the needs of industry digitization in agile connection, real-time business, data optimization, application intelligence, Key requirements for security and privacy protection. There is a very vivid analogy in the industry. Edge computing is like the nerve endings of the human body, which can process simple stimuli by itself and feed back characteristic information to the cloud brain. With the implementation of AIoT, in the intelligent connection of all scenario, devices will be interconnected, forming a new ecology of data interaction and sharing. In this process, the terminal not only needs to have more efficient computing power, in most scenarios, it must also have local autonomous decision and response capabilities.

6.4.3.1 Overview of Alibaba Cloud Edge Computing

Alibaba IoT edge computing is a PaaS layer software product that integrates cloud and edge. It sinks the capabilities of the cloud to the edge and solves the problems encountered in real-time, reliability and operation and maintenance economics of the edge. Southbound provides a communication protocol framework for software and hardware developers to provide convenient communication protocol development capabilities, and Northbound provides SaaS developers with the ability to quickly build cloud applications through Open API. For operation and maintenance, the cloud provides integrated operation and maintenance tools, which can centralize operation and maintenance in the cloud, reduce operation and maintenance costs and improve operation and maintenance efficiency.

Technically speaking, the edge capability of the IoT is the expansion of the capabilities of the IoT platform at the edge, inheriting the capabilities of the IoT platform in security, storage, computing and AI. It can be deployed in smart devices and computing nodes of different levels. By defining device models to connect devices with different protocols and different data formats, it provides safe, reliable, low-latency, low-cost and easily scalable local computing services. At the same time, it can combine big data, AI learning, voice, video and other capabilities to create a three-in-one computing system at the edge of the cloud, providing a wealth of edge solutions that integrate software and hardware.

IoT edge computing products mainly involve three parts: device side, edge computing side and cloud:

- Cloud

  After device data are uploaded to the cloud, it can be combined with Alibaba Cloud functions, such as big data and AI learning, to achieve more functions and applications through standard API interfaces.

- Edge computing end

  After the device is connected to the gateway, the gateway can collect, transfer, store, analyze and report device data to the cloud. At the same time, the gateway provides a rule engine and a function calculation engine to facilitate scene planning and business expansion.

- Device side

  Developers use the device to connect to the SDK, convert non-standard devices into standard object models and connect to the nearest gateway to realize device management and control.

For project managers, they can complete all business management locally. For project operation and maintenance personnel, it supports unified operation and maintenance of edge applications and cloud-side collaboration, supports high-reliability architecture, and satisfies stable operation of edge applications, and can be obtained in real time for group administrators. The data in the project assist cloud analysis and achieve refined operations.

6.4.3.2 Cloud

Edge computing includes two parts: cloud management and control services and edge runtime. The edge is released in the form of SDK and runs on the edge gateway or industrial computer. The cloud provides services to the outside in the form of SaaS service through API and console.

As shown in Figure 6.9, the cloud management and control part includes the creation of edge rules, the deployment and configuration of functions, the arrangement of rules for streaming data and the addition, deletion

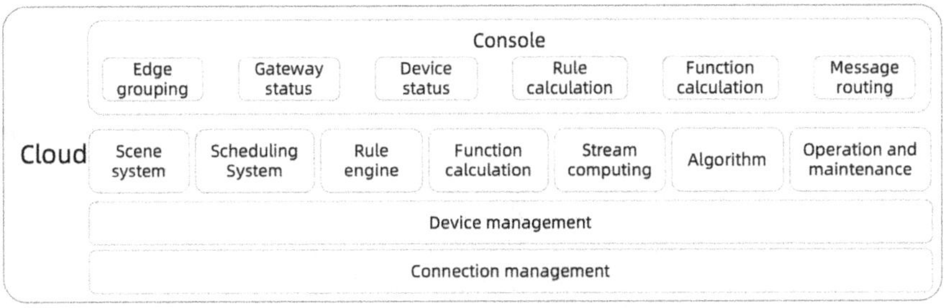

FIGURE 6.9 Cloud edge collaborative cloud product architecture diagram.

and modification of message routing. All configurations are encrypted and deployed to the edge gateway in the form of resource packages. The gateway downloads, decrypts, parses, loads and runs resources and feeds back data or results to the cloud platform. Cloud and edge integration, cloud management and control, edge operation, taking advantage of the real-time edge of the edge, local processing of raw data, reducing traffic costs and feedback of essential data to the cloud data processing platform for secondary data analysis and processing.

The edge capabilities of the IoT platform support edge application hosting and support the following functions:

- Unified resource scheduling: By virtualizing the computing resources of all edge computing nodes and deploying applications in containers, the system can make fuller use of all resources.
- Unified application operation and maintenance: All applications are deployed through Kubernetes, and users can perform unified management of all applications for capacity expansion, monitoring and log capture and login operations.
- High reliability of applications: The high reliability of applications is manifested in multi-copy deployment of stateless nodes and active/standby deployment of stateful nodes. The former guarantees the continuity of the service, and the latter guarantees the reliability of the data.
- Resource level expansion: The platform provides the ability to expand, that is, users can increase the number of devices as needed, and users can allocate resources to applications as long as they use expansion tools.

- Edge autonomy: Clusters can realize edge management in a weak or disconnected network environment.
- Free login for cluster applications: Unified docking of applications in the cluster, and a set of account system enables password-free login between multiple applications.
- AI capabilities: Edge gateways can support AI capabilities, support integrated scheduling of cloud and edge terminals, support functions such as the recognition, tracking and detection of images of people, vehicles and objects, and provide more intelligent and real-time analysis and prediction capabilities.

6.4.3.3 Edge Video All-in-One AI-Box

Alibaba Cloud IoT Edge Video Intelligent All-in-One Machine (hereinafter referred to as AI-BOX) is an edge computing product deployed in the customer's data center. It is connected to the local camera device through a local switch or router. It has product features such as wide adaptation, low deployment and operation costs and edge computing cloud-side collaboration.

AI-BOX in the customer data center has the functions of connecting various front-end cameras, adapting to various video protocols, doing local video AI analysis and calculation and operation and maintenance management.

AI-BOX provides standard API services in the cloud to support cloud applications based on video data or structured data after local video AI analysis for secondary application development. At the same time, it provides services such as remote control, remote configuration and remote algorithm update and distribution of AI-BOX.

AI-BOX can not only support the typical cloud-side-end application architecture, but also provide standard local API services to support pure local offline operation in special scenarios. Offline operation cannot provide various value-added services provided by remote cloud.

The main functions of the edge intelligent video service all-in-one include:

1. Local AI computing capabilities

   The local edge intelligent video service all-in-one machine completes the video image analysis and reasoning calculation, and uploads the required result data and structured data to the cloud. Greatly reduce the upstream bandwidth and traffic costs; at the same time, the cloud supports secondary big data calculations.

2. Use old access, strong compatibility

   It supports local camera rejuvenation, supports GB/T 28181 or ONVIF standard protocol, and also supports private protocol docking of mainstream manufacturers such as Haikang, Dahua and Univision. As long as it has a camera with a standard universal video protocol, or can provide a private protocol device with an SDK, AI-BOX can be adapted to connect.

3. Out of the box, easy to configure

   The router can be used when it is powered on and the network cable is connected. At the same time, a console is provided locally to complete device connection configuration and algorithm tasks.

4. Remote upgrade, algorithm can be upgraded

   Local algorithm container deployment supports remote upgrade and replacement of algorithms. In the cloud, remote equipment operation, maintenance and management, algorithm upgrades and iterations can be implemented locally, and algorithms can be remotely tuned, reducing the cost of subsequent on-site operations. Supports algorithms such as face recognition, human traffic statistics, helmet recognition and open flame recognition.

   - Typical video intelligent scene

     The local camera is connected to AI-BOX and AI-BOX to complete the video AI calculation locally, and upload the calculated results or structured data to the cloud through the public network or private line network. The cloud application service obtains the video stream or video analysis through the standard API. The result data are used for secondary development. (Cloud applications are generally customized and developed by customers.) In the subsequent operation and maintenance stage, the remote operation and maintenance tools provided by the cloud and the cloud algorithm competence center can be used to support the update and upgrade of the cloud to the local algorithm.

- Suggested application scenarios in various industries

  AI-BOX is mainly a video image connection and analysis product, which can basically be used in the scene requirements of various solutions for video intelligence. Including but not limited to specific application scenarios such as smart safe communities, smart parks, smart buildings, smart campuses and smart industries.

  At the same time, the core value of AI-BOX is more cost-effective and efficient edge video intelligent services, providing various recognition and detection algorithms running in local data centers, so that ordinary cameras can also have 'brains'.

6.4.3.4 Protocol Gateway EdgeBox

Alibaba Cloud Protocol Gateway (hereinafter referred to as EdgeBox) is an edge computing product deployed in customer data centers. It connects to local equipment, facilities and systems through local switches or routers. It provides complete application life cycle management for local applications. Security, disaster tolerance and fault tolerance guarantee, cloud-side collaboration, remote operation and maintenance management.

EdgeBox is a transformation of on-site application deployment mode. Through Kubernetes technology, it virtualizes the scheduling of CPU, memory, storage, network and other resources of the edge host, as well as the containerized orchestration of applications, so as to realize the life cycle management and safe operation support of edge applications, high availability and disaster recovery guarantees, and remote operation and maintenance capabilities. At the same time, it effectively improves the resource utilization of the host.

In layman's terms, EdgeBox virtualizes multiple physical hosts into a resource pool (CPU, memory, etc.). Applications are packaged into containers and arranged into the total resource pool. At the same time, we are in this resource pool. Established a set of capabilities to ensure the operation of these applications.

The protocol gateway provides functions such as device access and microservices. It connects to the intelligent subsystem in the south direction and connects to the Alibaba Cloud IoT platform in the north direction.

The protocol gateway architecture diagram is as follows (Figure 6.10):

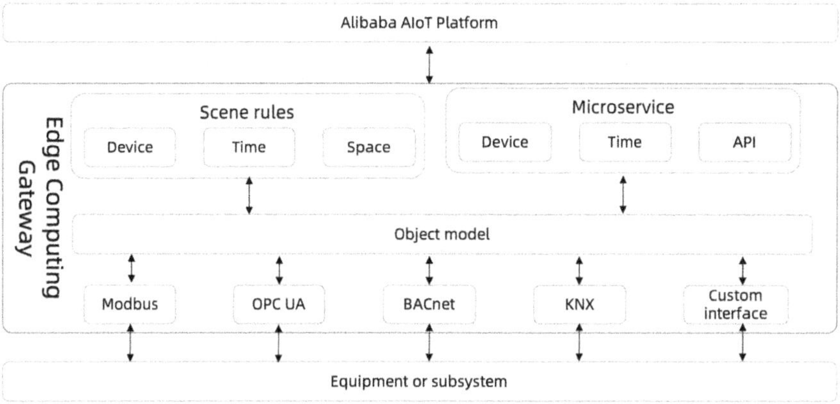

FIGURE 6.10 Edge protocol gateway product architecture diagram.

The main functions of the protocol gateway are as follows:

1. Equipment access capability

   Device access is a basic capability provided by Alibaba Link IoT Edge. The device access module is called a driver or device access driver in Link IoT Edge. All devices connected to Link IoT Edge need to be connected through drivers.

   The device access driver is an independent service module in Link IoT Edge. You can develop a custom device access driver according to the requirements of the business protocol. Figure 6.11 shows the function and data flow of the custom driver, and points out the development work that needs to be done to develop a custom driver.

2. Scene linkage

   Scene linkage is a visual programming method for developing automated business logic in the rule engine. You can visually define linkage rules between devices and deploy the rules to the cloud or edge. Each scene linkage rule consists of three parts: trigger (Trigger), execution condition (Condition) and execution action (Action). This rule model is called the TCA model.

   When the event or attribute change event specified by the trigger occurs, the system determines whether to execute the execution action defined in the rule by judging whether the execution condition has been met. If the execution conditions are met, the defined execution action is directly executed; otherwise, it is not executed.

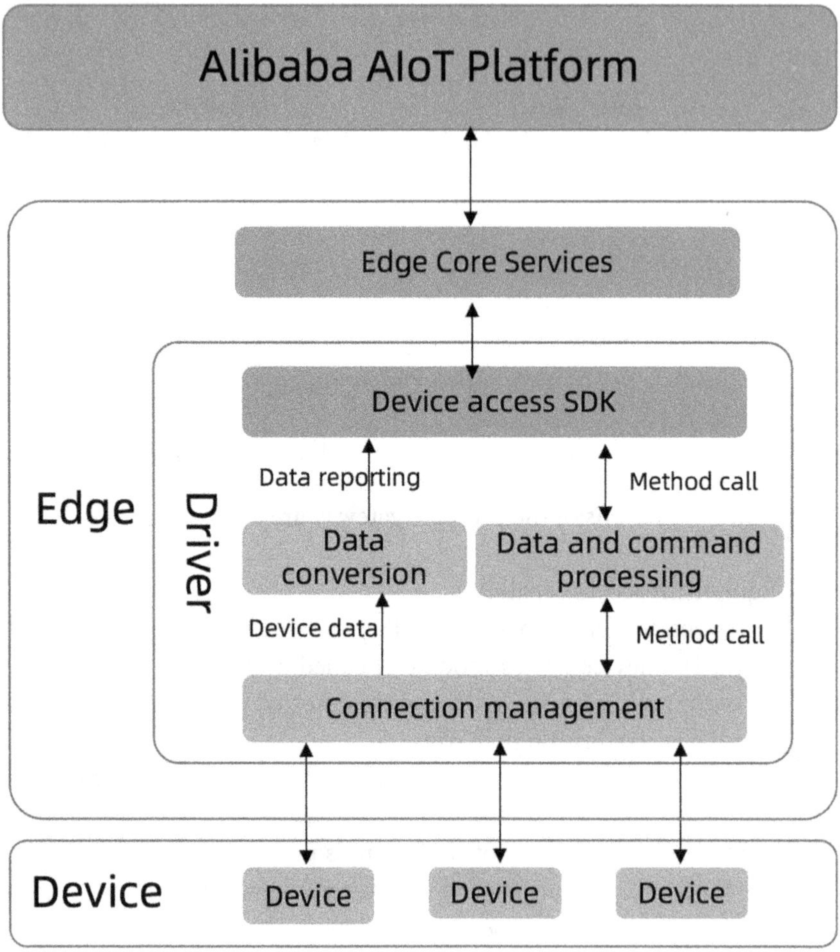

FIGURE 6.11 Edge device access driver development architecture diagram.

6.4.4 IoT Devices

The terminal device is composed of sensors, smart terminals and other devices that are responsible for sensing information. The terminal device has functions such as sensing signals and identifying objects.

Link SDK is provided by Alibaba Cloud for device manufacturers to integrate into the device, securely access the device to the Alibaba Cloud IoT platform, and allow the device to be controlled and managed by the Alibaba Cloud IoT platform. The device needs to support the TCP/IP

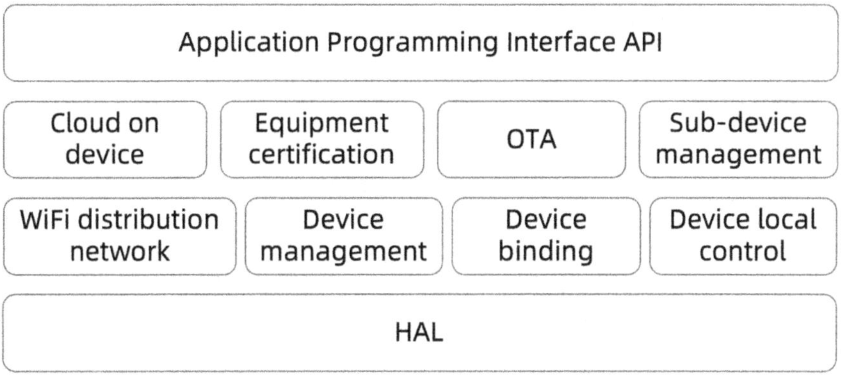

FIGURE 6.12 Full diagram of device access SDK function.

protocol stack to integrate the Link SDK. For non-IP devices such as zigbee and KNX, it needs to be connected to the Alibaba Cloud IoT platform through a gateway device, and the gateway device needs to integrate the Link SDK.

For terminal device access, it is necessary to meet the functional modules designed in Figure 6.12: device cloud connection, device authentication, OTA, sub-device management, WiFi network distribution, device management, user binding support and device local control.

For the design of terminal access SDK, full consideration is given to Alibaba Cloud Computing as the supporting basic cloud platform for IoT middle stations. The terminal access SDK provided by the Ali product Link Kit meets the technical design requirements required by the project scenario. Therefore, consider recommending the use of Alibaba's IoT cloud platform product Link Kit product. The following specifically introduces the support of related products for terminal access.

The Link Kit SDK is provided by Alibaba Cloud to the device manufacturer. The device manufacturer integrates the sensor device into the Alibaba Cloud IoT cloud platform through the SDK, so that the sensor devices can be managed by the Alibaba Cloud IoT cloud platform. Its software structure and functions are as follows:

1. Application Programming Interface (API)

   The Link Kit SDK provides APIs for device calls to control various functional modules provided by the SDK.

2. Function module

   The Link Kit SDK provides a series of functional modules for the device to call:

   - Connecting devices to the cloud: Provide MQTT, CoAP, HTTP/S and other methods to connect to the Alibaba Cloud IoT cloud platform.
   - Device identity authentication: Provide one device with one password, one type and one password for device identity authentication.
   - OTA: Provide device firmware upgrade.
   - Sub-device management: Access to sub-device.
   - WiFi distribution network: Transmit the SSID and password of the wireless router AP to the WiFi device.
   - Equipment management: Provide attributes, services and events to manage and control equipment.
   - User binding: Provide a secure binding token to support the binding between users and devices.
   - Local device control: For devices that use WiFi and Ethernet access, if the mobile phone or gateway is located in the same local area network as the device, the device can be controlled through the local area network instead of the cloud, so that the control is faster and more reliable.

3. Hardware Adaptation Interface (Hardware Abstraction Layer, HAL)

   Some functional modules require the device manufacturer to provide some information or processing functions. The Link Kit SDK defines HAL for these interfaces for the device manufacturer to implement the access to the 'end'/'side' device.

   The Alibaba Cloud IoT IoT platform provides multiple industry solutions/services in the cloud such as smart life, smart manufacturing and smart living. After devices are connected to the Alibaba Cloud IoT IoT platform using Link SDK, they can be managed by these industry solutions. In order to simplify the difficulty of device access or add special business logic on the device side, some industry solutions will integrate the Link SDK and output the device-side SDK for the industry, such as the Feiyan device SDK for life scenarios and LinkEdge SDK for industrial scenarios. In this case, please use the device-side SDK provided by industry solutions/services for device development. The relationship between Alibaba Cloud's IoT industry solutions and Link SDK is shown in Figure 6.13.

The Link SDK is suitable for all products connected to the Alibaba Cloud IoT platform. Here are some solutions for connecting to the Alibaba Cloud IoT platform by integrating the Link SDK:

- Lifestyle products

  Most of these devices connect to the wireless router in the home through WiFi, and then connect to the Internet and communicate with the Alibaba Cloud IoT platform. Common equipment includes: electronic peepholes, smart locks, fans, sweeping robots, air conditioners, refrigerators, wiring boards, air purifiers, heaters, curtains, lamps, electric water heaters, range hoods, microwave ovens and ovens.

- Gateway products

  Some devices do not support the TCP/IP protocol and cannot directly integrate the Link SDK. Such devices need to connect to the Alibaba Cloud IoT platform through the gateway integrated with the Link SDK, including ZigBee gateway, Bluetooth gateway, 433 gateway and KNX gateway.

- Cellular network access products

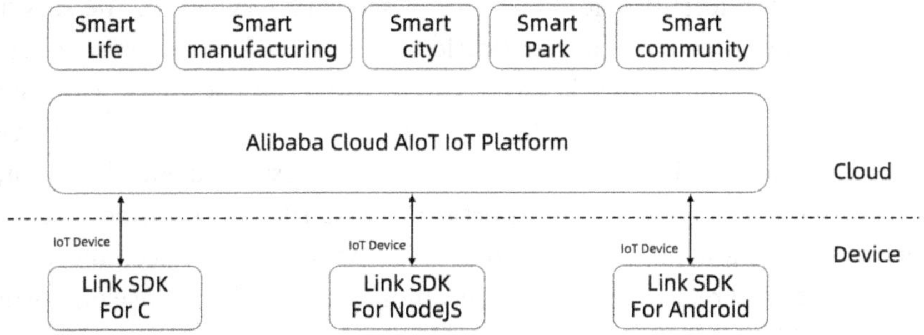

FIGURE 6.13 Link SDK solution architecture diagram.

IoT products connected using the cellular network of telecom operators are mostly used in agriculture, cities and other scenarios with a wide coverage area, or in scenarios where the equipment is mobile, such as logistics trucks, vending machines, weather collection systems, and hydrological collection systems, Smart meters and smart water meters.

6.4.5 Internet of Things Security Services

The Alibaba Cloud Link Security IoT security platform builds a full-link multi-level security defense system for the full life cycle of IoT devices, and comprehensively builds a trusted IoT platform from three levels of trusted terminals, trusted access and trusted services.

Trusted terminal: The foundation of IoT security is terminal security. The Link Security platform builds a trusted terminal based on trusted hardware, trusted computing framework and security SDK. Supports the trusted root security suite that provides multiple security levels and multiple types of security carriers, provides a trusted computing framework for different hardware and software operating systems, and a building-block multi-terminal security development kit to ensure end-side storage, operation, computing and networking Certified local security.

Trusted access: Supports end-to-end secure two-way identity authentication, provides multi-level and multiple types of IoT security encryption protocols, guarantees data confidentiality, integrity and availability during transmission, and can be applied to IoT terminals with light resource occupation. For scenarios where business data are sensitive, additional and independent business data encryption transmission capabilities are provided to ensure business data privacy.

Trusted services: Provide a variety of security services on the server side, including in-depth authentication of trusted devices, trusted service management, trusted key services and firmware security scanning services. It also provides full-cycle, one-stop and visualized security operation capabilities. Through continuous analysis of device behavior, a unified device digital security portrait is constructed, continuous security detection and security situation awareness of the device, and one-stop, visualized real-time Security risk monitoring and response platform (Figure 6.14).

The Link Security IoT security platform builds a full-link multi-level security defense system for the full life cycle of IoT devices. The IoT IoT

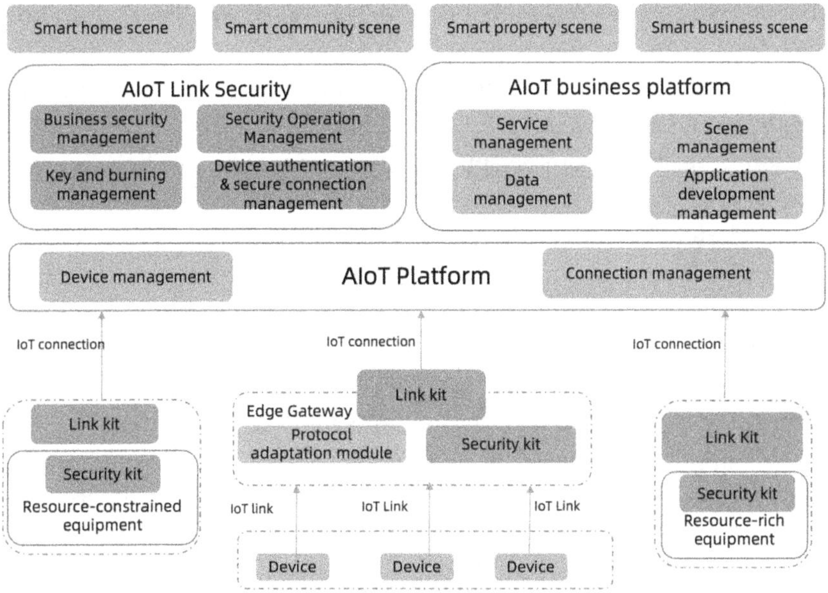

FIGURE 6.14 Link Security architecture diagram.

platform's business can integrate security capabilities on demand at different levels.

Terminal perception layer: Provides a building-block multi-terminal security development kit, including a trusted root security kit with multiple security levels and multiple security carriers, so as to support the rapid development and integration of multiple types of terminal equipment.

Edge gateway layer: Provides a building-block multiple-edge gateway security suite, including a trusted root security suite with multiple security levels and multiple types of security carriers, with trusted identity authentication, secure storage environment, secure execution environment, container security protection and other security protections ability.

Terminal and edge gateway burning: For the keys and sensitive data of terminals and edge gateways, it provides multi-level safe production burning and safe air distribution schemes, combined with unified key management services to realize the process of production, transportation and burning. The security, integrity and availability of identity information and keys.

Network connection layer: Provides full-terminal security connection capabilities, based on the trusted digital identity authentication of the

IoT, through end-to-end secure one-way/two-way identity authentication, provides a variety of secure encrypted transmission protocols, including TLS and lighter security encryption. The transmission protocol guarantees the confidentiality, integrity and availability of data during the transmission process, and can be applied to IoT terminals with light resource occupation.

IoT platform layer: Full-cycle, one-stop, visualized security operation, through continuous monitoring and continuous analysis of equipment behavior, provide a one-stop, visualized security risk monitoring platform.

6.4.6 Internet of Things Application Service Platform

The IoT application service platform is committed to quickly building intelligent solutions for IoT scenarios. Combined with a large number of industry intelligent hardware and software pre-integrated on the platform, and with the power of Alibaba Cloud and the ecology, it quickly connects OT and IT, and realizes the automated deployment and delivery of applications and solutions through containerization technology. Through pre-integration of software and hardware capabilities with industry-leading scenario solutions, combined with Alibaba Cloud's advantages in AI and big data, it provides scenario-oriented intelligent data analysis services to achieve an out-of-the-box activation experience (Figure 6.15).

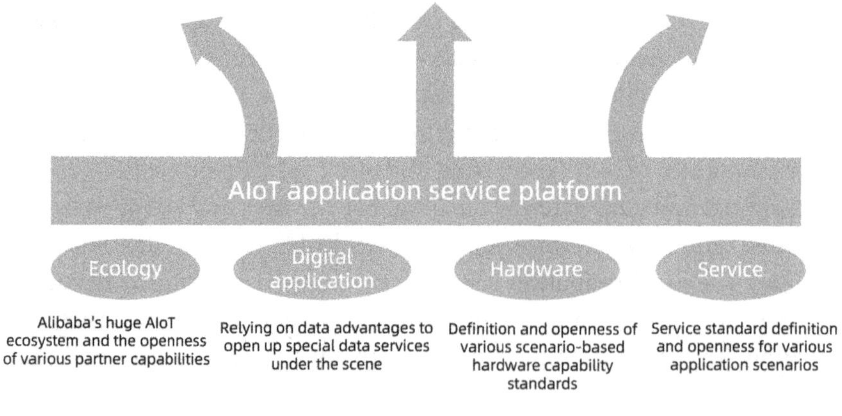
AIoT application service platform
Ecology
Digital application
Hardware
Service
Alibaba's huge AIoT ecosystem and the openness of various partner capabilities
Relying on data advantages to open up special data services under the scene
Definition and openness of various scenario-based hardware capability standards
Service standard definition and openness for various application scenarios

FIGURE 6.15 AIoT application service platform.

6.4.6.1 *Equipment Management*

Device management provides functions such as creation and management of IoT devices.

Product: A product is a collection of equipment, usually a set of equipment with the same function definition. For example: A product refers to a product of the same model, and a device is a certain device under that model.

Equipment: A product refers to a certain type of equipment. After creating the product, you need to create an identity for the device. You can create a single device or a batch of devices. This article describes the creation of a single device for you.

Gateway: A gateway is a piece of hardware that is deployed on site with users and quickly connects various local wired and wireless devices through different link protocols to make them quickly networked.

Authorization: Authorization is the processing of device permissions. You can authorize the device you produce to other tenants, so that other tenants' applications can obtain the data reported by the device.

6.4.6.2 *Project Management*

Project management provides the basic management process for the implementation of IoT projects and the management of various capabilities that the project depends on.

Service integration: The service integration function can be used in the IoT project to select the services that need to be added (such as services in the pedestrian field), and unified support.

Data integration: The business data involved in IoT projects, such as vehicle barrier data, are supported through the data integration function.

Rule engine: According to the data reported by the object, rules can be defined according to the TCA model to trigger new scenarios.

Alarm service: According to the data reported by the object, rules can be defined according to the TCA model to trigger an alarm.

Personnel management: Set up personnel information in the IoT project.

Organizational management: Set up organizational relationships within IoT projects.

Permission management: Set permissions in the IoT project.

Space management: Set up and manage space information.

Account opening: Support various account docking.

6.4.6.3 Field Service

Domain service is a unique vertical service provided by Alibaba Cloud IoT. The goal is to solve the problems of unified logic integration and unified data integration related to devices. Functionally, it is divided into service bus and data bus. For example, in the smart community project, Alibaba Cloud IoT will formulate and provide services in four areas: pedestrian services, automotive services, security services and equipment and facilities services to support the rapid docking of business centers and equipment-related content.

The design goal of the service bus is to standardize the behavioral expressions between applications and the expected results. For service providers, through service standardization, it is possible to clearly and concisely express which interfaces, definitions and the specific functions that the service provides, and any application that depends on the capabilities provided by the service. The service supply must be implemented in accordance with this set of interfaces; for the service relying party (that is, the application that uses the service) can clearly and concisely express which interfaces it depends on, and what specific functions are expected to be completed by these interfaces, and any application that provides services for it, as long as it follows the same service model, can achieve the replacement of service providers.

The design goal of the data bus is to standardize the transfer and expression of data between applications. The platform provides four APIs for adding, deleting, modifying and checking data, as well as a message subscription mechanism in HTTP2 mode.

1. Intelligent human settlement open platform

   Whether it is a family, a community or a long-term rental apartment, the living environment carries the longest living time in people's lives, and also carries people's pursuit and yearning for a better life. Alibaba Cloud IoT Smart Habitat carries the vision of Alibaba's new retail mission. Through the integration of online and offline, life and services are connected to make the living experience of all people richer, more convenient and enjoyable.

   The intelligent human settlement open platform provides customers with a complete set of basic IoT-related capabilities that can be used to quickly build their own whole-house intelligence, rental apartments, smart communities and other business systems. The platform supports the construction of whole-house smart solutions, smart community solutions, data security, and provides after-services such as data analysis.

Developers can use the mobile SDK&API and Open API to implement mobile applications and the cloud to network and control Alibaba Cloud IoT ICA standard devices, create scenarios (device linkage) and local control (devices, scenarios). At the same time, community services (community access control, vehicle systems and property systems) can be opened through cloud APIs, and services such as smart convenience stores and health centers can be provided.

The Smart Habitat Open Platform is based on a variety of Alibaba Cloud cloud products, including OTS (distributed storage), OSS (file storage), MQ (distributed message queue), API gateway, log service and mobile push. Developers can use other cloud products of Alibaba Cloud to build their own smart living solutions (Figure 6.16).

The intelligent human settlement open platform provides basic business entities, including users, equipment, spaces and scenes, as well as service APIs that maintain the relationship between them.

The smart human settlement platform can be connected to the devices in the Alibaba Cloud IoT system, including those connected to the 'Internet of Things Platform' and the 'Internet of Things

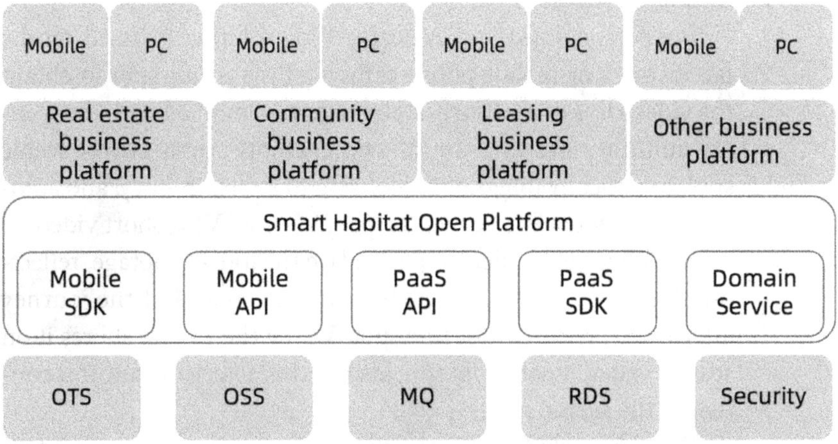

FIGURE 6.16 Architecture diagram of the intelligent human settlement development platform.

Platform'. Connect the device to the smart habitat platform for management and operation through network distribution or binding.

Devices connected to the intelligent human settlement platform can realize voice control by connecting to the Tmall Genie, which can provide users with a more natural way of using the device.

After accessing the intelligent human settlement platform, various standard field services in the extended service can be used. Choose a service provider that matches your needs from the existing service providers to achieve cloud capabilities docking.

2. Cultural and tourism services

The scenic Vlog smart tour system is centered on tourists, through customizing high frame rate cameras, shooting the unique scenery of the scenic area, automatically capturing faces, supporting algorithmic capabilities such as beauty and background blur, based on face recognition, combined with scenic theme templates, and automatically realized. Multi-camera play moves in series to create a star-like Vlog. The product uses Alibaba Cloud's latest IoT and AI technology based on the video materials of scenic spots and tourists to automatically generate Vlog short videos of 'people+scenery', allowing tourists to easily obtain exclusive video travel notes, and visitors can share them. Large social platform, at the same time realize the interactive communication of scenic beauty and culture.

I. Business Process

The personalized and exclusive Vlog solution is based on the face camera or mobile phone camera of the scenic spot to obtain the video clips of the tourists' characters, combined with the scenery, cultural relics, activities and other empty mirrors in the scenic spot, organically integrates according to a certain story, and automatically generates a section of 'people +Jing' Vlog short video.

The Vlog short video is centered on the tourist footage, reflecting the tourist's 'C position', leaving a memorial of the journey after he/she leaves the scenic spot. While the tourist shares it on social media, it naturally spreads the characteristic cultural content of the scenic spot.

The plan includes two product gameplay methods, one is face capture based on scenic cameras, and the other is AR interactive experience based on mobile terminals.

- Tourist footage: Camera capture, mobile AR interactive experience.
- Empty footage: Including opening and ending, landscape time-lapse photography or aerial photography, amusement activities, animal activities and other video footage, with additional lens effects and design aesthetics, creating visual effects that are difficult for tourists to simply shoot with their mobile phones.

II. Product Architecture (Figure 6.17)

Intelligent real-time generation of Vlog based on multiple AI algorithms.

- Intelligent lens analysis: Analyze scenes, faces and other information, and intelligently extract video clips according to certain collection scripts to realize real-time synthesis.
- Abundant video packaging: Organically combine materials such as the beginning, ending and empty lens to produce diversified templates and lens languages.
- Use various AI technologies: Based on AI algorithms such as face recognition, human body recognition, beauty and material selection, highlight individual collections and enhance beauty.

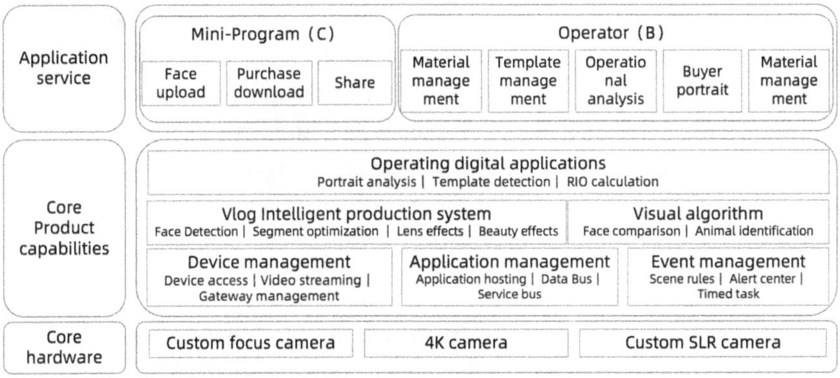

FIGURE 6.17 Vlog system product architecture diagram.

3. Smart life IoT platform

Life IoT platform (Feiyan platform) is Alibaba Cloud IoT's IoT platform for consumer smart devices. It is a set of public cloud platforms built on the basis of Alibaba Cloud IaaS and PaaS layer cloud products. The platform helps developers and solution providers in the service life field, providing functional design, embedded development and debugging, equipment security, cloud development, App development, operation management, data statistics, etc., from the full life cycle of product development to post operation service.

In July 2020, the Tmall Genie IoT open platform and Alibaba Cloud Life IoT platform will be integrated. The integrated life IoT platform has become a unified consumer-grade smart device IoT platform within Alibaba Group, which can support the following two business forms:

- Cloud products for the empowerment of the smart life industry provide global and intelligent paid cloud services for global customers. Connecting to the IoT products of the private brand project of the Internet of Things of Life, you can sell your device worldwide. At the same time, you can also support the App with your own brand to enhance your brand image.
- For the device access of the Tmall Elf IoT ecosystem, we will work with domestic equipment manufacturers to build an IoT ecosystem around the Tmall Elf. Products that choose to connect to the Tmall Genie IoT ecosystem can be controlled by all Tmall Genie's ecological terminals, including various types of Tmall Genie speakers, Tmall Genie App, Tmall Genie car machine and AliGenie Inside smart devices, which can realize voice, Touch screen and other multi-modal interactions to provide consumers with control, query, broadcast, scene and active services. At present, the Tmall Genie IoT ecosystem has access to more than 1,000 brands, more than 200 categories, and more than 4,000 models.

4. Urban IoT platform

The base of urban digitization and intelligence, the necessary infrastructure for 'new infrastructure', intelligentize every urban digital space, and help build a new type of smart city with 'good governance, benefits for the people, and prosperity'.

Realize the digitization of core elements such as people and things in urban space with the help of IoT sensing equipment, and realize the intelligentization of urban governance by constructing an event-driven urban resource scheduling system.

The Alibaba Cloud IoT CityLink smart city platform is based on Alibaba's years of technology accumulation and is built on the base of Alibaba Cloud Feitian distributed system. It supports the city's intelligent operation and management based on the IoT, cloud computing, big data, and spatial geographic information technology.

The platform is positioned as the base and infrastructure for urban intelligence and digital transformation. It is hoped that the platform will provide unified access and life cycle management of massive sensing devices, multi-domain standard data model precipitation and unified authority control, urban physical space and physical components. Digital mirroring construction, flexible configuration of data visualization presentation, event-driven scheduling of urban resource scheduling rules, unified application (stock and new) access and management and other core modules to help city integrators quickly complete the integration of smart city projects, Delivery, to help city operators build continuous operation capabilities based on the platform, and ultimately achieve a new smart/smart city operation mode of 'full time and space awareness', 'full field online', 'full element linkage' and 'full-cycle iteration'.

- Full time and space perception

  Relying on the real-time monitoring of multi-type sensing devices deployed in the entire city and connected to the city platform to complete the digital mirror construction of people, things, objects and fields in the city, thereby completing the digitalization of the physical world.

- All areas online

  The smart city platform supports the access of stock devices, data and applications. Through new additions and stock collection, the platform can accumulate resources in all fields and give full play to the value of the stock system.

- All elements linkage

  The standard model constructed by the smart city platform realizes the unified connection of equipment data, domain model data and digital component data. Through the event-driven urban resource scheduling engine, people, organizations, equipment and applications are connected in series to achieve cross-organization, Cross-system process management, process linkage and event closed loop.

- Full-cycle iteration

  Intelligent operation capabilities built through the smart city platform, using equipment assets, data assets, event operation centers, application integration and other configuration systems to achieve rapid iteration and management optimization of various IoT application services in the city to meet business changes during operation. Realize the improvement of operational efficiency.

6.5 ALIBABA CLOUD AIoT PROJECT CASES

Alibaba Cloud AIoT provides a horizontal general IoT platform, focusing on industry areas such as cities, real estate, parks, life, manufacturing and agriculture; in the face of the diverse and fragmented needs of IoT customers, AIoT is formulated according to customer needs outstanding solutions for different scenarios.

One is to be able to find substantial demand in the subdivision field; the other is to be able to conduct continuous research in the subdivision field.

1. Deeply cultivate subdivided industries

   Alibaba Cloud AIoT IoT platform is technically a low-level pan-platform concept. Its maturity and availability are actually very high. Of course, it can be applied to many industries from a horizontal perspective, but it is seen by many companies. In the future, what the IoT platform should do more is actually vertical, and it should fully integrate the knowledge of sub-industry to solve customer problems in specific scenarios and industries.

2. Construction benchmark case

   An important feature of benchmark cases is that they can be copied quickly, which can promote the conversion of user needs into landing projects more quickly. Alibaba Cloud AIoT launched the

'IoT Choice' channel. Alibaba Cloud IoT selects various industry standard solutions to provide customers with high-quality software and hardware products, ensuring that solutions can be quickly replicated and giving the market more trust and confidence.

3. Selectively enter the new incremental market

    Because different industries have different maturity levels, there must be a sequence of intelligentization processes in different industries. For example, the actual transformation needs of industries, parks, communities and buildings are now large, attracting many companies to participate, and there may be the next blue ocean market in the future. When realizing that all traditional industries will have opportunities for intelligent upgrades, Alibaba cannot rule out the possibility that companies will expand their business scenarios at a certain time. Just as many companies choose industry as their new market development direction in 2019. For the underlying pan-platform with strong market adaptability, it is a rare business expansion opportunity.

4. Develop industry ecological partners

    Every vertical industry requires deep experience and technology accumulation. Alibaba Cloud AIoT is impossible to do any vertical industry. Therefore, Alibaba develops industry ecosystem partners to achieve broad technology empowerment. It is worth noting that we can understand ecological cooperation from two levels: one is the ecology of the business model, which is like general contracting and subcontracting, where companies work together in business activities; the other is a technology-based ecology, where enterprises have a unified technical framework at all levels, so various products will be more technically coupled, and the things they make will be more consistent. This is actually a type of ecology that can truly promote the large-scale development of the industry.

    Alibaba Cloud AIoT faces the different needs of different customers at all levels and provides services to customers quickly and efficiently (Figure 6.18).

6.5.1 Case of Smart Community

With the continuous improvement of the technical conditions for the construction of smart communities, the development foundation has been continuously strengthened. The first is the increasing popularity of

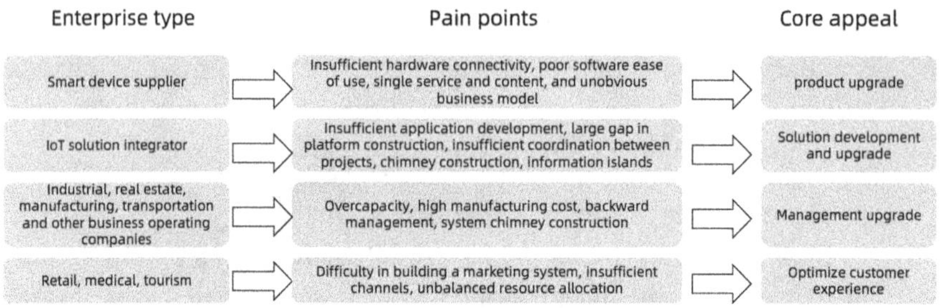

Enterprise type
Pain points
Core appeal
Smart device supplier
Insufficient hardware connectivity, poor software ease of use, single service and content, and unobvious business model
product upgrade
IoT solution integrator
Insufficient application development, large gap in platform construction, insufficient coordination between projects, chimney construction, information islands
Solution development and upgrade
Industrial, real estate, manufacturing, transportation and other business operating companies
Overcapacity, high manufacturing cost, backward management, system chimney construction
Management upgrade
Retail, medical, tourism
Difficulty in building a marketing system, insufficient channels, unbalanced resource allocation
Optimize customer experience

FIGURE 6.18 Alibaba Cloud AIoT enterprise service model.

diversified Internet terminals, the penetration rate of mobile phone users exceeds 100%, the rapid development of IoT terminals, and the application entities of smart communities fully touch all aspects of urban social and economic life; the second is the rapid development of mobile networks and high-speed broadband access. The development of the IoT further expands the depth and breadth of urban networks, and realizes all-round network access between people and things; third, the cloud computing/big data industry is booming, and the application of smart communities has laid the foundation for data storage, computing and analysis processing. Basic capabilities, machine learning and neural networks have brought about a leap forward in AI, which has further created the basic conditions for the development of intelligence; fourth, the technology solutions for application development and platform operation of smart communities have become increasingly mature, and a variety of digital information technologies have gradually formed integrated smart community construction solution.

6.5.1.1 Market Status

The IoT platform is the foundation of device access management, edge computing and AI capabilities in the smart community. It should have protocol specifications that connect smart vendors, build unified device models and service models, and support unified smart application systems Building and other capabilities.

The IoT is a network that realizes the comprehensive interconnection between people, people and things, things and things through information sensing equipment according to agreed agreements. Its main feature is to obtain various information about the physical world through information sensing equipment. Information, combined with the Internet, communication network and other networks for information transmission and interaction, using intelligent computing technology to analyze and process information. So as to improve the perception of the material world and realize intelligent decision-making and control. As a typical representative of a new generation of information technology, it is called the pillar of the 'smart community' along with emerging hotspot technologies such as cloud computing and big data, and its applications are increasing and becoming more and more important.

With the rapid development of the IoT technology, smart devices such as sensors, data acquisition devices and smart home appliances have been widely used in all aspects of society, resulting in a variety of application scenarios such as smart communities, smart life and smart industries, generating massive amounts of data. Behind these massive amounts of data are the laws governing user behavior, energy management, production efficiency, etc., which have great value.

Through the integration of big data computing platforms, cloud computing service capabilities and AI technology, the IoT will greatly promote the discovery of the huge value contained in IoT devices and data, so as to realize the intelligence of various IoT application fields. At the same time, there are still many challenges in the IoT industry.

1. Lack of basic standards for the IoT leads to industry fragmentation

   There are many application scenarios of the IoT, covering a wide range, and the basic standards in the construction process are not uniform, resulting in insufficient resource sharing and utilization, difficulties in data and information exchange, and the application of the IoT showing a fragmented state of isolation. On the one hand, the investment in the IoT project is large, the cycle is long and the industry manufacturers involved are scattered. In the absence of unified standards, the system is closed for self-built and self-use, basic capabilities are difficult to share and use, and cloud computing and big data analysis platforms cannot effectively connect with things. Internet-connected devices seriously affect resource efficiency; on the other hand, the application scenarios of the IoT involve all aspects of social and economic life, and the collected data information is complex and diverse. Without a unified standard, it is difficult to communicate and share data information, cloud computing and big data analysis. The platform cannot fully perform in-depth analysis of the data generated by the IoT, the subsequent application value is difficult to realize, and the intelligent management is difficult to realize.

2. Insufficient resource capacity restricts the construction of the IoT.

   The construction of the IoT is a systematic project that requires the integration and application of multiple digital information technologies, and the application scenarios are diversified, the investment in infrastructure construction is large, and the construction period is long. Judging from the current construction practice, due to the lack of a complete IoT industry chain platform, various manufacturers in the industry chain are separated from each other, which has an impact on demand interoperability, and it is difficult to effectively integrate the IoT industry ecology, which makes it difficult to meet the huge IoT construction Application requirements.

The IoT platform is designed to solve the problem of resource integration in the IoT in the smart community. The IoT platform serves as an important bridge to communicate the underlying perception layer equipment and the upper application layer business, encapsulates the heterogeneity of the underlying equipment and network, provides a unified and universal access interface and realizes the reuse of data and computing resources. The IoT platform is the center of data management, equipment management and event management, and is the core component of IoT application integration.

The smart community solution will build an IoT platform, support diversified protocols and devices and develop application middleware that supports multiple application functions and services on the platform, meet the needs of smart community application expansion, and provide system flexibility and scalability. The main problems to be solved are as follows:

1. Establish a monitoring system for the comprehensive perception of smart communities

   The IoT platform has the ability to access a large number of heterogeneous IoT sensing devices, and realizes the independence of hardware manufacturers and the independence of communication methods, making it more flexible and fast to connect to various IoT sensing devices, thereby achieving higher efficiency and lower cost-effective use of IoT technology to conduct comprehensive and real-time perception of behaviors and actions such as basic implementation equipment management, security monitoring, crowd statistics and network public opinion of various projects in the smart community. Extensive comprehensive perception data will provide a solid data foundation for scientific, sophisticated and digital community operation and management.

2. Improve the emergency response capability and operation management level of smart communities

   With the continuous accumulation of comprehensive perception data and continuous integration and fermentation with business data, a large number of events can be defined to define and explain various daily, urgent and cumulative changes and dynamics that occur in the project. With the support of a powerful intelligent linkage engine, a large number of events no longer require manual intervention, but will be automatically handled, and the methods and means of disposal will be continuously optimized and updated.

3. Build a safe and reliable platform that is not afraid of attacks and intrusions

   In the face of various information attacks and viruses that continue to be rampant around the world, the IoT platform can respond calmly and protect its comprehensively aware network and intelligent processing engine from being invaded due to data tampering, device hijacking or channel monitoring. Ensure the credibility of the data on the platform and the safe operation of various linkage disposal processes.

4. Ecologically compatible and at the forefront of innovation

   The IoT platform has set a benchmark for the entire community's intelligence and information industry, and has also opened up a broad business prospect for many unique vertical applications or smart device manufacturers. More and more innovative and black technology products continue to emerge, and the original equipment and applications are constantly being introduced with the development of business. The compatibility and open standards of the IoT can easily and quickly access these Innovate solutions and continuously improve the level of community informatization, operational efficiency and user experience.

From the perspective of economic benefits, building a unified IoT management platform can reduce the overall cost of smart application construction for the enterprise. Each department can collect the perception data of each application through a unified IoT platform, and at the same time manage the corresponding perception equipment, greatly shorten the development and application time of the business system, use the platform's public service capabilities to call, and save system development costs.

1. Direct benefits are based on the construction of a unified IoT management platform, and expected benefits are basically distributed in the incubation of the IoT industry, community operations, or low-cost integrated service response and cost input. The unified IoT management platform can provide public platform services instead of building a system separately. The potential benefits of the unified IoT system have high expectations. Their expectations include: improving the quality of products or services, improving corporate productivity and improving community management, efficiency and service quality, as well as improving the reliability of operations. Benefits come from the cost savings of improving operating efficiency, new or better data flow, helping to make better decisions, improving corporate productivity, and improving the transparency of the entire enterprise organization's assets, enabling better monitoring, new or new good customer experience, comprehensive investment in community management and cost savings for efficiency improvement, through unified and reusable collection and perception services, reduce investment costs and maximize service efficiency. Based on the formation of new perceptions or more refined and multi-faceted perception data streams based on the IoT, it helps companies and community managers make scientific decisions, guarantees the scientificity and feasibility of decision-making inputs and saves implementation time and cost inputs.

2. Indirect benefits are aimed at business and community operation and management personnel, through the IoT management platform, to achieve transparent management of related facilities and assets for corporate products and community public areas, which can better monitor, ensure asset availability and improve operational efficiency.

Reduce maintenance costs. At the same time, by opening up the interaction between people and things, new or better service experiences can be provided to enterprises, consumers and citizens.

6.5.1.2 Customer Needs

According to market research and visits to the property companies of domestic leading real estate groups, most of the property companies' management community systems are relatively independent, and the degree of autonomy of the project's property management is very high, resulting in pedestrian, vehicle and security in the community management unit. Hardware brands are different, which makes after-sales operation and maintenance unable to be managed uniformly, and there is no maintenance after equipment is damaged. Especially the parking system is relatively large in scale, rich in business types and large in number. Property management personnel lack unified construction specifications and business standard processes; EBA system: The management is complex, the threshold for rule configuration is high, and each project is different, and the group cannot manage and control it; each parking lot information and personnel traffic information are independent of each other, and each other is an information island, and the group level cannot be effectively supervised. There are many suppliers of parking systems and personnel access systems, which are difficult to manage, high training costs and difficult to standardize. Among them, the parking lot toll booth guards at least many people and multiple shifts, and the personnel costs are increasing year by year. On-site maintenance, operation and maintenance of equipment, engineering operation and maintenance personnel have high operating costs. The pedestrian access control management model is relatively outdated, the owner's experience is not good, the entry and exit of the personnel cannot be controlled, the express delivery, meal delivery and marketing can enter and exit the community casually, and the old monitoring cannot save the video surveillance for a long time.

The domestic smart community scenarios mainly focus on personnel traffic management, vehicle dealership management, AI security management and property basic service management, community smart services, etc. In the domestic real estate group and the existing management model of the property group, each property community has its own The scene management system is in a state of autonomous use, and the group cannot accurately and timely grasp the detailed information of each project's management status, financial status and security status. It has caused great difficulties and risks to group management.

The major real estate groups across the country have begun to plan and promote smart upgrade plans, and carry out the construction of the group's information system in accordance with the 'unified planning and step-by-step implementation' approach to promote the overall upgrade of smart communities in an orderly manner. Construct an information system investment guarantee mechanism. Gradually promote the construction of the group's information and data integration platform, realize the integration of data from various business, financial and other information systems within the group, and provide scientific data support, analysis and decision-making methods for business decisions. Based on the actual situation of the group, it will gradually advance in an orderly manner, eliminate information islands and provide support for integrated and intelligent management.

6.5.1.3 Alibaba's Solution

Based on current customer status, Alibaba Cloud builds a unified IoT platform for smart communities, provides ubiquitous connectivity and cloud services for various devices in the community, and forms an organic network with comprehensive perception, extensive interconnection and mutual collaboration. Realize the collection of various types of data, precipitate a unified community data center and build a unified user identity authentication, a unified payment system and a unified operation management platform based on this, support community security, inspection and other management applications, and pass the community one. The multi-code combination of Code (Face) connects people, places and things; realizes intelligent management of people, places and things; supports community owners' consumer services and other user services; and builds community smart applications and experience portals. Through digital modeling, a digital twin with all-element digitization and virtualization, real-time status and visualization can be constructed to realize the digital and visual display of community management and operation.

The smart community platform adopts the microservice architecture commonly used in the industry, uses the mainstream open source technology framework, separates front and back ends, supports distributed cluster deployment, and follows a unified technical route. The architecture design focuses on loose coupling between services and high cohesion within services. Business abstraction and mapping realize business object componentization and unified service invocation, and fully consider the system's scalability, reusability and configurability, reduce development and maintenance costs, so that the system can change as needed, and quickly and flexibly meet the business needs.

Based on the comprehensive and multi-channel analysis of the business needs of multiple customers' smart community projects in the real estate industry, in-depth understanding and grasping of the construction goals and expected effects of local customer projects, as well as grasping the needs in multiple business scenarios, and refined smart community projects. The overall architecture system design is as follows:

From the overall architecture system in Figure 6.19, we can see that the overall architecture is divided into three layers. Below we will elaborate on the three layers in the figure:

1. Application front end

   - C-side mini program: Committed to providing community owners with a fast and convenient visualized operation interface, helping them quickly complete payment services, reporting for repairs, pedestrian registration, visitor invitations, community

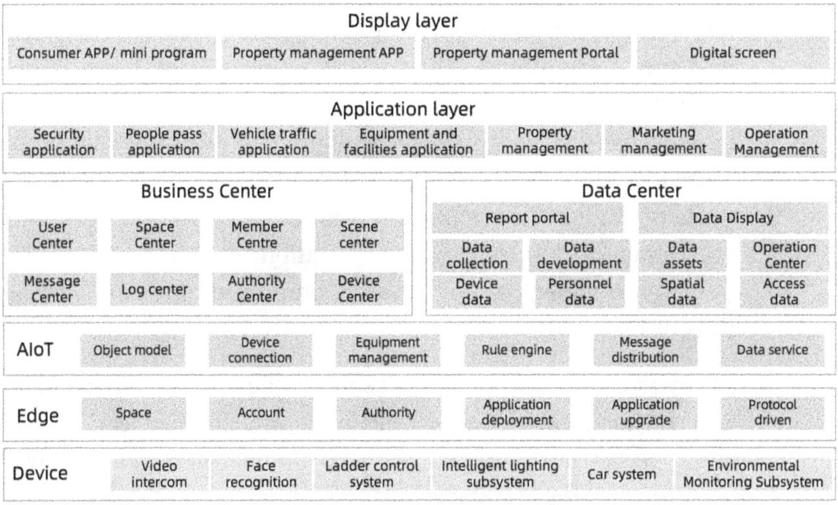

FIGURE 6.19 Alibaba Cloud AIoT smart community business architecture diagram.

life and other services, and intimately providing owners with online help. Create a safe, comfortable and convenient community management environment for owners, and realize convenient property management for community 'smart life' and 'resident autonomy'.

- B-end mobile management terminal: To provide property management personnel with more convenient property collection fees, convenient meter reading, work order management, daily inspections, vehicle management and other property services, while better providing services to owners and improving property service owners Satisfaction.
- B-side web management background: Property management personnel web background visual operation interface, unified application management platform, management applications include car dealerships, pedestrians, security, toll, basic files, work order management and EBA rule management, to improve property management Efficiency, to help the property quickly complete the informatization and intelligence of community management.

2. Group cloud

- AIoT middle station: AIoT middle station serves as an important bridge to communicate the underlying perception layer equipment and upper application layer services. It encapsulates the heterogeneity of the underlying equipment and network, provides a unified and universal access interface and realizes the reuse of data and computing resources. AIoT middle station is the center of data management, equipment management and event management, and is the core component of IoT application integration.
- Business center: The business center is the unified management and authorization of multi-business applications (car dealerships, pedestrians, security, property management), and realizes platform sharing and loose business coupling.
- Data center: Through the integration of existing ERP enterprise data with real estate groups and property companies, and at the

same time pull through the core master data of multiple formats (car dealerships, pedestrians, security and property management), improve data quality, and provide data decision-making basis for horizontal operation and management of multiple formats.

- Vehicle application: Core business applications such as parking space management, vehicle entry and exit management, toll management, exception management and financial management in the smart community project.
- Pedestrian applications: Core business applications such as owner access, visitor access, face recognition, all-in-one card, QR code access and visual cloud intercom in smart community projects.
- Video applications: AI security applications such as personnel control, face access control, visitor management, perimeter alarms, parabolic objects and public area occupancy in smart community projects can be effectively managed through security event analysis and picture and video viewing.
- Property management services: Core business applications such as community basic file management, property charges and work order management in smart community projects.

3. Local property project
    - Application containerized deployment: The edge applications of car dealerships, pedestrians and videos are all deployed in the docker container in the local property all-in-one machine of the project. Containerized deployment is more convenient and safe, can effectively compartmentalize various services, and is easy to maintain and troubleshoot problems.
    - Unified control of hardware devices: All hardware devices in the community management unit are managed uniformly by the device control interface service, which can centrally manage each device.
    - The project's local car dealerships, pedestrians, video applications, etc. are connected with the equipment through the equipment control unified service. After the local business logic is

processed, the Alibaba Cloud AIoT edge field services are transferred to the Alibaba Cloud IoT platform.

- Through the edge autonomous application management capabilities and the distribution and deployment capabilities based on the cloud-side collaboration system, it provides highly available, operable, and iterable remote operation and maintenance capabilities for the edge applications of the property all-in-one machine. At the same time, it provides cloud-side collaboration components to realize standardized cloud-side system integration between the project and the headquarters. Through the implementation of the smart community construction program based on Alibaba Cloud AIoT 'Edge Application Management Platform', it helped the client property group to achieve rapid replication and rapid integration of the project system, which greatly reduced the cost of smart project transformation (Figure 6.20).

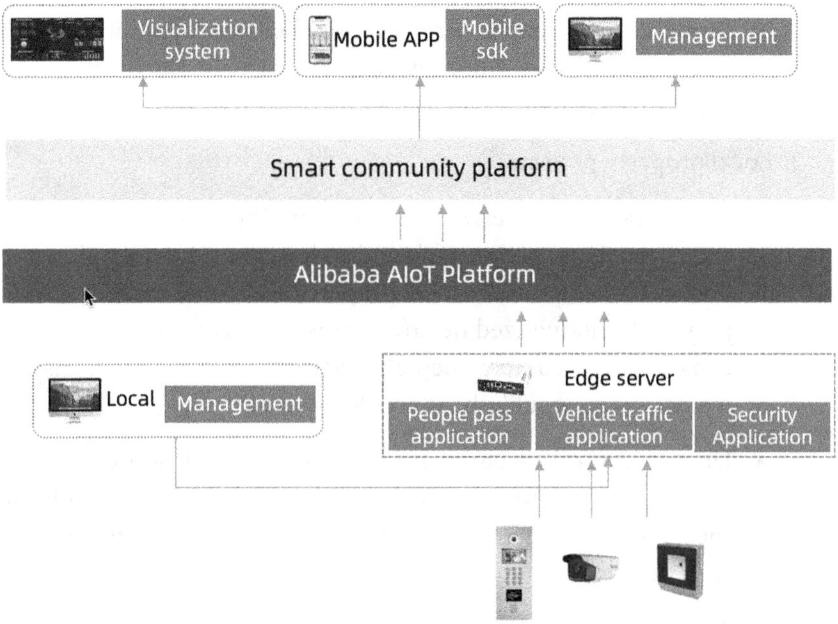

FIGURE 6.20 Alibaba Cloud AIoT smart community solution architecture diagram.

Through the implementation of a smart community construction program based on cloud-side collaboration, it has helped client real estate group properties achieve rapid replication and rapid integration of community systems, greatly reducing the cost of smart community transformation.

6.5.2 Case of Smart Park

The smart park is an important position for building a new era of industrial economic development. It is an important starting point for timely capture of emerging technologies and continuous integration into the park's innovative development. It is an important driving force for the organic integration of park production, life and ecology. It is a multi-format park construction, an important guarantee for modern management systems and capabilities.

As a leading cloud service provider for the new infrastructure of the digital economy, Alibaba Cloud has built multiple smart parks across the country in the construction and operation of market-oriented smart parks. The cooperation of outstanding groups in their respective fields will surely create industry representatives with a new era, a new benchmark for meaning.

Through the comprehensive use of the IoT, big data, cloud computing, AI, 5G and other new technologies and new applications, integrated into the construction of the smart park platform system, it will effectively improve the park management group's standardized management and control of parks located in different geographical locations. Utilize the rapid information collection, high-speed information transmission, highly centralized computing, intelligent transaction processing and ubiquitous service provision capabilities of the smart park to realize timely, interactive and integrated information perception, transmission and processing from the headquarters to the park. Improve the competitiveness and sustainable development of the park, and realize a first-class, smart, safe, comfortable, convenient and efficient soft environment for economic development in the park.

In the application and construction of the smart park, the IoT plays the role of the central nervous system and is the cornerstone of the 'smart park'. The IoT sensing front-end deployed in the park can agilely perceive the problematic infrastructure through data, quickly, accurately and effectively process all kinds of collected information through a unified management and control platform, and make intelligent management and control of various facilities, to realize the linkage between the systems, to realize the automatic perception and intelligent control of the infrastructure.

6.5.2.1 Market Status

In recent years, with the development of new-generation information technologies such as mobile communications, the IoT, big data and cloud computing, digital parks have further evolved to intelligentization on the basis of digitalization. Relying on the IoT can realize the intelligent perception, identification, positioning, tracking and supervision of physical entities; with the help of cloud computing and intelligent analysis technology, it can realize the processing and decision support of massive information. Through the development of digital parks, each park has accumulated a large amount of basic and operational data, but it also faces many challenges. The main challenges include processing issues such as the collection, analysis, storage and utilization of massive information at the park level, various complex issues in the integration of multiple systems, and the alienation of park development brought about by technological development. More importantly, the current digital parks only realize the one-way transmission of data and information (upstream transmission of data), and record and measure the data of entities. It is recorded in the 3D geographic information system and displayed through 3D virtual reality technology, but neglected, or in other words, the downlink transmission and application of data (and the transmission of control data to physical entities) could not be realized under the technical conditions at that time.

6.5.2.2 Customer Needs

The focus of the smart park platform lies in integrated construction. Historical experience has shown that system segmentation and data chimneys are common phenomena in the information age. Traditional building smart construction requires a set of application systems for hardware from different manufacturers. For the management personnel, the operation pressure is very high, and sometimes even the login password is always forgotten or misunderstood. The system value is very single and cannot form the linkage between the systems. In the digital age, the integration of data, the unification of data assets, and the integration and application of data are the source of sustainable development. Therefore, the difficulty of the smart park platform lies in:

1. How to realize the convergence of the whole area of the park, including the data automatically collected by the IoT, the collection of online and offline management service data and the data connection between horizontal heterogeneous systems.
2. How to realize the openness and advancement of platform construction, the rapid development of technology and the need for ecological integration of resources, all of which require the construction and expansion of the platform to be continuous, and it is a system that can 'grow'.

The specific construction goals of the park include:

1. Build an IoT network covering the park to realize comprehensive perception of power plant data

   The fourth technological revolution based on big data, the IoT, AI and the IoT is bringing us into an intelligent world of perception, interconnection and intelligence of all things. The rise of technologies such as the IoT and AI have provided mature conditions for the construction of a truly smart park in terms of device interconnection and network transmission.

   In the application and construction of the smart park, the IoT plays the role of the central nervous system and is the cornerstone of the 'smart park'. The construction of a smart park connects the core technologies of the IoT based on sensing and radio frequency technology, and connects the ubiquitous smart sensors implanted in the infrastructure of the park through the IoT to realize a comprehensive perception of the park.
2. Analyze and make decisions on perception data to realize intelligent management and control of park facilities

   The IoT sensing front-end deployed in the park can agilely perceive the problematic infrastructure through data, quickly, accurately and effectively process all kinds of collected information through a unified management and control platform, and make intelligent management and control of various facilities, to realize the linkage between the systems, to realize the automatic perception and intelligent control of the infrastructure.

3. High-tech application scenarios are accurately implemented, and the humanized service experience is fully improved

   Based on big data applications, extensive use of AI technologies such as voice interaction, face recognition, and machine learning, combined with various front-end smart interactive screens and other terminal equipment deployed in the park, make the mutual assistance between humans and machines more concise, enhance the human–computer interaction experience, and make work. Life is more convenient.

6.5.2.3 Alibaba's Solution

In November 2020, Alibaba launched the 'Alibaba Cloud Digital Intelligence Park White Paper', clearly defining that the Digital Intelligence Park is a brand new development stage of the park in the new environment, with data as the core production element and digital operation as the core production relationship system reform. The new type of park is a new type of park with a comprehensive increase in productivity and a new type of park with a comprehensive linkage of people, things and space. It not only covers a complete industrial ecology, but also an organic life unit with self-growth ability.

In fact, the intelligence of the park can be said to be reflected in each stage of the park form, and the characteristics and focus of the intelligence of each stage are different. In the 1.0 stage of the park, through the application of automated office systems, the management department of the park only has simple data collection work for the park, and uses a small amount of data to adjust the industrial layout and development policies of the park; in the 2.0 stage, not only the buildings in the park have realized the intelligent weak current system, but with the improvement of inter-industry synergies, the amount of data in the park is gradually enriched, and emerging information technology methods are gradually applied to the planning and management of the park, and the inter-department Simple data connectivity began to appear. The operation of the park begins to reflect the characteristics of intelligence; in the 3.0 stage, with the expansion of the production and living functions of the park, the data flow of the park became more and more abundant, and the management and operation departments of the park began to accumulate a large amount of data. A smart park is based on a unified integrated framework. The prototype is beginning to emerge, but there is still no suitable answer to how to fully

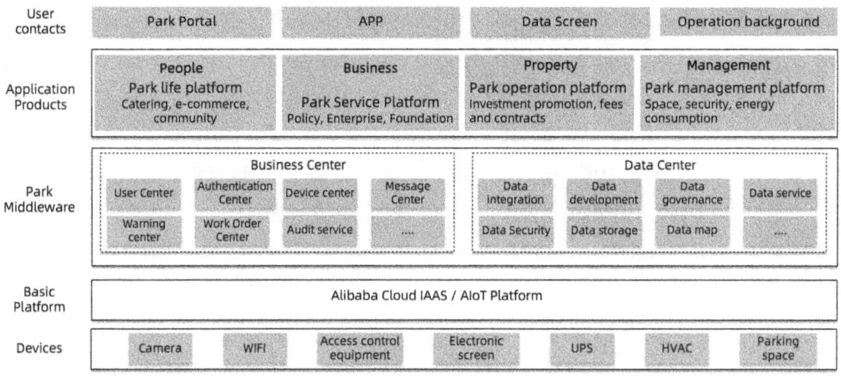

FIGURE 6.21 The overall architecture of Ali AIoT Digital Intelligence Park.

transform the data assets into the driving force for the development of the park; in the 4.0 stage, with the further deepening of industry–city integration, the functions and formats of the park will become more precise and more integrated. With the evolution, people's increasing production and living needs have caused serious conflicts with the backward operation and management status of the park. The Digital Intelligence Park has become a key measure to solve this problem (Figure 6.21).

The effective experience that Alibaba Cloud's solution in the campus brings to customers is practice:

1. Through the continuous fall to do the digital operation of the park, to truly perceive and recognize the pain points and needs of the park, in order to push back the construction of the smart park, and form the continuous rolling optimization of construction and operation to refine and refine Develop an effective smart park system and capabilities.

2. Build a digital 'new weak current' based on the capabilities of Alibaba Cloud's IoT, realize the interconnection of everything, global intelligent control, and truly amplify the intelligent value of things, facilitate unified management and maintenance, reduce usage costs, and improve response to emergencies. Ability to enable supervisors to make decisions quickly and reduce losses caused by emergencies.

3. The smart park platform is a tool, and people's livelihood is the key. For this reason, the construction of the operation team first needs

to maintain a high degree of ideological consistency with the group, and then gradually promote the park management and service from empirical judgment to data analysis and passive response to active service, and strive to achieve the goal of 'managing people, managing houses and managing affairs' is transformed into 'managing platforms, managing data and managing events'.

CHAPTER 7

Comprehensive Cases of Digital Transformation

Zhuang Longsheng, Li Chengqiang,
Hu Zhiqiang, and Caolin

7.1 CASE OF AN AIRLINE COMPANY

7.1.1 Customer Background

At the present stage, the civil aviation industry in the newly developed regions is still in an important period of opportunity and will continue to grow rapidly for a long time. G Airlines (hereinafter referred to as G Airlines) has made clear the goal and development blueprint of building the world's leading air transport enterprise. G navigation in the field of informatization construction will keep up with the trend of the industry and technology development, the implementation of the 'leading construction of information system, provide the best IT service' for the construction of the goal, continued its digital transformation, to strengthen the new technology, aviation core algorithms and big data analysis research and applications, to lay a good foundation for G intelligent navigation new era development.

G Airlines adheres to the general keynote of seek improvement in stability in the current digital construction, hoping to complete the IT technology transformation through the implementation of cloud computing and big data technology, realize unified data management, system interconnection, interconnection, establish a safe, efficient, advanced and intelligent microservice platform.

DOI: 10.1201/9781003272953-7

To this end, G Airlines introduced a series of products and services such as consultation, design, deployment, training and maintenance from Alibaba Cloud after bidding. By drawing on the advanced experience of Internet companies and based on the IT development vision of G Airlines and the current situation of IT facility management and construction, the company planned, designed and gradually assisted G Airlines to implement the digital transformation based on the mid-station architecture.

The digital transformation of G Airlines could be divided into several phases to timely for deployment in different periods and fields: as of this book is written, the double Middle Office architecture of G airlines only completed the design and implementation of the Data Middle Office, and the scheme of its Business Middle Office has not yet been finalized. Therefore, this case will mainly introduces the Data Middle Office.

7.1.2 Alibaba Cloud Solution

Considering the business and IT system status of G Airlines, we mainly implemented relevant transformation from the following three directions.

- Build resilient, highly available and secure data centers for the future.
- Establish a unified data management system and tool platform.
- To build a business-oriented data capability sharing platform: Data Middle Office.

The following will be a brief introduction from the above three aspects.

7.1.2.1 Construction of Hybrid Cloud Data Center

The construction of 'Hybrid Cloud Data Center' of G Airlines is a strong base and cornerstone to support the digital transformation (Figure 7.1).

G Airlines hybrid cloud construction objectives can be summarized as: Enhance business availability, improve business response capacity, accelerate business innovation, reduce the total cost of IT ownership, so as to build a flexible and elastic IT infrastructure architecture.

The overall capacity building framework includes six parts: strategic direction, resource pool and cloud relocation planning, operation and maintenance planning, disaster recovery management system planning, safety management planning, and governance and control planning.

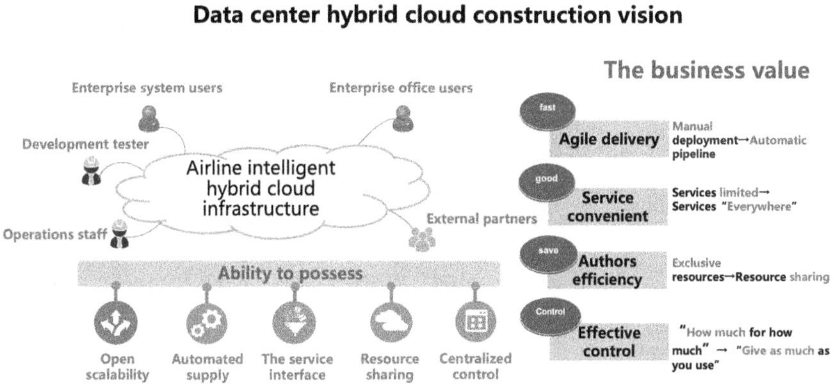

FIGURE 7.1 G airline data center construction vision.

With the rise of cloud computing, big data and artificial intelligence applications, the whole Data Middle Office market presents four major trends from general to heterogeneous, from coupling fixed to pooling combination, from center to edge, and from single cloud to hybrid cloud. The strategic direction of G aviation cloud platform construction can also be constructed through four stages of evolution according to this trend. The overall IT infrastructure development will gradually break the boundaries of CPU, server, data center and enterprise network, and go through four stages from universal to heterogeneous, from coupling fixed to pooling combination, from center to edge, and from single cloud to hybrid cloud, to practice the strategy of boundary-less data center (Figure 7.2).

Data center resource pool and migration cloud planning mainly includes server resources, storage backup resources, network resources, database resources, cloud desktop and computer room planning. As well as the application of cloud evaluation model, migration path.

Taking the mature infrastructure as a service (IaaS) platform in the industry as the benchmark, and combining with the trend development, the gap of G aero cloud is analyzed moderately and in advance. The gap analysis is carried out from the six aspects of computer room, network, storage, server, database and application migration involved in the two stages of construction and use (Figure 7.3).

Overall architecture of server resources and ability to replenish the cloud of optimized structure and the FPGA and the GPU ability form gradually retire on minicomputers, convert most of the traditional virtual machine host and ECS cloud and keep some physical machine and

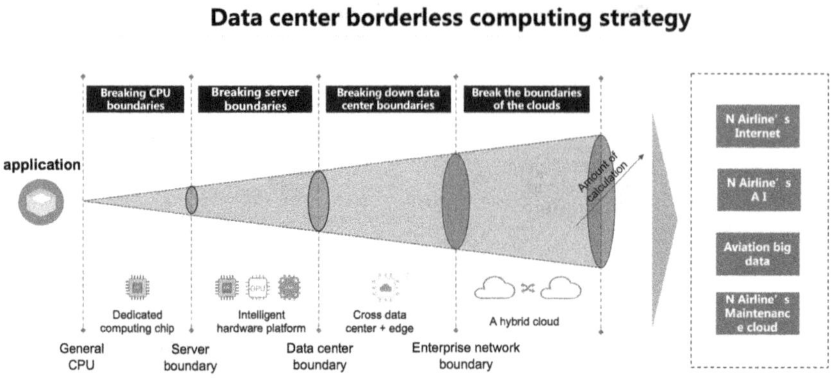

FIGURE 7.2 G airline borderless computing strategy.

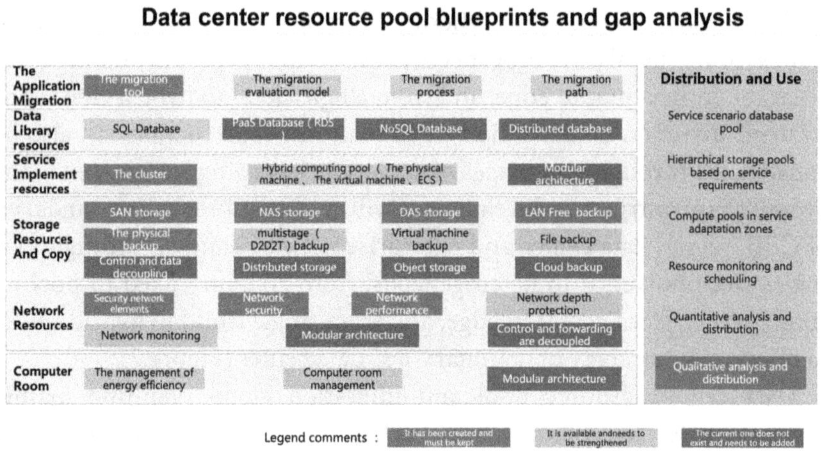

FIGURE 7.3 Blue chart and gap analysis of data center resource pool.

machine mixed state computing pool formation, and functional partition of computing pool performance, business demand assignment in the computing pool and migration. Atomic state of resources and cloud scheduling technology to achieve quantitative allocation and flexible scheduling capabilities are to provide support for computing full service.

Storage backup planning includes the overall architecture design of storage backup, resource type and hierarchical design, usage allocation principles, storage and backup architecture, storage and backup capacity estimation, as well as recommendations for the use of cloud backup.

The network planning mainly includes the overall functional architecture planning, G Airlines proprietary cloud available area, network security, data center interconnection, network operation and maintenance, disaster recovery network design and double active network design.

Database planning is mainly guided by business development to build multi-type and multi-architecture database services. In the future, G Airline Database Ecology focuses on supplementing the distributed architecture, introducing non-relational databases, providing them to users through mirroring, interfaces and containers, and providing users with user-side database management services at the same time.

Migration cloud planning should have a scientific evaluation model, which is similar to funnel to screen out applications that need to go to the cloud and design the most appropriate method for these applications, evaluate the resources needed for migration cloud and give migration strategy and route.

The main objectives of G Airline hybrid cloud operation and maintenance operation are the improvement of the support system (PPTR), the process design and the construction of tools in the two aspects of intelligent operation and maintenance and business operation (including service support). Operation and maintenance operation construction tasks include asset and configuration management domain, monitoring operation and maintenance domain, service support domain, business operation domain and automatic management five capability domains. Key points of construction include:

- Hybrid cloud asset configuration: Configuration database, configuration baseline management and automated configuration.
- Hybrid cloud operation and maintenance monitoring: Implement comprehensive and unified monitoring, integrate monitoring data, develop automation engines and adapters, deploy process data bus services, create and test process choreography for tasks such as installation and configuration.
- Hybrid cloud operation: Service catalog design, publication, delivery and support, user usage metering, usage analysis and user experience management.

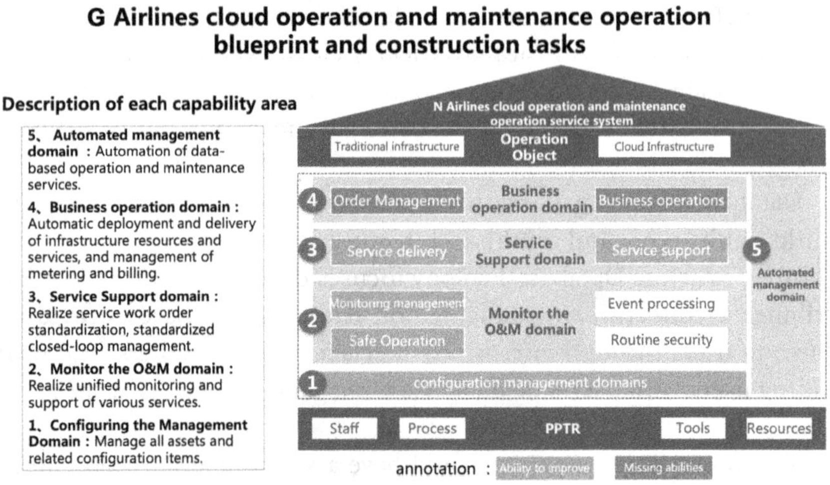

FIGURE 7.4 G airline cloud operation blueprint and construction tasks.

- Intelligent operation and maintenance: System log data access and data governance, mining and analysis, data visualization, supervised machine learning, automated problem analysis and processing (Figure 7.4).

Disaster recovery system planning includes the setting of disaster recovery metric, the design of disaster recovery strategy, the preparation of disaster recovery technology and architecture, and the design of disaster recovery management system. The overall view is as follows (Figure 7.5):

Based on the construction principle of maintaining development in the same city and survival in different places, the disaster recovery capability of 'two centers in the same city' and 'three centers in different places' of G Airline hybrid cloud is built. At the initial stage, most data-level disaster tolerance and a small part application-level disaster tolerance are realized.

The construction of disaster recovery system needs to ensure the integrity and recoverability of information system data as well as the continuity of critical business processing when disaster occurs, and it needs to be constructed and adjusted in combination with the construction pace of G Airline Data Middle Office's proprietary cloud. In the later stage, it is necessary to realize full data-level disaster tolerance and most application-level disaster tolerance.

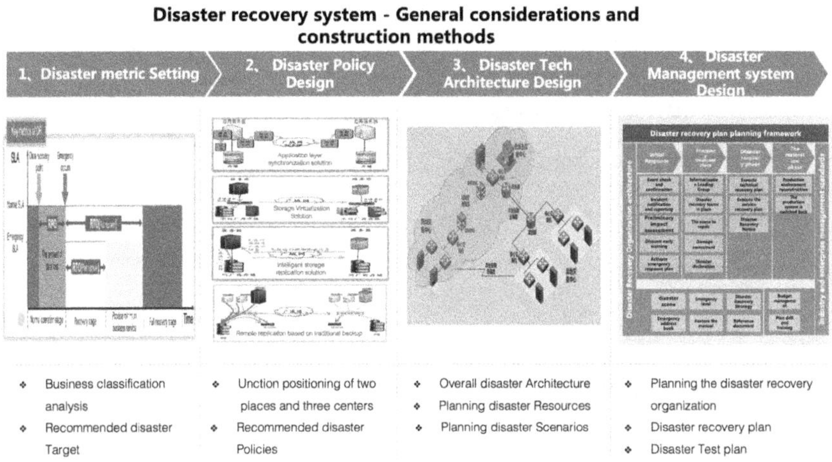

FIGURE 7.5 G airline disaster recovery system.

Specifically, the old and new airport machine rooms in Guangzhou jointly form the data center environment, undertake the data backup and application recovery of the core application and realize real-time mutual standby and double active disaster recovery. There are multiple exits between the disaster preparedness center of the new airport and the disaster preparedness center of the stock building, so as to ensure non-destructive disaster preparedness of the business. Beijing Daxing Computer Room builds an off-site backup center to undertake comprehensive off-site data backup functions.

When the new computer room (main center) cannot be used normally, the old computer room can be taken over as the disaster recovery center in the same city. When both the new computer room and the old computer room cannot be used, the core application should be rebuilt in the Beijing disaster recovery center (Figure 7.6).

The security management of the data center is mainly combined with the security issues that G Airlines focuses on, clarifying the key security tasks of cloud platform construction, and implementing the landing security operation plan. The security framework provides a top-down view of the components that enable G Airlines to build a full range of high-performance security systems.

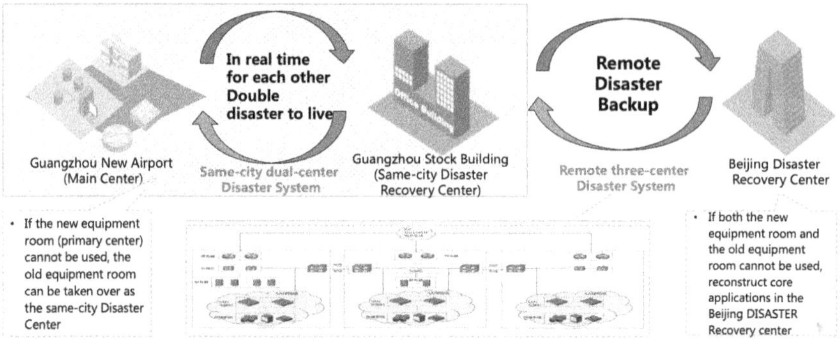
In real time
for each other
Double
disaster to live
Remote
Disaster
Backup
Guangzhou New Airport
(Main Center)
Same-city dual-center
Disaster System
Guangzhou Stock Building
(Same-city Disaster
Recovery Center)
Remote three-center
Disaster System
Beijing Disaster
Recovery Center
• If the new equipment room (primary center) cannot be used, the old equipment room can be taken over as the same-city Disaster Center
• If both the new equipment room and the old equipment room cannot be used, reconstruct core applications in the Beijing DISASTER Recovery center

FIGURE 7.6 Overall disaster recovery plan of G airline.

7.1.2.2 Data Governance System and Tool Platform Construction

As a huge enterprise, G Airlines have pretty complex businesses, also its systems. More than 400 sets of various application systems are being used. The system construction after the replacement of the multiple period or iterative transformation, and the construction goal of each period and construction standards and specifications are inconsistent, the company of the whole system data not only failed to become a value-added assets, but a variety of inconvenience to the performance of a business, 'Pull through, standardization, safety and value realization' need to be addressed. For example:

1. Passenger marketing data have the following problems:

    - Insufficient data perception: Internal and external data are not connected, such as customer information, competitors and search data; all business departments are managed by chimneys, and there is no unified planning for system construction in the early stage, and the standards are inconsistent.

    - Lack of customer insight: Lack of scene marketing, such as B-side management lack of structured product matrix, no product packaging or delivery for specific scenes; product homogenization is serious, product management does not combine customer experience process and preference combination of products.

    - Product management tradition: Lack of product management, lack of product landing ability, lack of supporting standards and assessment metrics, such as insufficient support for product delivery terminals, and lack of ability to recover from abnormal situations.

- Low efficiency of organizational structure: Lack of effectiveness of organizational structure, unclear responsibilities, complex decision-making and management, lack of coordination among businesses, such as lack of horizontal connection among seven expert working groups for marketing system construction, resulting in insufficient market perception and corresponding deficiencies.

2. The customer service data have the following problems:

- Data incomplete: For each service node of the service module and chain, there is a lack of chain information; for example, customer return visit data are distributed in multiple channels.
- Insufficient fine management: Lack of unified service system, marketing management does not provide corresponding rules.
- Lack of customer research: Lack of in-depth research and insight into customer experience process and key nodes, lack of integration of customer experience process.
- Lack of differentiation: Customer experience has not achieved the core value points, the core touch experience and leading airlines gap is large; failing to provide a signature highlights experience that clearly sets it apart from its competitors to impress the passenger airline...

  Other modules are not enumerated.

Overall requirements for data governance based on intelligent transformation, G intelligent navigation transformation goal vision, set the following data governance goal: To build a fusion, standard unified standardization, data timely and credible, safe and efficient supply chain data, realize the value of data-driven business and promote liquid, enhance the production and operation ability (Figure 7.7).

According to the practical experience of Alibaba, the framework of the data asset management system includes four levels of work content: The whole supply chain of the data layer is the core; the governance layer is

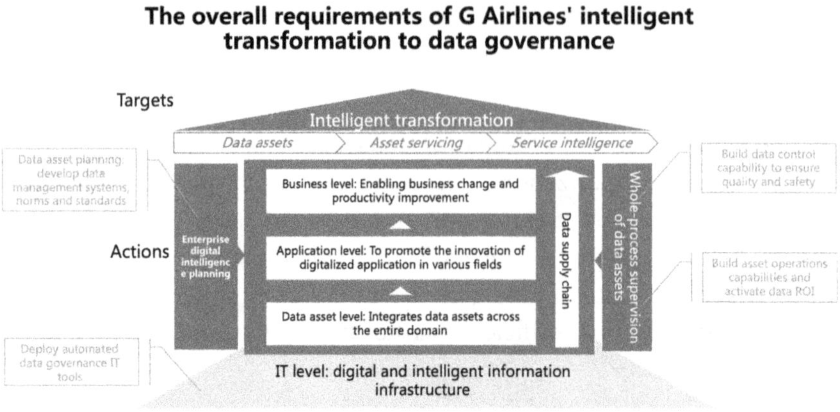

FIGURE 7.7 G airline general requirements of data governance.

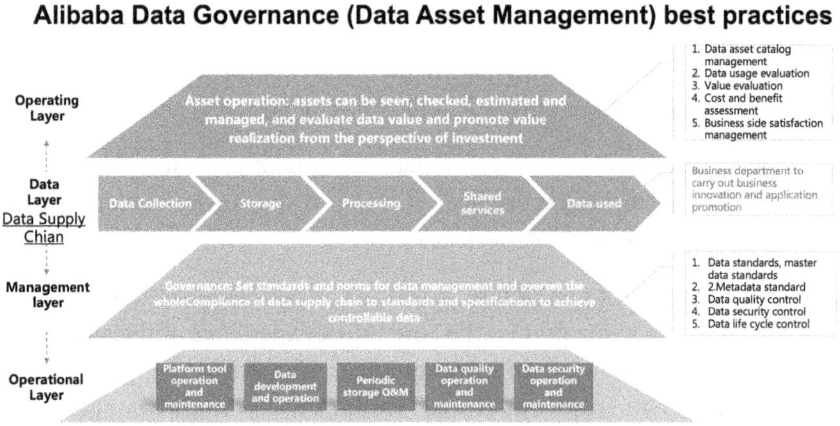

FIGURE 7.8 Data asset management system.

responsible for setting standards and monitoring to achieve controllable data; the operation and maintenance layer is responsible for ensuring normal operation; and the operation layer is responsible for making assets visible, traceable, estimable and manageable (Figure 7.8).

By combining the status quo of data governance of G Airlines and the historical practice of Ali, we designed a more detailed governance scheme.

- Governance purpose: To ensure the accuracy, timeliness, security and high ROI (Return on Investment) of the data supply chain.

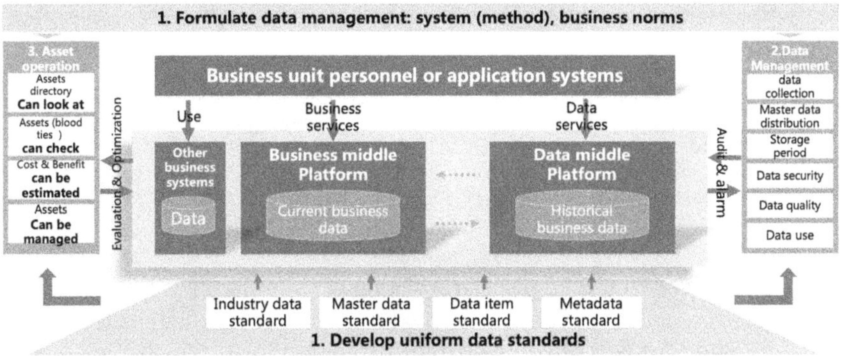

FIGURE 7.9 Data governance mechanism.

- Governance mechanism: Formulate unified standards and norms in advance, and all systems comply with the unified standards; establish the management and control system, audit the data, find the problems and deal with the alarm to ensure the consistency with the standard; after the operation of assets, evaluate the full amount of data, find problems and optimize them (Figure 7.9).

We first need to develop standards to guide the definition of global data for the company, including industry reference standards, master data standards, data item standards and metadata standards. After having the common data definition standard, it is necessary to formulate the standard and constraint means for data storage, flow, access rules, secondary utilization and value realization. A very important thing to consider is the security of the data; based on the above premise, Dataphin, Alibaba's data construction platform, is strictly synchronized and complies with relevant standards, and the company's global data are integrated, timely and agile to share various data capabilities with the front desk and the third party in the way of data middle desk, so as to support the discovery and solution of problems, trial and error and innovation of the business.

Formulated a series of systems such as (Figure 7.10):

The Dataphin platform of Ali itself is a set of solidified data governance norms for the use and construction of data. Combined with the overall system of the company, the goal of data governance can be achieved in a better and more efficient way.

① Data management system (i.e. general provisions)
② Data accountability management system
③ Master data management system
④ Data standard management system
⑤ Metadata management system
⑥ Data sensitivity classification specification
⑦ Personal privacy data management system
⑧ Data acquisition management system
⑨ Data storage management system
⑩ Data processing management system
⑪ Data usage management system
⑫ Data quality management system
⑬ Data security management system
⑭ Data life cycle management system
⑮ Data assessment management system
⑯ Data demand management system
⑰ Data supervision and inspection system

FIGURE 7.10 G airline data-related regulations.

7.1.2.3 Data Middle Office Construction

The first two parts mainly introduce some preconditions from the infrastructure, data management standards and tool platform, which are both auxiliary and necessary work for the digital transformation project of G Airlines in the data field. The following will introduce some contents of G air Data Middle Office.

The so-called Data Middle Office is the shared capability center of the whole domain data, including data collection, data modeling, data asset management, external data services and various data capabilities (mainly metrics and labels). The basic method is based on the middle desk methodology of Ali. First, it surveys the client's business picture, business process and data asset status quo, and then extracts the key business scenarios under the theme by business theme to design the corresponding data capability (MRD). Then according to the business needs, the shared data ability could be used in a variety of ways on Alibaba or other applications to show the value of data platform (PRD) to reflect the value of data middle office. Data in the data middle office after preliminary system construction are more important and are the optimized operation that takes a long time, continuous improvement and promote data middle office ability and value.

The following is a schematic diagram of the G air Data Middle Office frame (Figure 7.11)

Since the Data Middle Office needs to integrate the data of various business segments of the company in time, the construction of the Data Middle Office is doomed to be a long-term and gradual process. We can only choose to first incorporate the data of the core business segments into the Data Middle Office, and then gradually expand to the whole region. For example, in the case of G Airlines, we first chose the business theme in the red box for construction (Figures 7.12 and 7.13).

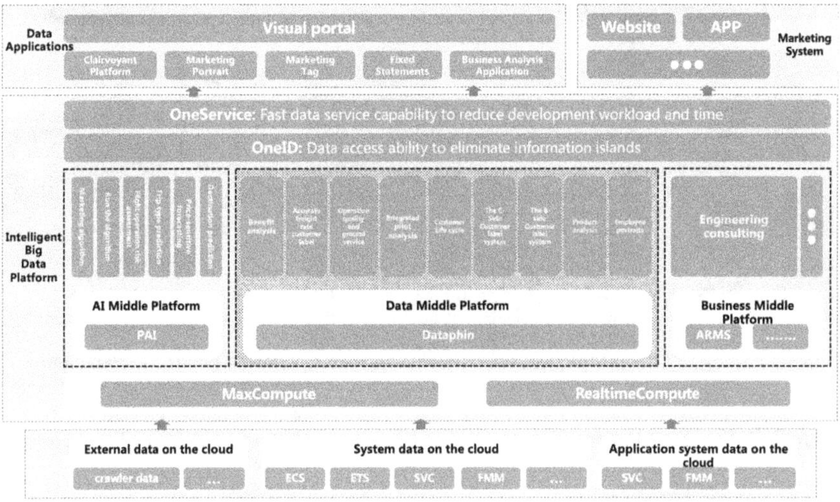

FIGURE 7.11 G airline overall architecture of Data Middle Office.

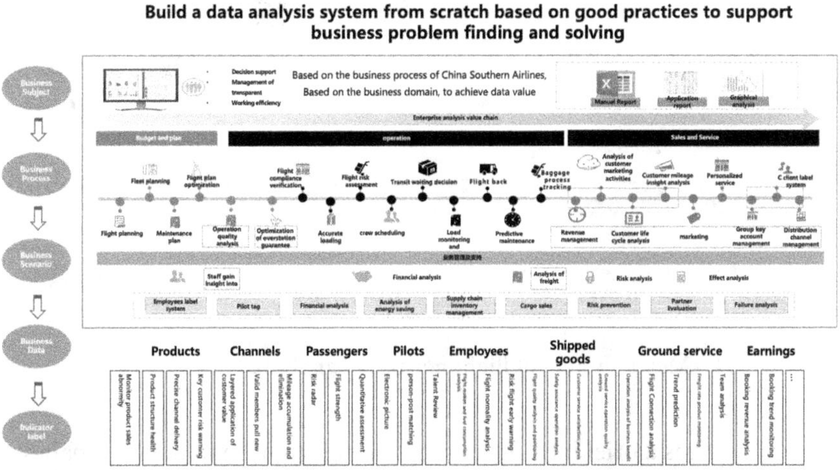

FIGURE 7.12 G airline overall design of data analysis system.

The core of the Data Middle Office is to provide the common data service capability to the front end, and form the application scenario solution of the demand through the free combination of applications. There are many business themes of airlines, and the scenarios under each business theme are inevitably complicated. In view of the limited length of cases,

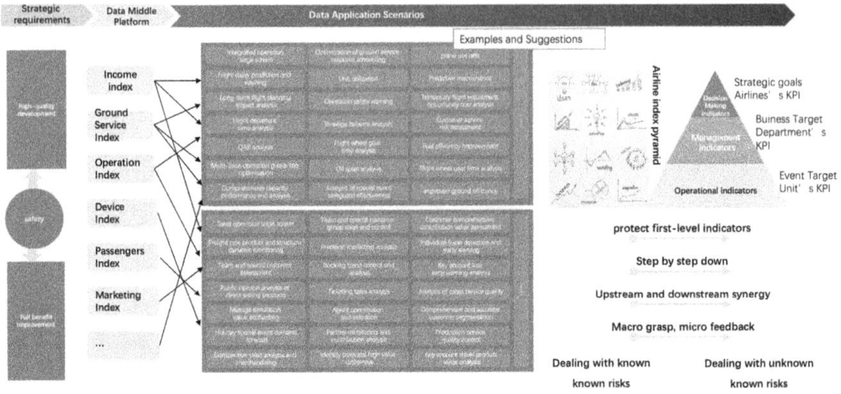

FIGURE 7.13 Big picture of data application scenarios.

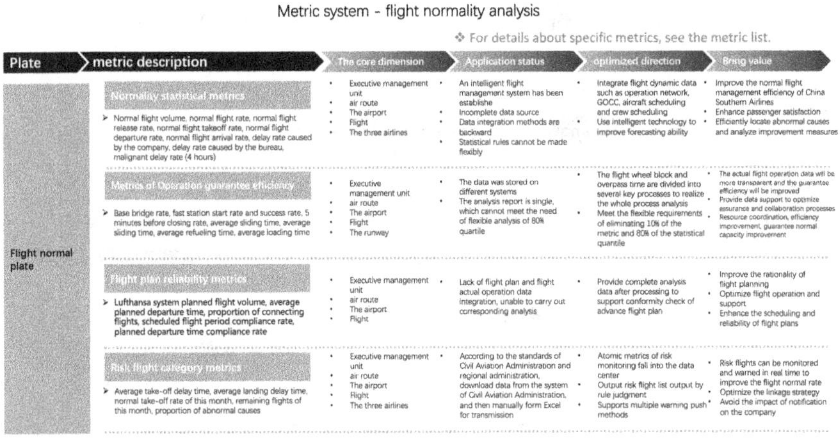

Metric system - flight normality analysis

❖ For details about specific metrics, see the metric list.

Plate	metric description	The core dimension	Application status	optimized direction	Bring value
Flight normal plate	Normality statistical metrics ➢ Normal flight volume, normal flight rate, normal flight release rate, normal flight takeoff rate, normal flight departure rate, normal flight arrival rate, delay rate caused by the company, delay rate caused by the bureau, malignant delay rate (4 hours)	• Executive management unit • air route • The airport • Flight • The three airlines	• An intelligent flight management system has been establishe • Incomplete data source • Data integration methods are backward • Statistical rules cannot be made flexibly	• Integrate flight dynamic data such as operation network, GOCC, aircraft scheduling and crew scheduling • Use intelligent technology to improve forecasting ability	• Improve the normal flight management efficiency of China Southern Airlines • Enhance passenger satisfaction • Efficiently locate abnormal causes and analyze improvement measures
	Metrics of Operation guarantee efficiency ➢ Base bridge rate, fast station start rate and success rate, 5 minutes before closing rate, average sliding time, average sliding time, average refueling time, average loading time	• Executive management unit • air route • The airport • Flight • The runway	• The data was stored on different systems • The analysis report is single, which cannot meet the need of flexible analysis of 80% quartile	• The flight wheel block and overpass time are divided into several key processes to realize the whole process analysis • Meet the flexible requirements of eliminating 10% of the metric and 80% of the statistical quantile	• The actual flight operation data will be more transparent and the guarantee efficiency will be improved • Provide data support to optimize assurance and collaboration processes • Resource coordination, efficiency improvement, guarantee normal capacity improvement
	Flight plan reliability metrics ➢ Lufthansa system planned flight volume, average planned departure time, proportion of connecting flights, scheduled flight period compliance rate, planned departure time compliance rate	• Executive management unit • air route • The airport • Flight	• Lack of flight plan and flight actual operation data integration, unable to carry out corresponding analysis	• Provide complete analysis data after processing to support conformity check of advance flight plan	• Improve the rationality of flight planning • Optimize flight operation and support • Enhance the scheduling and reliability of flight plans
	Risk flight category metrics ➢ Average take-off delay time, average landing delay time, normal take-off rate of this month, remaining flights of this month, proportion of abnormal causes	• Executive management unit • air route • The airport • Flight • The three airlines	• According to the standards of Civil Aviation Administration and regional administration, download data from the system of Civil Aviation Administration, and then manually form Excel for transmission	• Atomic metrics of risk monitoring fall into the data center • Output risk flight list output by rule judgment • Supports multiple warning push methods	• Risk flights can be monitored and warned in real time to improve the flight normal rate • Optimize the linkage strategy • Avoid the impact of notification on the company

FIGURE 7.14 Metric system of flight normality analysis.

the following is a brief introduction to the design of shared data capability in some scenarios under some business themes:

Flight normalcy analysis scenario: Flight normalcy analysis is used to measure the acceptance rate of aviation products, evaluate and assess the operational efficiency by using data metrics, track the flight dynamics of each node of operational support, compare the differences between flight plan and actual implementation, identify the unreasonable arrangement of flight plan or whether there is room for optimization, and ensure the normality of follow-up support and connection, etc. (Figures 7.14 and 7.15).

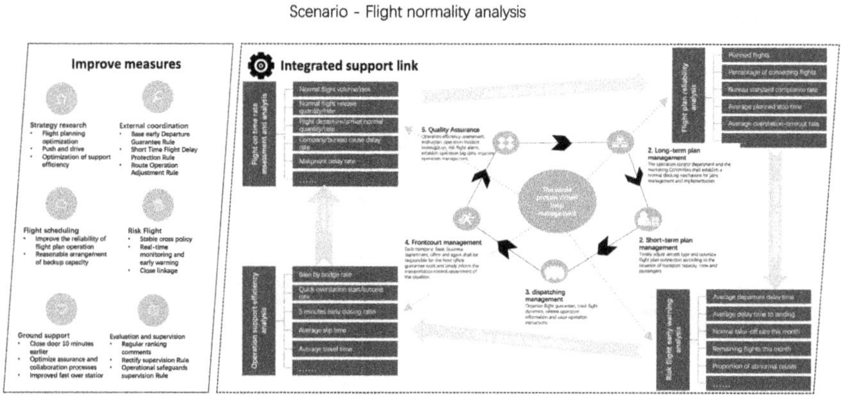

FIGURE 7.15 Analysis scenarios of flight normality.

The Ground service metric

❖ For details about specific metrics, see the metric list.

Plate	Metric description	The core dimension	Application status	optimized direction	Bring value
Operation quality analysis	**Closed door normal metric** ➢ Guangzhou Southern airlines closed X minutes ahead of normal class numbe ➢ Guangzhou Southern Airlines X minutes ahead of the number of late closing ➢ Close 5 minutes earlier _ domestic outbound flights	• air force station • Region, hub • air route • Flight plan • On the flight	• This metric is counted in daily and monthly reports, and there is no monitoring of flights corresponding to key events	• Increase the analysis dimension of planned flights and key flights, and provide means for business personnel to carry out monitoring of key events	• Can understand the general closing normal situation, provide decision support for daily operation
	Fast station pass efficiency analysis metric ➢ Quick over station can start the number of classes ➢ Number of effective start classes for fast overpass ➢ Quick over station start rate ➢ Quick overpass success rate	• Hub, region • Domestic and international • air route • air force station • Assessment metrics KPI	• This metric is counted in daily and monthly reports, and no corresponding KPI is associated for evaluation	• The analysis of fast overstation assessment metric KPI is added to improve the efficiency of fast overstation execution	• Through rapid station passing analysis, the efficiency of station passing execution can be evaluated, and the standard level of KPI of closing normal assessment of aviation bureau can be improved
	Metric of delayed disposal ➢ Number of flights, delayed flights, ground service responsibility delayed flights ➢ Length of delay (Max., Min., average) ➢ Assessment of disposal efficiency KPI, cost and expense	• Hub, station • Domestic and international • The delay reason • Molecular company	• This metric has been counted in daily and monthly reports. For the assessment KPI, it has not been separately refined, and the cost item needs to be refined	• By analyzing the assessment KPI, it can provide guidance for the management work of delay support and insight into the cost composition through cost analysis	• Through the analysis of delay disposal, we can understand the service efficiency and insight into cost expenditure to control cost input
	Process time assessment metrics ➢ Average process time ➢ On-time process completion rate ➢ Process time up to standard rate	• Hub • Air force station • Domestic and international • Process	• This metric is counted in daily and monthly reports. The process statistics are not complete at present, so it needs to be refined	• The process is more detailed, and the total time consumption of parallel and sequential processes is counted	• By subdividing the process, we can deeply understand the minimum time-consuming of the overall process, which is convenient for the subsequent support of faster decision-making
	Operation service evaluation metrics ➢ Process time up to standard rate ➢ Reduced the total number of people ➢ Closing downsizing metric, downsizing rate	• Hub • Domestic and international • Important Flight • The key models	• There are statistics of this metric in daily and monthly reports, which lack insight into key flights	• Increase key flights, as well as key models of special analysis, insight into key flights to ensure service governance ability	• To carry out the overall evaluation of guarantee service can find the shortcomings of guarantee capability and improve customer satisfaction

FIGURE 7.16 Metric system of ground service.

Application scenario of data analysis in the field of ground service–customer service satisfaction operation analysis: Carry out the satisfaction analysis of 'full touch' experience centered on passengers to promote and enhance the brand influence (Figures 7.16 and 7.17).

In the revenue management business, analytics are complex and important. Taking freight rate adjustment scenario as an example, it is necessary to evaluate the effectiveness and timeliness of pricing and promotion strategies based on the sales results of various promotional products, so as to further evaluate the effectiveness of promotional products and timely adjust the strategy of freight rate products (Figures 7.18 and 7.19).

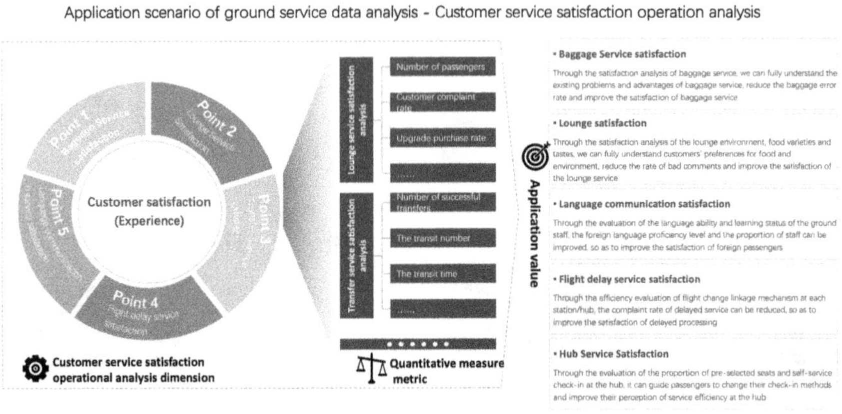

FIGURE 7.17 Analysis scenarios of customer service.

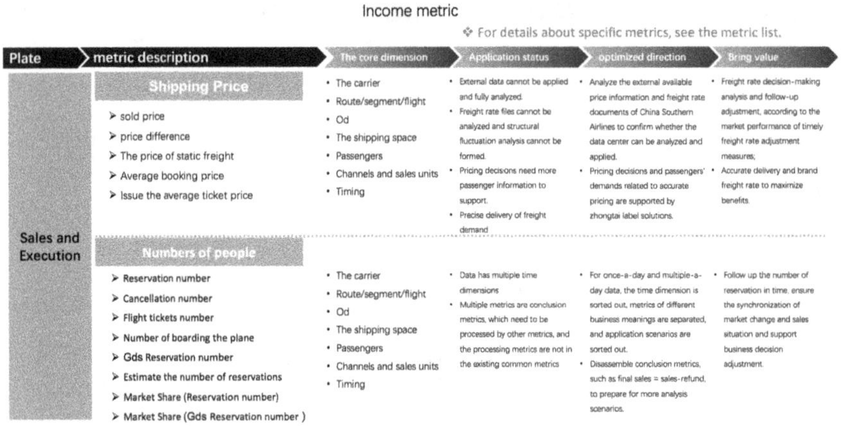

FIGURE 7.18 Metric system of income.

In the passenger marketing business, the B-end channel and C-end passengers should not only do the analysis and portrait, but also take into account the subsequent marketing activities to do targeted target group selection, so it is necessary to design a variety of labels for these business objects. With regard to the C-terminal passenger theme, we combined the practice of Alibaba New Retail and made some capacity planning and design for the data middle station according to Ali AIPL theory, customer life cycle and the current situation of G Airlines' user operation (Figures 7.20 and 7.21):

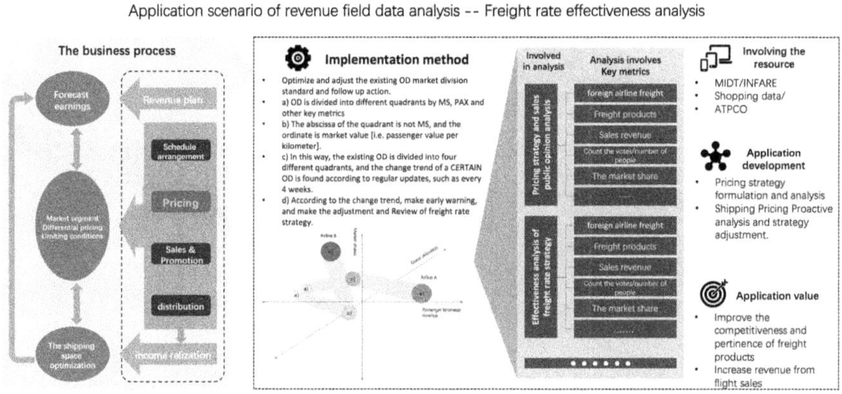

FIGURE 7.19 Analysis scenarios of income.

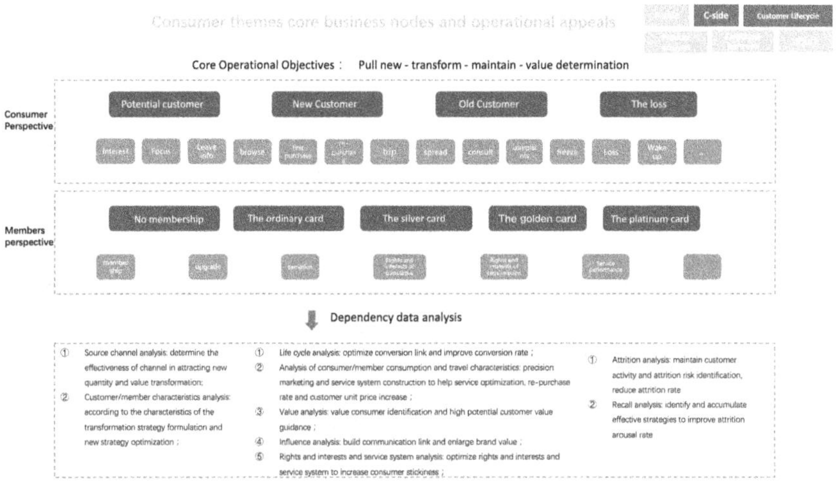

FIGURE 7.20 Customer-side passenger core business nodes and operation demands.

For example, in the application scenario design of customer value stratification: Due to the limit marketing resources, G Airlines should establish customer value evaluation system, combined with the specific business needs/objectives. When selecting target customers, the Airline could summarize customers' feature and preference, and then confirm products, services, sales, and events to be pushed. With ROI being guaranteed, the Airline could improve existing customer viscosity, and implement efficient membership development and maintenance.

(Figures 7.22).

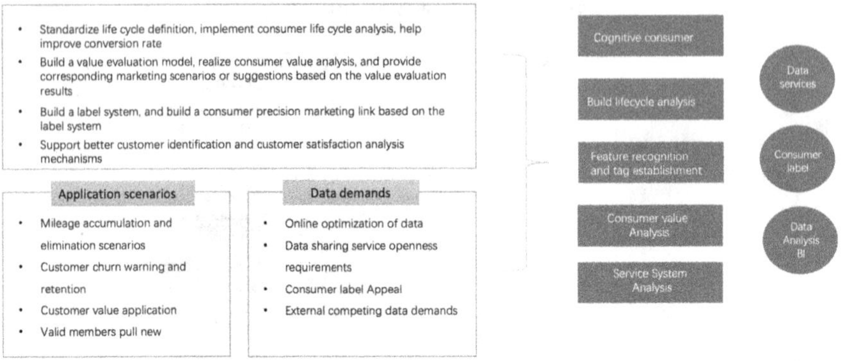

FIGURE 7.21 Customer-side passenger application scenario design.

FIGURE 7.22 Customer value-layered application design thinking.

Then take the early warning and retention scenarios of customer loss as examples: Navigation department business personnel by the loss of early warning model, identification has loss tends to customer in advance, on this basis, further value object, passenger in-depth, multi-dimensional analysis for high value (including the loss of reason and analysis of characteristics of customers), to develop and deliver targeted retention strategy, implementation based on the control of marketing costs, minimize customer churn rate and promote customer activity (Figure 7.23).

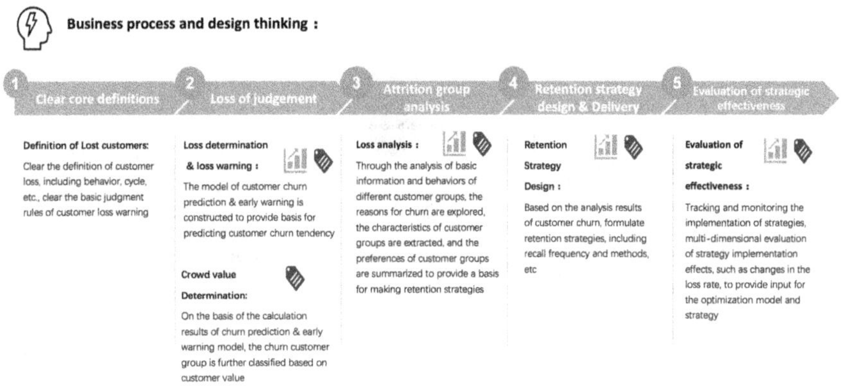

FIGURE 7.23 Loss of customer analysis application design thinking.

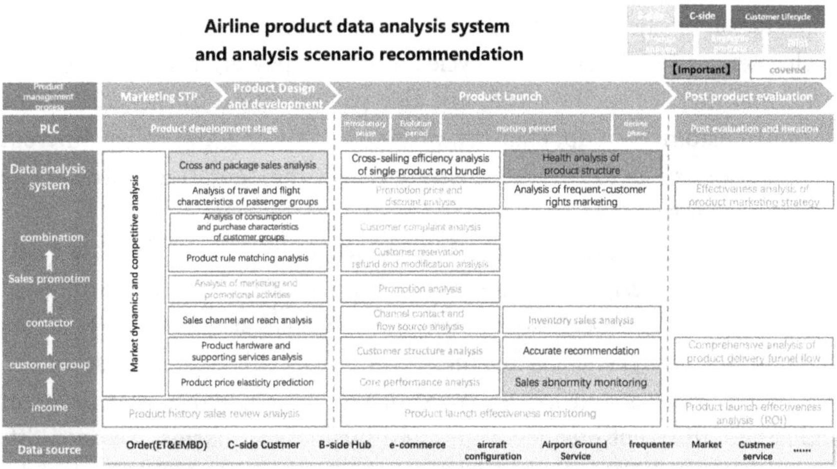

FIGURE 7.24 Airline company product data analysis system and scenario.

In the product analysis system, some supplements are made on the basis of some existing analysis scenarios: In addition to the main air ticket products, the business processes of additional, auxiliary and value-added products operate according to the product life cycle, and the corresponding data analysis also provides supportive insights for different product life cycles. The analysis is not only limited to sales volume and revenue, but also needs to pay attention to three closely related factors: customer base, channel and touch point and promotion (Figure 7.24).

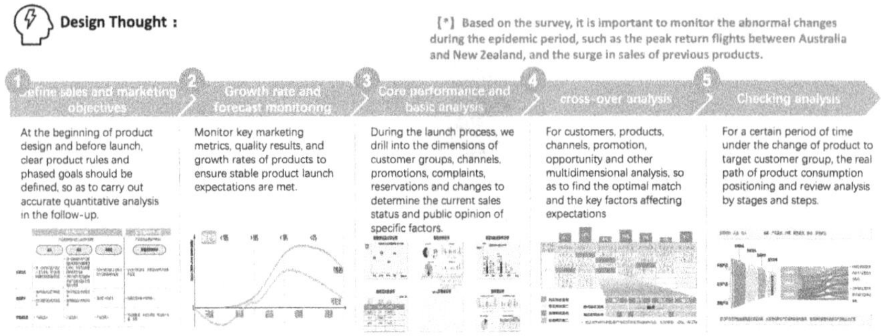

FIGURE 7.25 Design thinking of sales change scene.

Take the scene of sales change as an example: On the basis of defined product rules, customer groups and channel delivery, clear the target and expectation of sales stage; after the launch, we should focus on the key life cycle metrics for different life cycles, not only pay attention to the conventional metrics such as ROI, but also pay attention to the sales status such as customer group, channel, promotion, timing and other different information, so as to comprehensively reflect the operation status of the product (Figure 7.25).

The above cases provide a brief introduction to some of the data capabilities we have designed for the airline Data Middle Office from different perspectives, and provide some reference examples at the level of business themes, business scenarios, business processes, critical analysis scenarios and logic, metrics and labels. What needs to be noted is that different industries, different business types and business scenarios, we need different data capabilities, cannot be applied mechanically. What needs to be remembered is that in MRD design of the Data Middle Office, according to the business model, the purpose of the Data Middle Office is to extract core data capabilities such as metrics and labels to help us find and solve pain points in the business in a timely, fast and convenient manner, and support trial-and-error innovation.

7.1.3 Project Delivery Process

To be honest, the G Airlines project is the first middle-platform project delivered by Alibaba in the aviation sector, and the whole delivery process is full of challenges. Both Airline and Alibaba have organized huge delivery teams, and the two sides have set up a joint delivery team to achieve the

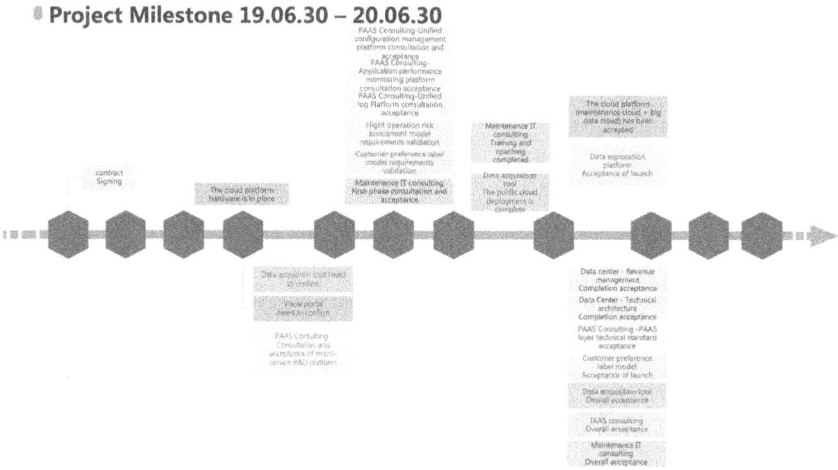

FIGURE 7.26 Project milestone I.

goal together. As soon as possible, we still found that in the daily communication and promotion of the project, various difficulties still exceeded our expectations. The airline company's business involves a large number of relationship departments and personnel. It is particularly important to be a sponsor, project director or committee that is authorized and able to make decisions and promote the coordination of resources and interests of various departments. In the research and co-creation stage of the project team, the airline company needs to invest a large amount of resources to cooperate. After the requirements and scheme are designed, the scheme needs to be recognized and accepted by those who dare to take responsibility. However, the later scheme implementation and verification can only be completed with the cooperation of many aviation departments and personnel. All these are far from enough to rely on the existing project management system of aviation companies.

Here are some of the project milestones (Figure 7.26 and 7.27):

The joint delivery team formed by both parties is as follows, and the more detailed organizational structure under each module will not be described (Figure 7.28):

Relatively speaking, the first two parts of the work, namely the construction of the Data Middle Office and the construction of the data governance system and platform, are similar to the traditional IT project delivery, so there is not much introduction. However, for a large Data

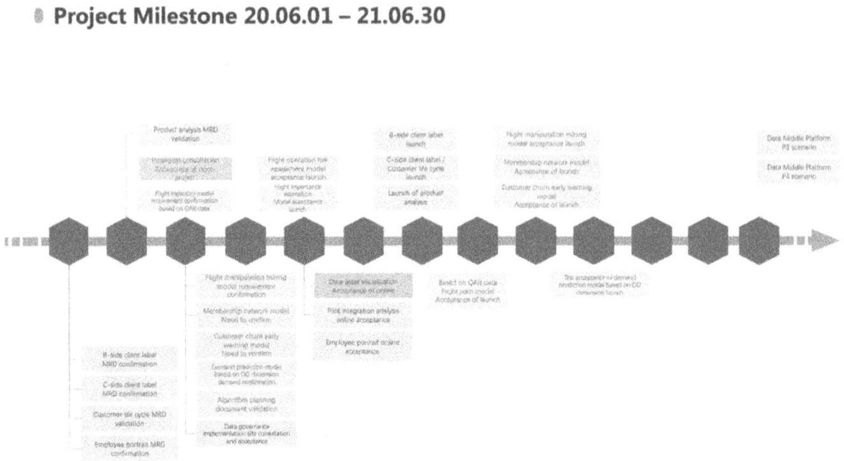

FIGURE 7.27 Project milestone II.

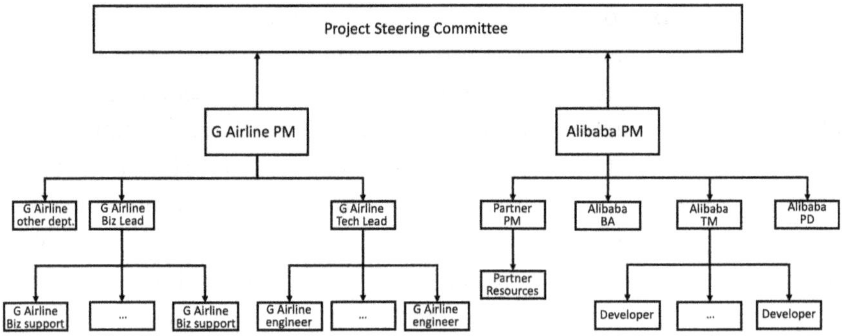

FIGURE 7.28 Project organization structure.

Middle Office project, the delivery process is usually as follows. The G Airlines project is basically promoted step by step in accordance with the following process (Figure 7.29):

It is also important to note that the derivation process of business scenario planning is as follows (Figure 7.30):

In general, the success of large-scale digital transformation projects depends not only on a mature business and technical solution, but also on experienced delivery teams, especially architects and project managers. At the same time, the client must have a strong and cooperative project management team and business and technical knowledge.

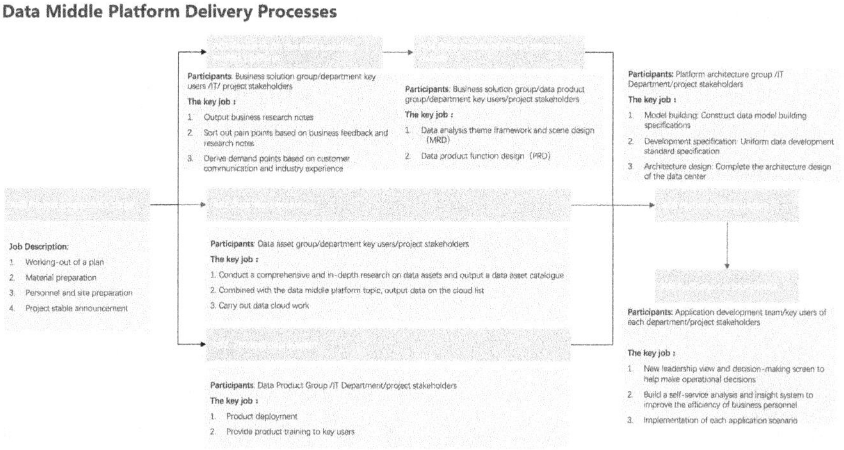

FIGURE 7.29 Data Middle Office delivery process.

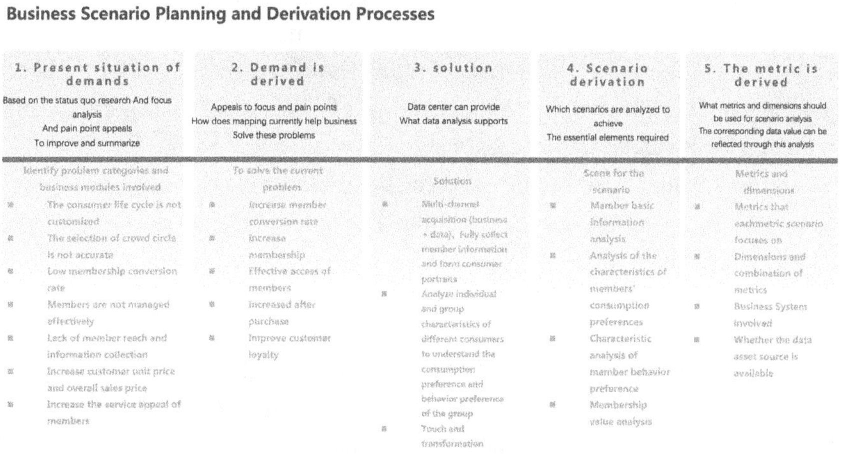

FIGURE 7.30 Business scenario derivation.

7.1.4 Customer Value

The G Airlines project shows many unique values before the project is fully completed, especially in the Data Middle Office before the station has gone through the normally necessary operation optimization stage:

Through the construction of the new hybrid cloud Data Middle Office, the whole group is equipped with the elastic capacity of IT infrastructure that can be allocated according to needs. Moreover, the disaster recovery system of the two places and three centers provides the high availability of the whole system, and the high security program provided by Alibaba makes the whole system can withstand the security test of depth.

In the first phase of the system construction, more than 40 cloud systems were completed, and relevant data with 76,000 fields on the data were completed cloud and business modeling. The data processing capacity of the system rose directly from TB to PB. After the adoption of ADB, data query response has improved from minutes to seconds compared to the previous MongoDB.

Through the construction of the Data Middle Office, the problem of data isolation between business departments has been solved, and all key data of RongTong Company are put on one platform and updated in a timely manner.

Newly established fast business analysis and labeling system based on global data: Including 16 data domains, 12 typical business topics (completed operation, revenue, ground service, marketing, pilots, employees, etc.), 50+ analysis and application scenarios, new design 500+ metrics, 500+ labels. There are 800+ users under the three themes of completed revenue and operations, saving an average of 2 hours per user per day, and reducing historical data processing from 1 month to 2 days.

Achieve visual insight into typical business scenarios such as operation, revenue, ground service and marketing; to help achieve flight on-time monitoring.

Newly established marketing system products, channels, users based on the whole life cycle business analysis strategy, such as early warning model, funnel model and correlation analysis model.

At the same time to help customers complete the intelligent transformation planning. Develop data governance plans and related systems from scratch.

In terms of platform and application construction, the deployment of Dataphin and other key platforms has been completed. The cloud tool saves 80% of the labor cost of cloud, and the efficiency of data development has been greatly improved.

By completing the deployment of QA (Quick Audience) and QBI (Quick BI) the efficiency of display, extraction and analysis of business data has been greatly improved.

Visualized portal and data asset manager application are highly appreciated by G Airline customers and have been successfully copied and promoted to Feihe, Beingmate and other projects and become standard equipment for data middle-platform projects.

The successful deployment of Trinidad Eye platform perfectly helps the airline to realize the decision support of seat warehousing transfer.

The customer portrait platform makes full use of the newly designed data metrics and labeling capabilities to provide new support capabilities for precision marketing business. The deployment of the PAI algorithm platform provides the basis for intelligent algorithms such as customer churn warning, such as identifying customers and passengers on business or private business trips with an accuracy rate of 90%.

More values are not listed one by one. At the same time, the Data Middle Office also needs a period of time to show its strong ability and value after operation precipitation.

7.2 CASE OF A TRAVEL AND HOTEL INDUSTRY COMPANY

7.2.1 Customer Background

R group is the largest comprehensive entertainment resort in Southeast Asia. Facilities include hotels, Universal Studios, aquariums, casinos and other facilities. The customer wants Alibaba Cloud to help them implement digital transformation based on Alibaba's best practices. The service utilizes Alibaba Cloud infrastructure and the dual Middle Office architecture to develop a unified system that provides services such as unified customer experience management, marketing engine, B2B business management, customer insight and Chatbot. In this project group, Alibaba Cloud provides customers with a complete cloud native development framework, such as IaaS, middleware, big data platforms, security, network and artificial intelligence services.

The progress of science and technology has changed the way consumers travel. Today's travelers, from airlines to hotels to travel plans, have many applications and platforms to provide technical solutions and restaurant recommendations. This trend also gives consumers higher expectations: from simple material satisfaction and demand to now, priority should be given to personalization, seeking different experiences, and at the same time ensuring convenience, efficiency and quickness.

The influence of technology on tourists' consumption behavior has also led to changes in the tourism industry. In addition to the overall business shift to online travel agencies and other technology providers, the tourism industry as a whole, all industry participants need to make a thorough change in their current business model to meet the growing demand and expectations. Through in-depth learning of the best practices of the Internet industry, supported by the concept of Middle Office, suggestions on improving B2B portals and open platforms as the starting point to better adapt to the above tourism industry trends, follow up the changes in the entire tourism industry and better meet the needs of consumers and partners.

This background also makes R group's digital transformation have more urgent needs, and Alibaba, which has rich and successful practice and technical ability in tourism digitalization and retail industry, has become the object of concern of R group. Through several months of discussion and evaluation, Alibaba will eventually cooperate strategically with Alibaba. Plan and implement a dual Middle Office strategy to support rapid business change and innovation.

7.2.2 Alibaba Cloud Solution

7.2.2.1 Project Objectives

R group supports the rapid growth of the organization's business by planning enterprise IT transformation to achieve a unified customer experience. Business Middle Office and Data Middle Office are Alibaba Group's practice methodologies for digital transformation of enterprises. Leveraging a set of business and data middleware, a Middle Office architecture refers to a powerful shared data asset layer and business logic layer that power agile and scalable applications throughout an organization.

Through the construction of Business Middle Office, a software application system with multi-channel business as the Innovation Point is built, and a multi-channel business operation support system with multi-channel collaborative development is achieved to increase sales scenarios and sales opportunities. Products, orders and inventory in existing online trading channels and distributors are interconnected to improve operational efficiency.

Build the following applications based on the Business Middle Office: Channel customer relationship management platform (CRM) and B2B trading platform. By supporting the new retail business at the C end and the channel business application at the B end, it helps R group to complete the transfer of 'service-centered mode' to 'customer-centered mode', effectively supports R to find prospective customers, establish trust relationships, tap customer needs, put forward customer plans, sign up commitments, improve customer satisfaction, etc.

Through the collaboration of Business Middle Office, Data Middle Office and digital factory, as well as a closed-loop data collection, storage, analysis and application based on big data technology, R group gradually improves its governance capabilities for business data and pan-member data, improves its R data application capability and provides strong data support for business development and platform operations.

The Business Middle Office uses the global digital operation infrastructure with 'Internet Plus' as its basic technology to achieve stable, efficient, secure and scalable goals. The Business Middle Office is responsible for aggregating, sharing and opening business capabilities, and effectively supporting the flexible expansion of front-end businesses.

7.2.2.2 Overall Design

The overall architecture design will be based on business needs research and analysis, to strengthen R group's business through marketing, sales and customer service as well as customer experience management (hereinafter referred to as System X) ecosystem.

The overall architecture design details the architectural decisions to implement marketing, sales and customer service functions, and provides an X system solution guide to meet RWS functional requirements.

Based on the concept of Middle Office, the project carries out business innovation of multiple formats and channels with the support of Business Middle Office and Data Middle Office (Figure 7.31).

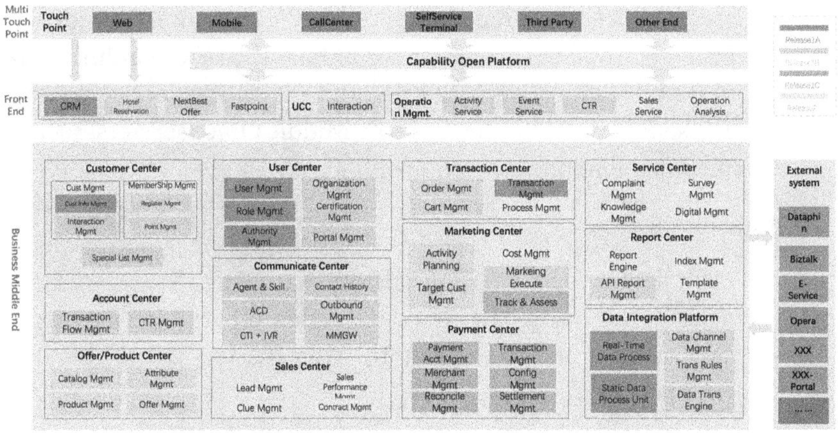

FIGURE 7.31 Overall architecture.

The overall architecture reflects the business requirements. It provides comprehensive decoupling to improve the flexibility and maintainability of the system. Overall system design principles:

1. Front-end and back-end decoupling
    - Each front-end is responsible for presentation and user interaction, and the business logic is implemented in the back-end Middle Office.
    - Both the front-end and back-end adopt their own technical routes for unified planning and parallel development.
    - Data interaction between the front-end and back-end through services.
2. Business decoupling
    - Therefore, the system is divided into independent service centers based on business domains. In the design of centers, the businesses must follow the principles of high cohesion and low coupling.
    - The service center provides external support for business capabilities.
    - In the design of the functional architecture, the domains, such as business domains, shared domains and technical domains, are classified as shared service centers. The services provided by business domains are generally unique to corresponding business applications, such as sales and marketing service centers. The shared domain provides core Middle Office services, such as customer and product centers.
3. Data decoupling
    - Each data center is responsible for managing its own business data. When using data in a data center, other centers or applications must access the data through the services of this data center.

7.2.3 Project Deliver Process

Through the design of the fourth phase of the project, the overall project will continuously deepen its enterprise application in R group with the idea of Middle Office. The advantage of the Middle Office architecture is that it can extract shared and reusable business logic to reduce repetitive development and sharing of global data, provide a foundation for business innovation and provide technical support for the sustained business growth of R group. The planning and implementation of the technology architecture design are as follows for your reference:

1. Layered system architecture

   The architecture of the back-end business system is designed based on domain modeling. Figure 7.32 shows the layers of the business system.

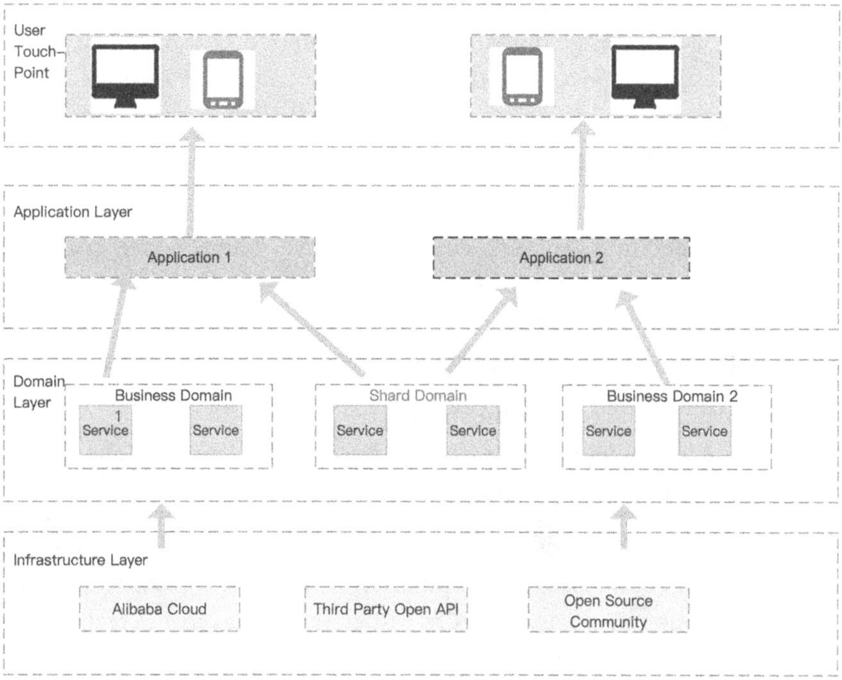

FIGURE 7.32 System layering.

- User touch-point layer

  The user layer provides an excellent user experience through the data and functions provided by the application layer. In terms of system architecture, this feature is reflected in the performance layer of PC and mobile client.

- Hybrid Application layer

  The application layer functions as the application system of each business. It manages services in the domain layer. For example, the four centers can serve as an application at the system architecture layer.

- Distributed domain layer

  The domain layer is a collection of all business services. The service provided by the business domain is usually unique to the corresponding business. The shared domain provides core Middle Office services, such as user and product centers.

  Based on the domain model, the shared Middle Office architecture supports the linear scalability required by business applications and reuses accumulated Middle Office services to adapt to rapid changes.

- Basic component layer

  The infrastructure component layer implements, manages and maintains services at the application layer and domain layer. It is usually composed of open source components, third-party platforms and middleware independent of business logic.

2. Overall technology architecture

   The technology architecture consists of the front-end, back-end, application O&M and integration modules (Figure 7.33).

   - The front-end is separated from the back-end, and the front-end architecture uses the popular front-end framework to support mobile terminals and PCs, respectively. Reasonable technology and humanized design meet users' sensory experience needs.

   - The distributed microservice architecture is the back-end architecture. Under the overall technology architecture, it is easy to scale, combine and deploy. It also supports dynamic scaling, precise monitoring and phased release.

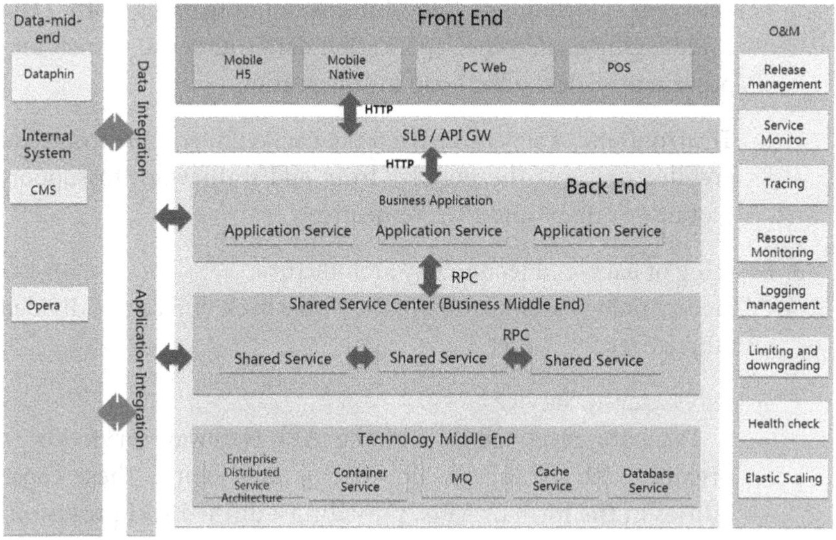

FIGURE 7.33 Overall technology architecture.

- The infrastructure back-end consists of the application layer and the basic services layer. Business Middle Office services and application services are implemented on the application layer with microservices model. The basic service layer provides the application layer with synchronous communication, asynchronous communication, object storage service, databases and distributed cache, based on the capabilities of basic middleware and cloud services.
- Currently, application O&M supports the O&M of all application services at the underlying layer. This system ensures high availability and stability by providing services such as basic monitoring, service upgrades and downgrades, phased release and auto scaling. The O&M layer provides capabilities such as Distributed Link tracking, online debugging and screen-based display of log collection to maximally reveal the service status.
- The integration includes data integration and application integration. Application Integration integrates system X with other systems using a Restful API (Application Programming Interface) to achieve efficient communication, calling, and collaboration

between applications. Data Integration integrates with Data Middle Office and other application data sources to extract and process historical data. Load for integration.

- The front-end is separated from the back-end, and the front-end architecture uses the popular front-end framework to support mobile terminals and PCs, respectively.

3. Layering of back-end technology architecture:

   The technology architecture of the whole back-end can be divided into three parts:

   - API Gateway Service

     The core responsibility of the API Gateway service is to organize and arrange business logic and expose these capabilities in the form of APIs, providing a unified API provisioning service that supports your application. This service is also provided to implement the service interface security-related control logic.

   - Deploy Stateless services

     Stateless services implement business features of each module. Stateless means that in the development of application code, only logic processing and numerical calculation, not data caching or persistence. Stateless services can be deployed to multiple server nodes according to specific business scenario requirements. This is also the basis for distributed high availability, high reliability and automatic scaling.

   - Supports Stateful services

     Stateful services mainly support cache-based acceleration, message queuing and database storage of each module. Stateful services employ technologies such as primary and secondary, read/write splitting and distributed storage to ensure high availability and high reliability.

Due to the finer-grained services of the microservices architecture, new applications can reuse former services when new technologies are needed. This is the most suitable framework for the business architecture of the shared Middle Office. Based on the microservices technology architecture,

Alibaba Cloud shields the technical complexity of microservices through basic service components of the technical Middle Office.

7.2.4 Customer Value

- Alibaba completed the construction of a closed-loop operation system based on Business Middle Office and Data Middle Office, and continuously precipitated core business service capabilities and data, and continuously explored new business models based on Middle Office capabilities.
- Alibaba Middle Office helps customers quickly build innovative businesses of multiple formats and build an enterprise-level capability open platform that flexibly supports B2C and B2B transaction modes.
- Through the Business Middle Office database, combined with Alibaba Cloud's big data capabilities, it provides user profiling, tag computing and other capabilities, guiding customers to enhance customer business value on the basis of better customer service.

7.3 CASE OF CHINA FEIHE LTD.

7.3.1 Customer Background

China Feihe Ltd. is one of the largest and most widely known infant formula milk powder brands in China. Counting from the second start-up in 2001, it started from a county in Northeast China. In 20 years, it has been the first chinese infant milk powder brand with revenues of over 10 billion, and won the leading position.

The reason why Feihe Dairy can stand alone in the 'vortex' of the industry is not due to good luck, but because it is 'planting grass and raising cattle' to build its own industrial clusters when all its counterparts are striving to conquer the city in the golden development period.

Feihe Dairy has built the first exclusive industrial cluster for infant milk powder in China for more than 10 years. From pasture planting, dairy cow breeding, fresh milk collection, to production and processing, and channel control, all links are fully controllable. At the same time, relying on the advantages of exclusive industrial clusters, a 2-hour industrial ecosystem has been formed. The fresh milk was squeezed from the milking parlor of the pasture will be transported to the world-class factory within 2 hours, in a fully enclosed transport vehicle with low-temperature. Then powdered milk was produced with nutrition and freshness.

Although Feihe Dairy has become the market leader, a reality that cannot be ignored is that chinese birth population has continued to decline since 2017. The total number of births continues to decrease, which means that the overall capacity of chinese infant milk powder market will gradually shrink.

As a result, Feihe Dairy has established a strategy to strengthen its head position in 2018. To this end, the cooperation with Alibaba has opened the road of digital intelligence transformation.

7.3.2 Alibaba Cloud Solution

Data islands, lack of corporate data standards and lack of data operations are pain points that need to be resolved urgently for Feihe Dairy, which is rapidly developing in business. Alibaba has proposed the construction of dual Middle Offices (Data Middle Office and Business Middle Office). From data presentation to data decision-making to data operation and then to business support, from capacity sharing to accelerating enterprise information construction, Feihe Dairy has built two phases of Data Middle Office, one phase of the Business Middle Office, to create an enterprise dual-middle-office structure, as shown in Figure 7.34.

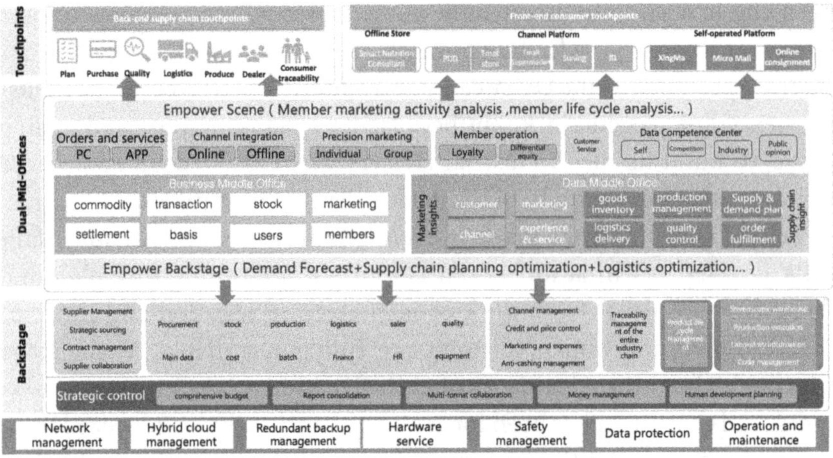

FIGURE 7.34 Feihe dual-middle-office structure.

The construction of the Dual-Mid-Offices interacts with basic data from the business back-end and empowers the back-end, supports the front end of the business from application to data, promotes full-link digital operations of enterprises, improves response efficiency and provides strong support for corporate decision-making and business development.

Therefore, we show the construction path of Feihe Dairy's digital transformation from the two dimensions of dual-Mid-offices construction.

7.3.2.1 Data Middle Office

Alibaba Data Middle Office is the accumulation and output of Alibaba Group's years of experience based on OneData methodology. Feihe Dairy, as the first batch of benchmark customers, successfully demonstrated the value of Data Middle Office.

Before the construction of the Data Middle Office, data analysis is centered on the ERP system in Feihe Dairy. It also built an e-commerce platform, a member operation platform and a marketing system, which laid the foundation for digital construction, but its accumulated huge data assets were scattered. The pain points of the C-end business are as follows:

- Consumers: Lack of comprehensive data support for operations.
- Shopping guide: Data metrics lack a unified caliber; APP tools and marketing management system data are not connected, data auditing is difficult; shopping guide marketing lacks comprehensive data support.
- Channels: Orders, marketing and other data are scattered in various systems and are difficult to integrate; sales management platform and terminal platform data are not communicated, and data in the entire link are difficult to insight.
- Customer experience and service: Lack of full life cycle management.

On the supply chain part, Feihe Dairy has invested decisively in the field of digital reform. It has successively reconstructed the supply chain systems from procurement, production, quality, inventory, logistics, etc., and promoted the digital transformation of the supply chain. Enterprises can quickly obtain various links. The module situation is further analyzed. After the digital transformation of the supply chain is completed, it is necessary to solve its pain points by the Data Middle Office:

- Procurement: Lacks analysis of raw material inventory age and timely monitoring and early warning.
- Plan: Rely heavily on human experience and judgment, and lack accurate data support for demand, supply, allocation and sales.
- Production: Lack of accurate management of capacity, inventory and production plan.
- Logistics: Inventory management mostly involves manually exporting data from multiple systems for analysis and processing, and lacks optimization analysis of logistics costs.
- Sales management: Lack of multi-dimensional analysis of order fulfillment rate, and manual processing of sales forecasts and demand planning.

Feihe Dairy has built two phases of Data Middle Office to break data islands. The first phase mainly solves the pain points of the C-end business, and the second phase mainly solves the pain points of the supply chain. The overall structure is shown in Figure 7.35.

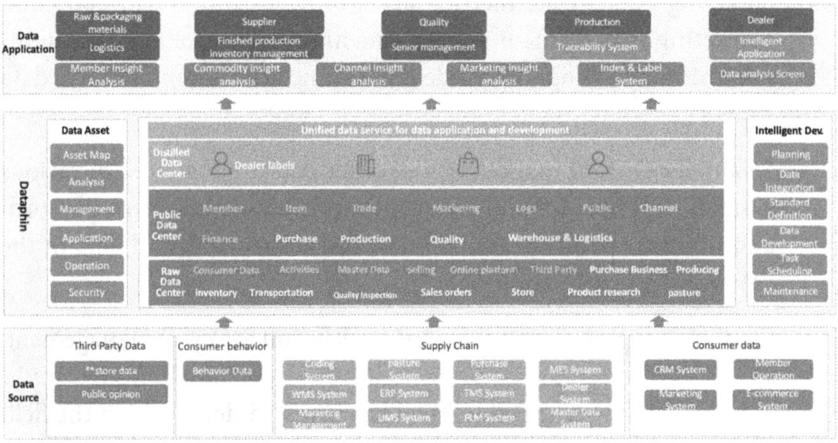

FIGURE 7.35 Overall architecture of Feihe Data Middle Office.

The first phase of the construction mainly covers consumer insights, sales promotion insights, channel insights, customer experience and services on the C end, among which consumer insights include:

- Based on OneID tag construction: Three-dimensional cross-analysis with consumer value as the main axis and behavioral characteristics and other information as the auxiliary axis for accurate identification.
- Consumer insight: Analysis based on the various stages of the consumer life cycle.
- Integral insight: Analyze the complete life cycle of the integral and early warning of abnormalities.

Camp sales insights include:

- Marketing activities: Covering content marketing, online sales promotion and offline sales promotion.
- Cross-industry alliance scenario analysis: For the cross-industry alliance cooperation scenario, analyze the three dimensions of product, activity and cooperation information;

Channel insights are analyzed from the four dimensions of purchase, sale, inventory and people, including:

- Analysis of terminal stores: Analyze the trend, comparison and structure of relevant metrics for terminal stores to facilitate the management of terminal stores.
- One party's LBS data analysis: Based on location information, regional characteristics analysis.
- Shopping guide analysis: For the trend, comparison and structure analysis of relevant metrics for shopping guides, it is convenient for the evaluation and management of shopping guides.

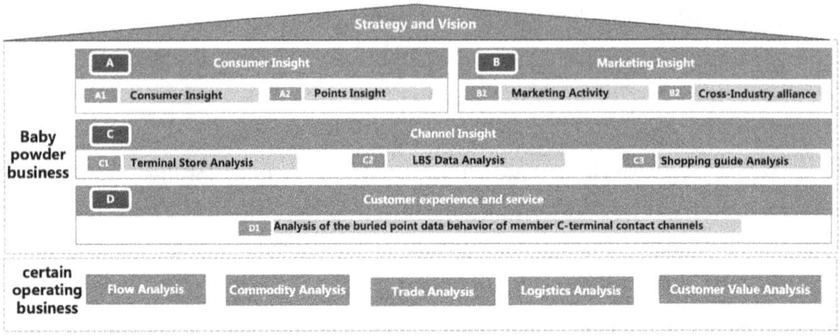

FIGURE 7.36 Overall scenario design of Feihe.

Customer experience and services include:

- Consumption data analysis: Relying on the C-end buried point data, according to the AIPL model, provide a basis for the analysis of the current situation of the enterprise and the direction of improvement.

The construction content of the first phase is shown in Figure 7.36.

Feihe Dairy is currently actively promoting the construction of the second phase of the Data Middle Office for the supply chain. The construction of the middle stage of the entire supply chain involves 8 sectors and 25 scenarios, as well as traceability systems and intelligent applications (consisting of sales forecasts, supply plans, logistics allocations and shipment plans). The project delivery process communicates with more than 10 departments on the customer side and clarifies the needs.

Main solutions for product inventory management mainly:

- Establish a product age management and early warning mechanism to control inventory turnover and consumption speed, and provide quality assurance for the freshness tracking of the entire supply chain.

- Establish a DOS early warning mechanism for available days, analyze the satisfaction of terminal requirements under different warehouse dimensions, and implement a replenishment mechanism for early warning of out-of-stock risks, and at the same time in-depth analysis of the root causes of out-of-stock status, laying a foundation for subsequent optimization of the stocking mechanism.

Main solutions for raw material packaging management and optimization:

- Pull through the raw material inventory at each node, establish a raw material shortage early warning mechanism, reasonably allocate existing resources and provide data support for the overall supply of raw materials.
- Establish a raw material age management and early warning mechanism, horizontally pull through the inventory data of each factory, combine the age of the goods and the speed of raw material consumption, implement self-consumption or inventory allocation and avoid the generation of older raw materials.

Main solutions for production and quality management:

- In terms of quality, cost, efficiency, etc., monitor the PCS time from the perspective of the head office, branch companies and parts.
- Break through data gaps, systematic quality monitoring, automatic early warning reminders to monitor the quality of raw and auxiliary materials.
- Monitoring of product detection and early warning, timely release and analysis of market complaints rate, so as to monitor product detection.

Main solutions for dealer management:

- From the dealer's back-end inventory to the front-end demand, tracking the order quantity.
- Monitor order delivery from the perspective of the head office, business departments and distributors.

Main solutions for logistics management:

- Multi-dimensional considerations in the application to make allocation and shipping decisions, effectively balance the supply and demand ends of the supply chain, and use logistics resources to reduce logistics costs.

- Monitoring the information of staying time in the warehouse through the distribution of the delivery age of the products.
- Through the establishment of the analysis scenarios of the stay time of each node of the product inventory and the monitoring of the goods in transit, the business department tracks the product inventory from the perspective of multi-dimensional and multi-node, and lays the foundation for subsequent decision-making.

In addition, the large screen provides corporate executives with an overview of global supply chain metrics, so that customer executives can learn about the supply chain in a timely manner and facilitate decision-making.

The second-stage Data Middle Office also delivered two applications, namely the traceability system and the intelligent application. The traceability system provides Feihe Dairy with the traceability of the entire process from pasture, production, quality, warehousing, circulation and sales, and provides external service interfaces to facilitate the inquiry and connection between consumers and regulatory authorities. The intelligent application uses intelligent algorithms to predict sales, so as to produce demand plans to support supply plans. In the intermediate environment of production and sales, intelligent logistics allocation and delivery modules provide logistics managers with reasonable logistics recommendations based on data output.

7.3.2.2 Business Middle Office

After the first phase of Data Middle Office construction, Feihe Dairy fully felt the value brought by digital transformation and the pain points of business system construction captured during the construction process, so it started the Business Middle Office construction. Before that, each business system has its own membership, order, product, inventory and other data. The master data uniformly maintain the content of corporate products, stores, distributors, prices, etc., and the marketing system maintains the relationship between shopping guides and members, and each system has its own technical system, resulting in high system operation and maintenance costs and new business undertakings are chimney construction, which is not convenient to conduct unified management. After the Business Middle Office runs, orders, commodities, inventory, membership, marketing, settlement and other content are uniformly maintained

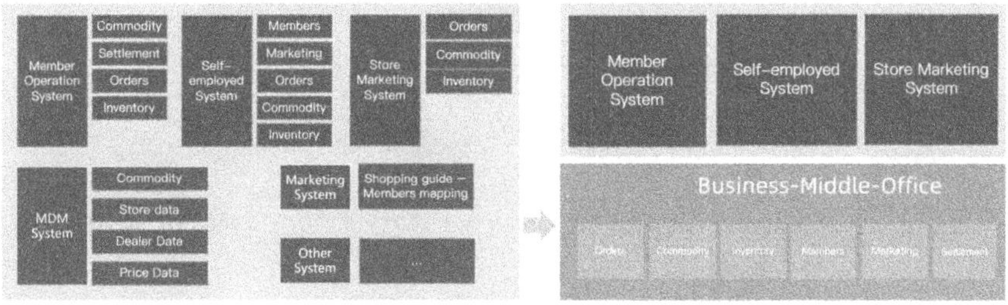

FIGURE 7.37 Business Middle Office iteration.

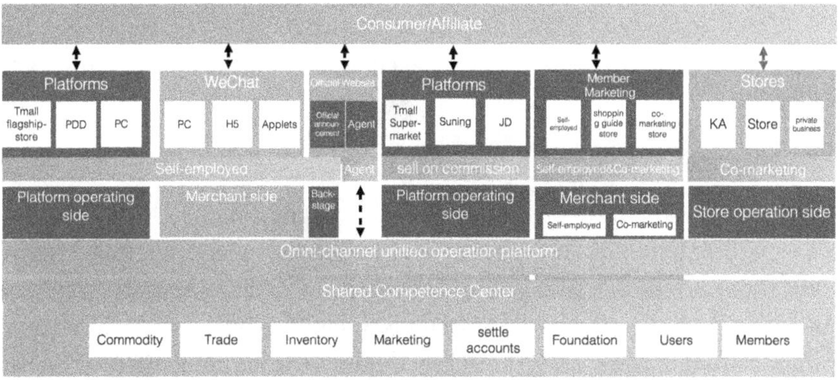

FIGURE 7.38 System architecture planning.

in the Business Middle Office, and the back-end of each business system is connected to the Business Middle Office for standard interface docking. Each business system only needs to pay attention to its own front-end system construction and personalized needs, speed up system iteration and quickly meet consumer needs, as shown in Figure 7.37:

At the system architecture level, we fully consider the customer's various channel platforms. The membership marketing system and the e-commerce system are used as pilots at first, and the omni-channel unified operation platform will be gradually adopted to connect the platform operation end, merchant end, and store operation end to support each channel business, the plan is shown in Figure 7.38.

The Business Middle Office undertakes to build commodity, transaction, inventory, marketing, settlement, basic, user and member sharing competence centers, and the omni-channel unified operation platform provides unified external standard services to support various forms

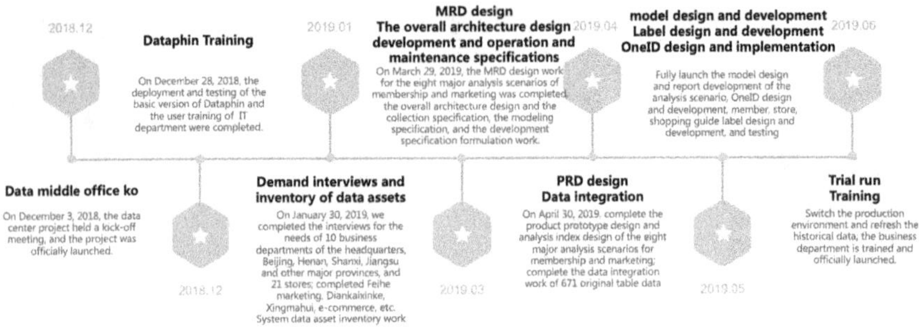

FIGURE 7.39 Key milestones of Data Middle Office phase I.

of business such as self-operated, agency sales, distribution and joint operations.

7.3.3 Project Delivery Process

Feihe Dairy is the head customer of Alibaba Data Middle Office, and it is also the first batch of landing customers. The first phase of Data Middle Office has undergone half a year of construction. The preliminary survey of more than 100 people covers 20 systems, including business departments, provincial managers and shopping guides, involving 8 data domains including member domain, marketing domain, logistics domain, channel domain, transaction domain, commodity domain, log domain and public domain, and 788 core business tables of 24 systems on the cloud, 125 of them Atomic metrics, 893 derived metrics and 221 member tags on the ground. The key milestone nodes are shown in Figure 7.39.

The second-phase Data Middle Office is still in the process currently (February 2020). There were 126 business pain points were sorted out and 80 improvement points were suggested, after 128 interviews with more than 145 employees from various departments and factories in the early stage of this project. We uploaded more than 1,000 tables from 13 supply chain business systems to the maxcompute (alibaba product), and built 627 supply chain metrics in data fields such as production domain, purchasing domain, warehousing logistics domain, and financial domain. Feihe expected to complete all the tasks in 14 months.

The key milestone nodes are shown in Figure 7.40.

The construction of the Business Middle Office began in early June 2020. The first phase completed the transformation of the point marketing system and launched in December 2020. The second phase will complete

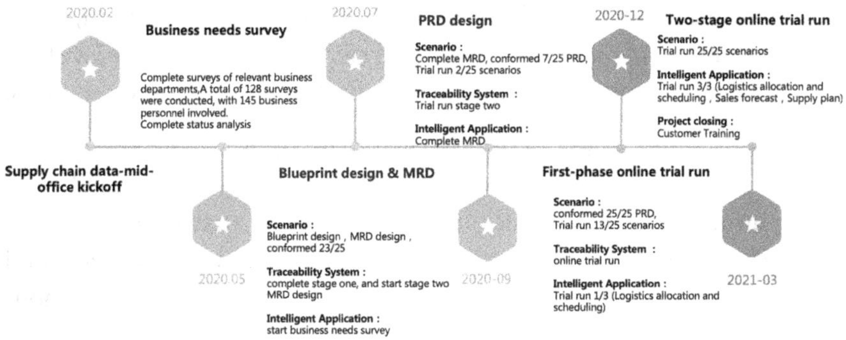

FIGURE 7.40 Key milestones of Data Middle Office phase II.

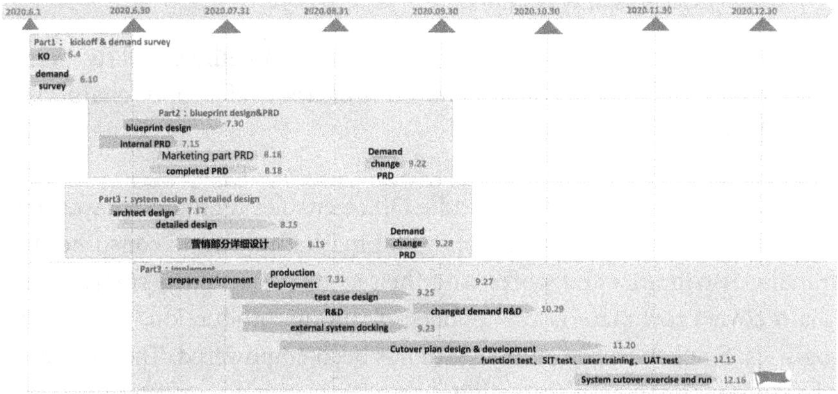

FIGURE 7.41 Business Middle Office project plan.

the e-commerce business undertaking, which is expected to be launched in March 2021 (Figure 7.41).

7.3.4 Customer Value

Feihe Dairy has achieved the transformation from empiricism to dataism after two phases of Data Middle Office construction, and achieved the following goals:

- Same source: Use the same source of data to provide comprehensive information for each business segment, from the management to the executive level, each business department uses the same analysis tool, unifies the logic and avoids communication misunderstandings.

- Two-way: Pull through the data of the entire industry chain, break the upstream and downstream barriers of the supply chain, effectively coordinate and improve the efficiency of business operations.
- Agility: Eliminate the need to manually collect data from various business systems and merge analysis work, use Alibaba Cloud computing power to conduct multi-dimensional analysis of big data, effectively use resources for decision-making and improvement and enhance the ability to respond to the supply chain and decision-making capabilities.
- Reverse: Through the construction of Data Middle Office, identify data problems and deficiencies, reverse the transformation of business processes and systems.
- Foresight: Through early warning logic, intelligent early warning of business problems and identification of supply chain risks (Figure 7.42).

The construction of the Data Middle Office enables digital empowerment of the business. The first phase of digital marketing covers consumer terminals, distributors and stores, and the second phase of the smart supply chain covers raw materials, suppliers, planning, production, quality and logistics. Since then, the entire chain has been empowered. The formation of a closed loop to accelerate digital transformation is consistent with the five steps of Alibaba's digital transformation: infrastructure cloudification, touch-point digitization, business online, operation digitization and decision-making intelligence.

The successful construction of the Data Middle Office has stimulated customers' expectations for the construction of the business center. Through the Business Middle Office landing in Feihe Dairy, it is expected to drive the business of the enterprise to better meet the needs of consumers:

- Based on the development of the Internet architecture to quickly respond to changes in business needs and achieve business innovation, platform data are online in real time; online and offline are connected to achieve membership, commodity, transaction and inventory.
- Real-time perception and support of business changes through enterprise-level Internet architecture.

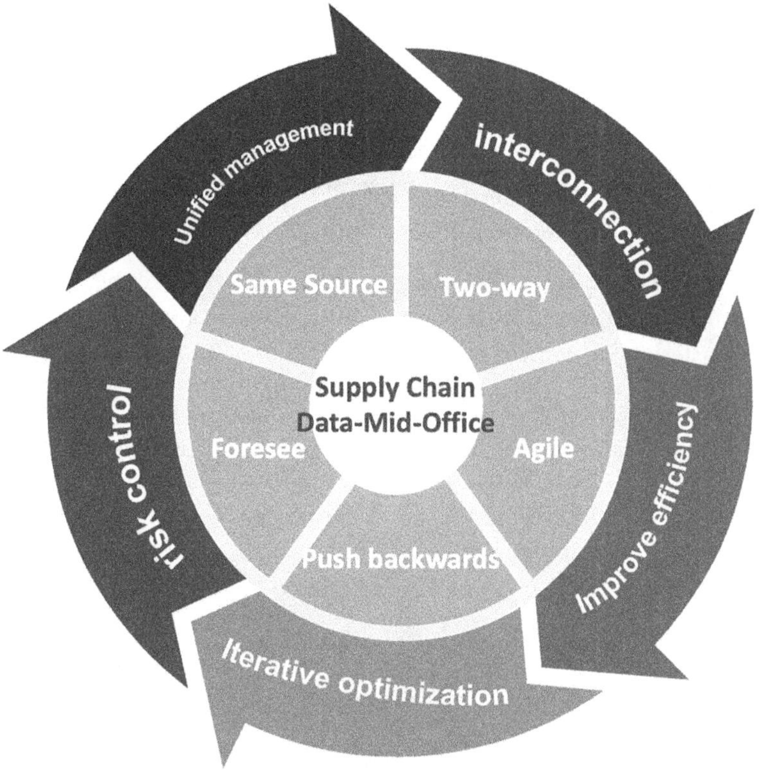

FIGURE 7.42 Value of Feihe Data Middle Office project.

- Business fragmentation leads to a direct decline in consumer experience, which directly affects corporate brand value. Through the shared Business Middle Office between China and Taiwan to achieve business interconnection, consumer experience is consistent, and brand value and user stickiness are improved.
- It is difficult to upgrade and iterate traditional business software requirements and to resolve faults. The Business Middle Office takes the core capabilities of informatization in the hands of the enterprise, and quickly customizes it on demand.
- The construction of Middle Offices has changed the chimney-style construction method of traditional enterprises, and also brought the optimization of enterprise process and organizational structure in the process of business analysis and analysis, and promoted

the continuous development and evolution of enterprises in market competition.

- Real-time online omni-channel data, real-time insight and analysis of current channel commodity inventory distribution, changes in sales trends in any cycle, and promote upstream supply chain planning and procurement optimization.
- Facing online and offline integration for global marketing, combining sales and product data, truly understanding the conversion process, and mastering the contribution of different channels, refined operations.
- Through the construction of middle stations, a digital ecological chain will be finally realized, and differentiated competitiveness will be created, including: consumer online, commodity online, transaction online, member online, inventory online and store online (Figure 7.43).

FIGURE 7.43 Value of Feihe Business Middle Office project.

Bibliography

Yu Changtao. New Infrastructure Accelerates the Digital Transformation of Enterprises and Promotes the Application of Industrial Internet of Things in a Multi-Dimensional Way. *Internet Economy*, 2020, (7): 66–69.

Su Chen. Aliyun Research Center – White Paper on New Digital Transformation, 2019.

Duan Chunlin, Lin Jiachun. Development of Enterprise Digital Transformation in the New Consumption Era. *Chinese Journal of Social Science*, 2020-08-20(006).

Chen Hao, Liu Yanbing, Zhu Haoran. Analysis and Research on key points of Enterprise Digital Benchmarking. *Enterprise Science and Technology and Development*, 2020, (8): 224–225+228.

≪Hologres+Flink流批一体首次落地4982亿背后的营销分析大屏≫ https://baijiahao.baidu.com/s?id=1684840580126060725&wfr=spider&for=pc.

Liu Junyan. Obstacles and Countermeasures for Digital Transformation and Upgrading of Traditional Foreign-Oriented SMEs – A Case Study of OEM Qingdao Home Textile Industry. *Science and Technology Bulletin*, 2020, 38(14): 126–133.

Chen Linchu, Pan Yanjun. A Brief Analysis of the Digital Transformation of Small, Medium and Micro Enterprises under the New Business Format. *Shopping Mall Modernization*, 2020, (14): 117–119.

Linmin Mining Group promotes enterprise digital transformation with big data governance. *China Information Industry*, 2020, (4): 95.

Liang Qigong. Industrial Big Data: Key Points of Digital Transformation of Manufacturing Enterprises. *China Information Weekly*, 2020-08-24(014).

Yang Meng. Actively Promoting the Digital Transformation and Upgrading of SMEs. *Entrepreneur Daily*, 2020-08-27(003).

Yang Meng. Embracing the Era of "Intelligent Manufacturing", Exploring the Way of Digital Transformation of Small and Medium-Sized manufacturing enterprises. *Science&Technology Daily*, 2020-08-31(008).

Zhang Qinglong. Analysis on the Digital Transformation Path of Financial Sharing Service. *Journal of Finance and Accounting*, 2020, (17): 12–18.

China Academy of Information and Communication Technology. White Paper on The Development of China's Digital Economy, 2020.

≪专访阿里姜伟华：实时数仓，如何能做得更好？≫ https://www.infoq.cn/article/6aksl76pmzruyymrtulf/.
≪阿里云为什么要做流批一体≫ https://weibo.com/ttarticle/p/show?id=2309404582106142474503&sudaref=www.baidu.com.
China Academy of Information and Communication Technology. Data Asset Management Practice White Paper 4.0, 2019.

For Product Safety Concerns and Information please contact our EU
representative GPSR@taylorandfrancis.com
Taylor & Francis Verlag GmbH, Kaufingerstraße 24, 80331 München, Germany

www.ingramcontent.com/pod-product-compliance
Lightning Source LLC
Chambersburg PA
CBHW060414220326
41598CB00021BA/2172
* 9 7 8 1 0 3 2 2 2 5 3 2 6 *